HYDROCARBON BIOREFINERY

HYDROCARBON BIOREFINERY

Sustainable Processing of Biomass for Hydrocarbon Biofuels

Edited By

SUNIL K. MAITY
Department of Chemical Engineering, Indian Institute of Technology
Hyderabad, Sangareddy, Telangana, India

KALYAN GAYEN
Department of Chemical Engineering, National Institute of Technology
Agartala, Agartala, Tripura, India

TRIDIB KUMAR BHOWMICK
Department of Bioengineering, National Institute of Technology
Agartala, Agartala, Tripura, India

ELSEVIER

Elsevier
Radarweg 29, PO Box 211, 1000 AE Amsterdam, Netherlands
The Boulevard, Langford Lane, Kidlington, Oxford OX5 1GB, United Kingdom
50 Hampshire Street, 5th Floor, Cambridge, MA 02139, United States

Copyright © 2022 Elsevier Inc. All rights reserved.

No part of this publication may be reproduced or transmitted in any form or by any means, electronic or mechanical, including photocopying, recording, or any information storage and retrieval system, without permission in writing from the publisher. Details on how to seek permission, further information about the Publisher's permissions policies and our arrangements with organizations such as the Copyright Clearance Center and the Copyright Licensing Agency, can be found at our website: www.elsevier.com/permissions.

This book and the individual contributions contained in it are protected under copyright by the Publisher (other than as may be noted herein).

Notices

Knowledge and best practice in this field are constantly changing. As new research and experience broaden our understanding, changes in research methods, professional practices, or medical treatment may become necessary.

Practitioners and researchers must always rely on their own experience and knowledge in evaluating and using any information, methods, compounds, or experiments described herein. In using such information or methods they should be mindful of their own safety and the safety of others, including parties for whom they have a professional responsibility.

To the fullest extent of the law, neither the Publisher nor the authors, contributors, or editors, assume any liability for any injury and/or damage to persons or property as a matter of products liability, negligence or otherwise, or from any use or operation of any methods, products, instructions, or ideas contained in the material herein.

Library of Congress Cataloging-in-Publication Data
A catalog record for this book is available from the Library of Congress

British Library Cataloguing-in-Publication Data
A catalogue record for this book is available from the British Library

ISBN: 978-0-12-823306-1

For information on all Elsevier publications
visit our website at https://www.elsevier.com/books-and-journals

Publisher: Candice Janco
Acquisitions Editor: Peter Adamson
Editorial Project Manager: Sara Valentino
Production Project Manager: Nirmala Arumugam
Cover Designer: Victoria Pearson

Typeset by STRAIVE, India

To our beloved parents.

Contents

Contributors	xiii
Editors biography	xvii
Preface	xix

1. Hydrocarbon biorefinery: A sustainable approach — 1

Alekhya Kunamalla, Swarnalatha Mailaram, Bhushan S. Shrirame, Pankaj Kumar, and Sunil K. Maity

1.1 Introduction	2
1.2 Biomass	3
1.3 Biorefinery	10
1.4 Biorefinery for traditional biofuels	13
1.5 Biorefinery for fuel-additives	18
1.6 Hydrocarbon biorefinery	24
1.7 Role of heterogeneous catalysis in the hydrocarbon biorefinery	30
1.8 Conclusions	35
References	35

SECTION 1 Chemical and thermochemical conversion of biomass

2. Fast pyrolysis of biomass and hydrodeoxygenation of bio-oil for the sustainable production of hydrocarbon biofuels — 47

Janaki Komandur and Kaustubha Mohanty

2.1 Introduction	47
2.2 Fast pyrolysis	50
2.3 Effect of operation conditions on fast pyrolysis products	56
2.4 Bio-oil upgrading	58
2.5 Conclusion and future perspective	71
References	73

3. Fischer-Tropsch synthesis to hydrocarbon biofuels: Present status and challenges involved — 77

Muxina Konarova, Waqas Aslam, and Greg Perkins

3.1 Introduction	78
3.2 Synthesis gas manufacture	79
3.3 Improved catalysts	82
3.4 Challenges and opportunities for biofuels	90

vii

viii Contents

3.5 Conclusions 94
References 94

4. Hydrodeoxygenation of triglycerides for the production of green diesel: Role of heterogeneous catalysis 97

Pankaj Kumar, Deepak Verma, Malayil Gopalan Sibi, Paresh Butolia, and Sunil K. Maity

4.1 Introduction 98
4.2 Reaction mechanism 101
4.3 Catalysts 105
4.4 Effect of process conditions 115
4.5 Coprocessing of triglycerides with the petroleum feedstock 116
4.6 Commercialization status 119
4.7 Process design and economics 120
4.8 Conclusions 122
References 122

5. Advances in liquefaction for the production of hydrocarbon biofuels 127

Gabriel Fraga, Nuno Batalha, Adarsh Kumar, Thallada Bhaskar, Muxina Konarova, and Greg Perkins

5.1 Introduction 128
5.2 Hydrothermal liquefaction 129
5.3 Liquefaction using organic solvents 152
5.4 Advances in commercialization 158
5.5 Process economics 165
5.6 Conclusions 168
References 169

6. Advances in the conversion of methanol to gasoline 177

Jyoti Prasad Chakraborty, Satyansh Singh, and Sunil K. Maity

6.1 Introduction 178
6.2 Production of methanol 179
6.3 Methanol to gasoline 184
6.4 Industrial development 194
6.5 Conclusions 196
References 197

SECTION 2 Biological and biochemical conversion processes

7. Biomass pretreatment technologies 203

Ayaz Ali Shah, Tahir Hussain Seehar, Kamaldeep Sharma, and Saqib Sohail Toor

7.1 Introduction 203
7.2 Lignocellulosic feedstock composition and pretreatment 204

Contents   **ix**

7.3 Pretreatment techniques for lignocellulosic feedstock   207
7.4 Sewage sludge composition and pretreatment   211
7.5 Pretreatment techniques for the sewage sludge   213
7.6 Challenges and future perspectives   222
7.7 Conclusions   223
References   223

8. Generation of hydrocarbons using microorganisms: Recent advances   229

Bhabatush Biswas, Muthusivaramapandian Muthuraj, and Tridib Kumar Bhowmick

8.1 Introduction   230
8.2 Biochemistry of hydrocarbon synthesis in microbes   231
8.3 Diverse microbial systems for hydrocarbon generation   236
8.4 Biosynthesis of hydrocarbons   244
8.5 Conclusions and future perspectives   249
Acknowledgment   249
References   249

9. Metabolic engineering approaches for high-yield hydrocarbon biofuels   253

Kalyan Gayen

9.1 Introduction   254
9.2 Microbial metabolic pathways involved in hydrocarbon biosynthesis   255
9.3 Metabolic engineering to improve yield of the hydrocarbon biofuels   258
9.4 Toxicity stress of hydrocarbons to microbial cells   261
9.5 Use of lignocellulosic materials as feedstock   262
9.6 Bioconversion of CO_2 to hydrocarbons   263
9.7 Challenges and future directions   265
9.8 Conclusions   265
References   266

10. Oligomerization of bio-olefins for bio-jet fuel   271

Joshua Gorimbo, Mahluli Moyo, and Xinying Liu

10.1 Introduction   271
10.2 Bio-jet-fuel production pathways   278
10.3 Oligomerization of olefins   283
10.4 Aromatization of hydrocarbons from oligomerization   287
10.5 Economics of bio-jet-fuel production   289
10.6 Conclusions   290
Acknowledgments   290
References   291

SECTION 3 Conversion of biomass-derived compounds to hydrocarbon biofuels

11. Carbon-carbon (C—C) bond forming reactions for the production of hydrocarbon biofuels from biomass-derived compounds — **297**

Olusola O. James and Sudip Maity

11.1 Introduction	298
11.2 Lignocellulose to hydrocarbon biofuels	299
11.3 Upgrading of fermentation fuels to drop-in fuels	301
11.4 Upgrading of sugar dehydration intermediates to drop-in fuels	306
11.5 Conclusions	321
References	323

12. Production of long-chain hydrocarbon biofuels from biomass-derived platform chemicals: Catalytic approaches and challenges — **327**

Sudipta De

12.1 Introduction	328
12.2 Strategies for C–C bond formation	329
12.3 Strategies for oxygen removal from oxygenated fuel precursors	336
12.4 Examples of long-chain hydrocarbon biofuels derived from different feedstocks	341
12.5 Perspectives and challenges	348
12.6 Conclusions	350
References	350

13. Bioeconomy of hydrocarbon biorefinery processes — **355**

Janakan S. Saral, R.S. Ajmal, and Panneerselvam Ranganathan

13.1 Introduction	355
13.2 Methodology of economic analysis and types of biorefinery	357
13.3 Conclusions and future prospects	380
Acknowledgment	380
References	381

14. Life-cycle analysis of a hydrocarbon biorefinery — **387**

Jasvinder Singh, Aman Kumar Bhonsle, and Neeraj Atray

14.1 Introduction	388
14.2 Biorefinery configurations	389
14.3 LCA of biorefineries	396
14.4 Conclusions	403
References	404

Contents **xi**

15. Hydrodeoxygenation of lignin-derived platform chemicals on transition metal catalysts **409**

Shelaka Gupta and M. Ali Haider

15.1 Introduction 409
15.2 Transition metals as hydrodeoxygenation catalysts 411
15.3 Rational design of bimetallic alloys for HDO reaction 421
15.4 Oxophilic metals as acidic sites for HDO reaction 424
15.5 Conclusions 425
Acknowledgment 426
References 426

Index *431*

Contributors

R.S. Ajmal
Department of Chemical Engineering, National Institute of Technology Calicut, Kozhikode, India

Waqas Aslam
Australian Institute for Bioengineering and Nanotechnology (AIBN), The University of Queensland, Brisbane, QLD, Australia

Neeraj Atray
Biofuels Division, CSIR-IIP, Dehradun, India

Nuno Batalha
School of Chemical Engineering, Faculty of Engineering, Architecture and Information Technology, The University of Queensland, Brisbane, QLD, Australia

Thallada Bhaskar
Material Resource Efficiency Division, CSIR-Indian Institute of Petroleum, Dehradun, Uttarakhand; Academy of Scientific and Innovative Research (AcSIR), Ghaziabad, India

Aman Kumar Bhonsle
Academy of Scientific and Innovative Research (AcSIR), Ghaziabad; Biofuels Division, CSIR-IIP, Dehradun, India

Tridib Kumar Bhowmick
Department of Bioengineering, National Institute of Technology Agartala, Agartala, Tripura, India

Bhabatush Biswas
Department of Bioengineering, National Institute of Technology Agartala, Agartala, Tripura, India

Paresh Butolia
School of Chemical Engineering, Sungkyunkwan University, Seoul, South Korea

Jyoti Prasad Chakraborty
Department of Chemical Engineering and Technology, Indian Institute of Technology (Banaras Hindu University), Varanasi, Uttar Pradesh, India

Sudipta De
KAUST Catalysis Center (KCC), King Abdullah University of Science and Technology, Thuwal, Saudi Arabia

Gabriel Fraga
School of Chemical Engineering, Faculty of Engineering, Architecture and Information Technology, The University of Queensland, Brisbane, QLD, Australia

Kalyan Gayen
Department of Chemical Engineering, National Institute of Technology Agartala, Agartala, Tripura, India

Joshua Gorimbo
Zhijiang College, Zhejiang University of Technology, Shaoxing, Zhejiang, China

Shelaka Gupta
Multiscale Modelling for Energy and Catalysis Laboratory, Department of Chemical Engineering, Indian Institute of Technology Hyderabad, Kandi, Telangana, India

M. Ali Haider
Renewable Energy and Chemicals Laboratory, Department of Chemical Engineering, Indian Institute of Technology Delhi, Hauz Khas, Delhi, India

Olusola O. James
Chemistry Unit, Faculty of Pure & Applied Sciences, Kwara State University, Malete, Kwara State, Nigeria

Janaki Komandur
Department of Chemical Engineering, Indian Institute of Technology Guwahati, Guwahati, India

Muxina Konarova
Australian Institute for Bioengineering and Nanotechnology (AIBN), The University of Queensland, Brisbane, QLD, Australia

Adarsh Kumar
Material Resource Efficiency Division, CSIR-Indian Institute of Petroleum, Dehradun, Uttarakhand; Academy of Scientific and Innovative Research (AcSIR), Ghaziabad, India

Pankaj Kumar
Department of Chemical Engineering, Birla Institute of Technology and Science (BITS), Pilani, Hyderabad Campus, Hyderabad, Telangana, India

Alekhya Kunamalla
Department of Chemical Engineering, Indian Institute of Technology Hyderabad, Kandi, Sangareddy, Telangana, India

Xinying Liu
Institute for the Development of Energy for African Sustainability (IDEAS), University of South Africa (UNISA), Florida Campus, Johannesburg, South Africa

Swarnalatha Mailaram
Department of Chemical Engineering, Indian Institute of Technology Hyderabad, Kandi, Sangareddy, Telangana, India

Sudip Maity
Gasification, Catalysis and CTL Research Group, Central Institute of Mining and Fuels Research (Digwadih Campus), Dhanbad, Jharkhand, India

Sunil K. Maity
Department of Chemical Engineering, Indian Institute of Technology Hyderabad, Kandi, Sangareddy, Telangana, India

Kaustubha Mohanty
Department of Chemical Engineering, Indian Institute of Technology Guwahati, Guwahati, India

Mahluli Moyo
Institute for the Development of Energy for African Sustainability (IDEAS), University of South Africa (UNISA), Florida Campus, Johannesburg, South Africa

Muthusivaramapandian Muthuraj
Department of Bioengineering, National Institute of Technology Agartala, Agartala, Tripura, India

Greg Perkins
School of Chemical Engineering, Faculty of Engineering, Architecture and Information Technology, The University of Queensland, Brisbane, QLD, Australia

Panneerselvam Ranganathan
Department of Chemical Engineering, National Institute of Technology Calicut, Kozhikode, India

Janakan S. Saral
Department of Chemical Engineering, National Institute of Technology Calicut, Kozhikode, India

Tahir Hussain Seehar
Department of Energy Technology, Aalborg University, Aalborg, Denmark; Department of Energy & Environment Engineering, Dawood University of Engineering & Technology, Karachi, Sindh, Pakistan

Ayaz Ali Shah
Department of Energy Technology, Aalborg University, Aalborg, Denmark; Department of Energy & Environment Engineering, Dawood University of Engineering & Technology, Karachi, Sindh, Pakistan

Kamaldeep Sharma
Department of Energy Technology, Aalborg University, Aalborg, Denmark

Bhushan S. Shrirame
Department of Chemical Engineering, Indian Institute of Technology Hyderabad, Kandi, Sangareddy, Telangana, India

Malayil Gopalan Sibi
School of Chemical Engineering, Sungkyunkwan University, Seoul, South Korea

Jasvinder Singh
Material Resource and Efficiency Division, CSIR-Indian Institute of Petroleum (CSIR-IIP), Dehradun; Academy of Scientific and Innovative Research (AcSIR), Ghaziabad, India

Satyansh Singh
Department of Chemical Engineering and Technology, Indian Institute of Technology (Banaras Hindu University), Varanasi, Uttar Pradesh, India

Saqib Sohail Toor
Department of Energy Technology, Aalborg University, Aalborg, Denmark

Deepak Verma
School of Chemical Engineering, Sungkyunkwan University, Seoul, South Korea

Editors biography

Sunil K. Maity is a professor in the Department of Chemical Engineering, Indian Institute of Technology, Hyderabad, India. He received his PhD from the Department of Chemical Engineering, Indian Institute of Technology, Kharagpur. His current research interests are
1. biorefinery for biofuels and organic chemicals
2. heterogeneous catalysis and chemical reaction engineering
3. process design and techno-economic analysis

He is currently working on hydrodeoxygenation of vegetable oils, steam and oxidative steam reforming, conversion of biobutanol to gasoline, butylenes, and aromatics, hydroxyalkylation-alkylation reaction, and hydrocarbon biofuels from platform chemicals. Prof. Maity has executed several externally funded research projects and guided five PhD students so far. Prof. Maity has published 35 research articles, 5 book chapters, 1 edited book, and delivered several presentations at national and international conferences.

Kalyan Gayen is an academician in the field of chemical engineering. He is currently working in the Department of Chemical Engineering, National Institute of Technology, Agartala, India. He received his PhD from Indian Institute of Technology Bombay and worked as a postdoctoral fellow at the University of California. His current research interests are
1. microalgae and cyanobacteria-based biofuels and bioproducts
2. conversion of lignocellulosic biomass into liquid biofuels and value-added chemicals
3. metabolic network analysis and systems biology
4. fermentation technology

He has executed five externally funded projects and guided three PhD students. Dr. Gayen has published more than 36 research articles, 1 edited book, 13 book chapters, and several presentations at national and international conferences.

Tridib Kumar Bhowmick is an Assistant Professor in the Department of Bioengineering, National Institute of Technology, Agartala, India. He received his PhD from Indian Institute of Technology Bombay, India. He has worked as a postdoctoral research assistant from 2008 to 2013 at the Institute for Bioscience and Biotechnology Research, University of Maryland, College Park, Maryland. His research was focused on traditional Indian medicine, nanomaterials, and targeted therapeutic delivery. Currently, he is involved in exploring the northeast region of India, representing a biodiversity hotspot with endemic flora and fauna, in identifying promising microalgal strains as alternative biofuel resources. His broad interest is to understand the characteristics of biomaterials at the nanoscale level and is looking forward to their novel applications. Dr. Bhowmick has published more than 25 research articles, 1 edited book, 4 book chapters, and attended several national and international conferences.

Preface

Petroleum is a finite, nonrenewable source of energy that is projected to be exhausted in the next few decades. Moreover, petroleum-based fuels are responsible for global warming and the deterioration of the Earth's environment by greenhouse and toxic gases. On the other hand, transportation fuels, such as gasoline, diesel, and jet fuel, play an extremely vital role in our society. Renewable transportation fuels are thus mandatory for the sustainability of our society. Biomass is a key renewable carbon source on Earth that is abundant. Therefore, tremendous emphasis has been put forward in the past few decades to convert biomass into transportation fuels, known as biofuels. Bioethanol and biodiesel have emerged as the most promising biofuels. These biofuels, however, contain oxygen in their structure, resulting in unmatched fuel properties and incompatibility with refinery infrastructures and internal combustion engines. These biofuels are thus mixed with transportation fuels only to a limited extent. Therefore, there is a huge demand to produce gasoline, jet fuel, and diesel-range hydrocarbons from biomass. These biofuels are commonly known as hydrocarbon biofuels. The hydrocarbon biofuels are analogous to existing transportation fuels and hence compatible with internal combustion engines and refinery infrastructures.

In this context, an attempt has been made to consolidate the existing knowledge for the production of hydrocarbon biofuels from biomass. This book covers three major areas for the conversion of biomass to hydrocarbon biofuels:

1. Chemical and thermochemical conversion processes
2. Biological and biochemical conversion processes
3. Conversion processes of biomass-derived compounds

This book further covers process design, economics, and life cycle analysis of various processes. The individual chapters are written by renowned professionals from different parts of the world. These chapters provide comprehensive and up-to-date information about the current state-of-the-art research and commercial initiatives. This information will be the foundation for the researchers and investors for possible investments. We firmly believe that this book will act as a catalyst for promoting both fundamental and applied research in this area that will eventually lead to the development of sustainable hydrocarbon biorefinery processes.

Edited by
Sunil K. Maity
Department of Chemical Engineering, Indian Institute of Technology Hyderabad,
Sangareddy, Telangana, India

Kalyan Gayen
Department of Chemical Engineering, National Institute of Technology Agartala, Agartala, Tripura, India
Tridib Kumar Bhowmick
Department of Bioengineering, National Institute of Technology Agartala, Agartala, Tripura, India

CHAPTER 1

Hydrocarbon biorefinery: A sustainable approach

Alekhya Kunamalla[a], Swarnalatha Mailaram[a], Bhushan S. Shrirame[a], Pankaj Kumar[b], and Sunil K. Maity[a]

[a]Department of Chemical Engineering, Indian Institute of Technology Hyderabad, Kandi, Sangareddy, Telangana, India
[b]Department of Chemical Engineering, Birla Institute of Technology and Science (BITS), Pilani, Hyderabad Campus, Hyderabad, Telangana, India

Contents

1.1 Introduction	2
1.2 Biomass	3
1.2.1 Types of biomass	3
1.2.2 Chemistry of biomass	6
1.2.3 Availability of biomass	9
1.3 Biorefinery	10
1.3.1 Sugar and starch-based biorefinery	11
1.3.2 Lignocellulose-based biorefinery	12
1.3.3 Triglyceride-based biorefinery	12
1.3.4 Biorefinery for biofuels	13
1.4 Biorefinery for traditional biofuels	13
1.4.1 Biodiesel	15
1.4.2 Bioethanol	16
1.4.3 Biobutanol	16
1.4.4 Dimethyl ether	17
1.5 Biorefinery for fuel-additives	18
1.5.1 γ-Valerolactone (GVL)	18
1.5.2 Alkyl levulinates	19
1.5.3 2-Methylfuran (2-MF)	22
1.5.4 2-Methyltetrahydrofuran (2-MTHF)	22
1.5.5 2,5-Dimethylfuran (2,5-DMF)	22
1.5.6 2,5-Dimethyltetrahydrofuran (2,5-DMTHF)	23
1.5.7 5-Ethoxymethylfurfural (5-EMF)	23
1.5.8 Glycerol acetals	24
1.6 Hydrocarbon biorefinery	24
1.6.1 Chemical and thermochemical conversion processes	26
1.6.2 Biological and biochemical conversion processes	28
1.6.3 Conversion of biomass-derived compounds to hydrocarbon biofuels	29
1.6.4 Petrochemical building-block chemicals from biomass	30
1.7 Role of heterogeneous catalysis in the hydrocarbon biorefinery	30
1.7.1 Supported metal catalysts	32

Hydrocarbon Biorefinery
https://doi.org/10.1016/B978-0-12-823306-1.00004-2

Copyright © 2022 Elsevier Inc.
All rights reserved.

1.7.2 Zeolite-type catalysts	33
1.7.3 Solid-acid and solid-base catalysts	34
1.8 Conclusions	35
References	35

Abbreviation

2,5-DMF	2,5-dimethylfuran
2,5-DMTHF	2,5-dimethyltetrahydrofuran
2-MF	2-methylfuran
2-MTHF	2-methyltetrahydrofuran
5-EMF	5-ethoxymethylfurfural
ABE	acetone–butanol–ethanol
FTS	Fischer–Tropsch synthesis
GVL	γ-valerolactone
HAA	hydroxyalkylation-alkylation
HDO	hydrodeoxygenation
HMF	5-hydroxymethylfurfural
LA	levulinic acid
MMT	million metric tons
MON	motor octane number
RON	research octane number

1.1 Introduction

Transportation fuels play an extremely vital role in present society. Gasoline, diesel, and jet fuel are major liquid transportation fuels at the moment. Diesel is the primary transportation fuel in India that accounts for more than 75% of transportation fuels consumed in the country. On the other hand, gasoline is the major transportation fuel in the United States. These transportation fuels are the major energy-consuming sector, accounting for about 28% of the world's energy consumption [1]. They are mainly produced from crude oil. However, some countries in the world blend a limited amount of biomass-derived fuels with these transportation fuels. Petroleum is, however, limited and nonrenewable. The current consumption rate indicates that existing petroleum reserves will be exhausted within the next 50 years [1]. However, reservoirs sometimes produce more petroleum than the estimated reserve due to the application of enhanced recovery techniques. Moreover, new petroleum reservoirs are discovered from time to time. Therefore, petroleum may last a little longer than 50 years. Moreover, these transportation fuels are responsible for emissions of carbon dioxide, NO_x, unburned hydrocarbons, and particulate matter into the atmosphere. These emissions are accountable for global warming and unhealthy air quality, causing various health issues. Petroleum reserves are also limited in many countries in the world. For example, India imports about 84% of the petroleum consumed in the country. Therefore, the production of transportation fuels from renewable sources is essential for the sustainability of human civilization and maintenance of clean air quality.

Biomass is the only carbon-based renewable energy source with the potential to provide transportation fuels, called biofuels. It may be emphasized that petroleum has been originated from biomass. The biofuels are carbon-neutral. The carbon dioxide emitted from these biofuels is reused for the growth of biomass, thereby preventing global warming and maintaining consistent air quality. The biofuels further offer advantages, such as the utilization of waste biomass and improvement in the rural economy. The biomass can also provide renewable organic chemicals, heat energy, and electricity. In contrast, other renewable energies, such as solar, wind, hydrothermal, etc., provide only energy in the form of electricity and heat. Biomass is thus considered the most promising renewable energy source. In ancient times, human civilization was dependent on biomass for its energy needs. Until now, biomass remains the primary energy source, mostly for cooking, in several underdeveloped or developing countries. Currently, biomass delivers around 10% of the world's energy or 50% of renewable energy. Biomass and waste account for about 22% of energy consumption in India [2]. The European Renewable Energy Council targeted 50% contribution of renewable energy in the world's energy consumption by 2040. Anticipating the potentials of biomass, the International Renewable Energy Agency proposed substituting about 22% of transportation fuels with biofuels by 2050 [3]. This chapter thus provides an overview of the various biomass sources with their availability and chemical structure and different types of biorefinery for the production of numerous biofuels. This chapter, specifically, introduces a novel hydrocarbon biorefinery concept for manufacturing hydrocarbon biofuels and building-block chemicals from biomass and the role of heterogeneous catalysis in this biorefinery.

1.2 Biomass

The definition of biomass has been widened since the evolution of bioenergy concepts. Biomass is a renewable organic matter that broadly includes plants, microorganisms, and animal wastes. Based on their source, biomass is classified as energy crops, wastes, and residues obtained from agriculture crops and forest management, wastes generated from industries and municipality, and aquatic plants (Table 1.1) [4, 5].

1.2.1 Types of biomass

Energy crops. Energy crops are cultivated wittingly for bioenergy production. These energy plantations are low-cost, short-rotation, and high-yielding varieties that need nominal maintenance. These dedicated crops encompass wood, perennial forage, sugar, starch, and oil-based energy crops. The woody energy crops have a wide geographical distribution with high potential yields. They are cultivated on a rotational basis for pulp production. The perennial forage energy crops are adaptable in the wasteland and require

4 Hydrocarbon biorefinery

Table 1.1 Different types of biomass based on their source.

Energy crops	
Oil crops	Oilseed rape, sunflower, castor oil, olive, coconut, groundnut, jatropha, soybean oil, palm oil, cottonseed oil, linseed oil, field mustard, palm oil, etc.
Sugar crops	Sweet sorghum, sugar beet, sugarcane, etc.
Starch crops	Wheat, sweet potato, corn, rice, barely, maize, etc.
Woody crops	Hybrid willow, silver maple, sweetgum, hybrid poplar, silver maple, eastern cottonwood, black walnut, green ash, sycamore, pine, miscanthus, etc.
Grass crops	Bamboo, switchgrass, kochia, wheatgrass, canary grass, coastal Bermuda grass, alfalfa hay, thimothy grass, common reed, Indian shrub, giant reed, immature cereals, etc.
Agriculture waste and residue	
Stalk	Bean, corn, cotton, kenaf, mustard, sunflower, triticale, etc.
Straw	Bean, corn, oat, rice, sesame, sunflower, wheat, mint, paddy, rape, rye, etc.
Shell and husk	Almond, cashewnut, coffee, olive, peanut, walnut, sunflower, cotton, etc.
Fiber	Flax, palm, kneaf bast, jute bast, coconut coir, etc.
Livestock waste	Chicken manure, pig manure, sheep manure, cattle manure, bones, meat bone meal, etc.
Forest waste and residue	
Forest waste	Wood blocks, wood chips from thinning, logs from thinning, barks, early thinning branches, leaves, bushes, etc.
Industry and municipality waste and residue	
Food industry residue	Cooking oil, proteins, tallow, wet cellulosic material (beet root tails), animal fat, oil ghee waste, fruit and vegetable scrap, fiber obtained from extraction of sugar and starch, etc.
Paper and wood industry residue	Fibrous waste from paper and pulp industries, waste wood from sawmills, waste wood from timber mills (bark, wood chips, slabs, and sawdust), paper and pulp sludge, etc.
Other industry waste	Tanning waste (leather particles, waste liquor, and fleshing residue), soap industry (oil and grease), wastewater sludge, etc.
Municipality waste	Sewage sludge, waste paper, yard waste, garden grasses and plants, dairy waste, wood pallets and boxes, fiberboard, plywood, paperboard waste, stale food (uneaten bread, rice, and vegetables), etc.
Marine biomass	
Marine biomass	Marine and freshwater algae (microalgae and macroalgae), cyanobacteria, marine microflora, giant kelp, other marine microorganisms (marine yeast), duckweed, water hyacinth, sweet water weeds, etc.

minimum water and nutrients for their cultivation. Typical examples of these energy crops are bamboo, switchgrass, wheatgrass, sweet sorghum, and miscanthus [4, 6]. The oil, sugar, and starch-based crops are cultivated across the globe for food purposes. These energy crops thus pose food versus fuel conflicts. Some countries, however, produce these energy crops in excess of human consumption for bioenergy applications. For example, Brazil produces sugarcane for bioethanol production. The Indian Government promoted jatropha cultivation in wastelands for biodiesel production. Jatropha oil is inedible and hence does not pose food versus fuel conflict.

Agriculture wastes and residues. A considerable amount of biomass wastes and residues is coproduced during the harvesting of crops. These agro-wastes have, however, limited utilization. They are either used as cattle fodder or fuel for cooking in remote villages. The majority of these biomasses are still either burned in the field or dumped into the wasteland. Rice husk is the most prominent agro-residue, and it accounts for about 25% of rice by mass [7]. Since corn is grown worldwide, corn stover is one of the major agro-residues. Animal manure is also included in agriculture wastes [4, 7]. Animal manure, such as cow dung, is currently used for biogas production. The agricultural wastes and residues are inedible and hence an attractive biomass for a biorefinery. This biomass has, however, low density and is generated in remote agriculture fields. The transportation of this biomass from a remote locality to a central biorefinery is thus expensive. Moreover, this biomass is seasonal. The storage facility for this biomass is thus obligatory for the uninterrupted operation of a biorefinery.

Forest waste and residue. Forest wastes and residues are another pertinent biomass. The logging residues generated from harvest operation, fuelwood derived from forestlands, stumps, and early thinning of branches are primary forest residues. In contrast, wood processing generates secondary residues. These operations are essential to maintain forest health and productivity. The bark and wood have a high heating value and are suitable for thermal energy applications. Solid wood is the main source of energy for the small-scale industries in underdeveloped and developing countries. The low recoverability and cost of transportation, logging, and collection activities are major hurdles for this biomass. The utilization of this biomass at or near source in a decentralized biorefinery is thus desirable.

Industry and municipality waste. The organic chemical industries and municipalities generate a large volume of organic wastes with only a small amount of inorganic materials [8]. Solid wastes, green wastes (nutshells, grass cutting, etc.), sewage sludge, and industry wastes are examples of industry and municipality waste. The municipality solid wastes comprise waste food, paper, plastic, and textiles. Food wastes are generated in various steps of the food processing chain, such as manufacturing, logistics, storage, and postconsumer stage [5, 9]. The wastes generated in the pulp industry, bagasse produced in the sugarcane industry, and coproducts (fatty acid distillate, protein-rich cake, gum, soap stock, etc.) of the oilseed processing industry are examples of industrial wastes.

Marine biomass. Marine biomass refers to diversified groups of aquatic species, photosynthetic algae, and cyanobacteria. The size of this biomass ranges from microscopic (cyanobacteria, microalgae, etc.) to macroscopic (macroalgae and brown, red, blue, and green seaweeds). Both micro and macroalgae have more than one million species [8, 10]. These species could be a potential source of triglyceride and cellulosic biomass. They can grow in wastewater, coastal seawater, saline water, and nonarable land. In nutrient-rich water, micro and macroalgae grow with a high cell density of up to 10^7 cells/mL [11]. The growth rate of algae is also very high with a short harvesting cycle (1–10 days).

1.2.2 Chemistry of biomass

The biomass is composed of a diverse range of organic compounds with an insignificant amount of inorganic substances. Carbon, hydrogen, and oxygen are major elements in biomass with a minor quantity of nitrogen and sulfur. Carbohydrates, lignin, lipid, proteins, and fats are major constituents of biomass. The amount of these constituents in biomass depends on their types, growth stage, source, and geographic location. The four categories of biomass are commonly used in a biorefinery: triglyceride, sugar, starch, and lignocellulose.

Triglycerides. The triglycerides include various inedible and edible vegetable oil, waste cooking oil, waste oils (trap grease, yellow grease, etc.), microalgae oil, and animal fats. A triglyceride molecule contains one glycerol unit and three same or different fatty acids linked by an ester bond (Fig. 1.1). The fatty acids contain a linear carbon chain in the range of C_8–C_{24}. Palmitic ($C_{16:0}$), stearic ($C_{18:0}$), oleic ($C_{18:1}$), and linoleic ($C_{18:2}$) acids are primary fatty acids in triglycerides. Both saturated and unsaturated fatty acids are present in triglycerides. The triglycerides also contain free fatty acids. The number of double bonds, the carbon chain length of fatty acids, and the extent of free fatty acids depend on the types of triglycerides (Table 1.2).

Sugar and starch. Sugar ($C_{12}H_{22}O_{11}$) is made of two six-carbon saccharides, namely α-glucose ($C_6H_{12}O_6$) and β-fructose ($C_6H_{12}O_6$.). They are bonded by α1 glucosidic and β2 fructosidic bonds (Fig. 1.1). Sugar is obtained from sugar beet, sugarcane, sweet sorghum, etc. On the other hand, in starch, α-glucose monomers are linked by α1,4 and α1,6 glucosidic bonds to form polymeric structures. Two principal components are present in starch: amylose (20%–25%, water-soluble) and amylopectin (75%–80%, water-insoluble). Amylose contains only α1,4 glucosidic bonds, while amylopectin contains both α1,4 glucosidic and α1,6 glucosidic bonds [4]. Starch is found in potato, rice, wheat, beans, etc.

Lignocellulose. Lignocellulose biomass is a composite of cellulose (40%–50%), hemicellulose (25%–35%), and lignin (15%–25%) with a small amount of protein, pectin, and extractives [4, 15] (Fig. 1.1). The carbohydrates (i.e., cellulose and hemicellulose) provide mechanical and structural strength to the plant, while the noncarbohydrate

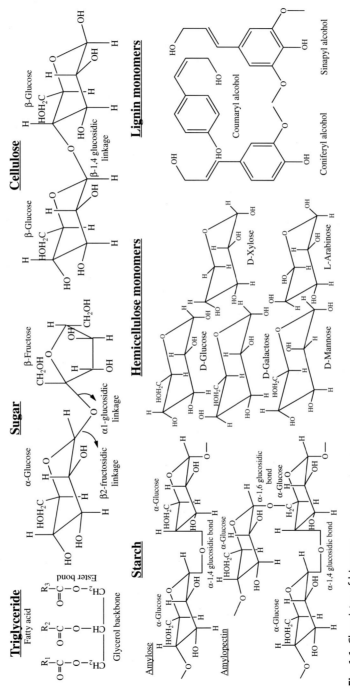

Fig. 1.1 Chemistry of biomass.

Hydrocarbon biorefinery

Table 1.2 Chemical composition of biomass [9, 12–14].

| Biomass | Lignocellulose biomass, wt% | | |
	Cellulose	Hemicellulose	Lignin
Bagasse	47.4	29.1	23.5
Bamboo	43.9	26.5	29.6
Corn cob	48.1	37.2	14.7
Cotton stalk	66.2	18.4	15.4
Pine sawdust	45.9	26.4	27.7
Rice straw	52.3	32.8	14.9
Eucalyptus	48	14	29
Switchgrass	40–45	31–35	6–12

| | Vegetable oil, wt% | | | | | | | | |
	$C_{12:0}$	$C_{14:0}$	$C_{16:0}$	$C_{18:0}$	$C_{18:1}$	$C_{18:2}$	$C_{18:3}$	$C_{20:0}$	$C_{22:1}$
Coconut[a]	44–51	13–20.6	7.5–10.5	1–3.5	5–8.2	1–2.6	0–0.2	–	–
Jatropha	–	1.4	14.6	7.4	41.4	35.4	0.2	0.3	–
Karanja	–	–	8.9	5.9	56.1	16.4	–	3.2	1.9
Palm	0.5–2.4	32–47.5	3.5–6.3	36–53	6–12	–	–	–	–
Rapeseed	–	0–1.5	1–6	0.5–3.5	8–60	9.5–23	1–13	–	5–56
Soybean	–		2.3–13.3	2.4–6	17.7–30.8	49–57.1	2–10.5	–	0–0.3
Sunflower	–	–	3.5–7.6	1.3–6.5	14–43	44–74	–	–	–

[a]$C_{8:0} = 4.6$–$9.5\,wt\%$, $C_{10:0} = 4.5$–$9.7\,wt\%$.

(i.e., lignin) gives stability to these structures. The composition of lignocellulose biomass is presented in Table 1.2. Cellulose is a long-chain polysaccharide with a high degree of polymerization (nearly 10,000). It is a linear polymer of D-glucopyranose formed by β1,4 glycosidic linkage. Cellulose has a crystalline structure with four different forms: Iα, Iβ, II, and III [16]. The crystalline structures are formed by intra and intermolecular hydrogen bonding. The tight packing of polymeric chains and crystalline structure makes cellulose recalcitrant for degradation and insoluble in water. In plants, it is present as elementary fibrils consisting of a single cellulose chain or bundles of elementary fibrils, i.e., microfibrils. These complexes offer strength and chemical stability to plants [17].

Hemicellulose is a branched polymer containing C_5 (xylose and arabinose, $C_5H_{10}O_5$) and C_6 (glucose, mannose, and galactose, $C_6H_{12}O_6$) sugars. These sugars are acetylated with uronic acid. The heterogeneous combination of monosaccharides in hemicellulose is formed by β1,4 glycosidic and β1,3 glycosidic bonds. It is an amorphous polymer that can readily hydrolyze to monomer sugars [17]. The abundant hemicellulose in hardwood is xylan, in which xylose sugars are linked at one and four positions. While softwood hemicellulose mostly contains glucomannan, a polymer of D-glucose and D–mannose bonded by β1,4 glucosidic bonds, lignin is abundant in lignocellulose biomass. It is a non-crystalline and three-dimensional polymer of phenylpropanoid units. The three units that form the lignin structure are coniferyl alcohol, sinapyl alcohol, and coumaryl alcohol. The lignin content and composition vary with the types of biomass. The hardwood lignin contains a combination of coniferyl (25%–50%) and sinapyl alcohol (50%–75%). The softwood lignin has a high percentage of coniferyl alcohol (90%–95%), while grass lignin has all three building blocks, i.e., coniferyl (25%–50%), sinapyl (25%–50%), and coumaryl (10%–25%) alcohol [18].

1.2.3 Availability of biomass

Biomass is distributed in vast territories on Earth with a large number of species. Therefore, it is difficult to assess the exact availability of biomass on a renewable basis. In general, waste biomass from agriculture, forest, industries, and municipalities is abundant without much impact on the environment and food products. This biomass is thus appropriate for a biorefinery. In 2014, the United States consumed 5.6 million metric ton (MMT) corn and 0.216 MMT biomass to produce bio-based products and gasoline blendstocks, respectively [19]. In 2013, the industrial sector in the United States utilized approximately 85.3 MMT (dry) wood and wood waste to generate 539 trillion Btu thermal and 15.4 trillion Btu electrical energy, while the residential sector produced 349.5 trillion Btu thermal energy from 44.8 MMT (dry) wood and wood waste. The United States further generated 254 MMT municipal solid waste in 2013. About 34% of these wastes were discarded for landfilling, and the remaining were used for either energy recovery or compost [19]. In 2016, the United States generated approximately 3.3 MMT wood pellets. Nearly 85% of these were residues originated from the sawmill, wood product manufacturing, and logging, while the remaining 15% were generated from logs harvested for industrial applications [20]. The United States proposed to generate 1.3 billion tons dry biomass by 2030 for bioenergy application. It includes 87 MMT grains, 106 MMT animal manure and food processing waste, 368 MMT forest waste, 377 MMT energy crops, and 428 MMT crop residue [21].

In 2017, the global bioenergy supply was 55.6 EJ with 48.2 EJ contribution from solid biofuels (wood pellets, wood chips, fuelwood, etc.). The contributions of liquid biofuels, municipality waste, biogas, and industrial waste were 3.65, 1.45, 1.33, and 1.07 EJ, respectively [22]. On the other hand, the contribution of renewable electricity was 0.33 EJ globally [22]. In 2017, biomass supplied 5% of total primary energy consumed in the United States with about 47%, 44%, and 10% contribution from wood, wood-derived biomass, and municipal waste, respectively [23]. In 2018, the United States produced 38.1 MMT of oil equivalent biofuel, which was the highest in the world (39.9%). Brazil was the world's second highest biofuel producer (21.4 MMT of oil equivalent, 22.4%) [24, 25].

In India, the estimated annual availability of biomass from agriculture, forest, and wasteland was 242 MMT in 2010–2011 and is anticipated to be 281 MMT by 2030–2031 [26]. Rice and wheat are the most cultivated crops in India which account for 41% of the cultivated area, while the remaining 15.9%, 13.8%, and 10.2% account for oilseed, pulses, and commercial crops, respectively [11]. In 2010–2011, the estimated production of rice straw was 172.8 MMT, of which 80.8% and 11.1% were used as fodder and fuel, respectively, and the remaining 8% were used for other purposes. About 139.2 MMT wheat straw was generated in 2010–2011, and it is estimated to be 193.7 MMT by 2030–2031 [26]. In 2015–2016, the total crop residues in India were estimated as 816.4 MMT. The sugarcane and cotton residues account for 282 MMT [11, 26]. Up to May 2019, the installed capacity of renewable power was 78.4 GW in India with 46%, 37%, and 12.8% contribution from wind, solar, and biomass, respectively. India estimates a projected 175 GW renewable power by 2022 with 10 GW from biomass. The principal feedstock is bagasse from sugar mills [27]. In 2015–2019, the average annual ethanol consumption growth was higher compared to the production growth (8%) in India [27]. The combined effect of enhanced fiscal deficit owing to the increase in crude oil import and depreciating rupee makes the Indian Government focus on biofuels. The primary fuel consumption in India was 809.2 MMT of oil equivalent, which is nearly 5.84% of global consumption in 2018 with 27.5 MMT of oil equivalent contribution from renewables [25]. The Indian annual biofuel production growth rate (19.7% in 2007–2017) was higher than the global growth rate (9%). In 2018, biofuel production in India was 1023 MMT of oil equivalent, which was 1.1% of global biofuel production [25]. The National Biofuel Policy in India proposed 20% mixing of bioethanol with gasoline and 5% mixing of biodiesel with diesel by 2030 [24, 27]. During 2013–2014, only 1.53% ethanol was blended with gasoline. It was increased to 3.5% in 2015–2016 and dropped to 2.07% in 2016–2017 [24].

1.3 Biorefinery

Coal is used mainly to produce electricity in thermal power plants. In contrast, the petroleum refinery refines crude oil to produce a spectrum of products, primarily fuels, such as liquefied petroleum gas, gasoline, kerosene, jet fuel, diesel, and fuel oil. These fuels find application in the transportation sector, cooking, illumination, and industry. Besides, the petroleum refinery produces naphtha as feedstock for petrochemical industries. On the other hand, natural gas is used as feedstock for petrochemical industries and road transportation fuel in the form of compressed natural gas. In the petrochemical industry, naphtha and natural gas are further processed to produce three basic building-block chemicals: (i) synthesis gas ($H_2 + CO$), (ii) olefins, such as ethylene, propylene, butylenes, and

Table 1.3 Platform chemicals.

Ethanol	Biohydrocarbons
Furfural	Succinic acid
5-Hydroxymethyl furfural	Hydroxypropionic acid/aldehyde
Furan dicarboxylic acid	Levulinic acid
Glycerol and derivatives	Sorbitol
Lactic acid	Xylitol
Isoprene	

butadiene, and (iii) aromatics, such as benzene, toluene, ethylbenzene, and xylenes. These building-block chemicals are the foundation for organic chemicals, fertilizers, polymers, and commodities for our society.

A novel manufacturing concept is developing throughout the world to substitute these fossil fuels with biomass. This concept is known as the biorefinery [3, 4]. The biorefinery should ideally produce a spectrum of products that are currently obtained from fossil fuels, such as biofuels, organic chemicals, polymers, materials, heat energy, and electricity. The biorefinery identified 12 primary biomass-derived organic chemicals, called platform chemicals (Table 1.3) [28]. These platform chemicals are produced from the carbohydrate fraction of biomass using a combination of chemical and biochemical processes. The glycerol is, however, coproduced in the transesterification process (10 wt% of biodiesel). These platform chemicals have vast derivative potentials to produce organic chemicals, polymers, and commodity products. The platform chemicals can also be transformed into various biofuels. The biorefinery is thus analogous to the integrated petroleum refinery and petrochemical industry. The biorefinery can be classified based on either feedstock, products, or the nature of processing technologies. In general, the specific biorefinery is expected to process biomass with a similar chemical nature. So the biorefinery is speculated based on the specific chemical nature of biomass, such as the sugar and starch-based biorefinery, lignocellulose-based biorefinery, and triglyceride-based biorefinery [3, 4]. These individual biorefineries can be further classified based on either targeted products or types of processing technologies.

1.3.1 Sugar and starch-based biorefinery

Sugar and starch-based biorefineries are well known because of bioethanol. The starch-based feedstock undergoes hydrolysis to form aqueous sugars. Aqueous sugars are extracted from sugar-based feedstock through physical separation methods. Aqueous sugars are fermented to vast ranges of products. Strains used in fermentation determine the types of products. Bacterial strains are used for acetone–butanol–ethanol (ABE) fermentation with coproduction of hydrogen, while yeast fermentation yields ethanol. Bioethanol and biobutanol are used as biofuel and solvent. They are also reformed to synthesis gas and catalytically converted to petrochemical building-block chemicals, such as ethylene, butylenes, and aromatics [29–31]. Fermentation of sugars also produces various platform chemicals, such as succinic acid, lactic acid, glutamic acid, itaconic acid, and 3-hydroxy propanoic acid (Table 1.3). A metabolically engineered strain produces linear or branched higher bio-alcohols from sugars. Aqueous-phase catalysis of sugars produces aromatics and alkanes [32]. Hydrogen is produced by the aqueous-phase reforming of sugars. 5-Hydroxymethylfurfural (HMF) is produced by dehydrocyclization of hexose sugars. The decomposition of HMF produces levulinic acid (LA). Microbial processing of hexose sugars produces liquid biofuels [33].

1.3.2 Lignocellulose-based biorefinery

A lignocellulose biomass provides a wide spectrum of products. However, the complex structure of the lignocellulose biomass is the main bottleneck. The combustion of the lignocellulose biomass produces heat and electricity. The gasification produces synthesis gas, which is further converted to ethanol and higher alcohols and liquid biofuels through the Fischer-Tropsch process. Fast pyrolysis and liquefaction of the lignocellulose biomass produce bio-oil. Bio-oil is upgraded to biofuels, aromatics, and organic chemicals using hydrodeoxygenation (HDO) and zeolite upgrading [34]. Pretreatment of lignocellulose biomass forms hydrolysate containing sugars. Anaerobic fermentation of hydrolysate produces biogas and hydrogen. Further hydrolysis of pretreated lignocellulose biomass reduces the structural complexity and produces lignin and sugars (hexose and pentose). Aqueous sugars are fermented to ethanol, butanol, acetone, and hydrogen. Aqueous-phase dehydration/hydrogenation of sugars produces C_1–C_6 alkanes. Zeolite upgrading of sugars produces aromatics and hydrocarbons. Aqueous-phase reforming and microbial processing produce hydrogen and liquid biofuels, respectively [33, 35]. Furfural and HMF are C_5 and C_6 furans and obtained by dehydration of C_5 and C_6 carbohydrates, respectively, in the presence of acid catalysts [36]. LA is obtained by subsequent dehydration of furans. Biofine Corporation produces LA on a large scale using a series of two reactors (70%–80% yield) [37]. In the first reactor, carbohydrate is converted to HMF at 483–503 K in the presence of mineral acids. HMF is then converted to LA in the second reactor at 468–488 K with more than 60% yield. Lignin is converted to aromatics through zeolite upgrading. Depolymerization, hydrogenation, and solvolysis of lignin give gasoline-range fuel and phenolics.

1.3.3 Triglyceride-based biorefinery

Triglycerides are transformed into biofuels, value-added products, and organic chemicals. The cake or residue obtained from oilseed is used as fertilizer, feed, or feedstocks to the lignocellulose biorefinery. Transesterification of triglycerides for biodiesel production has gained enormous attention in this biorefinery. Hydrolysis of triglycerides produces

Table 1.4 Classification of biorefinery based on types of biofuels.

Biorefinery	Biofuels
Biorefinery for traditional biofuels	Biodiesel, bioethanol, biobutanol, dimethyl ether, etc.
Biorefinery for fuel-additives	γ-Valerolactone, furanic compounds (2-methylfuran, 2-methyltetrahydrofuran, 2,5-dimethylfuran and 2,5-dimethyltetrahydrofuran), 5-ethoxymethylfurfural, alkyl levulinates, glycerol ether/acetal, etc.
Hydrocarbon biorefinery	Green liquefied petroleum gas, green gasoline, green kerosene, green jet fuel, green diesel, etc.

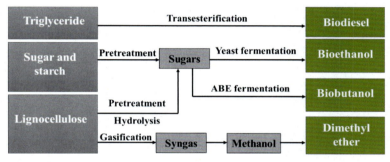

Fig. 1.2 Biorefinery for traditional biofuels.

fatty acids that are hydrogenated and isomerized to green diesel [38]. Fatty acids are also used to produce many value-added products like soaps, emulsifiers, surfactants, etc. Green diesel and green gasoline are also produced by direct HDO, catalytic cracking over solid-acid catalysts, and pyrolysis of triglycerides [39]. Steam and dry reforming of triglycerides produce synthesis gas [40, 41].

1.3.4 Biorefinery for biofuels

The biorefinery can be broadly classified into three types based on the nature of biofuels produced: (i) biorefinery for traditional biofuels, (ii) biorefinery for fuel-additives, and (iii) hydrocarbon biorefinery (Table 1.4) [42].

1.4 Biorefinery for traditional biofuels

The biorefinery for traditional biofuels is shown in Fig. 1.2. The fuel properties of these biofuels are compared with gasoline and diesel, as presented in Table 1.5.

Table 1.5 Fuel properties of traditional biofuels [43–46].

Property	Diesel	Gasoline	Biodiesel	Bioethanol	Biobutanol	Dimethyl ether
Composition	C_{12}–C_{25}	C_5–C_{12}	C_{12}–C_{24} Fatty acid methyl esters	C_2H_5OH	C_4H_9OH	C_2H_6O
Molar mass, g/mol	198.4	95–120	–	46.07	74.12	46
Water, ppm	–	–	0.05% max	–	–	–
C, wt%	–	86	77	52.2	65	52.2
H, wt%	–	14	12	13.0	13.5	13
O, wt%	0	0	11	34.8	21.5	34.8
S, ppm	<10	–	<1	–	–	
Specific gravity at 293 K	0.84	0.739	0.88 (288 K)	0.789	0.809	0.661
Boiling point, K	513–633	308–413	455–610	350.3	390.7	248.1
Flash point, K	338–341	230	373–443	285.5	308	–
Cloud point, K	268	–	268–288	–	–	–
Pour point, K	–	–	258–289	–	–	–
Auto ignition temperature, K	483–523	298.7	409	696	638	508
Air/fuel ratio, wt/wt	–	14.8	12.5	9	11.1	9
Lowering heating value, MJ/kg	42–46	44–46	28.12	26.7	33.1	27.6
Calorific value, MJ/L	38.6	33.5	31–33	21.2	29.2	–
Heat of vaporization, MJ/kg	0.25–0.6	0.26–0.36	–	0.83	0.71	0.47
Cetane number	40–55	5–20	50–60	5–8	~25	55–60
Research octane number	20–30	95	–	107	105–113	–
Motor octane number	20–30	90	–	89	80	–
Reid vapor pressure, kPa	–	45–90	<0.6	17	2.3	–
Kinematic viscosity, mm^2/s at 293 K	1.9–4.1	0.4–0.8	3.5–5.2 (313 K)	1.5	3.6	<1

1.4.1 Biodiesel

Transesterification of triglycerides with alcohol forms fatty acid alkyl ester, commonly known as biodiesel (Fig. 1.3). Biodiesel contains no aromatics and sulfur and shows a high cetane number (50–60). The cold flow properties of biodiesel vary with the types of alcohol used in the reaction. Methanol is low-cost alcohol and is widely used in this reaction. The biodiesel is produced using either acid, alkali, or enzyme as the catalyst [47]. Conventional processes mostly use alkalis, such as NaOH and KOH, as catalysts. In the presence of alkali catalysts, the reaction occurs at a faster rate compared to acid and enzyme catalysts and is operated under mild reaction conditions ($303 - 333\,K$). A methanol to triglycerides molar ratio of 6:1 is maintained for the alkali-catalyzed transesterification reaction. Complete conversion with more than 98% biodiesel yield is achieved in the alkali-catalyzed transesterification reaction. However, the presence of water in methanol and free fatty acids in triglycerides results in the formation of soap that poses difficulty in product separation. In this reaction, vegetable oil, methanol, and catalyst are fed into the transesterification reactor [48]. The product stream containing biodiesel, glycerol, excess methanol, and catalyst is sent to the settling tank where the glycerol phase (dense) is separated from the biodiesel phase (light). The glycerol phase contains the catalyst, traces of methanol, and soap. Phosphoric acid is used to break soap into fatty acids and salts. It also neutralizes the catalyst and forms phosphate, which is separated by a centrifuge. Traces of methanol are removed from the glycerol phase by evaporation and glycerol is purified using vacuum distillation. The biodiesel phase contains excess methanol, which is separated and recycled to the reactor. Biodiesel is sent to the water washing column, followed by neutralization to remove fatty acids and catalysts. The alkali-catalyzed transesterification process involves the complex separation of products, catalyst, soap, unreacted oil, and excess methanol. Supercritical and microwave-assisted transesterification reactions are alternative processes to overcome these complex separation processes. Feedstock contributes 60%–75% of the biodiesel cost. The revenue generation from glycerol is about 10% of the biodiesel cost [49]. The break–even price of the alkali and acid-catalyzed continuous biodiesel process from vegetable oil and waste cooking oil was reported to be in the range of 2.10–2.90 USD per gal [50].

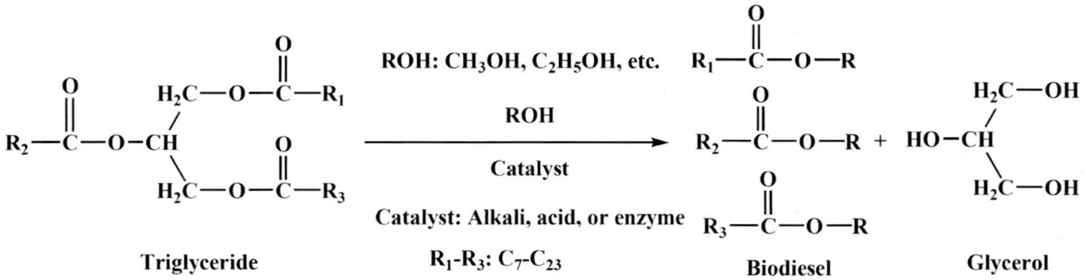

Fig. 1.3 Transesterification reaction.

1.4.2 Bioethanol

Bioethanol is the most dominant biofuel with about 85.1% contribution toward the world's total biofuel production in 2017 [22]. Bioethanol is largely derived from maize in the United States (60 billion liters annually) and sugarcane in Brazil (20 billion liters annually) [51]. These two countries together contributed nearly 87% of the world's bioethanol production in 2017. The high octane number (107 RON and 89 MON) makes bioethanol an attractive biofuel. The use of bioethanol in an internal combustion engine is also advantageous compared to that of gasoline [3]. The broader flammability limit, better flame speed, and higher heat of vaporization of bioethanol increases the compression ratio and shortens burn time. The presence of structural oxygen with negligible amounts of sulfur in bioethanol reduces emissions, facilitates clean-burning, and improves combustion efficiency. The lignocellulose biomass contains both hexose and pentose sugars, which are not easily digestible by a single enzyme. These sugars are fermented either together using modified enzymes (co-fermentation) or separately using their respective enzymes. Baker's yeast is the most common microorganism used in the fermentation of sugars to ethanol (operated at 306 K) [52]. However, a sugar concentration of more than 60 g/L and bioethanol concentration of larger than 10 wt% is toxic to the microorganism. The maximum yield is 0.48 g bioethanol per g sugar. The aqueous bioethanol is dehydrated using a series of distillation columns. Less than 95 wt% bioethanol, below its azeotropic point, is achieved in these columns [53]. This azeotropic mixture is subsequently refined to fuel-grade anhydrous bioethanol by azeotropic distillation using an entrainer, such as ethylene glycol and benzene. Bioethanol dehydration is an energy-intensive process. The fermentation integrated with hybrid separation techniques reduces the energy consumption. There are many bioethanol producers all over the world. Beta Renewables in Italy produces 0.075 million m^3 per year bioethanol from agricultural residues [3]. The estimated bioethanol production cost is 0.7 USD/L and 0.73 USD/L of gasoline-equivalent from corn and lignocellulose biomass, respectively. Bioethanol seems to be cheaper than gasoline (0.78 USD/L) for 120 USD/bbl oil price. If the oil price drops to 60 USD/bbl, bioethanol (0.6 USD/L), however, becomes more expensive than gasoline (0.42 USD/L).

1.4.3 Biobutanol

n-Butanol has emerged as a potential biofuel after David Ramsey drove his butanol-fueled car in 2005 [3]. Biobutanol has better fuel properties compared to bioethanol, such as higher energy density, lower water solubility, lower volatility, lesser corrosiveness, and lesser hygroscopicity. It has an octane number (105–113 RON, 80 MON) similar to that of gasoline. It can be mixed with petrol in any proportion or used in pure form without engine modification. The aqueous sugars undergo anaerobic fermentation in the presence of bacterial clostridia species to produce ABE in the mole ratio of 3:6:1. ABE fermentation is carried out at 304–308 K in a controlled pH of 5.4 [54]. ABE fermentation follows acidogenesis and solventogenesis metabolic pathways. *Clostridium beijerinckii* and *Clostridium acetobutylicum* are the most effective strains to produce ABE from both pentose and hexose sugars. Fermentation is operated in batch, fed-batch, and continuous modes. However, the biobutanol production process is facing various techno-economic challenges [55]. The low biobutanol yield (0.28–0.33 g/g), productivity (<0.3 g/L/h), and solvent concentration (<20 g/L) are the primary challenges in this process. The toxicity of solvents to microorganisms limits the concentration of ABE. Genetic modification of microorganisms is an important tool to increase tolerance toward solvents and to improve yield and productivity. The high energy consumption for the separation of biobutanol from ABE and water is another major problem. Conventional distillation consumes 79.5 MJ/kg, which is nearly two times that of biobutanol's (36 MJ/kg) energy content [56]. In situ solvent removal techniques are thus extensively studied to improve biobutanol concentration and to decrease energy consumption [57]. The liquid–liquid extraction has been reported to be the least energy-consuming (about 6 MJ/kg) technique among all ABE separation processes. The leading producers of biobutanol in the world are Butamax, Gevo, US Technology Corporation, and Green Biologics [58]. BP and DuPont started 30,000 tons of biobutanol production per year in Great Britain [59]. The ABE technology developed by China in the 1950s is regaining its interest by restoring units [60]. In this process, feedstock contribution is more than 65% of the biobutanol production cost. An additional pretreatment step in the lignocellulose biomass adds 37% to capital investment compared to sugar-based feedstock. The production cost is in the range of 0.59–0.75 USD, 0.62 USD, and 1.29 USD per kg of biobutanol for various cellulosic feedstock, sugarcane, and corn, respectively, for the 10,000 tons of biobutanol per year plant capacity [61]. Currently, more than 4.5 billion liters of biobutanol are produced per year, and the market is estimated to be USD 247 billion by 2020 [62].

1.4.4 Dimethyl ether

Dimethyl ether is highly volatile, and it is liquefied above 0.5 MPa and can be handled like liquefied petroleum gas. It has a cetane number (55–60) similar to that of diesel. It is thus a potential fuel in diesel engines. It has a low boiling point (248 K) that helps quick evaporation inside engines [63]. The presence of structural oxygen (35%) and absence of the C—C bond helps clean-burning with reduction of NO_x, SO_x, and particulate matters. It does not form explosive peroxides, unlike other ethers. Hence it is very safe for storage and handling. But it has a lower viscosity, combustion enthalpy, and modulus of enthalpy than diesel. It can also be used as a fuel additive, aerosol, and green refrigerant, a substitute for chlorofluorocarbons. Two different methods are followed for the synthesis of dimethyl ether: direct and indirect [64]. In indirect synthesis, methanol is first produced from synthesis gas, and it is then converted to dimethyl ether in a separate reactor. Methanol is manufactured from synthesis gas using an alumina-supported copper–zinc oxide catalyst at 50–100 bar and 523 K. Dimethyl ether synthesis from methanol is carried out at 523–593 K over solid-acid catalysts, such as γ-Al_2O_3, ion exchange resins, and zeolites. HZSM-5 with Si/Al of 15–25 showed good catalytic activity and stability but strong acid sites are responsible for the formation of hydrocarbons that reduced dimethyl ether selectivity. Metal supported HZSM-5 like Na-HZSM-5 and K-HZSM-5 demonstrated higher activity, stability, and selectivity. In the direct synthesis, methanol formation and its dehydration to dimethyl ether occur in the same reactor. The separation of dimethyl ether and CO_2 is difficult when methanol is present in the system. The direct process has a higher CO conversion compared to indirect synthesis. Catalysts used in this method require metal for methanol synthesis and solid-acid for dimethyl ether formation. Solid-acid catalysts modified with metal oxides like CuO, ZnO, Al_2O_3, and Cr_2O_3 are used in direct synthesis.

1.5 Biorefinery for fuel-additives

Furan compounds (furfural and HMF), glycerol, and LA are the most attractive platform chemicals for the production of biofuels (Table 1.3). The interesting solicitation of these platform chemicals is their evolution into fuel-additives (Table 1.4). Fuel-additives are blended with transportation fuels to increase fuel efficiency and mitigate environmental issues. A biorefinery for the production of these fuel-additives is shown in Fig. 1.4. Fuel properties of these biofuels are compared with gasoline, as shown in Table 1.6.

1.5.1 γ-Valerolactone (GVL)

GVL is a C_5 cyclic ester with a high boiling point (480 K), low vapor pressure (0.066 kPa), and low melting point (242 K). Its vapor pressure is lower than that of ethanol with a possibility of blending up to 10% into gasoline. In a comparative study, 10 (v/v)% GVL-gasoline and 10 (v/v)% ethanol–gasoline blends exhibited similar MON and RON [68]. GVL also finds applications as a solvent, food ingredient, and feedstock for manufacturing various value-added chemicals, such as valeric esters, 2-methyltetrahydrofuran (2-MTHF), 5-nonanone, and 1,4-pentanediol. GVL is obtained from LA by the combination of hydrogenation and dehydration reactions following two different pathways (Fig. 1.5). The former pathway involves dehydration of LA, followed by hydrogenation to GVL. The latter pathway follows hydrogenation of LA, followed by dehydration of 4-hydroxy pentanoic acid intermediate. The hydrogen donor solvents, such as formic acid and alcohols, are also used instead of direct hydrogenation [69, 70]. An equimolar amount of formic acid is co-produced during the production of LA from cellulose. Hence the utilization of formic acid as a hydrogen source has received extensive attention. In the 1950s, Quaker Oats synthesized GVL from LA by the continuous hydrogenation process at 473 K over a copper chromite catalyst [71]. A Ru-based catalyst is most effective and offers good stability over 10 days without loss of catalytic activity [72, 73]. The nonprecious metal catalysts, such as Ni, Cu, and, Co, are also

Hydrocarbon biorefinery: A sustainable approach   19

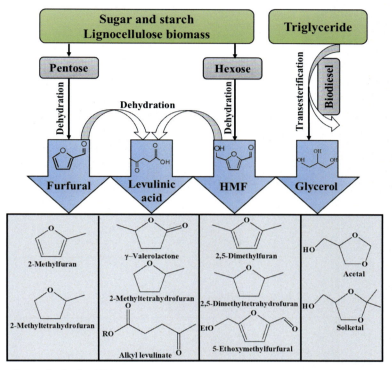

Fig. 1.4 Biorefinery for fuel-additives.

developed for industrial applications. A mixed metal catalyst, Ni/Cu/Al/Fe, showed the highest GVL yield of 98.1% at 415 K and 2 MPa hydrogen pressure [74]. The Ni$_{4.59}$Cu$_1$Mg$_{1.58}$Al$_{1.96}$Fe$_{0.7}$ catalyst showed stability for more than five cycles.

1.5.2 Alkyl levulinates

The properties of alkyl levulinates are almost identical to those of fatty acid methyl esters (biodiesel). Alkyl levulinates are produced through three different pathways: one-step conversion of cellulose, alcoholysis of furfuryl alcohol, and esterification of LA (Fig. 1.5). The first pathway was investigated using a cellulose-rich feedstock (*Eucalyptus nitens* wood) with butanol using sulfuric acid as the catalyst under microwave at 463 K for 15 min [75]. The obtained product mixture was directly used in an engine and showed a calorific value similar to that of diesel. Various metal oxides were applied to convert glucose to ethyl levulinate directly in a batch reactor at 473 K [76]. However, the activity of metal oxides was poor with a low yield of ethyl levulinate (around 30%). Recently, a sulfonic acid functionalized titanium nanotube was successfully developed to produce butyl levulinate with 94.6% yield. The catalyst

Table 1.6 Properties of various fuel-additives [42, 65–67].

	Gasoline	GVL	EL	2-MF	2-MTHF	2,5-DMF	2,5-DMTHF	5-EMF
Molecular formula	C_5–C_{12}	$C_5H_8O_2$	$C_7H_{12}O_3$	C_5H_6O	$C_5H_{10}O$	C_6H_8O	$C_6H_{12}O$	$C_8H_{10}O_3$
Molecular-weight, g/mol	72–150	100.12	144.17	82.1	86.13	96.13	100.16	154.16
Melting point, K	–	242	298	184	137	211	228	–
Boiling point, K	308–413	480	479.2	337	351	367	365	507
Flash point, K	230	369	364	251	261	272	300	368
Vapor pressure @293K, kPa	–	0.066	0.033	13.9	13.6	7.07	6.64	0.007
Lower heating value, MJ/L	44–46	26.2	24.8	27.6	28.2	30.1	31	–
Kinematic viscosity (mm^2/s) @313K	(MJ/kg) 0.4–0.8	2.1	1.5	0.39	0.70	0.48	0.47	
Density, g/mL	0.739	1.046	1.014	0.91	0.854	0.89	0.83	1.1
Auto ignition temperature, K	693	797	698	723	543	559	–	
Research octane number	95	100	110	103	86	119	82	–
Motor octane number	90	100	102	86	73	88.1	–	–
Cetane number	5–20	<10	<10	–	–	9	17.2	–
Carbon, wt%	86	60	58.3	73.2	69.7	75.0	72	62.3
Hydrogen, wt%	14	8	8.3	7.3	11.7	8.3	12	6.5
Oxygen, wt%	0	32	33.3	19.5	18.6	16.7	16	31.2
Enthalpy of vaporization, kJ/kg	260–360	442.4	306.7	357	364.4	332	348	–
Water miscibility, mg/mL	0	>=1	45.6	3	150	1.47	Immiscible	8.3
Energy density, MJ/L	33.5	–	–	28.5	–	29.3	31.8	30.3

Hydrocarbon biorefinery: A sustainable approach

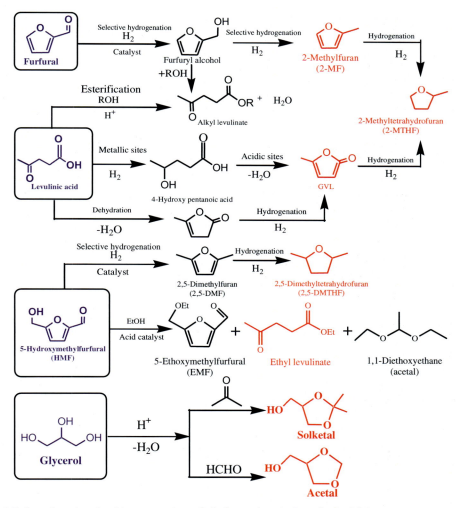

Fig. 1.5 Reactions involved in conversion of platform chemicals to fuel-additives.

showed excellent Brønsted and Lewis acid properties after reuse of six cycles. The ethanolysis of furfuryl alcohol was carried out using an Fe/USY catalyst and 90.6% ethyl levulinate yield was obtained. The loading of Fe species on USY suppressed the acidity of the catalyst with diminution of by-products [77]. The acid-catalyzed esterification of LA with alcohol is the simplest pathway. The supported heteropolyacids [78], zeolites [79], sulfonated carbon [80], silica [81], and MOFs [82] have been employed for LA esterification. Biofine and Texaco tested a mixture of 20% ethyl levulinate, 79% diesel, and 1% coadditive in a diesel engine. The fuel properties of this blend were found to be enhanced compared to those of diesel [83].

1.5.3 2-Methylfuran (2-MF)

2-MF is a selective hydrogenated product of furfural. Cu-based catalysts, such as Cu/Al_2O_3, Cu/SiO_2, and Cu/Cr, have been widely used in the production of 2-MF at 473–673 K and low pressure [84]. However, the Cr-based catalyst is toxic and causes catalyst deactivation and coke formation. Recently, Cu/ZnO showed a high yield of 2-MF (94.5%) in a fixed bed reactor. The effect of Cu loading was studied using a Cu/SiO_2 catalyst with a Cu loading of 2.5–20 wt% at 483 K [85]. 10 wt% Cu/SiO_2 resulted in a nearly 100% yield of 2-MF due to the smaller particle size and higher dispersion of Cu. A study was carried out to compare 2-MF with benchmark fuels like ethanol and 95 RON gasoline [86]. The evaporation and combustion duration of 2-MF were reported to be similar to these fuels and NOx emission in the lean-burn combustion system is lower than that of 95 RON gasoline. These results advocate the use of 2-MF as a fuel additive.

1.5.4 2-Methyltetrahydrofuran (2-MTHF)

2-MTHF has been recognized as a "P-Series fuel" by the US Department of Energy owing to its promising fuel properties. The hydrophobic nature, high heating value, and high specific gravity make 2-MTHF suitable for mixing with gasoline up to 70% [15]. 2-MTHF is produced through two different reaction routes: sequential hydrogenation of furfural and hydrogenation of LA (Fig. 1.5). The former pathway consists of hydrogenation of furfural to 2-MF, followed by hydrogenation of 2-MF to 2-MTHF. Recently, the one-pot synthesis of 2-MTHF was reported at 453 K and 1 MPa hydrogen pressure in a single reactor packed with two different layers of nonprecious metal catalysts [87]. The first layer consists of a Co-based catalyst (Co/Al_2O_3, Co/SiO_2, Co/TiO_2, Co/MgO_2, Co/CeO_2, and Co/ZrO_2) that converts furfural to 2-MF. The second layer consists of Ni-based catalysts (Ni/Al_2O_3, Ni/SiO_2, and Ni/CeO_2) that convert 2-MF to 2-MTHF. The high temperature favors a high yield of 2-MF in the first catalyst layer, while the low temperature is suitable for the high selectivity of 2-MTHF in the second catalyst layer. Hence, an optimum temperature of 453 K was used to obtain a high yield (87%) of 2-MTHF using Ni/Al_2O_3. The latter pathway involves GVL and 1,4-pentanediol intermediates. A combination of metal and acid catalysts, such as NiCu/Al_2O_3-ZrO_2, was developed to activate hydrogenation and ring-opening reactions and to promote cyclodehydration of 1,4-pentanediol [88]. The multimetallic properties of the catalyst resulted in high activity with 99.8% selectivity toward 2-MTHF.

1.5.5 2,5-Dimethylfuran (2,5-DMF)

It has fuel properties similar to those of gasoline. Thus, an engine test was performed to compare its performance with that of gasoline [89]. It was reported that 2,5-DMF can be used either directly or as an additive in gasoline (13–16 wt%) [90]. Selective hydrogenation of HMF gives 2,5-DMF. Noble metals, such as Ru, Pd, and Pt, have been studied extensively for hydrogenation of HMF. Ru/Co_3O_4 exhibited excellent catalytic activity and achieved 93.4% 2,5-DMF yield due to multifunctional properties of Ru and CoO_x species [91]. Later on, researchers shifted to copper-based catalysts to overcome the high cost of noble metals. Various supports (Al_2O_3, Fe_2O_3–Al_2O_3, and Nb_2O_5–Al_2O_3) were investigated for the Cu catalyst [92]. Cu/Al_2O_3 and Cu/Nb_2O_5–Al_2O_3 resulted in 90% 2,5-DMF yield with 100% HMF conversion at 423 K. An integrated process was developed to convert fructose to 2,5-DMF directly through the combination of dehydration and hydrogenation reactions [93]. Dehydration of fructose gave 93% HMF yield using Amberlyst-15 and the obtained HMF was converted to 2,5-DMF with 99% yield over Ru–Sn/ZnO.

1.5.6 2,5-Dimethyltetrahydrofuran (2,5-DMTHF)

2,5-DMTHF has properties similar to those of 2,5-DMF, in particular immiscibility with water, high energy density (31.8 MJ/L), and high boiling point (365 K). The nitrogen–doped, carbon-decorated, copper-based catalyst exhibited high catalytic performance and converted HMF to 2,5-DMTHF with 94.6% yield at 423 K [94]. A dual catalytic system composed of HI and $RhCl_3 \cdot xH_2O$ salt was implemented for direct conversion of fructose to 2,5-DMTHF with 70% yield [95]. HI is responsible for dehydration of fructose to HMF, and Rh favors C=C and C=O bond hydrogenation. The multifunctional Cu-doped, porous metal oxide catalyst was investigated for in situ hydrogenation of HMF using methanol in the temperature range of 513–593 K [96]. The elevated temperature promoted undesirable side products, and the highest yield of 2,5-DMF and 2,5-DMTHF mixture (57%) was observed at 543 K.

1.5.7 5-Ethoxymethylfurfural (5-EMF)

5-EMF has notable fuel properties, such as a high boiling point (508 K). The energy density (30.3 MJ/L) is similar to that of gasoline and higher than that of ethanol [97]. Etherification of HMF with ethanol using solid-acid catalysts yields 5-EMF through ethyl levulinate and 1,1-diethoxyethane intermediates (Fig. 1.5). A highly efficient catalytic system of multilayer polyoxometalates was developed with controlled Nb/Mo composition to regulate the acidity of the catalyst [98]. The niobium molybdate catalyst achieved 100% HMF conversion with more than 99% selectivity toward 5-EMF at 60 min. The production of 5-EMF directly from corn stover was reported using mixed acids (0.1% H_2SO_4 and 1.0% zeolite USY) as the catalyst [99]. About 23.9% 5-EMF yield was observed at 483 K and 125 min. Tungstophosphoric acid supported on SnO_2 was used for etherification of HMF and one-step conversion of fructose to 5-EMF [100]. About 90% and 68% 5-EMF yield was obtained through etherification and one-pot conversion processes, respectively.

1.5.8 Glycerol acetals

The acetalization of glycerol with aldehyde or ketone produces acetals. The acetalization of glycerol with various aldehydes, such as butanal, pentanal, hexanal, octanal, and decanal, was studied over the Amberlyst-15 catalyst [101]. The conversion of glycerol decreased with increasing aldehyde chain length and the highest conversion of 85% was observed using butanal. HPMo@Y zeolite resulted in 83% selectivity toward solketal at 303 K in 5 h [102]. Zeolites (USY, BEA, and ZSM-5) were found to be highly active for this reaction and achieved 88% conversion of glycerol at low temperatures [103]. The acetal obtained from acetalization of glycerol with acetone was slightly modified by the addition of the acetyl group to OH [104]. The inclusion of modified acetal in diesel improved fuel properties.

1.6 Hydrocarbon biorefinery

The exiting liquid transportation fuels, such as gasoline (C_5–C_{12}), jet fuel (C_{10}–C_{16}), and diesel (C_{12}–C_{22}), are composed of hydrocarbons (linear and branched paraffins, naphthenes, and aromatics). Similarly, liquefied petroleum gas (propane and butane), kerosene (C_{10}–C_{14}), and fuel oil are hydrocarbon-based fuels. On the other hand, traditional biofuels and fuel-additives are oxygenated compounds (Table 1.4). The presence of oxygen in their structure causes the mismatch of fuel properties with existing transportation fuels. It also offers lower calorific value and lesser fuel mileage than existing transportation fuels. Moreover, these biofuels are incompatible with refinery infrastructures including internal combustion engines. Therefore, these biofuels are limited to blending with existing transportation fuels. For example, biodiesel contains about 10%–15% oxygen in its structure. The cold flow properties of biodiesel, such as kinematic viscosity and cloud/pour point, are much higher compared to those of diesel (Table 1.5). Biodiesel is thus mixed with diesel up to 20 (v/v)% for application in an unmodified combustion engine. Likewise, bioethanol contains about 35% structural oxygen. Further, it is fully miscible with water and slightly corrosive to metallic parts of the engine. Bioethanol is thus mixed with gasoline up to 15 (v/v)% for application in an existing combustion engine. Similarly, biobutanol contains 21.6% oxygen in its structure with lower fuel mileage. The biofuels analogous to existing fuels are thus essential to overcome the bottlenecks of traditional biofuels and fuel-additives. These biofuels are termed hydrocarbon biofuels. The energy density or calorific value of hydrocarbon biofuels is much higher than that of oxygenated biofuels and similar to that of petroleum-derived fuels. The fuel mileage of hydrocarbon biofuels is thus much higher than that of oxygenated biofuels and comparable to that of current transportation fuels. Hydrocarbon biofuels are also compatible with refinery infrastructures, such as storage, production, distribution, fueling stations, and existing engines.

On the other hand, platform chemicals in the biorefinery are entirely different than petrochemical building-block chemicals (Table 1.3). These platform chemicals are oxygenated compounds and reduce the oxidation step for producing oxygenated chemicals. However, the chemistry involved in the production and downstream transformation of these platform chemicals to value-added chemicals or commodity products is completely different from the current practices of the petrochemical industry. The biorefinery concept involving these platform chemicals thus implicates huge capital investments for building new petrochemical production facilities. The production of petrochemical building-block chemicals from biomass, such as synthesis gas, olefins, and aromatics, is thus highly desirable for making use of existing production facilities of petrochemical industries. This novel concept of manufacturing hydrocarbon biofuels and petrochemical building-block chemicals from biomass is known as hydrocarbon biorefinery [3, 105].

Hydrocarbon biofuels are produced from biomass through a combination of conversion processes, such as (i) chemical and thermochemical, (ii) biological and biochemical, and (iii) conversion of biomass-derived compounds (Fig. 1.6). HDO of vegetable oils, integrated fast pyrolysis and HDO of bio-oil, Fischer-Tropsch synthesis (FTS), combined hydrothermal liquefaction and HDO of bio-oil, and methanol-to-gasoline are the potential chemical and thermochemical conversion processes. The biological and biochemical conversion processes include microbial routes and metabolic engineering, bioethanol-to-gasoline, biobutanol-to-gasoline, and oligomerization of biomass-derived olefins. On the other hand, the removal of oxygen from platform chemicals is needed to obtain hydrocarbon biofuels. HDO of these low molecular-weight compounds, however, results in volatile hydrocarbons that are inappropriate as liquid

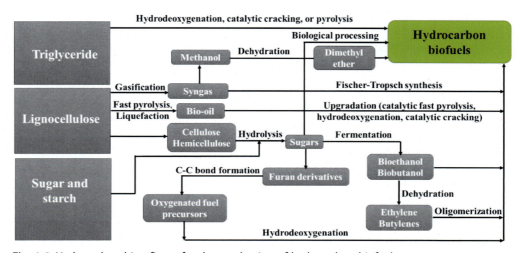

Fig. 1.6 Hydrocarbon biorefinery for the production of hydrocarbon biofuels.

26 Hydrocarbon biorefinery

transportation fuels. Various C—C bond forming reactions, such as aldol condensation, ketonization, and hydroxyalkylation-alkylation (HAA) reaction, are important strategies to increase molecular-weight. HDO of these high molecular-weight fuel precursors results in hydrocarbon biofuels. Fuel properties of hydrocarbon biofuels produced by various routes are compared with those of diesel, as summarized in Table 1.7.

1.6.1 Chemical and thermochemical conversion processes

HDO of triglycerides is carried out over supported metal catalysts at high hydrogen pressure (up to 150 bar) in the temperature range of 523–773 K [114, 115]. In this process, $C=C$ bonds are saturated and oxygen is removed as water, CO_2, and CO with propane as a coproduct. The diesel-range hydrocarbon obtained in this process is known as green diesel. Green diesel is economically competitive with the current diesel price. The manufacturing cost of green diesel by direct HDO of karanja oil was reported as USD 0.84 per kg compared to USD 0.798 per kg for two-step HDO (hydrolysis of vegetable oil, followed by HDO of fatty acids) [38]. Fast pyrolysis and hydrothermal liquefaction are two complementary technologies in the hydrocarbon biorefinery. In fast pyrolysis, lignocellulose biomass is thermally decomposed to bio-oil in the absence of oxygen at a moderate temperature (573–873 K), a shorter residence time of 1–2 s, and a high heating rate of 10^3–10^4 K/s in a fluidized-bed reactor [116, 117]. This technology is suitable for dry biomass. This technology involves low capital investment and is economical both on a small and large scale. It is thus an ideal technology for a decentralized biorefinery. On the other hand, hydrothermal liquefaction is carried out at slightly below the supercritical temperature of water (647 K) and high pressure (100–250 bar) using water as the solvent and reactant. This process produces mainly bio-oil with a small amount of char, gas, and water-soluble compounds. This technology is appropriate for wet biomass, such as microalgae, municipal sludge, etc. Bio-oil obtained from hydrothermal liquefaction contains a lesser amount of oxygen with a higher heating value compared to bio-oil derived from fast pyrolysis (Table 1.7) [118]. However, bio-oil contains a large amount of water and a vast range of chemical compounds, such as acids, alcohols, aldehydes, ketones, guaiacols, phenolics, and esters. Consequently, bio-oil is facing stability issues and is unsuitable as a biofuel. Bio-oil is thus upgraded to hydrocarbon biofuels by HDO. HDO of bio-oil is typically carried out at high pressure (up to 200 bar) and moderate temperatures (573–873 K) over supported metal, metal oxide, and metal sulfide catalysts. Bio-oil is also upgraded using zeolite catalysts to obtain aromatics [34]. The gasification of biomass to synthesis gas $(CO + H_2)$ is carried out in the presence of air at high temperatures (1000–1500 K). The synthesis gas is then transformed into methanol. Methanol is, however, unsuitable as a biofuel. Methanol is thus transformed into gasoline-range biofuel, known as methanol-to-gasoline. The synthesis gas is also converted to hydrocarbon biofuels directly using FTS [119]. The production of hydrocarbon biofuels via the

Table 1.7 Properties of various hydrocarbon biofuels [12, 39, 106–113].

| | Diesel | HDO of vegetable oil | Fischer-Tropsch diesel | Fast pyrolysis | | Hydrothermal liquefaction of microalgae | | Methanol-to-gasoline |
				Bio-oil	HDO	Bio-oil	HDO	
Water, wt%	–	–	–	20–30	1.5	2.8–5.6	–	
Aromatics, vol%	–	–	–	–	5–30	–	–	26
Benzene, vol%	–	–	–	–	0.3–0.8	–	–	0.3
pH	–	–	–	2–3	5.8	–	–	–
Specific gravity	0.84	0.78	0.77–0.78	1.05–1.25	1.2	0.94–1.04	0.76	–
Viscosity at 313 K, cP	1.9–4.1	3.81	3.2–4.5	40–100	1–5	15–330	1.3–4.5	–
Sulfur, ppm	<10	<1	<10	<500	<50	3000–6500	<50	nil
Heating value, MJ/kg	42–46	44	43	16–19	42–45	26–36	38–44	–
Cloud point, K	268	253–293		–	–	–	–	–
Cetane number	40–55	70–90	73–81	–	22	–	–	–
Research octane number	–	–	–	–	64–88	–	–	92
RVP, psi	–	–	–	–	–	–	–	9
Flash point, K	338–341	411	–	326–341	–	–	–	–
Fire point, K	529	418	–	237–264	–	–	–	–
Pour point, K	–	282	–	248	–	–		
Stability	Good	Good	–	Poor	Good	–	–	–
O, wt%	Nil	0.53	0	28–40	<5	6–8	1.85–2.2	–
H/C	2.0	1.97	–	0.9–1.5	1.3–2.0	1.2–1.6	1.92–1.94	–

gasification route is popularly known as biomass-to-liquid. FTS is carried out using 2:1 H_2/CO synthesis gas at 10–60 bar and 473–623 K. FTS was applied industrially in Germany in 1938 with a capacity of 660×10^3 tons per year [120]. FTS is performed at low temperature (473–523 K) in a slurry reactor over a Co catalyst to obtain a diesel-range hydrocarbon biofuel [121, 122]. On the other hand, FTS is performed at high temperature (573–623 K) in a fluidized-bed reactor over an Fe catalyst to obtain gasoline-range hydrocarbon biofuel. However, gasification has constraints of feeding highly pressurized reactors, the formation of tar, fouling of reactors, and expensive process plants. These factors make this route economically unviable.

1.6.2 Biological and biochemical conversion processes

Hexose and pentose sugars are converted to hydrocarbon biofuels and aromatics using microorganisms [123, 124]. Natural microorganisms convert sugars to C_{14}–C_{25} range hydrocarbons, while genetically engineered microorganisms produce C_{15} hydrocarbons from sugars. Hydrocracking and hydroisomerization of these high molecular-weight hydrocarbons produce hydrocarbon biofuels. Dehydration of bio-alcohols (bioethanol and biobutanol) in the presence of acid catalysts forms its respective olefins, i.e., ethylene and butylenes. Oligomerization of these olefins produces hydrocarbon biofuels in the range of diesel, gasoline, and jet fuel. Oligomerization is an existing process in the petroleum refinery which is performed over a solid-acid catalyst at 423–773 K and high pressure up to 50 bar [125]. Diesel-range hydrocarbon biofuels are formed at high pressure (30–50 bar) and low reaction temperatures (less than 573 K). However, gasoline-range hydrocarbon biofuel dominates at lower pressure and higher temperatures. Downstream hydrogenation of these oligomers produces gasoline or diesel-range hydrocarbon biofuels. Isomerization and cracking reaction also occur over the acid sites forming branched olefins with better fuel properties. ExxonMobil, Lurgi, and UOP are the pioneers in the olefin oligomerization process. Both ExxonMobil and Lurgi use zeolite types of catalysts, while UOP developed the process using the solid phosphoric acid catalyst.

Bioethanol and biobutanol can also be converted to gasoline-range hydrocarbon biofuel directly [31]. Ethanol-to-gasoline is operated at 573–723 K in a fixed-bed reactor using zeolites and modified zeolites as catalysts. Mobil Oil established the ethanol-to-gasoline process using the HZSM-5 catalyst with the formation of C_5^+ alkanes and aromatics. High aromatic content is, however, a major problem in this process. Downstream hydrogenation of aromatics using the Pt—Re catalyst is usually performed to reduce benzene content below 1% and aromatic content below 20%. Conversion of butanol or ABE mixture to gasoline-range hydrocarbon biofuels is, however, in the early stage of research. The reaction is also performed over the zeolite type of catalyst, such as HZSM-5, and produces aromatic-free, gasoline-range hydrocarbons at around 523 K and low weight-hourly space velocity [31]. Reaction proceeds by a combination of dehydration and dehydrogenation routes. Butylene and propylene formed by this reaction are converted to C_5–C_{12} range olefins by oligomerization reaction. Gasoline-range hydrocarbon biofuels are obtained by downstream hydrogenation of these olefins. Separation of ethanol or butanol from the aqueous solution is, however, an energy-consuming process. Research efforts are needed for the direct conversion of these aqueous alcohols or sugars to hydrocarbon biofuels. Aqueous-phase catalysis is an alternate route for the conversion of aqueous sugars to hydrocarbon biofuels. Aqueous-phase dehydration/hydrogenation produces alkanes that instantaneously separate from the aqueous phase. Huber et al. and Xing et al. recently developed conversion of sugars to hydrocarbon biofuels [126, 127].

1.6.3 Conversion of biomass-derived compounds to hydrocarbon biofuels

Aldol condensation involves furfural or HMF and a carbonyl compound (aldehydes or ketones) containing at least one hydrogen atom at the α-position (Fig. 1.7) [128]. Traditionally, a base catalyst and mild reaction conditions (303–393 K and atmospheric pressure) are suitable for aldol condensation. HAA reaction most commonly occurs between furfural and 2-MF in the presence of solid-acid catalysts under mild reaction temperature (313–363 K) and atmospheric pressure [129]. This reaction occurs between a furan with at least one α-hydrogen and a carbonyl compound. In the ketonization reaction, two carboxylic acids are condensed in the presence of base catalysts with the elimination

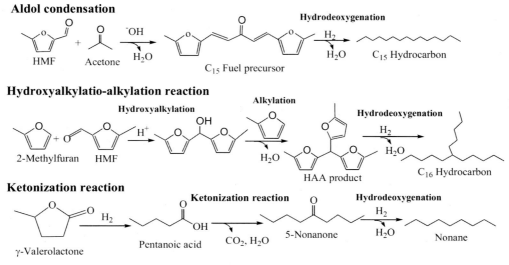

Fig. 1.7 Hydrocarbon biofuels from platform chemical.

of water and CO_2. These high molecular-weight oxygenated fuel precursors are upgraded to hydrocarbon biofuels by catalytic HDO [105]. The types of hydrocarbon biofuels obtained in these processes depend on the molecular-weight of reactants used. LA is a highly reactive molecule due to its dual functionality (C=O and COOH). It is reduced to GVL, which is sequentially converted to pentanoic acid by ring-opening and hydrogenation reactions (Fig. 1.7) [130]. The pentanoic acid is then converted to C_9 ketone (5-nonanone) by ketonization reaction. Finally, C_9 ketone is converted to 5-nonane (gasoline-range hydrocarbon biofuel) by HDO reaction.

1.6.4 Petrochemical building-block chemicals from biomass

In the hydrocarbon biorefinery, synthesis gas is produced by either gasification of lignocellulose biomass or steam reforming of bio-oil, triglycerides, and various biomass-derived compounds, such as biomethanol, bioethanol, biobutanol, etc. [40, 131] (Fig. 1.8). The dehydration reaction of bioethanol and biobutanol in the presence of solid-acid catalysts produces ethylene and butylenes, respectively [31]. Methanol-to-olefin provides another opportunity to produce a wide range of olefins, including ethylene and propylene [132]. The reaction of bioethanol and biobutanol in the presence of the zeolite catalyst produces aromatics [31]. On the other hand, bio-oil upgrading in the presence of zeolite or the metal modified zeolite catalyst produces both aromatics and light olefins [133]. The lignin depolymerization and HDO produce aromatics [18].

1.7 Role of heterogeneous catalysis in the hydrocarbon biorefinery

The hydrocarbon biorefinery is facing many engineering, scientific, and economic challenges. Heterogeneous catalysis provides remarkable opportunities in breaking these

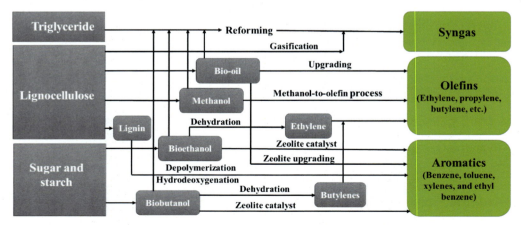

Fig. 1.8 Hydrocarbon biorefinery for the production of petrochemical building-block chemicals.

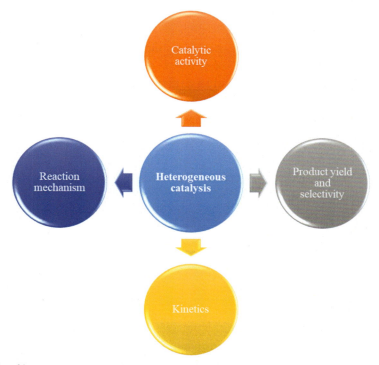

Fig. 1.9 Role of heterogeneous catalysis in the hydrocarbon biorefinery.

barriers to make hydrocarbon biofuels economically feasible. Either homogeneous, heterogeneous, or enzyme catalysts are used in the hydrocarbon biorefinery. Heterogeneous catalysts are, however, advantageous over homogeneous catalysts. It is easy to separate from the reaction mixture and resilient to harsh reaction conditions, such as high pressure (350 bar) and temperature (1600 K). It is noncorrosive to metal or alloy reactors, unlike homogeneous catalysts, e.g., sulfuric acid. Heterogeneous catalysis plays a vital role in the hydrocarbon biorefinery in terms of catalytic activity, product yield and selectivity, reaction mechanism, and kinetics (Fig. 1.9).

Heterogeneous catalysts are porous/nonporous solid materials. They are either bulk catalysts or supported catalysts. The bulk catalysts consider the entire solid material as the catalyst, whereas active metals or metal oxides are deposited on a high surface area material (called support) in supported catalysts. The high surface area ensures a better dispersion of active components. The reaction occurs on active sites of the catalyst. Two or more different active sites are sometimes introduced in supported catalysts, known as bi or multifunctional catalysts. Heterogeneous catalysis involves the diffusion of reactants from the bulk to catalyst surface, diffusion of reactants through pores, adsorption of reactants on active sites, reaction on active sites, and desorption of products from active sites.

These steps depend on the catalyst design, reactor types, and reaction conditions. The performance of heterogeneous catalysts is judged by three factors: catalytic activity, product yield and selectivity, and time-on-stream stability. Hydrocarbon biorefinery, however, involves the processing of oxygenated compounds. The development of selective, durable, and active catalysts is thus challenging for the hydrocarbon biorefinery. Three different types of catalysts are generally used in the hydrocarbon biorefinery: supported metal or metal oxides, zeolites or modified zeolites, and solid-acid and solid-base catalysts.

1.7.1 Supported metal catalysts

HDO of triglycerides. HDO of triglycerides is carried out over noble metal (Pt, Pd, etc.) and transition metal (Ni, Co, NiMo, CoMo, and NiW) catalysts. The nature of metal and support influence catalytic activity, product selectivity, time-on-stream stability, and the reaction mechanism [114]. The decarbonylation reaction dominates over supported monometallic catalysts, such as Ni and Co [134] (Fig. 1.10). The reaction, however, follows the HDO route over bimetallic (NiMo and CoMo) and metal supported on reducible oxides (CeO_2 and ZrO_2) and solid-acid (zeolite) catalysts [114, 135]. The synergetic effect of metals in bimetallic catalysts enhances catalytic activity [134, 135]. The mole ratio of individual metals in bimetallic catalysts influences the structure, reducibility, and formation of active species [136–139]. For example, increasing the Ni/Mo mole ratio increases the NiMo alloy content in the catalyst with high HDO activity [138]. Similarly, high metal loading (Ni and Mo) improves catalytic activity due to the increase in active metal sites (Ni and NiMo alloy) in the catalyst [138]. The calcination temperature of the catalyst also influences catalytic activity. The catalytic activity of Ni, NiMo, and CoMo catalysts enhances with increasing calcination temperature up to 973 K [135]. The improved catalytic activity is due to the enhanced metal-support interaction, resulting in increased metal dispersion.

The high support surface area provides better metal dispersion, resulting in higher catalytic activity [140]. The acidic supports, such as ZSM-5, also increase catalytic activity [114]. The higher catalytic activity of the Ni/ZSM-5 catalyst is due to its bifunctionality: acid and metal sites. ZSM-5 favors catalytic cracking as well. Therefore, mild acidic

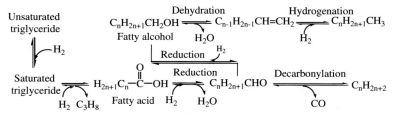

Fig. 1.10 Reaction mechanism for HDO of vegetable oil.

γ-Al_2O_3 is considered a promising support [114]. The acidity of supports also plays an important role in the metal-support interaction, resulting in the formation of different active species with a distinct reaction mechanism [141]. The pore size of support influences the catalytic activity, selectivity, and cetane number of green diesel [142]. The appropriate pore size and moderate acidity enhance the isomerization reaction, resulting in improved fuel properties [143]. The mesoporous support facilitates metal dispersion and easy diffusion of bulky reactants or products [143, 144]. Mesoporous supported metal catalysts are thus appropriate for the production of jet fuel-range hydrocarbon biofuel [145]. The mesoporous Ni-alumina composite catalyst further displayed a higher catalytic activity than alumina and mesoporous alumina supported Ni catalysts due to the incorporation of Ni in the alumina framework [146].

HDO of bio-oil and oxygenated fuel precursors. Metal sulfides (MoS_2, $NiMoS_2$, and $CoMoS_2$), noble metals, and transition metals (Ni, Co, Cu, Fe, and Mo) are normally used for HDO of bio-oil [147]. Cheap transition metal catalysts, especially the sulfided form of $NiMo/\gamma$-Al_2O_3 and $CoMo/\gamma$-Al_2O_3, are commonly used in the HDO of bio-oil. Sulfur vacancy in these catalysts works as active sites [148]. Ni and Co metals in NiMo and CoMo catalysts act as a promoter [149, 150]. Choice of support is important in the HDO of bio-oil. γ-Al_2O_3 is somewhat acidic and participates in the HDO reactions. However, γ-Al_2O_3 is not stable under the prevalent bio-oil water environment and converts to boehmite (AlO(OH)). Hence, it is not suitable as a support for HDO of bio-oil. Carbon (C), silica (SiO_2), ceria (CeO_2), and zirconia (ZrO_2) are thus considered as alternative water-resistant supports. Carbon and silica are neutral and show a lower tendency of coke formation compared to γ-Al_2O_3 [151–153]. Furthermore, oxophilic ZrO_2 and CeO_2 activate oxy-compounds and hence are considered a promising support [106]. Monometallic and bimetallic catalysts are commonly used for HDO of oxygenated fuel precursors to hydrocarbon biofuels. For example, Pt supported on SiO_2, Nb_2O_5, SiO_2–Al_2O_3, and ZrP was investigated for HDO of fuel precursors obtained from HAA of furfural with 2-MF [154].

Fischer-Tropsch synthesis (FTS). This reaction is performed using Fe and Co-based catalysts [121]. These metals are supported on Al_2O_3, TiO_2, and SiO_2 with Ru, Pt, and Re as promoters. The Co catalyst promotes a hydrogenation reaction. An alkali is generally used in the Fe catalyst, and it prevents methane formation and boosts catalytic activity [155]. The Co catalyst shows a higher catalytic activity than the Fe catalyst [119].

1.7.2 Zeolite-type catalysts

Catalytic fast pyrolysis. Fast pyrolysis is sometimes carried out using zeolite catalysts for in situ removal of oxygen, known as catalytic fast pyrolysis [156]. It is carried out at 773–873 K and atmospheric pressure and produces mainly aromatics [156]. Zeolites are shape-selective catalysts, and hence aromatics yield and product selectivity depend on their pore size, acidity, and types of zeolites used. The shape selectivity of zeolites is caused by both the mass transfer effect and transition state effect, linked to the pore size and internal void space, respectively. The medium pore size (ZSM-5, 5.5 Å) is suitable for high aromatic yield, while the large pore size favors coke formation [157, 158]. The one-dimensional zeolites with large pore sizes, such as ZSM-23, SSZ-20, and SSZ-55, show high selectivity to naphthalene. Two- and three-dimensional zeolites with large pore sizes, such as Y and Beta, show high selectivity to benzene-toluene-xylenes [156]. Silicalite has a pore structure similar to that of HZSM-5 but it does not contain Brønsted acid sites. Silicalite thus produces more coke and lesser aromatics compared to HZSM-5. ZSM-5 with an Si/Al ratio of 30 showed the maximum aromatic yield with a negligible amount of coke [159]. The mesopores provide easy diffusion of bulky oxygenates, resulting in improved catalytic activity [160]. The incorporation of Ga in ZSM-5 enhances the aromatic yield due to the presence of dual catalytic centers [161].

Methanol, ethanol, and butanol to gasoline. Zeolite-type catalysts play an important role in these processes. Product quality, expressed in terms of chemical composition and yield of gasoline, is mainly governed by the characteristic parameters of zeolites, such as the Si/Al ratio, acidity, pore structure, and doping of metals. The broad pore diameter is suitable for hydride transfer with the formation of aromatics. The product is, however, limited to linear olefins for small pore zeolite. Similarly, mesopore zeolites with a short diffusional path length, nanocrystal zeolites with a high external surface area, and hierarchical zeolites improve the catalytic activity with reduced coke formation. The modification of zeolite by metal oxides, such as ZnO and CuO, enhances acidity with improved catalytic activity and stability of the catalyst.

1.7.3 Solid-acid and solid-base catalysts

Oligomerization of olefins. This reaction is performed over solid-acid catalysts, such as sulfonic acid-functionalized silica, solid phosphoric acid, zeolites, and mixed metal oxides [125, 162, 163]. The microporous structure of zeolites is responsible for deactivation. In the case of zeolites, oligomerization reaction generally happens on Brønsted acid sites, while branching occurs on external acid sites [164]. The zeolite catalyst thus prevents branching due to its microporous structure [165]. The side reactions dominate over Lewis acid catalysts with the formation of branched olefins, aromatics, and coke.

Carbon–carbon bond forming reactions. Aldol condensation and HAA reactions are performed over solid-base and solid-acid catalysts, respectively. The role of the base catalyst in aldol condensation is to abstract a α-hydrogen atom to form an enolate ion for condensation. The mixed metal oxides, also known as hydrotalcites, are widely used for aldol condensation due to their high basicity, high surface area, and pore structure. $MgO-ZrO_2$ mixed oxide, basic zeolite NaY, and nitrogen substituted NaY (Nit–NaY) were employed for aldol condensation of HMF with acetone [166]. $MgO-ZrO_2$ showed almost 100% selectivity to aldol products. Further, Nit–NaY exhibited excellent catalytic activity compared to NaY owing to the enhancement of basic site strength by the incorporation of Nit. Solid–acid catalysts, such as resins, zeolites, sulfonated carbon, and mixed metal oxides, were studied for HAA of furfural and 2-MF. Among these catalysts, commercial Nafion-212 showed excellent activity (75% yield) due to the enhancement of acidity by electron–withdrawing fluorine [154]. Recently, sulfonated nanostructured carbon showed 90% 2-MF conversion and more than 95% selectivity toward the HAA product [129]. Pt/ZrP showed high activity (94% carbon yield) due to the high Brønsted to Lewis acid ratio. Bifunctional catalysts, such as $Pd/MgO–ZrO_2$ and Pd/Nb_2O_5, are also used for simultaneous HAA and HDO reactions in one step [130, 167].

1.8 Conclusions

Sustainable hydrocarbon biofuels have the potential to meet the world's ambitious goal for the transition away from fossil fuels. Potential assets for this ambition are agriculture residue, energy crops, forest residue, industrial and municipality waste, and algal biomass. Traditional biofuels, such as bio–alcohols and biodiesel, are well established and produced by fermentation and transesterification reactions, respectively. Platform chemicals, such as HMF, furfural, LA, and glycerol, are converted to various fuel-additives, such as 2-MF, 2-MTHF, 2,5-DMF, 2,5-DMTHF, 5-EMF, GVL, alkyl levulinates, and glycerol acetals. However, the manufacturing pathway of these fuel-additives comprises multiple reaction steps with a high production cost. Moreover, traditional biofuels and fuel-additives are incompatible with present refinery infrastructures including combustion engines. Hydrocarbon biofuels are, however, analogous to current transportation fuels and compatible with existing facilities. This chapter discussed the perception of the hydrocarbon biorefinery for the production of hydrocarbon biofuels. Hydrocarbon biofuels are produced by the combination of three different routes: chemical and thermochemical, biological and biochemical, and conversion of biomass–derived compounds. Heterogeneous catalysis plays a key role in the hydrocarbon biorefinery. This chapter further elucidated the role of heterogeneous catalysts in this biorefinery.

References

[1] IEA, Data & Statistics, https://www.iea.org/data-and-statistics?country=WORLD&fuel=Energysupply&indicator=Totalprimaryenergysupply(TPES)bysource (accessed June 5, 2020).
[2] T. Niblock, BioEnergy outlook, Biocycle 58 (2012) 37–39.
[3] S.K. Maity, Opportunities, recent trends and challenges of integrated biorefinery: part II, Renew. Sust. Energ. Rev. 43 (2015) 1446–1466, https://doi.org/10.1016/j.rser.2014.08.075.
[4] S.K. Maity, Opportunities, recent trends and challenges of integrated biorefinery: part I, Renew. Sust. Energ. Rev. 43 (2015) 1427–1445, https://doi.org/10.1016/j.rser.2014.11.092.

[5] L. Cao, I.K.M. Yu, X. Xiong, D.C.W. Tsang, S. Zhang, J.H. Clark, et al., Biorenewable hydrogen production through biomass gasification: a review and future prospects, Environ. Res. 186 (2020) 109547, https://doi.org/10.1016/j.envres.2020.109547.

[6] M.A. Carriquiry, X. Du, G.R. Timilsina, Second generation biofuels: economics and policies, Energy Policy 39 (2011) 4222–4234, https://doi.org/10.1016/j.enpol.2011.04.036.

[7] H. de Lasa, E. Salaices, J. Mazumder, R. Lucky, Catalytic steam gasification of biomass: catalysts, thermodynamics and kinetics, Chem. Rev. 111 (2011) 5404–5433, https://doi.org/10.1021/cr200024w.

[8] K.C. Badgujar, B.M. Bhanage, Dedicated and waste feedstocks for biorefinery: an approach to develop a sustainable society, in: Waste Biorefinery Potential Perspect, Elsevier, 2018, pp. 3–38, https://doi.org/10.1016/B978-0-444-63992-9.00001-X.

[9] M. Carmona-Cabello, I.L. Garcia, D. Leiva-Candia, M.P. Dorado, Valorization of food waste based on its composition through the concept of biorefinery, Curr. Opin. Green Sustain. Chem. 14 (2018) 67–79, https://doi.org/10.1016/j.cogsc.2018.06.011.

[10] D. Yue, F. You, S.W. Snyder, Biomass-to-bioenergy and biofuel supply chain optimization: overview, key issues and challenges, Comput. Chem. Eng. 66 (2014) 36–56, https://doi.org/10.1016/j.compchemeng.2013.11.016.

[11] N. Kumar, A. Sonthalia, H.S. Pali, Sidharth, Next-generation biofuels—opportunities and challenges, Green Energy Technol. (2020) 171–191, https://doi.org/10.1007/978-981-13-9012-8_8. Springer Verlag.

[12] G.W. Huber, S. Iborra, A. Corma, Synthesis of transportation fuels from biomass: chemistry, catalysts, and engineering, Chem. Rev. 106 (2006) 4044–4098, https://doi.org/10.1021/cr068360d.

[13] S.V. Vassilev, D. Baxter, L.K. Andersen, C.G. Vassileva, T.J. Morgan, An overview of the organic and inorganic phase composition of biomass, Fuel 94 (2012) 1–33, https://doi.org/10.1016/j.fuel.2011.09.030.

[14] B. Sajjadi, A.A.A. Raman, H. Arandiyan, A comprehensive review on properties of edible and non-edible vegetable oil-based biodiesel: composition, specifications and prediction models, Renew. Sust. Energ. Rev. 63 (2016) 62–92, https://doi.org/10.1016/j.rser.2016.05.035.

[15] D.M. Alonso, J.Q. Bond, J.A. Dumesic, Catalytic conversion of biomass to biofuels, Green Chem. 12 (2010) 1493–1513, https://doi.org/10.1039/c004654j.

[16] M.S. Singhvi, D.V. Gokhale, Lignocellulosic biomass: hurdles and challenges in its valorization, Appl. Microbiol. Biotechnol. 103 (2019) 9305–9320, https://doi.org/10.1007/s00253-019-10212-7.

[17] S.H. Hazeena, R. Sindhu, A. Pandey, P. Binod, Lignocellulosic bio-refinery approach for microbial 2,3-butanediol production, Bioresour. Technol. 302 (2020) 122873, https://doi.org/10.1016/j.biortech.2020.122873.

[18] C. Li, X. Zhao, A. Wang, G.W. Huber, T. Zhang, Catalytic transformation of lignin for the production of chemicals and fuels, Chem. Rev. 115 (2015) 11559–11624, https://doi.org/10.1021/acs.chemrev.5b00155.

[19] Resources AD, Billion-Ton Report 2016, 2016.

[20] EIA, New EIA Survey Collects Data on Production and Sales of Wood Pellets—Today in Energy, U.-S. Energy Information Administration (EIA), https://www.eia.gov/todayinenergy/detail.php?id=29152 (accessed May 29, 2020).

[21] M. Guo, W. Song, The growing U.S. bioeconomy: drivers, development and constraints, New Biotechnol. 49 (2019) 48–57, https://doi.org/10.1016/j.nbt.2018.08.005.

[22] World Bioenergy Association, Global Bioenergy Statistics, 2019. https://worldbioenergy.org/uploads/191129%20WBA%20GBS%202019_LQ.pdf.

[23] EIA, Biomass Explained—U.S. Energy Information Administration (EIA), https://www.eia.gov/energyexplained/biomass/ (accessed June 12, 2020).

[24] S. Das, The National Policy of biofuels of India—a perspective, Energy Policy 143 (2020), https://doi.org/10.1016/j.enpol.2020.111595.

[25] C.H.W. Ruhe, Statistical review, JAMA J. Am. Med. Assoc. 225 (2019) 299–306, https://doi.org/10.1001/jama.1973.03220300055017.

[26] P. Purohit, V. Chaturvedi, Biomass pellets for power generation in India: a techno–economic evaluation, Environ. Sci. Pollut. Res. 25 (2018) 29614–29632, https://doi.org/10.1007/s11356-018-2960-8.

[27] A. Aradhey, India Biofuels Annual, USDA Foreign Agric Serv, 2019. https://apps.fas.usda.gov/newgainapi/api/report/downloadreportbyfilename?filename=Biofuels%20Annual_New%20Delhi_India_8-9-2019.pdf.

[28] T. Werpy, G. Petersen, Top Value Added Chemicals from Biomass Volume I—Results of Screening for Potential Candidates from Sugars and Synthesis Gas Produced, Staff at Pacific Northwest National Laboratory (PNNL), National Renewable Energy Laboratory (NREL), Office of Biomass Program (EERE) for the Office of the Energy Efficiency and Renewable Energy.

[29] V. Dhanala, S.K. Maity, D. Shee, Steam reforming of isobutanol for the production of synthesis gas over Ni/γ-Al$_2$O$_3$ catalysts, RSC Adv. 3 (2013) 24521–24529, https://doi.org/10.1039/c3ra44705g.

[30] V.C.S. Palla, D. Shee, S.K. Maity, Production of aromatics from n-butanol over HZSM-5, H-Beta, and γ-Al$_2$O$_3$: role of silica-alumina mole ratio and effect of pressure, ACS Sustain. Chem. Eng. 8 (2020) 15230–15242, https://doi.org/10.1021/acssuschemeng.0c04888.

[31] V.C.S. Palla, D. Shee, S.K. Maity, Conversion of n-butanol to gasoline range hydrocarbons, butylenes and aromatics, Appl. Catal. A Gen. 526 (2016) 28–36, https://doi.org/10.1016/j.apcata.2016.07.026.

[32] T. Wang, S. Qiu, Y. Weng, L. Chen, Q. Liu, J. Long, et al., Liquid fuel production by aqueous phase catalytic transformation of biomass for aviation, Appl. Energy 160 (2015) 329–335, https://doi.org/10.1016/j.apenergy.2015.08.116.

[33] M.R. Connor, J.C. Liao, Microbial production of advanced transportation fuels in non-natural hosts, Curr. Opin. Biotechnol. 20 (2009) 307–315, https://doi.org/10.1016/j.copbio.2009.04.002.

[34] F. Li, Y. Yuan, Z. Huang, B. Chen, F. Wang, Sustainable production of aromatics from bio-oils through combined catalytic upgrading with in situ generated hydrogen, Appl. Catal. B Environ. 165 (2015) 547–554, https://doi.org/10.1016/j.apcatb.2014.10.050.

[35] G.W. Huber, J.A. Dumesic, An overview of aqueous-phase catalytic processes for production of hydrogen and alkanes in a biorefinery, Catal. Today 111 (2006) 119–132, https://doi.org/10.1016/j.cattod.2005.10.010.

[36] A. Mittal, H.M. Pilath, D.K. Johnson, Direct conversion of biomass carbohydrates to platform chemicals: 5-Hydroxymethylfurfural (HMF) and furfural, Energy Fuels 34 (2020) 3284–3293, https://doi.org/10.1021/acs.energyfuels.9b04047.

[37] J.J. Bozell, L. Moens, D.C. Elliott, Y. Wang, G.G. Neuenscwander, S.W. Fitzpatrick, et al., Production of levulinic acid and use as a platform chemical for derived products, Resour. Conserv. Recycl. 28 (2000) 227–239, https://doi.org/10.1016/S0921-3449(99)00047-6.

[38] S. Mailaram, S.K. Maity, Techno-economic evaluation of two alternative processes for production of green diesel from karanja oil: a pinch analysis approach, J. Renewable Sustainable Energy 11 (2019) 025906, https://doi.org/10.1063/1.5078567.

[39] S.R. Yenumala, S.K. Maity, D. Shee, Hydrodeoxygenation of karanja oil over supported nickel catalysts: influence of support and nickel loading, Catal. Sci. Technol. 6 (2016) 3156–3165, https://doi.org/10.1039/C5CY01470K.

[40] S.R. Yenumala, S.K. Maity, Thermodynamic evaluation of dry reforming of vegetable oils for production of synthesis gas, J. Renewable Sustainable Energy 4 (2012) 043120, https://doi.org/10.1063/1.4747819.

[41] S.R. Yenumala, S.K. Maity, Reforming of vegetable oil for production of hydrogen: a thermodynamic analysis, Int. J. Hydrog. Energy 36 (2011) 11666–11675, https://doi.org/10.1016/j.ijhydene.2011.06.055.

[42] S. Mailaram, P. Kumar, A. Kunamalla, P. Saklecha, S.K. Maity, 3. Biomass, biorefinery, and biofuels, in: Sustain. Fuel Technol. Handb, Academic Press, 2021, pp. 51–87, https://doi.org/10.1016/B978-0-12-822989-7.00003-2.

[43] A. Pugazhendhi, T. Mathimani, S. Varjani, E.R. Rene, G. Kumar, S.H. Kim, et al., Biobutanol as a promising liquid fuel for the future—recent updates and perspectives, Fuel 253 (2019) 637–646, https://doi.org/10.1016/j.fuel.2019.04.139.

[44] A. Demirbas, Competitive liquid biofuels from biomass, Appl. Energy 88 (2011) 17–28, https://doi.org/10.1016/j.apenergy.2010.07.016.

[45] J.C. Serrano-Ruiz, E. Ramos-Fernández, A. Sepúlveda-Escribano, From biodiesel and bioethanol to liquid hydrocarbon fuels: new hydrotreating and advanced microbial technologies, Energy Environ. Sci. 5 (2012) 5638–5652, https://doi.org/10.1039/c1ee02418c.

[46] C. Arcoumanis, C. Bae, R. Crookes, E. Kinoshita, The potential of di-methyl ether (DME) as an alternative fuel for compression-ignition engines: a review, Fuel 87 (2008) 1014–1030, https://doi.org/10.1016/j.fuel.2007.06.007.

[47] L.C. Meher, D. Vidya Sagar, S.N. Naik, Technical aspects of biodiesel production by transesterification—a review, Renew. Sust. Energ. Rev. 10 (2006) 248–268, https://doi.org/10.1016/J.RSER.2004.09.002.

[48] J.V. Gerpen, Biodiesel processing and production, Fuel Process. Technol. 86 (2005) 1097–1107, https://doi.org/10.1016/j.fuproc.2004.11.005.

[49] L. Tao, A. Aden, The economics of current and future biofuels, Biofuels, Springer, New York, NY (2011) 37–69, https://doi.org/10.1007/978-1-4419-7145-6_4.

[50] Y. Zhang, M.A. Dubé, D.D. McLean, M. Kates, Biodiesel production from waste cooking oil: 2. Economic assessment and sensitivity analysis, Bioresour. Technol. 90 (2003) 229–240, https://doi.org/10.1016/S0960-8524(03)00150-0.

[51] M. Balat, H. Balat, Recent trends in global production and utilization of bio-ethanol fuel, Appl. Energy 86 (2009) 2273–2282, https://doi.org/10.1016/j.apenergy.2009.03.015.

[52] B.O. Abo, M. Gao, Y. Wang, C. Wu, H. Ma, Q. Wang, Lignocellulosic biomass for bioethanol: an overview on pretreatment, hydrolysis and fermentation processes, Rev. Environ. Health 34 (2019) 57–68, https://doi.org/10.1515/reveh-2018-0054.

[53] L.M. Vane, Separation technologies for the recovery and dehydration of alcohols from fermentation broths, Biofuels Bioprod. Biorefin. 2 (2008) 553–588, https://doi.org/10.1002/bbb.108.

[54] A.P. Mariano, M.O.S. Dias, T.L. Junqueira, M.P. Cunha, A. Bonomi, R.M. Filho, Butanol production in a first-generation Brazilian sugarcane biorefinery: technical aspects and economics of greenfield projects, Bioresour. Technol. 135 (2013) 316–323, https://doi.org/10.1016/j.biortech.2012.09.109.

[55] H.J. Huang, S. Ramaswamy, Y. Liu, Separation and purification of biobutanol during bioconversion of biomass, Sep. Purif. Technol. 132 (2014) 513–540, https://doi.org/10.1016/j.seppur.2014.06.013.

[56] C. Xue, X.-Q. Zhao, C.-G. Liu, L.-J. Chen, F.-W. Bai, Prospective and development of butanol as an advanced biofuel, Biotechnol. Adv. 31 (2013) 1575–1584, https://doi.org/10.1016/j.biotechadv.2013.08.004.

[57] A. Friedl, Downstream process options for the ABE fermentation, FEMS Microbiol. Lett. 363 (2016), https://doi.org/10.1093/femsle/fnw073.

[58] Y. Wang, S.-H. Ho, H.-W. Yen, D. Nagarajan, N.-Q. Ren, S. Li, et al., Current advances on fermentative biobutanol production using third generation feedstock, Biotechnol. Adv. 35 (2017) 1049–1059, https://doi.org/10.1016/j.biotechadv.2017.06.001.

[59] DuPont and BP disclose advanced biofuels partnership targeting multiple butanol isomers, Focus. Catal. 2008 (2008) 3, https://doi.org/10.1016/s1351-4180(08)70165-x.

[60] Y. Ni, Z. Sun, Recent progress on industrial fermentative production of acetone-butanol-ethanol by *Clostridium acetobutylicum* in China, Appl. Microbiol. Biotechnol. 83 (2009) 415–423, https://doi.org/10.1007/s00253-009-2003-y.

[61] M. Kumar, Y. Goyal, A. Sarkar, K. Gayen, Comparative economic assessment of ABE fermentation based on cellulosic and non-cellulosic feedstocks, Appl. Energy 93 (2012) 193–204, https://doi.org/10.1016/j.apenergy.2011.12.079.

[62] E. Heracleous, E.S. Vasiliadou, E.F. Iliopoulou, A.A. Lappas, A.A. Lemonidou, Conversion of lignocellulosic biomass-derived intermediates to hydrocarbon fuels, in: Biorefineries: An Introd, De Gruyter, 2015, pp. 197–218.

[63] T.H. Fleisch, A. Basu, M.J. Gradassi, J.G. Masin, Dimethyl ether: a fuel for the 21st century, Stud. Surf. Sci. Catal. 107 (1997) 117–125, https://doi.org/10.1016/S0167-2991(97)80323-0.

[64] Z. Azizi, M. Rezaeimanesh, T. Tohidian, M.R. Rahimpour, Dimethyl ether: a review of technologies and production challenges, Chem. Eng. Process. 82 (2014) 150–172, https://doi.org/10.1016/j.cep.2014.06.007.

[65] K. Yan, C. Jarvis, J. Gu, Y. Yan, Production and catalytic transformation of levulinic acid: a platform for specialty chemicals and fuels, Renew. Sust. Energ. Rev. 51 (2015) 986–997, https://doi.org/10.1016/j.rser.2015.07.021.

[66] K. Yan, G. Wu, T. Lafleur, C. Jarvis, Production, properties and catalytic hydrogenation of furfural to fuel additives and value-added chemicals, Renew. Sust. Energ. Rev. 38 (2014) 663–676, https://doi.org/10.1016/j.rser.2014.07.003.

[67] X. Liu, R. Wang, Upgrading of carbohydrates to the biofuel candidate 5-ethoxymethylfurfural (EMF), Int. J. Chem. Eng. 2018 (2018), https://doi.org/10.1155/2018/2316939.

[68] I.T. Horváth, H. Mehdi, V. Fábos, L. Boda, L.T. Mika, γ-Valerolactone—a sustainable liquid for energy and carbon-based chemicals, Green Chem. 10 (2008) 238–242, https://doi.org/10.1039/b712863k.

[69] Y. Kuwahara, W. Kaburagi, T. Fujitani, Catalytic transfer hydrogenation of levulinate esters to γ-valerolactone over supported ruthenium hydroxide catalysts, RSC Adv. 4 (2014) 45848–45855, https://doi.org/10.1039/c4ra08074b.

[70] C. Ortiz-Cervantes, M. Flores-Alamo, J.J. García, Hydrogenation of biomass-derived levulinic acid into γ-valerolactone catalyzed by palladium complexes, ACS Catal. 5 (2015) 1424–1431, https://doi.org/10.1021/cs5020095.

[71] A.P. Dunlop, J.W. Madden, Process of Preparing Gamma Valerolactone, United States Patent 2,786,852, 1957.

[72] P.P. Upare, J. Lee, D. Wang, S.B. Halligudi, Y.K. Hwang, J.-S. Chang, Selective hydrogenation of levulinic acid to γ-valerolactone over carbon-supported noble metal catalysts, Ind. Eng. Chem. Res. 17 (2011) 287–292, https://doi.org/10.1016/j.jiec.2011.02.025.

[73] C. Li, N.I. Xiao-juan, X. Di, C. Liang, Aqueous phase hydrogenation of levulinic acid to γ-valerolactone on supported Ru catalysts prepared by microwave-assisted thermolytic method, J. Fuel Chem. Technol. 46 (2018) 161–170, https://doi.org/10.1016/S1872-5813(18)30008-2.

[74] J. Zhang, J. Chen, Y. Guo, L. Chen, Effective upgrade of levulinic acid into γ-valerolactone over an inexpensive and magnetic catalyst derived from hydrotalcite precursor, ACS Sustain. Chem. Eng. 3 (2015) 1708–1714, https://doi.org/10.1021/acssuschemeng.5b00535.

[75] C. Antonetti, S. Gori, D. Licursi, G. Pasini, S. Frigo, M. López, et al., One-pot alcoholysis of the lignocellulosic *Eucalyptus nitens* biomass to n-butyl levulinate, a valuable additive for diesel motor fuel, Catalysts 10 (2020) 509, https://doi.org/10.3390/catal10050509.

[76] L. Peng, L. Lin, J. Zhu, J. Shi, S. Liu, Solid acid catalyzed glucose conversion to ethyl levulinate, Appl. Catal. A Gen. 397 (2011) 259–265, https://doi.org/10.1016/j.apcata.2011.03.008.

[77] X. Kong, X. Zhang, C. Han, C. Li, L. Yu, J. Liu, Ethanolysis of biomass based furfuryl alcohol to ethyl levulinate over Fe modified USY catalyst, Mol. Catal. 443 (2017) 186–192, https://doi.org/10.1016/j.mcat.2017.10.011.

[78] S.S. Dharne, V.V. Bokade, Esterification of levulinic acid to n-butyl levulinate over heteropolyacid supported on acid-treated clay, J. Nat. Gas Chem. 20 (2011) 18–24, https://doi.org/10.1016/S1003-9953(10)60147-8.

[79] C.R. Patil, P.S. Niphadkar, V.V. Bokade, P.N. Joshi, Esterification of levulinic acid to ethyl levulinate over bimodal micro–mesoporous H-BEA zeolite derivatives, Catal. Commun. 43 (2014) 188–191, https://doi.org/10.1016/j.catcom.2013.10.006.

[80] B.L. Oliveira, V. da Silva, Sulfonated carbon nanotubes as catalysts for the conversion of levulinic acid into ethyl levulinate, Catal. Today 234 (2014) 257–263, https://doi.org/10.1016/j.cattod.2013.11.028.

[81] J.A. Melero, G. Morales, J. Iglesias, M. Paniagua, B. Hernández, S. Penedo, Efficient conversion of levulinic acid into alkyl levulinates catalyzed by sulfonic mesostructured silicas, Appl. Catal. A Gen. 466 (2013) 116–122, https://doi.org/10.1016/j.apcata.2013.06.035.

[82] F.G. Cirujano, A. Corma, F.X.L.i. Xamena, Conversion of levulinic acid into chemicals: synthesis of biomass derived levulinate esters over Zr-containing MOFs, Chem. Eng. Sci. 124 (2015) 52–60, https://doi.org/10.1016/j.ces.2014.09.047.

[83] D.J. Hayes, S. Fitzpatrick, M.H.B. Hayes, J.R.H. Ross, The biofine process—production of levulinic acid, furfural, and formic acid from lignocellulosic feedstocks, in: Biorefineries-Industrial Processes and Products: Status Quo and Future Directions, WILEY-VCH Verlag GmbH & Co. KGaA, 2006, pp. 139–164.

[84] J.G.M. Bremn, R.K.F. Keey, The hydrogenation of furfuraldehyde to furfuryl alcohol and sylvan (2-methylfuran), J. Chem. Soc. (1947) 1068–1080.

[85] C.P. Jiménez-Gómez, J.A. Cecilia, R. Moreno-Tost, P. Maireles-Torres, Selective production of 2-methylfuran by gas-phase hydrogenation of furfural on copper incorporated by complexation in mesoporous silica catalysts, ChemSusChem 10 (2017) 1448–1459, https://doi.org/10.1002/cssc.201700086.

[86] M. Thewes, M. Muether, S. Pischinger, M. Budde, A. Brunn, A. Sehr, et al., Analysis of the impact of 2-methylfuran on mixture formation and combustion in a direct-injection spark-ignition engine, Energy Fuel 25 (2011) 5549–5561, https://doi.org/10.1021/ef201021a.

[87] P. Liu, L. Sun, X. Jia, C. Zhang, W. Zhang, Y. Song, et al., Efficient one-pot conversion of furfural into 2-methyltetrahydrofuran using non-precious metal catalysts, Mol. Catal. 490 (2020) 110951, https://doi.org/10.1016/j.mcat.2020.110951.

[88] Z. Xie, B. Chen, H. Wu, M. Liu, H. Liu, J. Zhang, et al., Highly efficient hydrogenation of levulinic acid into 2-methyltetrahydrofuran over Ni-Cu/Al$_2$O$_3$-ZrO$_2$ bifunctional catalysts, Green Chem. 21 (2019) 606–613, https://doi.org/10.1039/c8gc02914h.

[89] S. Zhong, R. Daniel, H. Xu, J. Zhang, D. Turner, M.L. Wyszynski, et al., Combustion and emissions of 2,5-dimethylfuran in a direct-injection spark-ignition engine, Energy Fuels 24 (2010) 2891–2899, https://doi.org/10.1021/ef901575a.

[90] E. Nürenberg, P. Schulze, F. Kohler, M. Zubel, S. Pischinger, F. Schüth, Blending real world gasoline with biofuel in a direct conversion process, ACS Sustain. Chem. Eng. 7 (2019) 249–257, https://doi.org/10.1021/acssuschemeng.8b03044.

[91] Y. Zu, P. Yang, J. Wang, X. Liu, J. Ren, G. Lu, et al., Efficient production of the liquid fuel 2,5-dimethylfuran from 5-hydroxymethylfurfural over Ru/Co$_3$O$_4$ catalyst, Appl. Catal. B Environ. 146 (2014) 244–248, https://doi.org/10.1016/j.apcatb.2013.04.026.

[92] M. Esteves Laura, M.H. Brijaldo, E.G. Oliveira, M.J. José, H. Rojas, A. Caytuero, et al., Effect of support on selective 5-hydroxymethylfurfural hydrogenation towards 2,5-dimethylfuran over copper catalysts, Fuel 270 (2020) 117524, https://doi.org/10.1016/j.fuel.2020.117524.

[93] P.P. Upare, D.W. Hwang, Y.K. Hwang, U.H. Lee, D.Y. Hong, J.S. Chang, An integrated process for the production of 2,5-dimethylfuran from fructose, Green Chem. 17 (2015) 3310–3313, https://doi.org/10.1039/c5gc00281h.

[94] Z. Gao, C. Li, G. Fan, L. Yang, F. Li, Nitrogen-doped carbon-decorated copper catalyst for highly efficient transfer hydrogenolysis of 5-hydroxymethylfurfural to convertibly produce 2,5-dimethylfuran or 2,5-dimethyltetrahydrofuran, Appl. Catal. B Environ. 226 (2018) 523–533, https://doi.org/10.1016/j.apcatb.2018.01.006.

[95] M.R. Grochowski, W. Yang, A. Sen, Mechanistic study of a one-step catalytic conversion of fructose to 2,5-dimethyltetrahydrofuran, Chem. Eur. J. 18 (2012) 12363–12371, https://doi.org/10.1002/chem.201201522.

[96] T.S. Hansen, K. Barta, P.T. Anastas, P.C. Ford, A. Riisager, One-pot reduction of 5-hydroxymethylfurfural via hydrogen transfer from supercritical methanol, Green Chem. 14 (2012) 2457–2461, https://doi.org/10.1039/c2gc35667h.

[97] X. Liu, R. Wang, 5-Ethoxymethylfurfural—a remarkable biofuel candidate, In: Biomass Biofuels Biochem.: Biomass, Biofuels, Biochemicals Recent Advances in Development of Platform Chemicals, Elsevier, 2020, pp. 355–375, https://doi.org/10.1016/b978-0-444-64307-0.00013-5.

[98] F. Yang, J. Tang, R. Ou, G. Zengjing, Y. Wang, X. Wang, et al., Fully catalytic upgrading synthesis of 5-ethoxymethylfurfural from biomass-derived 5-hydroxymethylfurfural over recyclable layered-niobium–molybdate solid acid, Appl. Catal. B Environ. 256 (2019) 117786, https://doi.org/10.1016/j.apcatb.2019.117786.

[99] B. Chen, G. Xu, C. Chang, Z. Zheng, D. Wang, S. Zhang, et al., Efficient one-pot production of biofuel 5-ethoxymethylfurfural from corn stover: optimization and kinetics, Energy Fuels 33 (2019) 4310–4321, https://doi.org/10.1021/acs.energyfuels.9b00357.

[100] K.P. Kumari, R.B. Srinivasa, D. Padmakar, N. Pasha, N. Lingaiah, Lewis acidity induced heteropoly tungustate catalysts for the synthesis of 5-ethoxymethyl furfural from fructose and

5-hydroxymethylfurfural, Mol. Catal. 448 (2018) 108–115, https://doi.org/10.1016/j.mcat.2018.01.034.

[101] P.H.R. Silva, V.L.C. Gonçalves, C.J.A. Mota, Glycerol acetals as anti-freezing additives for biodiesel, Bioresour. Technol. 101 (2010) 6225–6229, https://doi.org/10.1016/j.biortech.2010.02.101.

[102] J.E. Castanheiro, J. Vital, I. Fonesca, A.M. Ramos, Glycerol conversion into biofuel additives by acetalization with pentanal over heteropolyacids immobilized on zeolites, Catal. Today 346 (2020) 76–80, https://doi.org/10.1016/j.cattod.2019.04.048.

[103] H. Serafim, I.M. Fonesca, A.M. Ramos, J. Vital, J.E. Castanheiro, Valorization of glycerol into fuel additives over zeolites as catalysts, Chem. Eng. 178 (2011) 291–296, https://doi.org/10.1016/j.cej.2011.10.004.

[104] E. García, M. Laca, E. Pérez, A. Garrido, J. Peinado, New class of acetal derived from glycerin as a biodiesel fuel component, Energy Fuels 22 (2008) 4274–4280, https://doi.org/10.1021/ef800477m.

[105] P. Kumar, M. Varkolu, S. Mailaram, A. Kunamalla, S.K. Maity, Biorefinery polyutilization systems: production of green transportation fuels from biomass, in: K.R. Khalilpour (Ed.), Polygeneration with Polystorage Chem. Energy Hubs, Elsevier Inc, 2019, pp. 373–407, https://doi.org/10.1016/B978-0-12-813306-4.00012-4.

[106] P.M. Mortensen, J. Grunwaldt, P.A. Jensen, K.G. Knudsen, A.D. Jensen, A review of catalytic upgrading of bio-oil to engine fuels, Appl. Catal. A Gen. 407 (2011) 1–19, https://doi.org/10.1016/j.apcata.2011.08.046.

[107] Z. Si, X. Zhang, C. Wang, L. Ma, R. Dong, An overview on catalytic hydrodeoxygenation of pyrolysis oil and its model compounds, Catalysts 7 (2017) 169, https://doi.org/10.3390/catal7060169.

[108] G. Perkins, T. Bhaskar, M. Konarova, Process development status of fast pyrolysis technologies for the manufacture of renewable transport fuels from biomass, Renew. Sust. Energ. Rev. 90 (2018) 292–315, https://doi.org/10.1016/j.rser.2018.03.048.

[109] A.H. Zacher, M.V. Olarte, D.M. Santosa, D.C. Elliott, S.B. Jones, A review and perspective of recent bio-oil hydrotreating research, Green Chem. 16 (2014) 491–515, https://doi.org/10.1039/c3gc41382a.

[110] S. Jones, Y. Zhu, D. Anderson, R.T. Hallen, D.C. Elliott, Process Design and Economics for the Conversion of Algal Biomass to Hydrocarbons: Whole Algae Hydrothermal Liquefaction and Upgrading, PNNL, 2014, pp. 1–69, https://doi.org/10.2172/1126336.

[111] F.M. Hossain, J. Kosinkova, R.J. Brown, Z. Ristovski, B. Hankamer, E. Stephens, et al., Experimental investigations of physical and chemical properties for microalgae HTL bio-crude using a large batch reactor, Energies 10 (2017) 467, https://doi.org/10.3390/en10040467.

[112] J.A. Ramirez, R.J. Brown, T.J. Rainey, A review of hydrothermal liquefaction bio-crude properties and prospects for upgrading to transportation fuels, Energies 8 (2015) 6765–6794, https://doi.org/10.3390/en8076765.

[113] A. Lappas, E. Heracleous, Production of Biofuels via Fischer-Tropsch synthesis: Biomass-to-Liquids, Woodhead Publishing Limited, 2011, https://doi.org/10.1533/9780857090492.3.493.

[114] P. Kumar, S.R. Yenumala, S.K. Maity, D. Shee, Kinetics of hydrodeoxygenation of stearic acid using supported nickel catalysts: effects of supports, Appl. Catal. A Gen. 471 (2014) 28–38, https://doi.org/10.1016/j.apcata.2013.11.021.

[115] B. Donnis, R.G. Egeberg, P. Blom, K.G. Knudsen, Hydroprocessing of bio-oils and oxygenates to hydrocarbons. Understanding the reaction routes, Top. Catal. 52 (2009) 229–240, https://doi.org/10.1007/s11244-008-9159-z.

[116] P.M. Lv, Z.H. Xiong, J. Chang, C.Z. Wu, Y. Chen, J.X. Zhu, An experimental study on biomass air-steam gasification in a fluidized bed, Bioresour. Technol. 95 (2004) 95–101, https://doi.org/10.1016/j.biortech.2004.02.003.

[117] P. Kumar, P. Kumar, P.V.C. Rao, N.V. Choudary, G. Sriganesh, Saw dust pyrolysis: effect of temperature and catalysts, Fuel 199 (2017) 339–345, https://doi.org/10.1016/j.fuel.2017.02.099.

[118] D.C. Elliott, Historical developments in hydroprocessing bio-oils, Energy Fuel 21 (2007) 1792–1815, https://doi.org/10.1021/ef070044u.

[119] S.S. Ail, S. Dasappa, Biomass to liquid transportation fuel via Fischer Tropsch synthesis—technology review and current scenario, Renew. Sust. Energ. Rev. 58 (2016) 267–286, https://doi.org/10.1016/j.rser.2015.12.143.

[120] M.J. Tijmensen, A.P. Faaij, C.N. Hamelinck, M.R. van Hardeveld, Exploration of the possibilities for production of Fischer Tropsch liquids and power via biomass gasification, Biomass Bioenergy 23 (2002) 129–152, https://doi.org/10.1016/S0961-9534(02)00037-5.

[121] M.E. Dry, The Fischer–Tropsch process: 1950-2000, Catal. Today 71 (2002) 227–241, https://doi.org/10.1016/S0920-5861(01)00453-9.

[122] M.E. Dry, Fischer–Tropsch reactions and the environment, Appl. Catal. A Gen. 189 (1999) 185–190, https://doi.org/10.1016/S0926-860X(99)00275-6.

[123] F. Zhang, S. Rodriguez, J.D. Keasling, Metabolic engineering of microbial pathways for advanced biofuels production, Curr. Opin. Biotechnol. 22 (2011) 775–783, https://doi.org/10.1016/j.copbio.2011.04.024.

[124] N. Ladygina, E.G. Dedyukhina, M.B. Vainshtein, A review on microbial synthesis of hydrocarbons, Process Biochem. 41 (2006) 1001–1014, https://doi.org/10.1016/j.procbio.2005.12.007.

[125] O. Muraza, Maximizing diesel production through oligomerization: a landmark opportunity for zeolite research, Ind. Eng. Chem. Res. 54 (2015) 781–789, https://doi.org/10.1021/ie5041226.

[126] G.W. Huber, J.N. Chheda, C.J. Barrett, J.A. Dumesic, Production of liquid alkanes by aqueous-phase processing of biomass-derived carbohydrates, Science 308 (2005) 1446–1450, https://doi.org/10.1126/science.1111166.

[127] R. Xing, A.V. Subrahmanyam, H. Olcay, W. Qi, G.P. van Walsum, H. Pendse, et al., Production of jet and diesel fuel range alkanes from waste hemicellulose-derived aqueous solutions, Green Chem. 12 (2010) 1933–1946, https://doi.org/10.1039/c0gc00263a.

[128] A. Bohre, B. Saha, M.M. Abu-Omar, Catalytic upgrading of 5-hydroxymethylfurfural to drop-in biofuels by solid base and bifunctional metal-acid catalysts, ChemSusChem 8 (2015) 4022–4029, https://doi.org/10.1002/cssc.201501136.

[129] D. Damodar, A. Kunamalla, M. Varkolu, S.K. Maity, A.S. Deshpande, Near-room-temperature synthesis of sulfonated carbon nanoplates and their catalytic application, ACS Sustain. Chem. Eng. 7 (2019) 12707–12717, https://doi.org/10.1021/acssuschemeng.8b06280.

[130] J.C. Serrano-Ruiz, D. Wang, J.A. Dumesic, Catalytic upgrading of levulinic acid to 5-nonanone, Green Chem. 12 (2010) 574–577, https://doi.org/10.1039/b923907c.

[131] V. Dhanala, S.K. Maity, D. Shee, Oxidative steam reforming of isobutanol over Ni/γ-Al_2O_3 catalysts: a comparison with thermodynamic equilibrium analysis, J. Ind. Eng. Chem. 27 (2015) 153–163, https://doi.org/10.1016/j.jiec.2014.12.029.

[132] P. Tian, Y. Wei, M. Ye, Z. Liu, Methanol to olefins (MTO): from fundamentals to commercialization, ACS Catal. 5 (2015) 1922–1938, https://doi.org/10.1021/acscatal.5b00007.

[133] P.S. Rezaei, H. Shafaghat, W.M.A.W. Daud, Production of green aromatics and olefins by catalytic cracking of oxygenate compounds derived from biomass pyrolysis: a review, Appl. Catal. A Gen. 469 (2014) 490–511, https://doi.org/10.1016/j.apcata.2013.09.036.

[134] L. Kaluža, D. Kubička, The comparison of Co, Ni, Mo, CoMo and NiMo sulfided catalysts in rapseed oil hydrodeoxygenation, React. Kinet. Mech. Catal. 122 (2017) 333–341, https://doi.org/10.1007/s11144-017-1247-2.

[135] P. Kumar, S.K. Maity, D. Shee, Hydrodeoxygenation of stearic acid using Mo modified Ni and Co/alumina catalysts: effect of calcination temperature, Chem. Eng. Commun. 207 (2020) 904–919, https://doi.org/10.1080/00986445.2019.1630396.

[136] E. Kordouli, B. Pawelec, K. Bourikas, C. Kordulis, J.L.G. Fierro, A. Lycourghiotis, Mo promoted Ni-Al_2O_3 co-precipitated catalysts for green diesel production, Appl. Catal. B Environ. 229 (2018) 139–154, https://doi.org/10.1016/j.apcatb.2018.02.015.

[137] E. Kordouli, L. Sygellou, C. Kordulis, K. Bourikas, A. Lycourghiotis, Probing the synergistic ratio of the $NiMo/\gamma$-Al_2O_3 reduced catalysts for the transformation of natural triglycerides into green diesel, Appl. Catal. B Environ. 209 (2017) 12–22, https://doi.org/10.1016/j.apcatb.2017.02.045.

[138] P. Kumar, S.K. Maity, D. Shee, Role of NiMo alloy and Ni species in the performance of NiMo/alumina catalysts for hydrodeoxygenation of stearic acid: a kinetic study, ACS Omega 4 (2019) 2833–2843, https://doi.org/10.1021/acsomega.8b03592.

[139] Y. Wang, G. Xiong, X. Liu, X. Yu, L. Liu, J. Wang, et al., Structure and reducibility of $NiO-MoO_3/$-$\gamma-Al_2O_3$ catalysts: effects of loading and molar ratio, J. Phys. Chem. C 112 (2008) 17265–17271, https://doi.org/10.1021/jp800182j.

[140] C. Hoang-Van, Y. Kachaya, S.J. Teichner, Y. Arnaud, J.A. Dalmon, Characterization of nickel catalysts by chemisorption techniques, x-ray diffraction and magnetic measurements: effects of support, precursor and hydrogen pretreatment, Appl. Catal. 46 (1989) 281–296, https://doi.org/10.1016/S0166-9834(00)81123-9.

[141] H. Chen, Q. Wang, X. Zhang, L. Wang, Effect of support on the NiMo phase and its catalytic hydrodeoxygenation of triglycerides, Fuel 159 (2015) 430–435, https://doi.org/10.1016/j.fuel.2015.07.010.

[142] R. Tiwari, B.S. Rana, R. Kumar, D. Verma, R. Kumar, R.K. Joshi, et al., Hydrotreating and hydrocracking catalysts for processing of waste soya-oil and refinery-oil mixtures, Catal. Commun. 12 (2011) 559–562, https://doi.org/10.1016/j.catcom.2010.12.008.

[143] Q. Liu, H. Zuo, T. Wang, L. Ma, Q. Zhang, One-step hydrodeoxygenation of palm oil to isomerized hydrocarbon fuels over Ni supported on nano-sized SAPO-11 catalysts, Appl. Catal. A Gen. 468 (2013) 68–74, https://doi.org/10.1016/j.apcata.2013.08.009.

[144] S.R. Yenumala, P. Kumar, S.K. Maity, D. Shee, Production of green diesel from karanja oil (Pongamia pinnata) using mesoporous NiMo-alumina composite catalysts, Bioresour. Technol. Rep. 7 (2019) 100288, https://doi.org/10.1016/j.biteb.2019.100288.

[145] T. Li, J. Cheng, R. Huang, J. Zhou, K. Cen, Conversion of waste cooking oil to jet biofuel with nickel-based mesoporous zeolite Y catalyst, Bioresour. Technol. 197 (2015) 289–294, https://doi.org/10.1016/j.biortech.2015.08.115.

[146] S.R. Yenumala, P. Kumar, S.K. Maity, D. Shee, Hydrodeoxygenation of karanja oil using ordered mesoporous nickel-alumina composite catalysts, Catal. Today 348 (2020) 45–54, https://doi.org/10.1016/j.cattod.2019.08.040.

[147] W. Mu, H. Ben, X. Du, X. Zhang, F. Hu, W. Liu, et al., Noble metal catalyzed aqueous phase hydrogenation and hydrodeoxygenation of lignin-derived pyrolysis oil and related model compounds, Bioresour. Technol. 173 (2014) 6–10, https://doi.org/10.1016/j.biortech.2014.09.067.

[148] D.A. Ruddy, J.A. Schaidle, J.R. Ferrell, J. Wang, L. Moens, J.E. Hensley, Recent advances in heterogeneous catalysts for bio-oil upgrading via "ex situ catalytic fast pyrolysis": catalyst development through the study of model compounds, Green Chem. 16 (2014) 454–490, https://doi.org/10.1039/c3gc41354c.

[149] Y. Romero, F. Richard, S. Brunet, Hydrodeoxygenation of 2-ethylphenol as a model compound of bio-crude over sulfided Mo-based catalysts: promoting effect and reaction mechanism, Appl. Catal. B Environ. 98 (2010) 213–223, https://doi.org/10.1016/j.apcatb.2010.05.031.

[150] F.E. Massoth, P. Politzer, M.C. Concha, J.S. Murray, J. Jakowski, J. Simons, Catalytic hydrodeoxygenation of methyl-substituted phenols: correlations of kinetic parameters with molecular properties, J. Phys. Chem. B 110 (2006) 14283–14291, https://doi.org/10.1021/jp057332g.

[151] D.C. Elliott, T.R. Hart, Catalytic hydroprocessing of chemical models for bio-oil, Energy Fuels 23 (2009) 631–637, https://doi.org/10.1021/ef8007773.

[152] H.Y. Zhao, D. Li, P. Bui, S.T. Oyama, Hydrodeoxygenation of guaiacol as model compound for pyrolysis oil on transition metal phosphide hydroprocessing catalysts, Appl. Catal. A Gen. 391 (2011) 305–310, https://doi.org/10.1016/j.apcata.2010.07.039.

[153] A. Popov, E. Kondratieva, J. Gilson, L. Mariey, A. Travert, F. Maugé, IR study of the interaction of phenol with oxides and sulfided CoMo catalysts for bio-fuel hydrodeoxygenation, Catal. Today 172 (2011) 132–135, https://doi.org/10.1016/j.cattod.2011.02.010.

[154] G. Li, N. Li, Z. Wang, C. Li, A. Wang, X. Wang, et al., Synthesis of high-quality diesel with furfural and 2-methylfuran from hemicellulose, ChemSusChem 5 (2012) 1958–1966, https://doi.org/10.1002/cssc.201200228.

[155] A.P. Steynberg, Introduction to Fischer-Tropsch technology, Stud. Surf. Sci. Catal. 152 (2004) 1–63, https://doi.org/10.1016/s0167-2991(04)80458-0.

[156] A. Zheng, L. Jiang, Z. Zhao, Z. Huang, K. Zhao, G. Wei, et al., Catalytic fast pyrolysis of lignocellulosic biomass for aromatic production: chemistry, catalyst and process, Wiley Interdiscip. Rev. Energy Environ. 6 (2017) 1–18, https://doi.org/10.1002/wene.234.

[157] T.R. Carlson, J. Jae, Y.-C. Lin, G.A. Tompsett, G.W. Huber, Catalytic fast pyrolysis of glucose with HZSM-5: the combined homogeneous and heterogeneous reactions, J. Catal. 270 (2010) 110–124, https://doi.org/10.1016/j.jcat.2009.12.013.

[158] J. Jae, G.A. Tompsett, A.J. Foster, K.D. Hammond, S.M. Auerbach, R.F. Lobo, et al., Investigation into the shape selectivity of zeolite catalysts for biomass conversion, J. Catal. 279 (2011) 257–268, https://doi.org/10.1016/j.jcat.2011.01.019.

[159] A.J. Foster, J. Jae, Y.-T. Cheng, G.W. Huber, R.F. Lobo, Optimizing the aromatic yield and distribution from catalytic fast pyrolysis of biomass over ZSM-5, Appl. Catal. A Gen. 423–424 (2012) 154–161, https://doi.org/10.1016/j.apcata.2012.02.030.

[160] J. Li, X. Li, G. Zhou, W. Wang, C. Wang, S. Komarneni, et al., Catalytic fast pyrolysis of biomass with mesoporous ZSM-5 zeolites prepared by desilication with NaOH solutions, Appl. Catal. A Gen. 470 (2014) 115–122, https://doi.org/10.1016/j.apcata.2013.10.040.

[161] Y.T. Cheng, J. Jae, J. Shi, W. Fan, G.W. Huber, Production of renewable aromatic compounds by catalytic fast pyrolysis of lignocellulosic biomass with bifunctional Ga/ZSM-5 catalysts, Angew. Chem. Int. Ed. 51 (2012) 1416–1419, https://doi.org/10.1002/anie.201107390.

[162] J. Skupinska, Oligomerization of α-olefins to higher oligomers, Chem. Rev. 91 (1991) 613–648, https://doi.org/10.1021/cr00004a007.

[163] C.T. O'Connor, M. Kojima, Alkene oligomerization, Catal. Today 6 (1990) 329–349, https://doi.org/10.1016/0920-5861(90)85008-C.

[164] X. Zhang, J. Zhong, J. Wang, L. Zhang, J. Gao, A. Liu, Catalytic performance and characterization of Ni-doped HZSM-5 catalysts for selective trimerization of n-butene, Fuel Process. Technol. 90 (2009) 863–870, https://doi.org/10.1016/j.fuproc.2009.04.011.

[165] A. de Klerk, Properties of synthetic fuels from H-ZSM-5 oligomerization of Fischer–Tropsch type feed materials, Energy Fuels 21 (2007) 3084–3089, https://doi.org/10.1021/ef700246k.

[166] W. Shen, G.A. Tompsett, K.D. Hammond, R. Xing, F. Dogan, C.P. Grey, et al., Liquid phase aldol condensation reactions with MgO-ZrO$_2$ and shape-selective nitrogen-substituted NaY, Appl. Catal. A Gen. 392 (2011) 57–68, https://doi.org/10.1016/j.apcata.2010.10.023.

[167] C.J. Barrett, J.N. Chheda, G.W. Huber, J.A. Dumesic, Single-reactor process for sequential aldol-condensation and hydrogenation of biomass-derived compounds in water, Appl. Catal. B Environ. 66 (2006) 111–118, https://doi.org/10.1016/j.apcatb.2006.03.001.

SECTION 1

Chemical and thermochemical conversion of biomass

CHAPTER 2

Fast pyrolysis of biomass and hydrodeoxygenation of bio-oil for the sustainable production of hydrocarbon biofuels

Janaki Komandur and Kaustubha Mohanty
Department of Chemical Engineering, Indian Institute of Technology Guwahati, Guwahati, India

Contents

2.1 Introduction	47
2.2 Fast pyrolysis	50
2.2.1 Reactor configuration	50
2.2.2 Catalytic fast pyrolysis	53
2.3 Effect of operation conditions on fast pyrolysis products	56
2.4 Bio-oil upgrading	58
2.4.1 Physical upgradation technique of bio-oil	58
2.4.2 Chemical upgradation technique of bio-oil	58
2.4.3 Hydrodeoxygenation of bio-oil	61
2.4.4 Reactor configuration of hydrodeoxygenation	62
2.4.5 In situ hydrogenation	64
2.4.6 Catalyst for hydrodeoxygenation process	64
2.4.7 Role of support	69
2.4.8 Effect of parameters on hydrodeoxygenation of bio-oil	70
2.5 Conclusion and future perspective	71
References	73

2.1 Introduction

Nowadays, there is a rapid increase in the world energy demand due to an increase in the population. Hence, advanced and sustainable technologies are required to fulfill energies as well as economic requirements [1]. According to an estimation, the world energy demand is expected to rise by 2.9% by 2018. Moreover, reducing the carbon emission from the burning of fossil fuels is also in focus. Also, biomass resources are the only renewable source for the production of liquid fuel and value–added chemicals. They are abundant and renewable, which makes them a viable option as an alternative to fossil fuels and also a wide range of value–added chemicals can be obtained from the pyrolytic oil [2]. The biomass conversion is possible by different technologies. The two crucial

Hydrocarbon Biorefinery
https://doi.org/10.1016/B978-0-12-823306-1.00003-0

Copyright © 2022 Elsevier Inc.
All rights reserved.

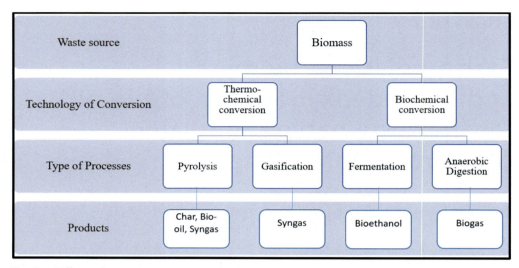

Fig. 2.1 Different biomass conversion technologies.

technologies employed for the conversion of biomass to biofuels and value-added chemicals are thermochemical and biological. Fig. 2.1 demonstrates the different types of technologies and the final useful products.

The biochemical conversion of the biomass takes place by different technologies such as anaerobic digestion, alcoholic fermentation, and photobiological hydrogen production. This process utilizes an enzyme or yeast in the conversion of biomass to bioethanol [3]. The limitations of biochemical conversion technology include size reduction, crystallinity, cellulase deactivation, and tolerance of ethanol by yeast [4]. The different thermochemical conversion processes include pyrolysis, gasification, combustion, etc. The gasification of biomass is operated at temperatures around 1000°C–1400°C. The main product of gasification process is syngas. Pyrolysis is a process in which the pyrolytic products include char, bio-oil, and gases. The pyrolysis processes are differentiated based on the heating rate of the process. The three different pyrolysis processes are slow pyrolysis, intermediate pyrolysis, and fast pyrolysis. The thermochemical technologies, such as fast pyrolysis, have many advantages over the traditional biochemical technologies (enzymatic conversion). The time taken for the conversion of biomass to the final product is considered as the key advantage. Also, this helps in the effective utilization of localized biomass, which indeed helps in the reduction of transportation cost of the raw material to the plant site [1]. Fig. 2.2 discusses the different types of thermochemical conversion technologies and their end products.

Pyrolysis of biomass contemplates being one of the feasible routes to produce valuable end products [5]. It is a process in which the biomass (usually a solid material) is rapidly heated in an inert atmosphere to a temperature of 450°C–600°C. A dark homogeneous liquid, commonly referred to as bio-oil, is formed, after cooling and condensing the

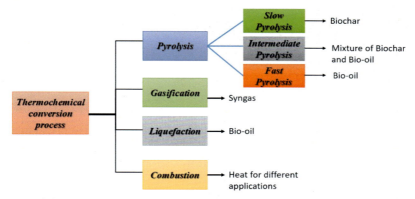

Fig. 2.2 Types of thermochemical technologies with their end products.

vapor, which is formed by fast heating of biomass [6]. Nevertheless, the oxygen content of bio-oil is very high (around 35%–55%). Higher oxygen content leads to a lower calorific value and instability. Thus, there is a need to improve the quality of the bio-oil. One of the other critical issues with pyrolysis oil is the water content. The bio-oil with minimal water content can be obtained by using dry biomass feed with less than 10% water content [7]. The products, such as char and gas, are used to generate the heat required for the process, which ensures that there are no or fewer waste streams coming out of the process. A maximum yield is achieved by essentially controlling the operating temperature and vapor residence time. Reasonable control of these two parameters can subsequently minimize the conversion of intermediates into undesired products and maximize the liquid yield. The yield of liquid depends on various factors, such as raw material, operational temperature, the residence time of hot vapor, separation of char, and ash contents [8].

Different upgradation techniques, such as hydrodeoxygenation, pyrolysis, and cracking in the presence of the catalyst, are employed to remove excess oxygen from the bio-oil. This process has different names, such as hydrogenolysis, hydrogenation, and hydrotreatment. Also, this process is similar to hydrodesulfurization and hydrodenitrogenation processes that are in use in petroleum refineries to treat crude oil [9]. Catalytic hydrodeoxygenation is a process in which catalyst is used [10]. The high oxygen content in the bio-oil restricts its application as a transportation fuel. Upgradation of bio-oil by hydrodeoxygenation essentially removes a significant amount of oxygen and also helps in improving the quality of the bio-oil [11]. The hydrodeoxygenation of bio-oil is a potential method in which the bio-oil is subjected to the high pressure of hydrogen in the presence of a catalyst. As the bio-oil contains different types of oxy-compounds, it is a challenge to determine a reaction pathway ultimately [12].

The motive of this book chapter is to present different types of reactors used in the fast pyrolysis process at different operating conditions. This chapter also emphasizes the catalyst used for fast pyrolysis and hydrodeoxygenation of the bio-oil. The effect of operating conditions on the final product is also focused.

2.2 Fast pyrolysis

The high heating rates and short vapor residence time (typically less than 2 s) distinguish the fast pyrolysis from other pyrolysis processes. A better liquid yield is obtained from this process. The heating value of the obtained liquid is half of the conventional fuel [7]. The transportation of liquid obtained by converting solid biomass is comparatively easy. One of the essential features of fast pyrolysis is the rate of heating. The rate of heating of the biomass particles is approximately equal to or more than 100°C/s. The other factor is the residence time of vapor. It should be maintained less than 2 s to avoid any secondary reactions. The secondary reactions effectively reduce the yield of bio-oil, and hence the vapor residence time should be minimal. The third feature is to maintain very high heating rates and high heat transfer rates at the interface of the reacting biomass particles. To achieve this, the size of the particle should be less than 3 mm. This is mainly attributed to the low thermal conductivity of the particle. The fourth feature is to obtain a maximum liquid yield. The reaction temperature is generally maintained at 500°C to obtain a maximum yield. Carefully controlling the reaction temperature at an optimum value can give maximum liquid yield. If the biomass particles are reacting at a temperature lower than the optimum temperature, then the reaction favors the formation of charcoal, which affects the liquid yield. One of the possible ways to overcome this issue is by using particles that are typically less than 3 mm. The last feature is the removal of char. It is imperative as it leads to the cracking of vapor and forms undesirable products [7].

2.2.1 Reactor configuration

Among the various thermochemical processes, much attention has been paid to the model and design of the reactors for pyrolysis processes. A reactor is considered as the heart of the pyrolysis process. The reactors are carefully designed to provide the essential features for the fast pyrolysis process. Some of the reactor types are given below:

  i. Ablative reactor

In ablative reactors, a high heating rate is obtained by pressing the biomass against a hot surface. An oil film is formed on the surface, which then evaporates and acts as a lubricant for the next layer of biomass particles. In ablative reactors, the heat transfer takes place from the wall of the reactor to the biomass particles that are in direct contact with each other under particular pressure. The outer surface of the biomass particles is rapidly heated, and the char formed from the particles is then shredded [6]. The pyrolysis reaction front moves in a uni-direction through the particle. The reaction rate is highly dependent on the pressure applied to the wood onto the hot surface, the relative velocity of the wood and surface of heat exchange, and the surface temperature of the reactor. Some of the critical

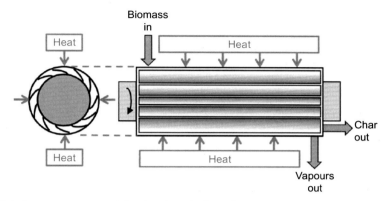

Fig. 2.3 Ablative reactor model developed by the Aston University. *(Reprinted from A.V. Bridgwater, Review of fast pyrolysis of biomass and product upgrading, Biomass Bioenergy 38 (2012) 68–94, doi:10.1016/j.biombioe.2011.01.048, with permission from Elsevier.)*

features for ablative reactors are as follows: the temperature of the reactor wall is less than 600°C, the relative motion between particles and the wall of the reactor is high, and the high pressure of the particle is exerted on the hot reactor wall [7]. As there is no limitation for the transfer of the heat to the biomass particles, larger-sized particles can be used in the ablative reactors. The advantage of ablative reactors is that it can be operated in an inert atmosphere [13]. An ablative plate reactor was developed by the Aston University in which pressure and motion are determined mechanically, preventing the need for carrier gas. The yield of liquid obtained is around 70–75 wt% (on dry basis) [7]. Fig. 2.3 demonstrates the ablative reactor model developed by the Aston University.

ii. Fluidized bed

It is a type of reactor which is partially filled with inert materials, such as sand. A fluidizing gas (generally nitrogen gas) and heat are supplied to the reactor (as pyrolysis is indigenously an endothermic process) [14]. The biomass feed is continuously supplied to the reactor, and the fluidizing gas is provided as an input that flows over the inert bed material. The entire bed is now in a fluidized state. The mode of heat transfer includes both conduction as well as convection. The transfer of heat is limited to the intra-particle region. Thus, the particle size required is minimal (typically less than 3 mm) [8]. The rate of heat transfer and mixing is considerably high in this type of reactor [6]. This reactor can be used for producing fuel of high moisture content or low-grade fuels. The fluidized bed is further classified into a bubbling fluid bed and circulating fluid bed depending upon the velocity of blowing air [13].

iii. Bubbling fluidized bed

The bubbling fluidized bed is a widely used reactor model owing to its simple construction and easy operation. Sand, which is often preferred as bed material, is supported on the perforated plate. The inert gas fluidizes the sand material. A high yield of oil

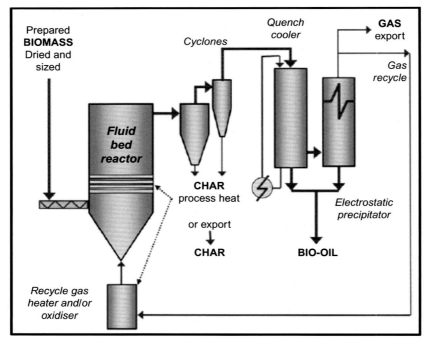

Fig. 2.4 Bubbling fluidized bed equipped with electrostatic precipitator. (Reprinted from A.V. Bridgwater, Review of fast pyrolysis of biomass and product upgrading, Biomass Bioenergy 38 (2012) 68–94, doi:10.1016/j.biombioe.2011.01.048, with permission from Elsevier.)

(70%–75%) is observed when this type of reactor is used [15]. The heat generated from the combustion of pyrolysis products, such as gas and char, is generally supplied to heat the fluid bed. Though the heat transfer from the inert bed material (sand) to the biomass is excellent, the resistance to the transfer of heat inside the coils limits the heat transfer from the coils to the fluid bed. This limitation should be overcome to obtain better heat transfer in the bubbling fluid bed reactors [13]. It is easy to control the residence time of both vapor as well as char. The char can be separated using the cyclones [6]. Fig. 2.4 displays the bubbling fluidized bed reactor equipped with electrostatic precipitator.

iv. Circulating fluid bed

The circulating fluidized bed follows the same principle as the bubbling fluidized bed. The residence time for char and sand is shorter in circulating fluid bed (0.5–1 s) than in bubbling fluid bed (2–3 s) [15]. Some of the many advantages include efficient heat and mass transfer, better production capacities, and the ease of operation. The velocity of the gas is comparatively high in the circulating fluidized bed reactor. The high gas velocities (5–15 m/s) lead to the attrition of biomass particles, which eventually make it difficult to separate the char particles from the liquid product. High throughput can be achieved using the circulating fluidized bed reactor. Hence, this reactor is used in

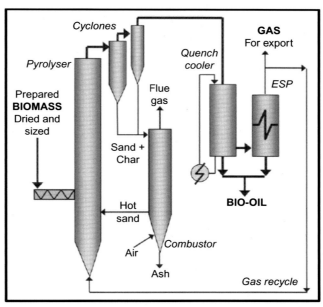

Fig. 2.5 Circulating fluidized bed reactor. (Reprinted from A.V. Bridgwater, Review of fast pyrolysis of biomass and product upgrading, Biomass Bioenergy 38 (2012) 68–94, doi:10.1016/j.biombioe.2011.01.048, with permission from Elsevier.)

large-scale requirements. A commercial-scale plant has been operated using circulating fluidized bed technology [6]. Fig. 2.5 shows a conventional design of the circulating fluidized bed reactor.

v. Rotating cone reactors

This technology was invented by researchers of the University of Twente and developed by Biomass Technology Group. It is a type of transport bed in which the transport of material takes place due to the centrifugal force (~10 Hz) rather than a gas. The char and sand are separated in a fluid bed reactor. The char is then combusted to reheat the sand, which is recycled back to the reactor. The integration of the three subsystems (rotating cone pyrolyzer, riser for sand recycling, and bubbling bed char combustor) is required [7]. Fig. 2.6 shows an initial model on the left and its role in an integrated fast pyrolysis process on the right.

It is known that pyrolysis always gives only three products (char, gas, and liquid). A better understanding of the process is helpful to enhance the yield of as per the required product.

2.2.2 Catalytic fast pyrolysis

Many chemical routes, including aldol condensation and ketonization, are employed to upgrade the bio-oil. Catalytic fast pyrolysis is a combination of the fast pyrolysis and chemical upgrading of the vapor [16]. This is a one-pot process that can be used to

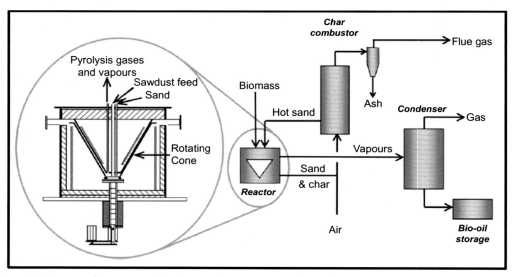

Fig. 2.6 Rotating cone reactor for fast pyrolysis. *(Reprinted from A.V. Bridgwater, Review of fast pyrolysis of biomass and product upgrading, Biomass Bioenergy 38 (2012) 68–94, doi:10.1016/j.biombioe.2011.01.048, with permission from Elsevier.)*

enhance the properties of the vapor as well as cut down the cost of the vapor upgradation step. Removal of oxygenates as carbon dioxide is advantageous as it ensures the minimal requirement of hydrogen as an external supply. It also enhances the hydrogen to carbon ratio in the end products. The different catalyst materials that are used in catalytic fast pyrolysis are soluble inorganics, metal oxides with and without support material, and microporous/mesoporous materials [16].

a. Mineral catalyst

The biomass contains materials that are effective in producing ash. These are generally formed as cations. These materials are catalytically active and present within the biomass, thus making it a self-driven catalyst. They are mainly made up of alkali and alkaline materials (K, Ca, Na, and Mg). These materials increase the catalytic cracking of the primary vapor and produce undesirable products [17]. To avoid the varying effect on the product distribution, the biomass is washed with water. This step essentially removes such elements to a greater extent. Washing the biomass with acid has a superior effect compared to water. Even though they are present in an infinitesimal amount (<0.5 wt%), they have a more significant impact on the chemical composition of the pyrolytic liquid. For example, the yield of levoglucosan can vary remarkably if the elements are present in the order of $K^+ > Na^+ > Ca^{+2} > Mg^{+2}$. The attrition of the catalyst can be one of the reasons for the decrease in catalytic activity. The effect of ash on the vapor phase is assumed to proceed in four different ways [18]:

  i. The effect of a catalyst on the primary pyrolytic vapor decreases the yield of condensable gases and increases char production.

Fast pyrolysis of biomass and hydrodeoxygenation of bio-oil   **55**

Table 2.1 Effect of inorganic material on the microcrystalline material.

Inorganic material	Effect on microcrystalline material	References
Magnesium chloride	No effect on weight loss or the temperature at which cellulose decomposes. Suppression of many organic functional group compounds, such as ketones, aldehydes, 2-furfuraldehyde, and furans.	[19]
Sodium chloride	Onset temperature tends to decrease. The total number of lower molecular weight compounds increases by three folds. The formation of levoglucosan significantly decreases (0.4 wt%) due to the suppression of the transglycosylation reaction. The presence of the sodium atoms inhibits this reaction.	[19]
Ferrous sulfate	Reduction in the decomposition temperature of cellulose by 50°C. Favorable for the formation of levoglucosan and levoglucosenone from wood.	[19]
Zinc chloride	There were two prominent peaks of which one represented a dehydration peak, and the other was associated with an untreated specimen.	[19]

    ii. The ash helps in cracking larger molecules into smaller molecules. This facilitates the adsorption of the smaller molecules in the pores of the catalyst.

   iii. Poisoning of the catalyst by the ash.

   iv. By the process of reforming of the cracked vapors.

The effect of soluble inorganics on the yield, distribution of the product, and operating conditions are reported in systemized studies. For example, the microcrystalline materials are impregnated with 1 mol% of each of magnesium chloride, sodium chloride, ferrous sulfide, and zinc chloride. This was studied using a thermogravimetric analyzer with mass spectroscopy. Table 2.1 shows the effect of these inorganic materials on the mass loss, distribution of the product, and characteristic decomposition temperature.

b. Metal oxides

Among various metal oxide catalysts, the transition metal oxides have been extensively applied in the catalytic fast pyrolysis process. The metal oxides, such as MgO, NiO, CeO_2, Al_2O_3, TiO_2, ZrO_2, etc., are examined as a catalyst. The metal oxides are mainly categorized as acid, base, and transition metal oxides. The acidic metal oxides [alumina (Al_2O_3), silica (SiO_2), sulfated titanium oxide (SO_4^{-2}/TiO_2), and sulfated–zirconia (SO_4^{-2}/ZrO_2)] and their composites (Al_2O_3–SiO_2) are widely explored in catalytic fast pyrolysis. Acidic metal oxides show a negative impact on the yield of the bio-oil, and they tend to increase the noncondensable gases and char. The catalytic fast pyrolysis of commercial wood biomass (Lignocel HBS 150–500) derived from beechwood is an exemplar of this process. The process was carried out in a fixed-bed reactor. In the case of a noncatalytic system, the highest yield of liquid and organics obtained are 58.76 and

37.37 wt%, respectively. The application of catalysts in the process has shown a negative effect on the yield of liquid, i.e., the yield of the liquid decreased while leading to an increase in the yield of organics [20]. Though alumina-based catalyst has a characteristic feature of high surface area, still the yield of the liquid has decreased (from 58.6 to 40 wt %), and that of organics have increased (33.85 wt%) when compared to the noncatalytic system (22.89 wt%). The increase in the solid product is attributed to the accumulation of the coke on the catalyst. The oxygen in the pyrolysis products is converted to carbon dioxide, carbon monoxide, and water. While the preferable form is CO_2, the less preferable form is H_2O as it removes hydrogen and reduces the hydrocarbon formation reaction. Acidic materials generally favor the decarbonylation reaction (conversion into CO) over ketonization reactions (conversion into CO_2). For instance, the basic materials attributed to the conversion of acidic compounds in the pyrolysis vapor to ketones. The nickel oxide (NiO) has high selectivity toward CO_2 [20]. The alkaline earth materials, such as MgO and CaO, are considered excellent catalysts for catalytic fast pyrolysis. The oxide ions act as a base, and the metal ions behave as a Lewis acid site. Using MgO catalyst, the bio-oil quality was significantly enhanced, but there was a decrease in the quantity of the bio-oil. The deoxygenation extent of the bio-oil was drastically reduced (from 9.56 to 4.9 wt%). The fragmentation of higher molecular weight compounds into a lower molecular weight compound was also achieved efficiently. The transition metal oxides, such as ZrO_2, TiO_2, composite materials (ZrO_2–TiO_2), Fe_2O_3, etc., have been looked over as a potential catalyst for catalytic fast pyrolysis. The ketonization reaction was favored in the presence of NiO and ZrO_2/TiO_2 catalyst. It is probably due to the steam reforming phenomenon over the catalyst. For instance, the catalytic fast pyrolysis of poplar wood biomass in a Py-GC/MS reactor with different catalysts was performed. The ZnO catalyst had a mild effect on the poplar wood biomass. Most of the pyrolytic vapor remained unchanged. However, there was a notable change in the pyrolysis product vapor when calcium oxide (CaO) was used as a catalyst. The primary product vapor was removed entirely, and the reaction favored the formation of linear ketones, cyclopentane, and hydrocarbons [21].

2.3 Effect of operation conditions on fast pyrolysis products

Even though the fast pyrolysis process is operated at a pilot plant scale, there are many aspects that are yet to be studied for the betterment of the process output. The quality (properties and composition) and quantity (yield) of the bio-oil obtained from the fast pyrolysis process is dependent on various process parameters, such as the composition of biomass, the residence time in the reactor, temperature, and insoluble inorganics [6]. Adding to the above factors, the ash is considered as one of the significant parameters, which affect the yield and composition of the bio-oil [22]. To fulfill the requirement of heat transfer rate, there is an imposition of particle size of biomass for some type of reactor configurations. The performance of the reactor should not be the only criteria to obtain an excellent liquid yield. The feed is typically dried to less than 10 wt% water. The bio-oil consistently consists of 15 wt% water in a typical yield of 60 wt% organics and 11 wt% reaction water. Separation of the organic and aqueous phases is economically impractical. A selective condensation process can be employed to do the same, but it can also be performed only up to a certain extent [8].

In the 1980s, it was proven that a maximum liquid (bio-oil) yield could be obtained by maintaining low pressure and pyrolysis temperature in the range of 400°C–550°C. Many other factors, such as heating rates and particle size of biomass, play a crucial role in the fast pyrolysis process. The cracking of the primary volatile product to form undesirable secondary products are also responsible for the declining quality of the bio-oil. While it was proposed that to avoid the secondary reactions, the residence time of vapor in the reactor should be less than 5 s. Nevertheless, studies have proven that the residence time can typically extend up to 10 s only if the reactor temperature is carefully controlled at a temperature of less than 500°C. The bio-oil is majorly composed of anhydrous-oligosaccharides and lignin-derived oligomers. These molecules are dominantly observed at a temperature of 400°C–500°C and at a residence time of 75 ms. Garcia-Perez et al. performed fast pyrolysis of mallee biomass and pine pellet. It was observed that as the temperature was increased, the yield of bio-oil gradually increased till a specific temperature (500°C) and decreased beyond that temperature. As the temperature was increased, the yield of char and gases decreased and increased, respectively. It was also observed that there is a strong correlation of water content in the bio-oil with the particle size of biomass for different types of reactors at the same temperature (500°C). As the particle size of biomass (microns) increases, the water content in the bio-oil increases. This indicates the effect of intra-particle dehydration secondary reactions. Earlier it has been shown that the secondary reaction catalyzed by fresh char is accountable for at least a 10 wt% increase in the char yield [23].

The effect of temperature was studied on pinewood fast pyrolysis in a fluidized bed [24]. As the temperature was increased, it was observed that the yield of gases increased while there was an acute decrease in the char yield and was observed to be constant beyond 500°C. The probable reason for the decrease in char yield can be attributed to the conversion of the lignin present in the biomass. Other parameters, such as viscosity, average molecular size, and heavy water-insoluble compounds, also increased notably as the temperature was increased. Although the amount of water-insoluble compounds in the bio-oil remain constant in the range of 500°C–530°C, viscosity and average molecular size continue to rise due to the polymerization reactions [24].

Particle size is an essential parameter in biomass fast pyrolysis performed in an entrained reactor. Partial conversion of the larger particles of biomass arises due to heat transfer limitations. Hence, the efficiency of the process is significantly decreased [25]. The particle size of less than 1 mm provides greater efficiency in an entrained flow reactor

[26]. High tar yield is obtained at higher heating rates in an entrained reactor. Fast pyrolysis of cypress sawdust with a particle size of 0.5 mm was performed at a temperature of 600°C–1400°C. It was observed that the soot generation was significantly high at 1000°C. At the same time, the tar yield and hydrocarbons were found to be low at the same temperature. A complete conversion of tar takes place at a temperature of 1400°C, leaving behind a meagre amount of gaseous hydrocarbons. The gaseous hydrocarbons majorly consisted of methane [27]. Instead, there was a significant increase in the yield of CO and H_2 [28].

One of the other crucial factors in the pyrolysis process is the vapor residence time. Higher residence time enhances the secondary vapor cracking, which eventually results in a decrease of liquid yield [27]. Since the vapor consists of many reactive species, an increase in vapor residence time will increase the size of molecules in due course of the polymerization reaction [29]. Table 2.2 represents a brief literature survey on the processes which invovled fast pyrolysis on various feedstock material.

2.4 Bio-oil upgrading

The properties of the crude bio-oil obtained from pyrolysis have many undesirable properties such as a high TAN number, high oxygen content, the potential for aging, etc. Hence, there is a need to upgrade the bio-oil for further use as a transportation fuel. The upgrading techniques include physical, chemical, and catalytic ways [15]. Fig. 2.7 demonstrates various bio-oil upgrading processes.

2.4.1 Physical upgradation technique of bio-oil

The bio-oil obtained from pyrolysis cannot be directly blended with fossil fuel because of its undesirable properties. Some of them are given below [7]:

Filtration: A hot vapor is essential to reduce the ash content in the bio-oil to less than 0.01%. It also helps in reducing the alkali content to less than 10 ppm. The char is catalytically active to promote the secondary reactions, which reduce the yield of oil up to 20%, and the average molecular weight also reduces. The feasibility of liquid filtration to a meagre particle size is difficult due to the physicochemical characteristics of the liquid [7].

Addition of solvent: Polar solvents are used to homogenize and decrease the viscosity of the bio-oil. Methanol as a solvent showed a remarkable effect on the stability of the bio-oil. Aging of bio-oil with a 10 wt% methanol was found to be 20 times less than the bio-oil without additives [7].

2.4.2 Chemical upgradation technique of bio-oil

The chemical techniques for upgrading the bio-oil include gasification, hydrothermal treatment, hydrodeoxygenation, and steam reforming [15]. Gasification of the produced

Table 2.2 Fast pyrolysis process performed on various feedstock materials.

Feedstock	Reactor	Catalyst	Operating Conditions		Catalyst: biomass ratio	Yield of products	Major findings	References
			Temperature	Residence time				
Forest pine wood chips	Auger reactor	Sand–CaO and Sand–CaO–MgO	500°C	7 min	3:1, 1:1, 1:2, 1:3	Oil yield: 45–50 wt%	Only at the ratio of 1:3, good oil yield was obtained. Addition of catalyst enhanced the oil quality.	[30]
Pine saw dust	Bubbling fluidized bed reactor	Spray dried ZSM-5	500°C–650°C	–	1:3	Olefins yield at 650°C is around 8.5%	Continuous production of olefins and aromatic. The catalyst was stable during the reaction-regeneration models. Catalyst coking occurred.	[31]
Corn cob	Fluidized bed reactor	HZSM-5 (bed height: 10 cm)	550°C	10 min	–	Liquid yield: 56.8% H/C: 1.511 HHV: 34.6 MJ/kg	Placing a second condenser increased the separation of oil fraction in the liquid yield.	[32]

Continued

Table 2.2 Fast pyrolysis process performed on various feedstock materials—cont'd

Feedstock	Reactor	Catalyst	Operating Conditions		Catalyst: biomass ratio	Yield of products	Major findings	References
			Temperature	Residence time				
Sewage sludge	Fluidized bed reactor	(CaO and La_2O_3)	450°C	–	–	Char yield: 37.1 wt% at 550°C. Oil yield: 57.5 wt% on a dry-ash free basis	Feed particle size <0.3 or >1.0 mm has negatively influenced the production of liquid yield. Chlorine content of the liquid is significantly removed in the presence of catalyst.	[33]
Acid leached bagasse and pine wood	Fluidized bed reactor	–	485°C	–	–	Wet oil yield: 0.62 kg/kg Char yield: 0.12 kg/kg	A longer hot vapor residence time decomposes the volatiles into gas and decreases the oil yield.	[34]

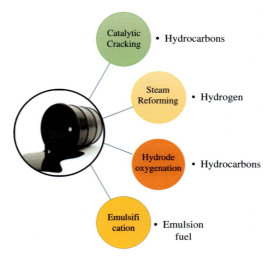

Fig. 2.7 Bio-oil upgradation techniques.

bio-oil to convert it into syngas has gained much attention these days. A small energy penalty is applied as compensation in plants operated commercially. This is due to the lesser energy-efficient pyrolysis process, transportation of the bio-oil to the centralized plants, and the gasification step [35]. This chapter emphasizes on hydrodeoxygenation technique for the upgradation of bio-oil properties.

2.4.3 Hydrodeoxygenation of bio-oil

It is a process in which oxygen is removed from the bio-oil in the presence of hydrogen. When a catalyst is used, the process is known as the catalytic hydrodeoxygenation process [36]. The catalyst for the hydrodeoxygenation process varies from that used in petroleum refineries. Several studies on catalytic hydrodeoxygenation were involved in the conversion of model compounds. Though the results provided the fundamental aid in understanding the mechanism/pathway of the reaction, these results are not sufficient enough to draw general conclusions for the hydrodeoxygenation of bio-oil [6]. The different types of reaction mechanisms include hydrogenation, hydrodeoxygenation, decarboxylation, hydrogenolysis, and dehydration [15]. The catalytic activity is different at each site on the catalyst as the reactions occurring are different. It increases the complexity of understanding the process. A wide variety of heterogeneous catalysts are available, and it is challenging to select an active site type [37]. The reaction chemistry is as follows:

$$\text{Oxygenated biomass} + \text{Hydrogen} \rightarrow \text{Deoxygenated compound} + \text{water}$$

Example: $CH_{1.33}O_{0.43} + 0.77\ H_2 \rightarrow CH_2 + 0.43 H_2O$.

Several undesired reactions occur at different sites on the catalyst that makes the selection of an efficient catalyst always challenging. The products in the above reaction generally tend to sinter and leach the metals. This often results in the requirement of precious metal catalysts that have better stability compared to other base metals. Sometimes this becomes economically unviable. Hydrodeoxygenation is classified into two categories: partial or mild hydrodeoxygenation (occurs at low temperature) and complete hydrodeoxygenation (occurs at high temperature). The mild hydrodeoxygenation process is used to stabilize the acid oxygen compounds. The operating conditions are typically maintained low (T: 150°C–250°C and P > 100 bar of hydrogen in the presence of a catalyst). In contrast, the complete hydrodeoxygenation process is used to remove oxygen in both high molecular weight alkanes and aromatic hydrocarbons. The operating temperature range is around 350°C–400°C, and typical pressures are around 200 bar of hydrogen in the presence of a catalyst. The common oxygenated compounds are aromatic compounds (phenols), carboxylic acids, aldehyde, and carbohydrates [11].

2.4.4 Reactor configuration of hydrodeoxygenation

Elliot et al. studied the hydrodeoxygenation of poplar wood bio-oil in the presence of a sulfided CoMo catalyst. They observed total deoxygenation of 23 wt% and also encountered an issue of catalyst deactivation due to coking and bed plugging. The issue of coke formation can be controlled by using a dual bed reactor [38]. The two-stage hydrotreatment process was initially conducted in the Pacific Northwest Laboratory (PNNL) in 1989 [39]. Of the two stages, the first stage is operated at a low temperature (553 K) in the presence of a Ni or sulfided CoMo catalyst to produce a more stable reactive species, such as ketone and furans aldehyde. In the second stage, the bio-oil from the first stage is subjected to extreme conditions in the presence of a hydrotreating catalyst used in petroleum refining. A significant increase in the yield of bio-oil up to 30%–55% could be possible using a two-stage reactor. The degree of deoxygenation can be up to 99% [40].

Noble metals, such as palladium (Pd), are employed in the two-stage hydrotreatment of bio-oil. The Pd/C catalyst was used to treat bio-oil from a mixed corn stover feedstock. The experiments were carried out in a bench and fixed-bed reactor. The first step is hydrotreating, while the second one is the hydrocracking step. The hydrocracking process produced a partially upgraded bio-oil. The temperature and pressure maintained in the reactor were around 400°C and 2000 psig, respectively. The process was run for 100 h. In the hydrotreating step, the bio-oil is refined in a typical range of operating conditions. The typical operating conditions are the temperature of 310°C–375°C and liquid hourly space velocity (LHSV) of 0.18–1.12 h^{-1}. In the first

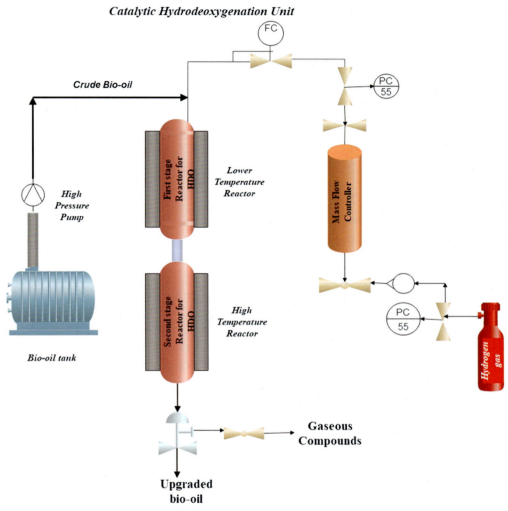

Fig. 2.8 Dual-stage reactor for hydrodeoxygenation process.

step, though the yield of gas increased, the yield of bio-oil significantly reduced at the higher temperature. The deoxygenation also improved. In the second step, usage of a conventional sulfide catalyst ensured the drop in oxygen content to below 1%. Integration of these process steps into a nonisothermal reactor enhanced the conversion to hydrocarbons. This also ensured the reduction of carbon loss in the water stream obtained as a by-product [38]. Fig. 2.8 shows the dual-stage reactor for the hydrodeoxygenation of bio-oil.

2.4.5 In situ hydrogenation

It is an alternate technique in which liquid hydrogen donors are used as a replacement of hydrogen gas. Some of the liquid hydrogen donors include acids and alcohols. The process of hydrogen donation is the reforming of the aqueous phase. Xiong et al. performed in situ hydrogenation of bio-oil provided by Yineng Bio-Energy Co. Ltd. Raney Ni catalyst, SBA-15, and Y zeolite was employed as the hydrodeoxygenation catalyst. The liquid hydrogen donors, such as methanol and formic acid, were used as a source for hydrogen. During the process, the liquid hydrogen donor (formic acid) decomposed to form carbon dioxide as a by-product. The CO_2 dissolved in the liquid medium, which increased the hydrogenation. They also observed high liquid yield with evident lesser coke formation [41].

2.4.6 Catalyst for hydrodeoxygenation process

The catalytic hydrodeoxygenation involves the use of various catalysts, which promote the reaction network. The performance of a different type of catalysts is given below [37].

a. Metal catalyst

Hydrodeoxygenation reactions occur at metal sites that involve the cleavage of C—C bonds. The selection of metallic catalysts often depends upon the target product. The main reactions involved are decarboxylation and decarbonylation. The functionalities that are essentially removed are —COOH and —CHO groups. Some of the metal-based catalysts are metal sulfided, Pd—Fe bimetallic catalyst, Nb-phosphate-supported Pd catalyst. The deoxygenation reactions are mostly favored by ruthenium (Ru)-based catalyst, while the hydrogenation reactions are mostly favored by the palladium (Pd)-based catalyst for the hydrodeoxygenation of eugenol [37]. Table 2.3 shows the effect of metal catalyst on the functional group removal and the reaction pathway followed.

Table 2.3 Effect of metal catalyst on the functional group removal and the reaction pathway followed.

Metal catalyst	Functional groups removed	Reaction mechanism
Pd—Fe bimetal	C=O of furans and nonfurans	Hydrogenation
Co	Carboxylic acids	
Nb-phosphate-supported Pd	CO and CO_2	Hydrogenolysis of esters
Ru	C—C	Hydrotreating chemical mixtures
Sulfided Ru		Hydrogenation of polyols and sugars
Ni—Mo sulfided supported on Al_2O_3		Hydrodeoxygenation of untreated microalgae
γ-Al_2O_3-supported Ni—Mo		Hydrodeoxygenation of aliphatic esters

b. Solid acid support

Among the various reactions, dehydration, hydration, and hydrolysis are some of the significant reactions that occur on the sites. The selectivity and activity of a catalyst for the C—O bond type via removal of the water molecule on a solid acid catalyst are dependent on the Bronsted and Lewis acid sites. They play an important role in the isomerization reactions also. The fructose produced from the isomerization of glucose predominantly occurs at a lower fraction of Bronsted sites. In comparison, the Lewis acid sites have a positive effect on the formation of fructose from glucose that is essentially carried out via isomerization reactions. [37].

c. Bifunctional metal–acid sites

Bifunctional metal–acid catalysts are often used in the hydrodeoxygenation process. These are produced by the dispersion of the metal particles over solid acid support. Among the various types of catalysts, the bimetallic catalysts containing a combination of reducing metal and an oxophilic metal have been considered as a catalyst for hydrodeoxygenation reactions. The hydrogenolysis of C—O—C bonds takes place via the opening of the ring structure through the removal of water molecules. Metal sites play a significant role in catalyzing these reactions. To attain a higher activity of the catalyst, both the reducing metal as well as the oxophilic metal should be present closer to each other on the support material. According to a recent report, they claimed that the proximity or closeness of the metals is optimum. However, the fundamental understanding of the metal–acid interactions is termed to be strenuous compared to that of a monofunctional catalyst for hydrodeoxygenation reaction [37]. There are different types of catalysts used for hydrodeoxygenation reactions. Some of them are sulfides, oxides, and noble metal catalysts. The essential features of the catalyst based on these materials are given below.

d. Sulfide-based catalyst

The biomass and the product bio-oil contain trace amounts of sulfur. It can poison the catalyst very easily. Hence, it is always advised to choose a catalyst that is resistant to sulfur. Generally, these catalysts are subjected to sulfidation in the presence of sulfiding agents [42]. The sulfided Co/Ni-Mo, such as $CoMoS_2$ and $NiMoS_2$ catalyst, has gained much attention for the hydrodeoxygenation process. The metals Ni and Co increase the catalytic activity and stability of the catalyst. Hence, they are used as a promoter [6]. There have been many models proposed for the promoted sulfided catalyst. Some of them are the monolayer model, intercalation model, catalytic cosite model, and the Co–(Ni)–MoS model, which is now a preferred model. The coordinatively unsaturated sites or the ions of exposed molybdenum with vacancies in sulfur at the edges and corners of the molybdenum sulfide crystallite are found to be active in typical reactions, such as hydrogenation and hydrogenolysis reactions. It was observed when a sulfided catalyst without any promoter is used. The Co-M-S structure is in stacked in such a way that the sulfur atoms are present at the edges in an alternating pattern [42]. Chemisorption of the target oxygen atom occurs at the sulfur vacancy. This vacancy is produced by the reduction reaction in the presence of hydrogen at the MoS_2 slab edge [6].

The MoS_2 catalyst was used in the hydrodesulfurization process. The metal sulfide catalyst significantly removes the sulfur present in the crude oil. The metal sulfide acts as an active phase that contains catalytically active sites, and numerous reactions are carried on them. This catalyst is mainly used in the hydrodeoxygenation of fatty acids and triglycerides. There are many disadvantages, such as deactivation of the catalyst and leaching of metal by water in the fluid phase. MoS_2 is prepared from the molybdenum trioxide (MoO_3), which acts as an active phase. The oxidation states of Mo play a crucial role in oxygen removal from the biomass products. Different types of support materials, such as alumina or zeolites, are used in the hydrodeoxygenation catalyst. The MoS_2 supported on $Al_2O_3-TiO_2$ has a better catalytic activity in the hydrodesulfurization process compared to that of the hydrodeoxygenation process [43]. The selection of support material for the catalyst is considered as one of the challenges in this process. The alumina is converted to boehmite or trioxide phase in the presence of water. This proves the instability of alumina as a support in the presence of any trace of water. In the case of the Ni–Mo–S catalyst, the nickel sulfide is converted to an oxide form. There is also blockage of the active sites of Ni and Mo occurs. The other disadvantage is the strong affinity of carbon precursors to the support material. This is attributed to the high acidity of the alumina support. The alternate materials considered for hydrodeoxygenation as support materials are carbon and silica (SiO_2). Both the support materials are neither acidic nor basic; they are neutral. Lower affinity of the support material toward carbon makes it a promising support material. The mesoporous silica can also be used as support owing to its many advantages, such as a large specific area (provides better dispersion of the active phase) and wider pore, diameter. The catalyst supported on the mesoporous silica shows a better performance compared to the catalyst supported on hexagonal mesoporous silica. Romero et al. performed hydrodeoxygenation of 2-ethyl phenol in the presence of MoS_2 catalyst and suggested a mechanism, as shown in Fig. 2.9. They demonstrated that the proton donation from the S molecule, which is chemisorbed with the target oxygen atom, provides the formation of a carbocation. The deoxygenated compound is obtained by the cleavage of $C=O$ in the carbocation [44].

e. Oxides-based catalyst

The metal oxides, such as MoO_3, have been used as a catalyst in the conversion of olefins to ketones and aldehydes by oxidation of olefins. These catalysts work on the principle of the Mars-van Krevelen mechanism. Recent studies show that these catalysts are used in the hydrodeoxygenation process in a reverse Mars-van Krevelen mechanism [6]. Fig. 2.10 shows the reverse Mars-van-Krevelen mechanism for the hydrodeoxygenation of alkyl-substituted phenol over the MoO_3 catalyst [45].

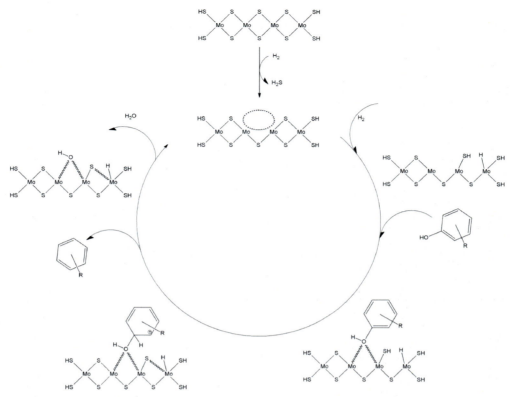

Fig. 2.9 Proposed hydrodeoxygenation mechanism of alkyl-substituted phenol over MoS$_2$ with sulfur vacancy as active site (R denotes alkyl group). *(Reprinted from T.M.H. Dabros, M.Z. Stummann, M. Høj, P.A. Jensen, J.D. Grunwaldt, J. Gabrielsen, P.M. Mortensen, A.D. Jensen, Transportation fuels from biomass fast pyrolysis, catalytic hydrodeoxygenation, and catalytic fast hydropyrolysis, Prog. Energy Combust. Sci. 68 (2018) 268–309, doi:10.1016/j.pecs.2018.05.002, with permission from Elsevier.)*

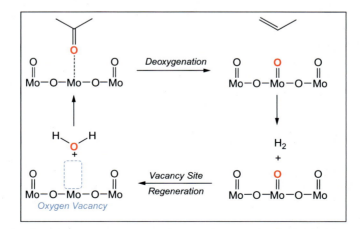

Fig. 2.10 Proposed reverse Mars-van Krevelen mechanism for hydrodeoxygenation of alkyl-substituted phenol over MoO$_3$ (R denotes alkyl group). *(Reprinted from T.M.H. Dabros, M.Z. Stummann, M. Høj, P.A. Jensen, J.D. Grunwaldt, J. Gabrielsen, P.M. Mortensen, A.D. Jensen, Transportation fuels from biomass fast pyrolysis, catalytic hydrodeoxygenation, and catalytic fast hydropyrolysis, Prog. Energy Combust. Sci. 68 (2018) 268–309, doi:10.1016/j.pecs.2018.05.002, with permission from Elsevier.)*

Reverse Mars–van Krevelen Mechanism

$$oxy - compound + vacant\ site\ [Adsorption]$$
$$\downarrow$$
$$Deoxygenated\ compound + oxygen\ adsorbed\ on\ catalyst\ [Desorption]$$
$$\downarrow$$
$$H_2 + oxygen\ absorbed\ on\ catalyst\ [Regeneration\ step]$$
$$\downarrow$$
$$oxygen\ vacancy + H_2O$$

Several metal oxides, including vanadium pentoxide (V_2O_5), ferric oxide (Fe_2O_3), cupric oxide (CuO), and molybdenum trioxide (MoO_3), are used in the hydrodeoxygenation of biomass. These catalysts are active in hydrodeoxygenation of acetone with a specific consumption rate of acetone in the order of $MoO_3 > V_2O_5 > Fe_2O_3 > CuO$. A vital characteristic of the MoO_3 catalyst is that the number of carbon atoms between the product and the parent atom is indistinguishable. It shows the selective cleavage of the C—O bond over C—C. From the DFT calculations, the heat of adsorption for cyclic ketones is found to be higher than linear ketones [45]. Although a high hydrogen pressure alleviates the coking phenomenon, the low hydrogen pressure is required to circumvent the reduction of the active phase. The catalytic activity of the metal oxides depends on the availability and strength of the acid sites. The Lewis acid sites are crucial for the chemisorption of oxygen lone pair of the oxygenated compound at the vacancy. Whereas, the Bronsted acid site affects the availability of hydrogen at the surface of the catalyst pertaining to the presence of hydroxyl groups [6]. A bifunctional catalyst, such as a metal oxide catalyst supported on metal, is considered as an effective catalyst for the hydrodeoxygenation process. The interaction of the metal oxide in its highest oxidation state is higher with the oxide surfaces of the support rather than the surface of the metal particles. Hence, selective scission of C—O is a major challenge [46].

f. Reduced transition metals

The reduced transition metals, such as nickel, platinum, palladium, ruthenium, and rhodium, are catalytically active in the hydrogenation and hydrodeoxygenation reactions. The supply of hydrogen sulfide (H_2S) is not required to maintain the activity of the catalyst [6]. The hydrodeoxygenation of phenol and 1-octanol in the presence of Ni-MoS_2/ZrO_2 indicated that a continuous addition of sulfur to the feed is required to preserve the catalyst stability for a prolonged period. This is also associated with a disadvantage of increased content of sulfur in the product. The unsaturated compounds formed from the dehydration reactions are inclined to react with H_2S to form thiols [47]. It is advisable to remove the sulfur in the product before upgrading it using the hydrodeoxygenation technique [6]. The advantages of using noble metal catalysts are that they can be used at lower temperatures. The lower temperature reactions inhibit the cracking of reactions to a great extent. They also reduce the coking and subsequent deactivation of the catalyst. However, these catalysts are exorbitant. They require a higher amount of hydrogen to get a better selectivity for the hydrogenation reactions [48]. These catalysts have better performance if they exist as a bifunctional catalyst. The bifunctional catalyst ensures the activation of oxygen-containing compounds and proton (hydrogen) donation to the oxygen-containing compound. In general, it is admitted that the hydrogen donation sites are composed of metal sites. However, the activation of the oxygen-containing compound is available either at the metal site or the metal–support interface. The above-described mechanism indicates the potential of such a catalytic system to have more pathways for the reaction to proceed [49].

g. Nonnoble metal

The nonnoble metals are considered as an alternative to the noble metals. The best example of a nonnoble metal is Ni. The bimetallic catalysts show a higher catalytic activity compared to a monometallic catalyst in C—O bonds scission at 100°C and 100 bar, respectively. The carbon-supported Ni-based catalyst exhibits superior performance. The advantages of using such support materials are low coke formation and higher stability. The mesoporous carbon and the activated carbon as a support material has gained a lot of interest for the hydrodeoxygenation reactions [50]. The hydrodeoxygenation of phenolic species in the presence of three different catalysts, namely Ni/HBeta, Fe/HBeta, and NiFe/HBeta was investigated by Shafaghat et al. [51]. A different catalyst provides different reaction pathways. For example, the Ni/HBeta facilitates the conversion of phenolics to cycloalkanes. While Fe/HBeta facilitated the production of aromatic hydrocarbons and the NiFe/HBeta produced both cycloalkanes and aromatic hydrocarbons probably due to the synergistic effect of Ni and Fe [51].

2.4.7 Role of support

Probe into the role of support for catalyst materials in the hydrodeoxygenation process provided scattered results for numerous catalytic systems as well as model compounds. The role of support is dependent on two factors mainly, the catalyst system and the type of oxygenates subjected to hydrodeoxygenation. The γ-Al_2O_3 support has been widely used in conventional hydrotreating. Though it has wide use in conventional processes, its use is limited in the hydrodeoxygenation process owing to its disadvantages. It reacts with water present in the bio-oil to form boehmite. The activity of the catalyst reduces due to the conversion into the boehmite. The probable reason for the decrease in the catalyst activity is that the crystals in the catalytic region might get trapped in the support lattice. Moreover, the γ-Al_2O_3 tends to higher coke formation due to its higher acid content [6]. Carbon-supported catalysts were also used in hydrodeoxygenation processes. These were found to be a good support for the catalyst as the acidity is relatively low, which therefore decreases the coke formation on the catalyst [52]. Silica is also considered as a promising support material for the catalyst. The order of the acidity of the support materials is given as follows:

$$Ni/_{Al_2O_3} > Ni/_{ZrO_2} > Ni/_{CeO_2} > Ni - V_2O_5/_{ZrO_2} > Ni/_{MgAl_2O_4}$$

$$> Ni/_{CeO_2 - ZrO_2} > Ni - V_2O_5/_{SiO_2} > Ni/_{SiO_2} > Ni/_{C}$$

A study concluded that both Ni supported on Al_2O_3 and activated carbon had a lower degree of deoxygenation than Ni supported on SiO_2 in the hydrodeoxygenation of anisole. This process was performed at a temperature of 180°C–220°C and a pressure of 5–30 bar H_2. Such a result can mainly be attributed to the interaction between the support and the model compound [53]. It is essential to address that the support material influences the selectivity and activity of the catalyst. Hence, the contribution of support should be included in the design of the catalyst. The three critical aspects in which the support material plays a major role are as follows [6]:

a. Tendency coke formation. This is dependent on the acidity of the catalyst.
b. The interaction between the support and the oxygenated compounds.
c. Resistance to water and the probability of regeneration.

2.4.8 Effect of parameters on hydrodeoxygenation of bio-oil

The operating conditions of hydrodeoxygenation reaction should be chosen in such a way that they minimize deactivation of the catalyst and increase the catalytic activity for a long operation time. Usually, the hydrodeoxygenation process is operated in the temperature range of 250°C–400°C. Hydrogenation of bio-oil is not favorable at a higher temperature. The exothermicity of the process could restrict the choice of temperature in different types of reactors (mostly in fixed-bed and batch reactors) due to the generation of hot spots. Also, the increase in temperatures (>450°C) leads to an increase in coke formation [6].

Elliott et al. carried out several experiments on the hydrodeoxygenation of bio-oil derived from wood biomass. Pd/C was used as a catalyst in a fixed-bed reactor at a temperature of 310°C–360°C and pressure of 140 bar. Their observations included a positive influence of temperature on the degree of deoxygenation in the temperature range of 310°C–340°C, and a negative impact on the yield of bio-oil and the degree of deoxygenation in the temperature range of 310°C–360°C. Simultaneously, they observed a threefold increase in the gas yield [6]. The dual-stage reactors have an advantage in this respect. The temperature of both reactors can be different since different oxygenated compounds require different temperatures to be stabilized. It facilitates an increase in the degree of deoxygenation. Also, the deactivation of the catalyst is addressed effectively in a dual-stage reactor [54]. Though the dual-stage reactor provides many benefits,

Routray et al. reported that severe reactor plugging was noticed at 55–71 h of the operation. It is due to the operational temperature being at 300°C–400°C. The main reason for the reactor plugging is the thermally induced repolymerization reaction at a temperature of <200°C. Hence, it is better to operate the first reactor at a lower temperature to decrease the effect of coke formation [55]. The liquid hourly space velocity should be maintained relatively lower to obtain better deoxygenation. It is preferable to maintain a lower range of LHSV to prevent the absorption of sulfur in the bio-oil [47].

The hydrogen pressure is one of the other vital parameters in the hydrodeoxygenation process. Very high pressure is maintained in the hydrodeoxygenation process. It ensures to decrease the coke formation, an increase in solubility of hydrogen, which eventually leads to the availability of hydrogen at the surface of the catalyst. It also favors the saturation of unstable compounds [56]. It also favors the saturation of unstable compounds. The hydrodeoxygenation of less reactive phenolic compounds requires a higher amount of hydrogen. Such a type of reaction follows the hydrogenation mechanism. A significant increase in the H/C ratio occurs due to the increase in the hydrogenation of bio–oil, which increases in HHV value of the bio–oil [57]. Table 2.4 presents the different operating conditions and feedstock used for the hydrodeoxygenation process.

2.5 Conclusion and future perspective

The world community is facing many challenges, such as the depletion of fossil fuels, and rapid expansion in population, leading to an increase in energy demand. Human life is dependent on energy in many ways. Hence, there is a requirement for sustainable and renewable energy sources. Lignocellulosic biomass is a renewable energy source that does not disturb the food cycle. Nevertheless, the usage of agricultural lands for the production of energy crops should be taken into account. Fast pyrolysis is an efficient process in which the solid biomass is converted into a liquid product. It is an effective and easy way to produce a liquid product suitable for transportation fuels. The fast pyrolytic liquid product is inherently associated with some issues, such as high oxygen content (around 30–50 wt%), high TAN, low heating value, storage instability, and handling. One of the methods to improve the quality of bio-oil is catalytic fast pyrolysis. The critical feature in the catalytic fast pyrolysis process is the development of the catalyst. Fundamental understanding of the pyrolytic chemistry during thermal decomposition of the biomass with catalyst must be known. It will help in developing desirable catalysts. The ideal condition in the upgradation process is to remove oxygen in the vapor product as a permanent gas (mostly CO_2) to conserve the hydrogen in the biomass. The vital feature of catalytic fast pyrolysis is to produce transportation fuels. However, it is not readily achievable in a single step.

Another suitable technology for upgrading fast pyrolysis oil is catalytic hydrodeoxygenation. The hydrodeoxygenation of bio-oil is an encouraging method in which the

Table 2.4 Hydrodeoxygenation of different fast pyrolytic oil.

Feedstock	Reactor	Catalyst used	Operating conditions		Hydrodeoxygenation	Major findings	References
			Temperature	Pressure (bar)			
BTG-BTL bio-oil produced from pine wood	Autoclave reactor with stirrer	Ru/C	450°C	350	15–26 g of $CO_2 \cdot 100\,g^{-1}$ of dry bio-oil) and (1.8–5.8 g of $H_2O \cdot 100\,g^{-1}$ of dry bio-oil	Up to a catalyst loading ratio of 5%, the removal of oxygen in the form of CO_2 and H_2O increased significantly	[58]
Straw bio-oil	Continuous fixed-bed reactor	Commercial sulfided NiMo/Al$_2$O$_3$	300°C–360°C	20–80	At T/P of 300/20, the oxygen content was found to be half of the original	Single-stage upgrading at 340°C and 4 MPa is desirable	[59]
Switchgrass oil	Batch autoclave	Pt/C, Ru/C, or Pd/C	320°C	144.78	–	Overall upgrading efficiency of switchgrass oil catalyzed by Pt/C is high	[60]
Oak wood oil	Dual bed reactor	Ru/C and Pt/ZrP	130°C and 300°C–400°C	139–149	Coke yield of 45% at 400°C	Addition of second-stage reactor improved the hydrocarbon content	[55]
Scotch pine wood oil	Two-stage reactor	Ni or CoMo and sulfided CoMo/Al2O3	150°C–250°C and 300°C–380°C	50–100	Oxygen content in first reactor was 26–27 wt% and in the second reactor is 2–3 wt%	Sulfided catalyst provided promising result	[61]

deoxygenation takes place in the presence of hydrogen. A lot of research and developments are focused on the hydrodeoxygenation of model compounds. The model compounds provide fundamental aid in understanding the hydrodeoxygenation process. It is essential to optimize the performance of the catalyst. The biomass feedstock in general is quite complex. This increases the complexity of the bio-oil produced. Hence, the implementation of hydrodeoxygenation of bio-oil at the industrial level requires a sound knowledge of the catalyst as well as biomass. The sulfided CoMo and carbon-supported catalysts provide optimistic result. The sulfided catalyst exhibited good hydrodeoxygenation results. Moreover, they are resistant to sulfur, can withstand high hydrogen pressure, and moderately economical. The continuous plugging of the single reactor can be avoided using a dual-stage reactor. It is recommended to check the stability of bio-oil during storage. The knowledge of fast pyrolysis and catalytic hydrodeoxygenation should be implemented in exploring new technologies. One such process is the hydropyrolysis of biomass. It is the combination of both fast pyrolysis and catalytic hydrodeoxygenation in a single process. It can also be performed in the presence of a catalyst.

References

[1] M.S. Mettler, D.G. Vlachos, P.J. Dauenhauer, Top ten fundamental challenges of biomass pyrolysis for biofuels, Energy Environ. Sci. 5 (2012) 7797–7809, https://doi.org/10.1039/c2ee21679e.

[2] K. Praveen Kumar, S. Srinivas, Catalytic co-pyrolysis of biomass and plastics (polypropylene and polystyrene) using spent FCC catalyst, Energy Fuel 34 (2020) 460–473, https://doi.org/10.1021/acs.energyfuels.9b03135.

[3] S.Y. Lee, R. Sankaran, K.W. Chew, C.H. Tan, R. Krishnamoorthy, D.-T. Chu, P.-L. Show, Waste to bioenergy: a review on the recent conversion technologies, BMC Energy 1 (2019) 1–22, https://doi.org/10.1186/s42500-019-0004-7.

[4] W.Y. Chen, T. Suzuki, J. Seiner, M. Lackner, Handbook of Climate Change Mitigation, Elesvier, 2012.

[5] B.B. Uzoejinwa, X. He, S. Wang, A. El-Fatah Abomohra, Y. Hu, Q. Wang, Co-pyrolysis of biomass and waste plastics as a thermochemical conversion technology for high-grade biofuel production: recent progress and future directions elsewhere worldwide, Energy Convers. Manag. 163 (2018) 468–492, https://doi.org/10.1016/j.enconman.2018.02.004.

[6] T.M.H. Dabros, M.Z. Stummann, M. Høj, P.A. Jensen, J.D. Grunwaldt, J. Gabrielsen, P.M. Mortensen, A.D. Jensen, Transportation fuels from biomass fast pyrolysis, catalytic hydrodeoxygenation, and catalytic fast hydropyrolysis, Prog. Energy Combust. Sci. 68 (2018) 268–309, https://doi.org/10.1016/j.pecs.2018.05.002.

[7] A.V. Bridgwater, Review of fast pyrolysis of biomass and product upgrading, Biomass Bioenergy 38 (2012) 68–94, https://doi.org/10.1016/j.biombioe.2011.01.048.

[8] A.V. Bridgwater, S. Czernik, J. Piskorz, An overview of fast pyrolysis, Prog. Thermochem. Biomass Convers. 30 (2008) 977–997, https://doi.org/10.1002/9780470694954.ch80.

[9] D. Hua, Y. Wu, Y. Chen, J. Li, M. Yang, X. Lu, Co-pyrolysis behaviors of the cotton straw/PP mixtures and catalysis hydrodeoxygenation of co-pyrolysis products over NI-MO/Al2O3 catalyst, Catalysts 5 (2015) 2085–2097, https://doi.org/10.3390/catal5042085.

[10] I. Johannes, L. Tiikma, H. Luik, Synergy in co-pyrolysis of oil shale and pine sawdust in autoclaves, J. Anal. Appl. Pyrolysis 104 (2013) 341–352, https://doi.org/10.1016/j.jaap.2013.06.015.

[11] A.O. Oyedun, M. Patel, M. Kumar, A. Kumar, The upgrading of bio-oil via hydrodeoxygenation, Chem. Catal. Biomass Upgrad. (2020) 35–60, https://doi.org/10.1002/9783527814794.ch2.

[12] H. Shafaghat, P.S. Rezaei, W.M. Ashri Wan Daud, Effective parameters on selective catalytic hydro-deoxygenation of phenolic compounds of pyrolysis bio-oil to high-value hydrocarbons, RSC Adv. 5 (2015) 103999–104042, https://doi.org/10.1039/c5ra22137d.

[13] T. Bhaskar, B. Bhavya, R. Singh, D.V. Naik, A. Kumar, H.B. Goyal, Thermochemical conversion of biomass to biofuels, Biofuels (2011) 51–77, https://doi.org/10.1016/B978-0-12-385099-7.00003-6.

[14] A. Bamido, Design of a fluidized bed reactor for biomass pyrolysis, A Thesis Submitted to the Graduate School of the University of Cincinnati, University of Cincinnati, 2018.

[15] M. Sharifzadeh, M. Sadeqzadeh, M. Guo, T.N. Borhani, N.V.S.N. Murthy Konda, M.C. Garcia, L. Wang, J. Hallett, N. Shah, The multi-scale challenges of biomass fast pyrolysis and bio-oil upgrading: review of the state of art and future research directions, Prog. Energy Combust. Sci. 71 (2019) 1–80, https://doi.org/10.1016/j.pecs.2018.10.006.

[16] C. Liu, H. Wang, A.M. Karim, J. Sun, Y. Wang, Catalytic fast pyrolysis of lignocellulosic biomass, Chem. Soc. Rev. 43 (2014) 7594–7623, https://doi.org/10.1039/c3cs60414d.

[17] G. Yildiz, F. Ronsse, R. Venderbosch, R. van Duren, S.R.A. Kersten, W. Prins, Effect of biomass ash in catalytic fast pyrolysis of pine wood, Appl. Catal. B Environ. 168–169 (2015) 203–211, https://doi.org/10.1016/j.apcatb.2014.12.044.

[18] P.R. Patwardhan, J.A. Satrio, R.C. Brown, B.H. Shanks, Influence of inorganic salts on the primary pyrolysis products of cellulose, Bioresour. Technol. 101 (2010) 4646–4655, https://doi.org/10.1016/j.biortech.2010.01.112.

[19] G. Várhegyi, M.J. Antal, E. Jakab, P. Szabó, Kinetic modeling of biomass pyrolysis, J. Anal. Appl. Pyrolysis 42 (1997) 73–87, https://doi.org/10.1016/S0165-2370(96)00971-0.

[20] S.D. Stefanidis, K.G. Kalogiannis, E.F. Iliopoulou, A.A. Lappas, P.A. Pilavachi, In-situ upgrading of biomass pyrolysis vapors: catalyst screening on a fixed bed reactor, Bioresour. Technol. 102 (2011) 8261–8267, https://doi.org/10.1016/j.biortech.2011.06.032.

[21] Q. Lu, Z.F. Zhang, C.Q. Dong, X.F. Zhu, Catalytic upgrading of biomass fast pyrolysis vapors with nano metal oxides: an analytical Py-GC/MS study, Energies 3 (2010) 1805–1820, https://doi.org/10.3390/en3111805.

[22] L. Axelsson, M. Franzén, M. Ostwald, G. Berndes, G. Lakshmi, N.H. Ravindranath, Perspective: Jatropha cultivation in southern India: assessing farmers' experiences, Biofuels Bioprod. Biorefin. 6 (2012) 246–256, https://doi.org/10.1002/bbb.1324.

[23] M. Garcia-Perez, X.S. Wang, J. Shen, M.J. Rhodes, F. Tian, W.J. Lee, H. Wu, C.Z. Li, Fast pyrolysis of oil mallee woody biomass: effect of temperature on the yield and quality of pyrolysis products, Ind. Eng. Chem. Res. 47 (2008) 1846–1854, https://doi.org/10.1021/ie071497p.

[24] R.J.M. Westerhof, D.W.F. Brilman, W.P.M. Van Swaaij, S.R.A. Kersten, Effect of temperature in fluidized bed fast pyrolysis of biomass: oil quality assessment in test units, Ind. Eng. Chem. Res. 49 (2010) 1160–1168, https://doi.org/10.1021/ie900885c.

[25] S. Septien, S. Valin, C. Dupont, M. Peyrot, S. Salvador, Effect of particle size and temperature on woody biomass fast pyrolysis at high temperature (1000-1400°C), Fuel 97 (2012) 202–210, https://doi.org/10.1016/j.fuel.2012.01.049.

[26] A. Van der Drift, H. Boerrigter, B. Coda, Entrained Flow Gasification of Biomass, vol. 58, ECN—Energy Cent, Petten, Netherlands, 2004, https://doi.org/10.1016/j.fuel.2011.10.063.

[27] G. Brem, E.A. Bramer, PyRos : a new flash pyrolysis technology for the production of bio-oil from biomass residues, in: Proc. Int. Conf. Exhib. Bioenergy Outlook, 2007, pp. 1–14.

[28] K. Qin, P.A. Jensen, W. Lin, A.D. Jensen, Biomass gasification behavior in an entrained flow reactor: gas product distribution and soot formation, Energy Fuel 26 (2012) 5992–6002, https://doi.org/10.1021/ef300960x.

[29] M.R. Hurt, J.C. Degenstein, P. Gawecki, D.J. Borton, N.R. Vinueza, L. Yang, R. Agrawal, W.N. Delgass, F.H. Ribeiro, H.I. Kenttämaa, On-line mass spectrometric methods for the determination of the primary products of fast pyrolysis of carbohydrates and for their gas-phase manipulation, Anal. Chem. 85 (2013) 10927–10934, https://doi.org/10.1021/ac402380h.

[30] A. Veses, M. Aznar, I. Martínez, J.D. Martínez, J.M. López, M.V. Navarro, M.S. Callén, R. Murillo, T. García, Catalytic pyrolysis of wood biomass in an auger reactor using calcium-based catalysts, Bioresour. Technol. 162 (2014) 250–258, https://doi.org/10.1016/j.biortech.2014.03.146.

[31] J. Jae, R. Coolman, T.J. Mountziaris, G.W. Huber, Catalytic fast pyrolysis of lignocellulosic biomass in a process development unit with continual catalyst addition and removal, Chem. Eng. Sci. 108 (2014) 33–46, https://doi.org/10.1016/j.ces.2013.12.023.

[32] H. Zhang, R. Xiao, H. Huang, G. Xiao, Comparison of non-catalytic and catalytic fast pyrolysis of corncob in a fluidized bed reactor, Bioresour. Technol. 100 (2009) 1428–1434, https://doi.org/10.1016/j.biortech.2008.08.031.

[33] H.J. Park, H.S. Heo, Y.K. Park, J.H. Yim, J.K. Jeon, J. Park, C. Ryu, S.S. Kim, Clean bio-oil production from fast pyrolysis of sewage sludge: effects of reaction conditions and metal oxide catalysts, Bioresour. Technol. 101 (2010) S83–S85, https://doi.org/10.1016/j.biortech.2009.06.103.

[34] P.S. Marathe, R.J.M. Westerhof, S.R.A. Kersten, Effect of pressure and hot vapor residence time on the fast pyrolysis of biomass: experiments and modeling, Energy Fuel 34 (2020) 1773–1780, https://doi.org/10.1021/acs.energyfuels.9b03193.

[35] T. Bridgwater, Challenges and opportunities in fast pyrolysis of biomass: part II, Johnson Matthey Technol. Rev. 62 (2018) 150–160, https://doi.org/10.1595/205651318X696738.

[36] N. Tran, Y. Uemura, S. Chowdhury, A. Ramli, A review of bio-oil upgrading by catalytic hydrodeoxygenation, Appl. Mech. Mater. 625 (2014) 255–258, https://doi.org/10.4028/www.scientific.net/AMM.625.255.

[37] S. Kim, E.E. Kwon, Y.T. Kim, S. Jung, H.J. Kim, G.W. Huber, J. Lee, Recent advances in hydrodeoxygenation of biomass-derived oxygenates over heterogeneous catalysts, Green Chem. 21 (2019) 3715–3743, https://doi.org/10.1039/c9gc01210a.

[38] Z. Si, X. Zhang, C. Wang, L. Ma, R. Dong, An overview on catalytic hydrodeoxygenation of pyrolysis oil and its model compounds, Catalysts 7 (2017) 1–22, https://doi.org/10.3390/catal7060169.

[39] Elliott, R.D.: United States Patent [19]. 5, 2–7 (1984).

[40] Wildschut, J., Mahfud, F.H., Venderbosch, R.H., Heeres, H.J.: Ie9006003.Pdf. 10324–10334 (2009).

[41] W.M. Xiong, Y. Fu, F.X. Zeng, Q.X. Guo, An in situ reduction approach for bio-oil hydroprocessing, Fuel Process. Technol. 92 (2011) 1599–1605, https://doi.org/10.1016/j.fuproc.2011.04.005.

[42] F. Schüth, Hydrogen: economics and its role in biorefining, Catalytic Hydrogenation for Biomass Valorization, in:, R. Rinaldi (Ed.), RSC Energy and Environment Series No. 13, Royal Society of Chemistry, 2015.

[43] D. Valencia, L. Díaz-García, L.F. Ramírez-Verduzco, A. Qamar, A. Moewes, J. Aburto, Paving the way towards green catalytic materials for green fuels: impact of chemical species on Mo-based catalysts for hydrodeoxygenation, RSC Adv. 9 (2019) 18292–18301, https://doi.org/10.1039/c9ra03208h.

[44] Y. Romero, F. Richard, S. Brunet, Hydrodeoxygenation of 2-ethylphenol as a model compound of bio-crude over sulfided Mo-based catalysts: promoting effect and reaction mechanism, Appl. Catal. B Environ. 98 (2010) 213–223, https://doi.org/10.1016/j.apcatb.2010.05.031.

[45] T. Prasomsri, T. Nimmanwudipong, Y. Román-Leshkov, Effective hydrodeoxygenation of biomass-derived oxygenates into unsaturated hydrocarbons by MoO_3 using low H_2 pressures, Energy Environ. Sci. 6 (2013) 1732–1738, https://doi.org/10.1039/c3ee24360e.

[46] K. Tomishige, Y. Nakagawa, M. Tamura, Design of supported metal catalysts modified with metal oxides for hydrodeoxygenation of biomass-related molecules, Curr. Opin. Green Sustain. Chem. 22 (2020) 13–21, https://doi.org/10.1016/j.cogsc.2019.11.003.

[47] P.M. Mortensen, D. Gardini, C.D. Damsgaard, J.D. Grunwaldt, P.A. Jensen, J.B. Wagner, A.D. Jensen, Deactivation of Ni-MoS2 by bio-oil impurities during hydrodeoxygenation of phenol and octanol, Appl. Catal. A Gen. 523 (2016) 159–170, https://doi.org/10.1016/j.apcata.2016.06.002.

[48] A. Berenguer, T.M. Sankaranarayanan, G. Gómez, I. Moreno, J.M. Coronado, P. Pizarro, D.P. Serrano, Evaluation of transition metal phosphides supported on ordered mesoporous materials as catalysts for phenol hydrodeoxygenation, Green Chem. 18 (2016) 1938–1951, https://doi.org/10.1039/c5gc02188j.

[49] P.M. Mortensen, J.D. Grunwaldt, P.A. Jensen, K.G. Knudsen, A.D. Jensen, A review of catalytic upgrading of bio-oil to engine fuels, Appl. Catal. A Gen. 407 (2011) 1–19, https://doi.org/10.1016/j.apcata.2011.08.046.

[50] E. Kordouli, C. Kordulis, A. Lycourghiotis, R. Cole, P.T. Vasudevan, B. Pawelec, J.L.G. Fierro, HDO activity of carbon-supported Rh, Ni and Mo-Ni catalysts, Mol. Catal. 441 (2017) 209–220, https://doi.org/10.1016/j.mcat.2017.08.013.

[51] H. Shafaghat, P.S. Rezaei, W.M.A.W. Daud, Catalytic hydrodeoxygenation of simulated phenolic bio-oil to cycloalkanes and aromatic hydrocarbons over bifunctional metal/acid catalysts of Ni/HBeta Fe/HBeta and NiFe/HBeta, J. Ind. Eng. Chem. 35 (2016) 268–276, https://doi.org/10.1016/j.jiec.2016.01.001.

[52] A. Centeno, E. Laurent, B. Delmon, Influence of the support of CoMo sulfide catalysts and of the addition of potassium and platinum on the catalytic performances for the hydrodeoxygenation of guaicol, J. Catal. 154 (1995) 288–298.

[53] S. Jin, Z. Xiao, C. Li, X. Chen, L. Wang, J. Xing, W. Li, C. Liang, Catalytic hydrodeoxygenation of anisole as lignin model compound over supported nickel catalysts, Catal. Today 234 (2014) 125–132, https://doi.org/10.1016/j.cattod.2014.02.014.

[54] H. Wang, J. Male, Y. Wang, Recent advances in hydrotreating of pyrolysis bio-oil and its oxygen-containing model compounds, ACS Catal. 3 (2013) 1047–1070, https://doi.org/10.1021/cs400069z.

[55] K. Routray, K.J. Barnett, G.W. Huber, Hydrodeoxygenation of pyrolysis oils, Energy Technol. 5 (2017) 80–93, https://doi.org/10.1002/ente.201600084.

[56] R.H. Venderbosch, A.R. Ardiyanti, J. Wildschut, A. Oasmaa, H.J. Heeres, Stabilization of biomass-derived pyrolysis oils, J. Chem. Technol. Biotechnol. 85 (2010) 674–686, https://doi.org/10.1002/jctb.2354.

[57] F. de Miguel Mercader, M.J. Groeneveld, S.R.A. Kersten, N.W.J. Way, C.J. Schaverien, J.A. Hogendoorn, Production of advanced biofuels: co-processing of upgraded pyrolysis oil in standard refinery units, Appl. Catal. B Environ. 96 (2010) 57–66, https://doi.org/10.1016/j.apcatb.2010.01.033.

[58] M. Benés, R. Bilbao, J.M. Santos, J. Alves Melo, A. Wisniewski, I. Fonts, Hydrodeoxygenation of lignocellulosic fast pyrolysis bio-oil: characterization of the products and effect of the catalyst loading ratio, Energy Fuel 33 (2019) 4272–4286, https://doi.org/10.1021/acs.energyfuels.9b00265.

[59] M. Auersvald, B. Shumeiko, M. Staš, D. Kubička, J. Chudoba, P. Šimáček, Quantitative study of straw bio-oil hydrodeoxygenation over a sulfided NiMo catalyst, ACS Sustain. Chem. Eng. 7 (2019) 7080–7093, https://doi.org/10.1021/acssuschemeng.8b06860.

[60] Y. Elkasabi, C.A. Mullen, A.L.M.T. Pighinelli, A.A. Boateng, Hydrodeoxygenation of fast-pyrolysis bio-oils from various feedstocks using carbon-supported catalysts, Fuel Process. Technol. 123 (2014) 11–18, https://doi.org/10.1016/j.fuproc.2014.01.039.

[61] V.A. Yakovlev, M.V. Bykova, S.A. Khromova, Stability of nickel-containing catalysts for hydrodeoxygenation of biomass pyrolysis products, Catal. Ind. 4 (2012) 324–339, https://doi.org/10.1134/s2070050412040204.

CHAPTER 3

Fischer-Tropsch synthesis to hydrocarbon biofuels: Present status and challenges involved

Muxina Konarova[a], Waqas Aslam[a], and Greg Perkins[b]

[a]Australian Institute for Bioengineering and Nanotechnology (AIBN), The University of Queensland, Brisbane, QLD, Australia
[b]School of Chemical Engineering, Faculty of Engineering, Architecture and Information Technology, The University of Queensland, Brisbane, QLD, Australia

Contents

3.1 Introduction	78
3.2 Synthesis gas manufacture	79
3.3 Improved catalysts	82
3.3.1 Iron catalysts	83
3.3.2 Cobalt catalysts	85
3.3.3 Reactor design	85
3.3.4 Fixed-bed multitubular	86
3.3.5 Fluidized bed	88
3.3.6 Slurry bubble column	88
3.3.7 Microchannel	89
3.3.8 3D printed	89
3.4 Challenges and opportunities for biofuels	90
3.4.1 Distributed processing	90
3.4.2 Gas loop configuration	91
3.4.3 Reactor/catalyst selection	91
3.4.4 Upgrading and refining	93
3.4.5 Plant modularization	93
3.4.6 Automated and unattended operations	93
3.4.7 Oxygenated fuels	93
3.5 Conclusions	94
References	94

Abbreviations

FT	Fischer-Tropsch
FTS	Fischer-Tropsch synthesis
GTL	gas to liquid
HTFT	high-temperature Fischer-Tropsch
LTFT	low-temperature Fischer-Tropsch
SMDS	shell middle distillate synthesis
TL	biomass to liquid
WTL	waste to liquid

Hydrocarbon Biorefinery
https://doi.org/10.1016/B978-0-12-823306-1.00006-6

Copyright © 2022 Elsevier Inc.
All rights reserved.

3.1 Introduction

Franz Fischer and Hans Tropsch first demonstrated the main gas-to-liquid conversion process in 1923, in which mixtures of linear hydrocarbons were produced by passing syngas over Fe, Ni, or Co catalysts at 180°C–250°C at 1 bar. Since then, other options for syngas conversion have been explored including the Fischer-Tropsch (FT) process, methanol, ethanol and higher alcohols, and some specialized derivatives (e.g., dimethyl ether, diethyl ether, and aromatics). FT synthesis produces a wide product spectrum mainly hydrocarbons requiring further processing, purification, and marketing. The FT process requires clean syngas and significant utilities, making the economic viability challenging for small-scale operations, e.g., less than 1000 t/day fuel. In the Fischer-Tropsch synthesis (FTS), carbon monoxide is hydrogenated over metallic catalysts producing linear hydrocarbons, according to the following overall reaction:

$$n\text{CO} + 2n\text{H}_2 \rightarrow n(-\text{CH}_2-) + n\text{H}_2\text{O} \tag{3.1}$$

The reaction involves polymerization of CH_x-alkyl radicals derived from C—O dissociation and the subsequent hydrogenation. Transition metals (Group VIII) have been reported to exhibit high catalytic activity toward long-chain hydrocarbons and numerous reports on the reaction mechanism and kinetics over transition metals can be found elsewhere [1–3]. Depending on the support material activity of catalysts can be varied as follows:

$$\text{Ru} > \text{Fe} > \text{Ni} > \text{Co} > \text{Rh on alumina} \tag{3.2}$$

$$\text{Co} > \text{Fe} > \text{Ru} > \text{Ni} > \text{Rh on silica} \tag{3.3}$$

Iron- and cobalt-based catalysts have been applied in FT synthesis at an industrial scale [4]. CO hydrogenation products consist of a mixture of hydrocarbons ranging from C_1 to long-chain compounds (waxes). The chemical composition of FT products depends on the kinetics of linear polymerization with chain growth probability (α). The Anderson-Shultz-Flory distribution (ASF distribution) predicts broad hydrocarbon distribution for the FT synthesis with dependency on α. The production of diesel-range components is favored at high values $\alpha > 0.9$, while the production of gasoline-range components is favored at moderate values $0.7 < \alpha < 0.8$, see Fig. 3.1.

Operating conditions, such as (i) temperature, (ii) pressure, (iii) H_2/CO ratio, and (iv) reactor design and the catalysts, all affect the range of carbon chain lengths (α) including olefin/paraffin ratio, carbon deposition, and methane selectivity, as shown in Table 3.1.

When the FT process is applied at industrial scales, there are three main processing steps that are required including the production of clean syngas (CO and H_2), FTS, and upgrading or refining of the FT products. While the FT process has been applied industrially using natural gas and coal as feedstocks, the conversion of biogenic feedstocks, such as biomass and wastes, is becoming the major focus of the recent studies and project

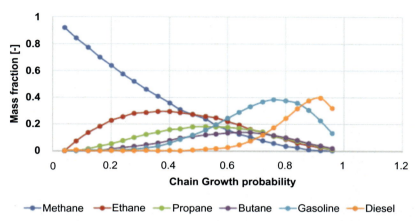

Fig. 3.1 Relationship between chain growth probability and the mass fraction of different components [1].

Table 3.1 Effect of FT operating conditions on chain growth probability, olefin/paraffin ratio, carbon deposition, and methane selectivity.

Parameter	Chain growth probability	Olefin/paraffin ratio	Carbon deposition	Methane selectivity
Temperature increase	Down	Down	Up	Up
Pressure increase	Up	Complex	Complex	Down
H_2/CO ratio increase	Down	Down	Down	Up
Conversion increase	Complex	Down	Up	Up
Space velocity increase	Complex	Up	Complex	Down

proposals. The major driver is to produce renewable fuels and chemicals that can decarbonize the transport sector—one of the largest single generators of greenhouse gases. In this chapter, the production of syngas for the FT process is briefly discussed, along with more in-depth summaries of the present status of catalyst development and reactor designs. The chapter concludes with a discussion of the challenges and opportunities of producing biofuels with the FT process.

3.2 Synthesis gas manufacture

The FT process requires clean syngas at pressures typically greater than 25 bar with a preferred H_2/CO ratio of around 2. The syngas should have major contaminants removed

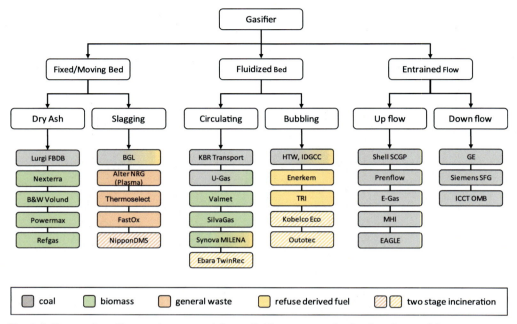

Fig. 3.2 Types of gasifiers and commercially available processes by feedstock type [9].

that would deactivate the catalysts, including sulfur (<10 ppb), HCN and NH_3 (<20 ppb), heavy metals (<10 ppb), chlorine (<10 ppb), and be free of tars and particulates [1]. It is also preferable that the syngas has a low concentration of components that are inert in the FT process, such as CO_2, N_2, and CH_4. The clean syngas may be generated by reacting carbonaceous material, such as natural gas, biogas, coal, biomass, and wastes with steam and/or oxygen in a reforming or gasification process. Biofuel production refers to using syngas derived from the biogenic origin, made from biogas, biomass, or wastes. Natural gas and biogas may be converted to syngas using tubular reformers, combined reforming reactors, autothermal reforming reactors, or partial oxidation reactors. Solid feedstocks are gasified by reacting them with a deficiency of oxygen at temperatures in the range of 800°C–1400°C [5–8]. Gasifiers may be classified by type as either fixed-bed, fluidized-bed, or entrained-flow, as shown in Fig. 3.2.

Biomass gasifiers are usually configured to be either fluidized-bed or fixed-bed reactors as entrained-flow reactors are less attractive due to the small particle sizes required [9]. Fixed-bed updraft or downdraft gasifiers can be suitable for small-scale projects, while circulating and bubbling fluidized-bed gasifiers can be better suited for larger projects [5, 10]. The overall block flow diagram for producing FT biofuels is shown in Fig. 3.3. The biomass is pretreated and gasified and then the raw syngas is cleaned to meet the requirements of the FT catalyst [7–9]. FT liquids are cooled and recovered and then

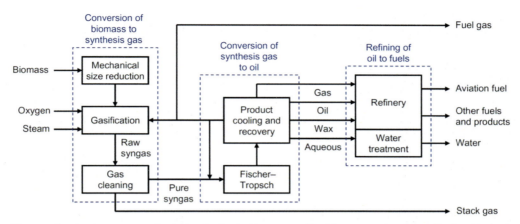

Fig. 3.3 Production routes of aviation fuels via biomass-to-liquids process involving Fischer-Tropsch synthesis [11].

Table 3.2 Classification of GTL plants based on processing capacity.

	World Scale	Small Scale	Mini-GTL	Micro-GTL
Gas feed rate (MMscfd)	>100	>10	>1	>0.1
Gas feed rate (m^3/d)	>3000,000	>300,000	>30,000	>3000
Capex ($MM)	>500	>100	>10	>1
Product make (bpd)	>10,000	>1000	>100	>10
Plant configuration	Fixed, 20+ years life	Fixed, 20+ years life	Moveable, modular	Unattended, modular
Examples	Oryx & Pearl GTL, Qatar	ENVIA, Oklahoma	INFRA Technology, Texas	GasTechno M300, North Dakota

hydrocracked/hydrotreated to meet final product specifications for diesel, kerosene, and aviation fuel.

Table 3.2 shows a classification of gas-to-liquid (GTL) plants based on processing capacity. Traditional GTL plants typically use natural gas or coal to achieve economies of scale and produce more than 10,000 bpd of liquid products. Examples include the Oryx GTL plant and Pearl GTL plants in Qatar and Bintulu, Malaysia. In recent years, the focus of innovation has been on the development of smaller plants designed to use stranded natural gas (flared gas), landfill gas, biogas, or biomass and residual wastes. Small plants produce over 1000 bpd of product, while mini and micro-GTL plants

Table 3.3 Examples of biofuel FT plants.

	ENVIA, Oklahoma, USA	Enerkem, Edmonton, CA	Fulcrum, Nevada, USA
Feedstock	Landfill biogas	RDF[a]	RDF
Product make	11 ML/year	38 ML/year	50 ML/year
Reforming/gasification	Haldor-Topsoe steam reformer	Enerkem bubbling fluidized bed	TRI reforming gasifier
Fischer-Tropsch technology	Velocys	Enerkem methanol and ethanol synthesis	BP
Status	Shut down	Operating	Under construction

[a]RDF, refuse-derived fuel.

produce between 1 and 100 bpd of liquid products. As the processing scale reduces, one or more of the following innovative features is generally incorporated into the plant design to achieve improved economics:

- Use of stranded natural gas or natural gas which is already being flared
- Use of a biogenic feedstock, such as biomass/biogas/waste, which enables the product to be classed as a renewable fuel, attracting a higher price than conventional fossil-derived synthetic fuels
- Application of compact, modular FTS units
- Improved catalysts aimed at increasing the value of the products
- Alternate products such as oxygenated fuels such as methanol and ethanol
- Plant modularization to reduce construction costs
- Automated or unattended operations to reduce operations costs

A small number of biofuel-to-liquid (BTL) plants have been developed. Table 3.3 shows examples of three different BTL plants. The ENVIA plant in Oklahoma converted landfill biogas using steam reforming and the Velocys FT microchannel technology. The plant was shut down in 2018 due to a leak identified in the FT coolant system. Enerkem has built a plant in Edmonton, Canada that gasifies refuse-derived fuel in a bubbling fluidized bed to produce syngas which is then reacted over catalysts to form methanol and ethanol [12]. Fulcrum Bioenergy is developing plants to convert refuse-derived fuel into FT liquids using a tri-reforming gasifier paired with BP's FT liquids technology, which uses a catalyst from Johnson-Matthey [13]. The main areas of innovation in FT over the past decade have involved improvement of catalysts and the development of novel reactor designs. We summarize these innovations in the following sections.

3.3 Improved catalysts

Depending on the required product, FT synthesis can be categorized as low-temperature Fischer-Tropsch (LTFT) and high-temperature Fischer-Tropsch (HTFT). LTFT is

Table 3.4 Distribution of products obtained from Fischer-Tropsch synthesis and crude oil [14].

Compound class	HTFT	LTFT	Crude Oil
Alkanes (paraffins)	>10%	Major product	Major product
Cyclo-alkanes (naphthenes)	<1%	<1%	Major product
Alkenes (olefins)	Major product	>10%	None
Aromatics	5%–10%	<1%	Major product
Oxygenates	5%–15%	5%–15%	<1% O
Sulfur compounds	None	None	0.1%–0.5% S
Nitrogen compounds	None	None	<1% N
Organometallics	Carboxylates	Carboxylates	Phorphyrines
Water	Major by-product	Major by-product	0%–2%

normally used for fuel synthesis, while HTFT is generally used for light olefin and oxygenate synthesis. Table 3.4 shows a comparison of the distribution of major compounds produced from HTFT and LTFT synthesis compared with crude oil.

Syngas to FT fuels conversion is limited by thermodynamics at certain process conditions, namely high temperature and low pressure. Catalyst development efforts for FT synthesis have been extensively reported. Most catalysts used in FT synthesis involve transition metals, such as ruthenium, iron, and nickel [15]. The catalyst development has been aimed to produce long-chain hydrocarbons and C_{14}–C_{20} range diesel hydrocarbons.

- In general, nickel promotes the methanation reaction, which leads to the formation of undesirable by-product methane [16]. However, methanation is thermodynamically favorable and significantly lowers hydrogen concentration inhibiting the formation of linear alkanes. By tuning catalyst composition, some of the methanation activity can be impeded.
- The low-cost iron catalyst is active for water-gas shift activity and alters feed composition (H_2/CO ratio) due to the formation of hydrogen and carbon dioxide. Feedstock with low H_2/CO ratios requires water-gas shift reaction if syngas is produced from coal and biomass, but undesirable for syngas originated from natural gas having a high H_2/CO ratio [16].
- Cobalt features more desirable catalytic properties and is often promoted with ruthenium [17]. Compared to iron, cobalt exhibited lower activity toward water-gas shift reaction and is considered an expensive metal.

In general, FT catalysts contain porous support and hydrogenating metals and can be grouped into two major platforms: iron (Fe)-based and cobalt (Co)-based catalysts.

3.3.1 Iron catalysts

Syngas produced from coal gasification contains a low H_2/CO ratio. Thus, catalysts based on iron are often used in CO hydrogenations to obtain FT fuels. Iron-based catalysts can be used at two regimes: the high-temperature regime (300°C–350°C) and low-temperature regime (220°C–270°C). LTFT synthesis produces wax-type linear hydrocarbons, and further processing is required to produce a shorter carbon chain length hydrocarbon, such as kerosene, diesel, and lubricant oil [18]. There are several preparation methods, such as precipitation and fusion, that are available to obtain iron-based FT catalysts. Commercial iron catalysts are iron oxides, hydroxides, or oxy-hydroxides, which must be activated before use. Typically, they are reduced with hydrogen or pre-treated with syngas before the operation. The desired properties of a good catalyst include low selectivity toward methane, high selectivity toward hydrocarbons, high activity, high stability, and a high degree of mechanical robustness, especially for catalysts utilized in fluidized beds and slurry bubble column reactors. The type of reactor employed for the FTS and the process is whether HTFT or LTFT put additional demands on the required catalyst performance. For example, in tube and shell fixed-bed reactors, the mass transfer barrier must be reduced to maximize the full potential of catalyst activity. In contrast, catalyst strength and attribution resistance are of less importance than catalyst shape and form. On the other hand, in a fluidized-bed reactor, catalyst robustness and attrition resistance, as well as particle size and density, are key performance criteria. Iron catalysts are comprised of nanosized Fe_2O_3 crystallites with transition metal promoters (e.g., Cu, Mn). Catalytic active phases of iron-based catalysts have often reported a mixture of iron carbide, iron oxide, and metallic iron, which are formed during activation (hydrogen) and reaction under syngas atmosphere [19]. Iron-catalyzed FT synthesis is among the most studied catalytic processes in the literature. Fe can not only catalyze light hydrocarbon product streams that are suitable as fuel and chemical feedstocks, but can also be used to synthesize long-chain hydrocarbons (C_{35}) suited for the processes involving wax. When compared with cobalt, iron is a relatively inexpensive material and so has been one of the first types of catalysts to be commercialized. Iron-based catalysts have been used since the late 1950s by Sasol (Table 3.1). Iron is reported to be more resistant to sulfur contaminants in the feed gas than cobalt catalysts. Iron-based catalysts have shown to be sensitive to the promoter addition and changes to process conditions, allowing fast manipulation of product selectivity.

The disadvantage of iron-based catalyst in the FTS process is its rapid deactivation rate (loss of active sites). By manipulating synthesis temperature in the iron-based FTS process, carbon chain length of products can be varied. At lower temperatures (220°C–250°C), catalyst chain growth probability (α) corresponds to 0.94, leading to the formation of hydrocarbons with chain lengths longer than C_{21}. At higher temperatures, 320°C–350°C, the value of α reduces to 0.7, forming light hydrocarbons employed in transportation fuels and petrochemicals. Fig. 3.1 shows the relationship between chain growth probability and carbon chain length. Although iron-based catalysts offer several advantages, the catalyst structure and active sites are very complex. Iron-based FT reactions require in situ, *operando* studies to understand their catalytic properties fully.

3.3.2 Cobalt catalysts

Cobalt catalysts are used only in the LTFT range. This is due to their high methanation activity at high temperatures. The cobalt catalysts are costlier than iron-based catalysts but using cobalt catalysts, FT synthesis can be operated at mild process conditions saving energy for downstream processing. Cobalt catalysts have a lower coke deposition rate than Fe catalysts, and consequently, Co catalysts have longer lifetimes. Operation at high temperature results in the gasoline/diesel ratio of 2:1, while operation at low temperature results in the gasoline/diesel ratio of 1:2 or less no matter whether the catalyst is Fe or Co. Short-chain hydrocarbons can be formed at high FT synthesis temperature and also shifting selectivity toward hydrogenated products. Higher temperatures also lead to increased branching and production of oxygenated compounds, including ketones and alcohols.

In the 1980s, Shell began development on the Shell Middle Distillate Synthesis (SMDS) technology, which is a two-stage process. In the first stage, called heavy paraffin synthesis (HPS), the FT process is applied with a cobalt catalyst ($Co/Zr/SiO_2$) operated at a low temperature of around 220°C and 2.5 MPa [20]. The intent was to achieve an α-value of at least 0.9 and with a stable and long catalyst lifetime [14]. To minimize development time and scale-up risk multitubular fixed-bed reactors were selected for the FT heavy paraffin synthesis. In the second stage, called heavy paraffin conversion, the heavy wax fraction is converted into middle distillates selectively which are then distilled into the desired boiling point ranges. The heavy paraffin conversion operates a mild trickle-flow hydrocracking process at 300°C–350°C and 3–5 MPa [20]. The two-stage process provides flexibility to adjust the product slate.

In the early 1990s, the first SMDS plant was built using natural gas as feedstock in Bintulu, Malaysia with an initial production capacity of 12,500 bpd, which was later increased to nearly 14,000 bpd. A second plant, called the Pearl GTL facility, was built in Qatar with two FT processing trains of 70,000 bpd each and started operations in 2011. The SMDS technology produces a syncrude that is very paraffinic and has a low content of alkenes and oxygenates. In 2007, the Oryx GTL plant was commissioned in Qatar, converting lean natural gas into 34,000 bpd of liquid products. The Sasol slurry phase distillate process was adopted for the liquid synthesis using a Co-based low-temperature FT catalyst ($Co/Pt/Al_2O_3$). Wax and cold condensate from the FT unit are partially upgraded in a single hydrocracker using a design from ChevronTexaco [14].

3.3.3 Reactor design

The choice of a suitable reactor for FT is intimately coupled with the selected catalyst and desired operating conditions. The very high exothermic heat of reaction requires effective dissipation routes to eliminate hot spots. Therefore, heat removal from the reaction zone is one of the most critical factors that need to be accounted for when designing a catalyst/reactor system for the FT process. Efficient reactor designs integrate temperature management strategies for optimal heat conduction and dissipation. Depending upon the

Table 3.5 Reactor technologies that have been used at industrial scale for Fischer-Tropsch synthesis [14].

Fe-HTFT	Fe-LTFT	Co-LTFT
Fixed fluidized bed (1951, Hydrocol) Circulating fluidized-bed (1955, Kellogg Synthol) Circulating fluidized bed (1980, Sasol Synthol) Fixed fluidized bed (1995, Sasol Advanced Synthol)	Fixed bed (1955, Ruhrchemie–Lurgi) Slurry bed (1993, Sasol slurry bed process)	Fixed bed (1936, German normal pressure) Fixed bed (1937, German medium pressure) Fixed bed (1993, Shell middle distillate synthesis) Slurry bubble column (2007, Sasol slurry bed process) Microchannel fixed bed (2013, Velocys)

reactor design, the per pass CO conversion is limited to reduce heat duty and to reduce the rate of catalyst deactivation [21]. The FT process has been applied industrially using multitubular fixed-bed reactors, microchannel fixed-bed reactors, circulating and fixed fluidized-bed reactors, and slurry-bed reactors, see Table 3.5 and Fig. 3.4.

3.3.4 Fixed-bed multitubular

In a multitubular reactor, a large surface area is achieved by utilizing many small diameter tubes filled with pelletized catalysts. The exothermic heat of the reaction is transferred from the tubes to boiling water to generate steam. Fixed-bed reactors are desirable because of the high density of catalyst per unit reactor volume and potential for high productivity. While studies show that external mass transfer limitations are generally negligible, the choice of catalyst size and shape is an optimization problem, requiring making trade-offs between pressure drop, heat transfer, and overall reaction kinetics [20]. To avoid high-pressure drop, the size of the catalyst pellets is limited to greater than 1 mm. Commercial fixed-bed reactors for the FT process use a concurrent trickle-flow arrangement whereby both gas and liquid flow from the top to the bottom of the reactor, see Fig. 3.4B [22].

The multitubular fixed-bed reactors were first applied in the 1930s in the Arge reactors that were jointly developed by Lurgi and Ruhrchemie. Today, Shell uses multitubular reactors in the SMDS plants built in Malaysia and Qatar. These multitubular reactors contain up to 3000 individual tubular reactors. Significant attention is required to ensure that the pressure drop over each tube is equal, so that the syngas and recycle liquid flow is distributed evenly over all of the tubes. A major advantage of fixed-bed reactors is that only a single tube is required in the laboratory setting to evaluate the commercial-scale performance [23].

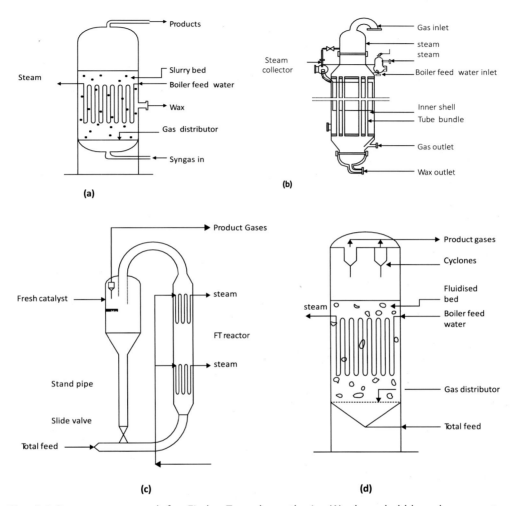

Fig. 3.4 Reactor types used for Fischer-Tropsch synthesis: (A) slurry bubble column reactor, (B) multitubular trickle fixed-bed reactor, (C) circulating fluidized-bed reactor, and (D) fixed fluidized-bed reactor [22].

BP and Johnson–Matthey have reported a technology based on "cans tech," which claims to increase productivity by a factor of three and reduce the cost by a factor of two compared with conventional FT reactors [24]. Currently, Fulcrum Bioenergy has a license for this technology (Sierra Biofuels Plant in Nevada). Another company using multitubular reactors for GTL is Emerging Fuels Technology. They have developed their catalyst and technology. By combining this technology with hydrotreatment, esters and fatty acids can be produced. Red Rock Biofuels together with EFT developed a second-generation TL8a catalyst/reactor requiring less catalyst volume [25, 26].

3.3.5 Fluidized bed

In a fluidized-bed reactor, the FT reactions take place in the gas phase at operating conditions, and the desired liquid products are formed by condensing the product gas outside of the reactor. Compared to fixed-bed reactors, fluidized-bed reactors offer uniform temperature distribution due to intensive mixing, which minimizes hot-spot formation. Heat transfer is achieved by circulating coolants within heat transfer surfaces placed inside the reactor. Small catalyst particle sizes ($<50\,\mu m$) are tolerable and the catalyst can be changed while the unit is online. However, the disadvantages are (i) low α-value operation to avoid catalyst agglomeration issues, (ii) catalyst particles that can withstand mechanical stress over a long period, often necessitating lower reactive surface area, and (iii) energy requirements to circulate and fluidize the catalyst [22].

As discussed by Sie et al., fluidized-bed reactors are not well suited for high α-value catalysts, wherein a heavy product slate is desired, as the product is prone to condensation on the catalyst surface leading to agglomeration problems [20]. All the commercialized technologies using fluidized-bed reactors operate as HTFT and have α-values of around 0.68. Fig. 3.4C shows the circulating fluidized bed of the Sasol Synthol technology that was originally installed at the Sasol Secunda site. These reactors were replaced with fixed fluidized-bed Advanced Synthol reactors in the 1990s, which were simpler to operate and more efficient [14].

3.3.6 Slurry bubble column

In the slurry bubble column (or fluidized-bed slurry) reactor, the catalyst is suspended in the heavier fraction of the liquid product from the FT process. A high α-value is, therefore, necessary to enable the formation of the liquid phase at reaction conditions. The catalyst particle size, being on the order of a few hundred microns, does not give rise to intraparticle diffusion limitations, and the slurry mode of operation allows effective heat removal and temperature control. The syngas is injected into the reactor through a proprietary distributor arrangement, rising through the liquid phase, forming gas bubbles. Mass transfer limitations are expected between the gas and liquid phases and are influenced by the hydrodynamic regime [22]. A large range of different reactor configurations is possible with the slurry bubble column, including the use of (i) a cascade bubble column with sieve trays, (ii) packed bubble column, (iii) multishaft bubble column, and (iv) static mixers. In the Sasol slurry bubble column reactor, see Fig. 3.4A, cooling coils are immersed within the reactor to control the temperature. Generally, slurry bubble column FT reactors are designed for very large capacities, and in the past have been most suitable for projects using natural gas or coal as the feedstock. Structured packings may also be used in slurry bubble column reactors to improve heat and mass transfer characteristics further [22].

3.3.7 Microchannel

Microchannel reactors are small and modular reactors with multiple millimeter channels where the reactions take place. Several companies have developed microchannel reactors suitable for the FT process. The microchannel reactors feature an innovative design where chemical reactions are intensified due to fast heat removal and enhanced mass transfer. In highly exothermic processes, such as FTS, the presence of many small coolant channels can significantly aid temperature control and heat removal capacity. Microchannels can also enhance gas flow distribution inhibiting the formation of side products and maximizing selectivity toward favorable products. This leads to the reduction of downstream processing units, thus can be considered as cost effective. Furthermore, fast heat transfer eliminates hot spots that could cause sintering of catalysts, thus losing activity [27]. The high rate of heat removal also increases catalyst lifetime.

Despite the obvious advantages mentioned above, commercialization of the microchannel technology is still facing significant obstacles. Guettel and Turek showed in a simulation-based study that the microchannel reactor has the highest productivity per unit of catalyst volume. However, the overall productivity of the reactor was comparable to conventional fixed-bed technology, possibly due to the low ratio of catalyst to reactor volume [28].

In 2012, Velocys Inc. first demonstrated a structured microchannel reactor in an operating environment for various types of syngas including coal-to-liquids, gas-to-liquids, and biomass-to-liquids. A few years later, Velocys' microchannel FT technology was used at the ENVIA plant that had a production capacity of around 19 million gallons per year.

Microchannel reactors contain low catalyst load per unit volume, which is considered to be one of the major disadvantages, requiring multiple parallel reactors in a plant, and other utility and catalyst processing units along with the requirement for clean boiler feed water. It is often claimed that microchannel FT processes are inexpensive and more productive than traditional reactors, but the cost effectiveness differs from project to project. Technology developers need to compare various technology options to assess the risks and opportunities properly. The FT unit typically only represents about 10%–20% of the capital cost of a typical plant [14]. Therefore, syngas conversion rates and selectivity to the favorable products are critically important. A disadvantage of having a large number of small capacity reactors is that the balance of plant costs can quickly mount up for syngas condensing, catalyst regeneration, and process monitoring and control. For these reasons, many recent project proposals have preferred fixed-bed multitubular reactors.

3.3.8 3D printed

In other microchannel designs, the complicated microchannel manufacturing (e.g., Velocys Inc.) can be hugely simplified by printing optimized catalyst channels using three-dimensional (3D) printing. In 3D printing (additive manufacturing), digital models are used to fabricate free-standing 3D structures. Target structure is initially drawn by computer-aided design software and then transferred into a digital file, which is used to instruct the 3D printer to print the material in layers [29]. 3D structures are built by depositing two-dimensional (2D) layers on top of each other using a computer-controlled nozzle feeder [29]. In 3D printing, the printer inks can be made of ceramics, polymers, colloidal systems, or semiconductor materials [5]. 3D printing is an exciting development for catalytic technology. The catalyst particles can be placed on the millimeter-sized reaction channels and catalytic paste can be externally manipulated to achieve the highest porosity needed for minimal pressure drop and maximal diffusion [29, 30].

Konarova et al. have shown catalytic properties of 3D-printed Ni—Mo metal nanoparticles within carbon scaffold [31], followed by a highly compact channeled reactor with printed catalyst $NiMoS_2$ on corrugated chevron heat-exchanger plates. The reactor consisted of two catalysts and two cooling plates and was fabricated by vacuum brazing (copper) [32]. More recently, Tsubaki and coworkers [33] have elaborated the concept of Fe and Co-based self-catalytic reactor (SCR) system via selective laser sintering (SLS). The structural studies of the Co-SCRs have shown that multiple catalytic functions for FT synthesis can be achieved by tuning different printing structures. The authors have claimed a simple, fast, and practical technology to optimize the synergies between reactors and catalysts and facilitate the design of future catalytic systems.

3.4 Challenges and opportunities for biofuels

The industrial development of FT process technology has mostly focused on the optimization of relatively large projects using natural gas or coal, where low feedstock costs and economies of scale are exploited to achieve acceptable economics [21]. The use of biogenic feedstocks, such as biomass and wastes, poses new opportunities and challenges that must be seized and addressed to enable FT to be a competitive technology. A major driver for considering biogenic feedstocks is that the hydrocarbon products will be classed as renewable [9]. However, the major disadvantages of using biogenic feedstocks as compared to natural gas are (i) much smaller feedstock availability, (ii) capital costs at least twice as high, and (iii) lower overall efficiencies. To overcome these disadvantages, many innovations have been proposed for BTL and waste-to-liquid (WTL) plants.

3.4.1 Distributed processing

Distributed processing of the biomass feedstock is one potential solution to address the challenge of processing a low-energy density material from a relatively wide area, such as farming lands. The intent is to first upgrade the biomass into a denser energy carrier via pyrolysis or torrefaction units located close to the feedstock source and transport the energy carrier to a central plant for upgrading into final products. Although many studies have been undertaken, the economics of using agricultural biomasses in this way does not look very attractive given current prices for the cost of supplying the biomass and the price paid for the fuel products [34, 35]. Distributed processing may be of some interest in the future for converting wastes, such as municipal solid waste, however, as hundreds of thousands of tons per annum of these wastes can already be brought to a single location and the feedstock attracts a waste gate fee, stand-alone waste-to-liquid plants using FT are already economically viable in some jurisdictions.

3.4.2 Gas loop configuration

Optimization of the FT gas loop is another area of interest in smaller projects—ostensibly to reduce plant complexity and capital costs [14]. In small-scale projects, internal recycling of some of the FT tail gas is likely. It is worth noting that in industrial plants, some 70%–80% of the cost is associated with producing clean syngas, so it is imperative that the high-cost syngas is efficiently converted into the desired products. Therefore, C_{5+} selectivity and overall product yield are critical factors that need to guide process design. An external recycle whereby some of the FT tail gas is recycled back to the gasifier is worth considering.

3.4.3 Reactor/catalyst selection

The considerations for reactor/catalyst selection for BTL and WTL plants are no different than for GTL and coal-to-liquid plants, except for the processing scale. It is desirable to have high C_{5+} selectivity and a high reactor volume productivity. This leads to favoring LTFT processes, especially Co-based catalysts, in fixed-bed multitubular and microchannel reactors or slurry bubble column reactors. Table 3.6 shows the main characteristics for each of the common FT reactor types and their suitability for BTL and WTL projects. Fluidized-bed reactors are not favored due to the requirement to operate at low α and their relatively high complexity. Multitubular reactors are considered to be one of the best choices due to simple reactor design, high α operation, upset tolerability, long catalyst lifetime, and easy catalyst replacement. Slurry bubble columns can also be considered, especially when the production capacity starts to become significant (>5000 bpd). Microchannel technology is potentially another option, however, catalyst replacement in some designs is very difficult, requiring new infrastructure to be built, which is unattractive. The feasibility of maintaining very clean coolant water quality for use in microchannels is also a major consideration that should not be overlooked when considering this reactor class. 3D printing of catalysts has promise, but the overall value it can bring to an industrial facility is not yet very clear. Catalyst replacement is one area that needs to be solved in commercial designs.

Table 3.6 Performance indicators of the main types of reactors used for Fischer-Tropsch synthesis and their suitability for smaller-scale BTL and WTL projects [14].

Description	Fixed-bed			Slurry bubble column	Fluidized bed	
	Multitubular	Microchannel	3D printed		Fixed	Circulating
Nature of reactor	PFR	PFR	PFR	CSTR	CSTR	CSTR
Reaction phase	g or g+1	g or g+1	g or g+1	g+1	g	g
Catalyst particle size (mm)	>2	<0.1	<0.1	<0.1	<0.1	<0.1
Mass transfer limitation	High	Low	Low	Medium	Medium	Medium
Heat transfer limitation	High	Low	Low	Low	Medium	Medium
Online catalyst replacement	No	No	No	Possible	Possible	Possible
Offline catalyst replacement	Easy	Hard	Hard	Easy	Easy	Easy
Catalyst mechanical strength	Low	Low	Low	Medium	High	High
Catalyst product separation	Easy	Easy	Easy	Hard	Medium	Medium
Scale-up risk	Low	Low	Low	Medium	Medium	Medium
Economy of scale	Medium	Low	Low	High	Very High	High
Feed poisoning	Local	Local	Local	Global	Global	Global
Turn-down limitations	None	None	None	Catalyst settling	Defluidization	Defluidization
Suitability for BTL/WTL	Excellent	Fair/good	Uncertain	Good	Poor	Very poor

3.4.4 Upgrading and refining

A major consideration for an FT plant is what products to sell to an end customer. There are many options, including upgrading, partial refining, and stand-alone refining [14]. Generally, the upgrading and refining steps have the lowest cost (<10% of plant capex), but can potentially create the most value. However, for smaller-scale plants upgrading is likely preferred, to create intermediate products that are destined for further refining. Partial refining may also be selected depending upon the plant location and local market conditions.

3.4.5 Plant modularization

Modularization is not a new concept and has been used for decades in the petrochemical industry. The main goal is to reduce the overall plant construction costs by moving site construction costs to a construction yard, where labor costs are much lower. The plant is built in prefabricated modules, which are hooked up and commissioned at the plant site.

3.4.6 Automated and unattended operations

For small-scale plants, it is necessary to reduce the operations costs as much as possible, and labor is one of the main cost items in running a plant. A combination of automation, IoT, and remote operations can be applied to reduce the need for onsite operators. A single remote team can be used to monitor and operate multiple plants and can be located in a low-cost jurisdiction. Today, remote operations are routinely used to monitor simple plants, such as landfill gas plants and solar PV plants. As IT systems become more intelligent, they will be increasingly used to monitor and operate more complex plants, such as FT liquids synthesis.

3.4.7 Oxygenated fuels

While not Fischer-Tropsch per se, catalytic conversion of syngas into oxygenates is also a potential option for BTL and WTL plants. The conversion of syngas into ethanol has been known for over a century but has not been commercialized due to poor ethanol yield and selectivity. Mixed alcohols were synthesized in the United States and Germany between 1927 and 1945 at an industrial scale but discontinued due to low petroleum prices. The main challenge is to synthesize robust and highly selective ethanol catalysts. Current research on synthetic ethanol is focused on improving ethanol catalysts.

(i) Rh doped with diverse different elements (Ir, Cu, Ni, Mn, Pt, and Ru). The highest ethanol selectivity reported has been 28% and CO conversion of 43%. The prohibitive cost and limited global supply (around 10 ton/year) of Rh makes this an implausible setup for large-scale operations.

(ii) Cu-based catalysts are cheaper but favor methanol and suffer from particle sintering/deactivation at temperatures of more than 280°C. A Cu–Co-based process has been patented using a multistep ethanol synthesis.

(iii) Recently, high CO conversion up to 84% with 38% ethanol selectivity was reported for Cu–(Pd/Pt) catalysts, in a multistep process. Multiple reactors/units, however, make the multistep synthesis commercially unattractive.

(iv) MoS_2-based catalysts are most attractive because they are (i) tolerant to sulfur impurities existing in syngas, (ii) resistant to coke formation, and (iii) active for water-gas shift reactions reducing water levels in products. Dow Chemicals failed to commercialize this route because of low ethanol selectivity.

Enerkem has constructed a WTL facility in Canada that converts refuse-derived fuel into methanol and ethanol [12].

3.5 Conclusions

The production of synthetic fuels from bioenergy sources can be achieved through the application of the FT process in BTL and WTL plants. While the FT process is challenged by low overall thermal efficiency and high capital costs, it does benefit from being one of the few options available to produce renewable transport fuels. Innovations in catalysts and microchannel and 3D printing reactor designs are emerging, though proven designs using multitubular fixed-bed reactors and slurry bubble columns continue to remain competitive depending upon the project scale. While many BTL projects have been announced, the economics remain challenging, especially in a low-price oil environment. WTL shows a lot of promise through the diversification of the plant's revenue streams and in reducing greenhouse gas emissions associated with wastes when they are sent to landfill. Several pioneer plants are under construction and in operation, and more are expected to be built in the coming decade.

References

[1] C. De Blasio, Fischer–Tropsch (FT) Synthesis to Biofuels (BtL Process), in: Fundamentals of Biofuels Engineering and Technology, Green Energy and Technology, Springer International Publishing, Cham, 2019, pp. 287–306, https://doi.org/10.1007/978-3-030-11599-9_20.

[2] A. Tavasoli, A.N. Pour, M.G. Ahangari, Kinetics and product distribution studies on ruthenium-promoted cobalt/alumina Fischer-Tropsch synthesis catalyst, J. Nat. Gas Chem. 19 (2010) 653–659, https://doi.org/10.1016/S1003-9953(09)60133-X.

[3] I.C. Yates, C.N. Satterfield, Intrinsic kinetics of the Fischer-Tropsch synthesis on a cobalt catalyst, Energy Fuel 5 (1991) 168–173, https://doi.org/10.1021/ef00025a029.

[4] A. de Klerk, D.L. King (Eds.), Synthetic Liquids Production and Refining, ACS Symposium Series, American Chemical Society, Washington, DC, 2011, https://doi.org/10.1021/bk-2011-1084.

[5] P. Basu, Biomass Gasification, Pyrolysis and Torrefaction, second ed., Academic Press, London, United Kingdom, 2013.

[6] C. Higman, M. van der Burgt, Gasification, second ed., Gulf Professional Publishing, Burlington, MA, 2008.

[7] G. Perkins, Production of electricity and chemicals using gasification of municipal solid wastes, in: Waste Biorefinery, Elsevier, 2020, pp. 3–39, https://doi.org/10.1016/B978-0-12-818228-4.00001-0.

[8] M. Shahabuddin, B.B. Krishna, T. Bhaskar, G. Perkins, Advances in the thermo-chemical production of hydrogen from biomass and residual wastes: summary of recent techno-economic analyses, Bioresour. Technol. 299 (2020) 122557, https://doi.org/10.1016/j.biortech.2019.122557.

[9] M. Shahabuddin, M.T. Alam, B.B. Krishna, T. Bhaskar, G. Perkins, A review on the production of renewable aviation fuels from the gasification of biomass and residual wastes, Bioresour. Technol. 312 (2020) 123596, https://doi.org/10.1016/j.biortech.2020.123596.

[10] V.S. Sikarwar, M. Zhao, P. Clough, J. Yao, X. Zhong, M.Z. Memon, N. Shah, E.J. Anthony, P.S. Fennell, An overview of advances in biomass gasification, Energy Environ. Sci. 9 (2016) 2939–2977, https://doi.org/10.1039/C6EE00935B.

[11] A. de Klerk, Aviation turbine fuels through the Fischer–Tropsch process, in: Biofuels for Aviation, Elsevier, 2016, pp. 241–259, https://doi.org/10.1016/B978-0-12-804568-8.00010-X.

[12] Enerkem, Enerkem, 2019. https://enerkem.com/. [WWW Document], accessed 3.31.19.

[13] Fulcrum Bioenergy, Fulcrum Bioenergy, 2019. http://fulcrum-bioenergy.com/. [WWW Document], accessed 3.31.19.

[14] A. de Klerk, Fischer-Tropsch Refining: DE KLERK: FISCHER-TROPSCH O-BK, Wiley-VCH Verlag GmbH & Co. KGaA, Weinheim, Germany, 2011, https://doi.org/10.1002/9783527635603.

[15] S.S. Ail, S. Dasappa, Biomass to liquid transportation fuel via Fischer Tropsch synthesis—technology review and current scenario, Renew. Sust. Energ. Rev. 58 (2016) 267–286, https://doi.org/10.1016/j.rser.2015.12.143.

[16] V. Subramani, S.K. Gangwal, A review of recent literature to search for an efficient catalytic process for the conversion of syngas to ethanol, Energy Fuel 22 (2008) 814–839, https://doi.org/10.1021/ef700411x.

[17] L.C. Almeida, O. Sanz, D. Merino, G. Arzamendi, L.M. Gandía, M. Montes, Kinetic analysis and microstructured reactors modeling for the Fischer–Tropsch synthesis over a co–re/Al2O3 catalyst, Catal. Today 215 (2013) 103–111, https://doi.org/10.1016/j.cattod.2013.04.021.

[18] Shell, Gas-to-Liquids, 2020. https://www.shell.com/energy-and-innovation/natural-gas/gas-to-liquids.html. [WWW Document]. accessed 11.18.20.

[19] E. de Smit, B.M. Weckhuysen, The renaissance of iron-based Fischer–Tropsch synthesis: on the multifaceted catalyst deactivation behaviour, Chem. Soc. Rev. 37 (2008) 2758, https://doi.org/10.1039/b805427d.

[20] S.T. Sie, M.M.G. Senden, H.M.H. Van Wachem, Conversion of natural gas to transportation fuels via the shell middle distillate synthesis process (SMDS), Catal. Today 8 (1991) 371–394.

[21] D.L. King, A. de Klerk, Overview of feed-to-liquid (XTL) conversion, in: A. de Klerk, D.L. King (Eds.), Synthetic Liquids Production and Refining, ACS Symposium Series, American Chemical Society, Washington, DC, 2011, pp. 1–24, https://doi.org/10.1021/bk-2011-1084.ch001.

[22] Z.I. Onsan, A.K. Avci (Eds.), Multiphase Catalytic Reactors: Theory, Design, Manufacturing, and Applications, Wiley, 2016.

[23] J.J.C. Geerlings, J.H. Wilson, G.J. Kramer, H.P.C.E. Kuipers, A. Hoek, H.M. Huisman, Fischer–Tropsch technology—from active site to commercial process, Appl. Catal. A Gen. 186 (1999) 27–40, https://doi.org/10.1016/S0926-860X(99)00162-3.

[24] BP, First Licence for New Waste-to-Fuel Technology, 2018. https://www.bp.com/en/global/corporate/news-and-insights/bp-magazine/new-waste-to-fuel-technology.html. [WWW Document]. First Licence New Waste—Fuel Technol accessed 2.9.20.

[25] EFT, Emerging Fuels Technology Licenses Its Fischer-Tropsch Technology to Red Rock Biofuels, 2018. https://www.prnewswire.com/news-releases/emerging-fuels-technology-licenses-its-fischer-tropsch-technology-to-red-rock-biofuels-300735494.html. [WWW Document]. Emerg. Fuels Technol. Licens. Its Fisch.-Tropsch Technol. Red Rock Biofuels. accessed 2.9.20.

[26] J. Lane, A New Technology Debuts for Renewable Jet Fuel, 2015. http://www.biofuelsdigest.com/bdigest/2015/03/11/a-new-technology-debuts-for-renewable-jet-fuel/. [WWW Document]. New Technol. Debuts Renew. Jet Fuel. accessed 2.9.20.

[27] B. Todić, V.V. Ordomsky, N.M. Nikačević, A.Y. Khodakov, D.B. Bukur, Opportunities for intensification of Fischer–Tropsch synthesis through reduced formation of methane over cobalt catalysts in microreactors, Cat. Sci. Technol. 5 (2015) 1400–1411, https://doi.org/10.1039/C4CY01547A.

[28] R. Guettel, T. Turek, Comparison of different reactor types for low temperature Fischer–Tropsch synthesis: a simulation study, Chem. Eng. Sci. 64 (2009) 955–964, https://doi.org/10.1016/j.ces.2008.10.059.

[29] C. Parra-Cabrera, C. Achille, S. Kuhn, R. Ameloot, 3D printing in chemical engineering and catalytic technology: structured catalysts, mixers and reactors, Chem. Soc. Rev. 47 (2018) 209–230, https://doi.org/10.1039/C7CS00631D.

[30] J. Gascon, J.R. van Ommen, J.A. Moulijn, F. Kapteijn, Structuring catalyst and reactor—an inviting avenue to process intensification, Cat. Sci. Technol. 5 (2015) 807–817, https://doi.org/10.1039/C4CY01406E.

[31] M. Konarova, W. Aslam, L. Ge, Q. Ma, F. Tang, V. Rudolph, J.N. Beltramini, Enabling process intensification by 3 D printing of catalytic structures, ChemCatChem 9 (2017) 4132–4138, https://doi.org/10.1002/cctc.201700829.

[32] M. Konarova, G. Jones, V. Rudolph, Enabling compact GTL by 3D-printing of structured catalysts, Results Eng. 6 (2020) 100127, https://doi.org/10.1016/j.rineng.2020.100127.

[33] Q. Wei, H. Li, G. Liu, Y. He, Y. Wang, Y.E. Tan, D. Wang, X. Peng, G. Yang, N. Tsubaki, Metal 3D printing technology for functional integration of catalytic system, Nat. Commun. 11 (2020) 4098, https://doi.org/10.1038/s41467-020-17941-8.

[34] F. Trippe, M. Frohling, F. Schultmann, R. Stahl, E. Henrich, Techno-economic assessment of gasification as a process step within biomass-to-liquid (BtL) fuel and chemicals production, Fuel Process. Technol. 92 (2011) 2169–2184, https://doi.org/10.1016/j.fuproc.2011.06.026.

[35] M.M. Wright, R.C. Brown, A.A. Boateng, Distributed processing of biomass to bio-oil for subsequent production of Fischer-Tropsch liquids, Biofuels Bioprod. Biorefin. 2 (2008) 229–238, https://doi.org/10.1002/bbb.73.

CHAPTER 4

Hydrodeoxygenation of triglycerides for the production of green diesel: Role of heterogeneous catalysis

Pankaj Kumar[a], Deepak Verma[b], Malayil Gopalan Sibi[b], Paresh Butolia[b], and Sunil K. Maity[c]

[a]Department of Chemical Engineering, Birla Institute of Technology and Science (BITS), Pilani, Hyderabad Campus, Hyderabad, Telangana, India
[b]School of Chemical Engineering, Sungkyunkwan University, Seoul, South Korea
[c]Department of Chemical Engineering, Indian Institute of Technology Hyderabad, Kandi, Sangareddy, Telangana, India

Contents

4.1 Introduction	98
4.1.1 Structure and composition of triglycerides	99
4.1.2 Cracking vs hydrodeoxygenation of triglyceride	101
4.2 Reaction mechanism	101
4.3 Catalysts	105
4.3.1 Noble metal catalysts	106
4.3.2 Sulfided transition metal catalysts	107
4.3.3 Reduced transition metal catalysts	108
4.3.4 Nitride, phosphide, and carbide catalysts	110
4.3.5 Composite catalyst	111
4.3.6 Role of acidity of support	111
4.3.7 Reducible oxides as support	113
4.3.8 Mesoporous support	113
4.3.9 Catalyst deactivation	115
4.4 Effect of process conditions	115
4.5 Coprocessing of triglycerides with the petroleum feedstock	116
4.6 Commercialization status	119
4.7 Process design and economics	120
4.8 Conclusions	122
References	122

Abbreviations

EXAFS	extended X-ray absorption fine structure
HDO	hydrodeoxygenation
HSACS	high acidity of high surface area semicrystalline
LSAC	low surface area crystalline
MMT	million metric tons
OMA	organized-mesoporous-alumina

Hydrocarbon Biorefinery
https://doi.org/10.1016/B978-0-12-823306-1.00013-3

Copyright © 2022 Elsevier Inc.
All rights reserved.

4.1 Introduction

At present, primary energy sources, such as crude oil, coal, and natural gas, supply about 80% of global energy. Such large-scale usages of these primary energy sources are responsible for their rapid depletion, associated air pollution, and drastic change in the world's climate owing to the release of greenhouse gases like CO_2 [1]. Renewable energy is thus considered actively throughout the world to reduce the dependency on fossil fuels and to ensure a healthy environment of our planet [2]. These resources provide energy for various sectors, such as residential, commercial, industrial, and transportation. Diesel, gasoline, and jet fuel are mainly used in the transportation sector. It is a major energy-consuming sector and it contributes about 28% of the energy consumption in the world. The demand for these transportation fuels is also rising continuously due to the growth of the world's population and advancement in the standard of living. The manufacturing of renewable transportation fuels is thus needed for the sustainability of human civilization.

Biomass is the only carbon-based renewable energy resource on the planet earth, and it is also capable of producing all kinds of transportation fuels, called biofuel. Biomass currently provides more than 10% of the world's energy. The contribution of renewables in India is about 27% of the energy consumption with around 22% contribution from biomass and waste [3]. As biomass has tremendous potentials, India has projected to blend 20% biofuels with transportation fuels [3]. Biomass is commonly categorized into three distinct types as per their chemical structure: triglycerides, lignocellulose, and sugar and starch. The triglyceride biomass is, however, gaining huge attraction in the biorefinery compared to other biomass. It is mainly due to its simple structure, the presence of long carbon chain length, lower oxygen content (around 15 wt%), and lower functionality (only ester bonds) than other biomass. The triglycerides are made up of C_8-C_{24} linear fatty acids with C_{16} and C_{18} being dominating ones [4]. The carbon chain length in the fatty acids falls in the range of diesel (C_{12}-C_{22} linear and branched paraffin, naphthenes, and aromatics). The triglycerides are thus considered as ideal biomass for the production of diesel-range biofuels. The triglycerides are, however, high molecular weight compounds with high viscosity and pour/cloud point. Pure triglycerides are thus not suitable as a biofuel. To overcome this problem, triglycerides are generally converted to fatty acid methyl esters, known as biodiesel, by transesterification reaction with methanol in the presence of alkali catalysts. Biodiesel, however, contains about 10%–15% structural oxygen. The presence of oxygen causes lesser calorific value and hence lesser fuel mileage of biodiesel than diesel. Moreover, this factor causes slightly higher viscosity and pour/cloud point of biodiesel than diesel. The mismatch in fuel properties makes pure biodiesel inappropriate for use in current-generation diesel engines [2, 5]. The biodiesel mixing with diesel is thus restricted to 15 (v/v)% for use in the present diesel engines.

Therefore, there is a strong need to remove oxygen from triglycerides to produce diesel-equivalent biofuel, known as hydrocarbon biofuel or green diesel. The green diesel has comparable fuel properties and calorific value with diesel and hence, neat green diesel can be used in existing diesel engines. For a competitive price scenario, the identical fuel mileage will eventually build the confidence of end users for green diesel. Therefore, the market penetration of these types of biofuels will become easier. This manufacturing concept of biofuels equivalent to current transportation fuels with identical calorific value is known as hydrocarbon biorefinery. Hydrodeoxygenation (HDO) is a promising technology in hydrocarbon biorefinery for eliminating oxygen from triglycerides and producing green diesel with a high yield and negligible loss of carbon. This chapter thus covers various aspects of HDO of triglycerides, such as reaction mechanism, the role of heterogeneous catalyst and supports, and the effect of process conditions on the fuel properties, composition, and yield of green diesel. The chapter further provides a glimpse of the commercial and techno-economic status of green diesel.

4.1.1 Structure and composition of triglycerides

The triglycerides, also known as triacylglycerols, share a common chemical structure, as shown in Fig. 4.1. Vegetable oils, algal oil, animal fats, waste cooking oils, etc. are the major sources of triglycerides. They are important biomass for hydrocarbon biofuels due to their high energy density. Triglycerides are the fatty acids ester of glycerol (1,2,3-propanetriol, often called glycerine) [6]. Sometimes, they contain a significant amount of free fatty acids. The triglycerides contain linear fatty acids in the range of C_8-C_{24} (major C_{16} and C_{18}). Table 4.1 presents the representative fatty acids composition of the commonly used animal fats and vegetable oils. Generally, vegetable oils contain a large quantity of unsaturated fatty acids, such as oleic acid (18:1), linoleic acid (18:2), and linolenic acid (18:3). In contrast, saturated fatty acids dominate animal fats, such as palmitic acid (16:0) and stearic acid (18:0) [7,8]. Vegetable oils are classified as edible and nonedible based on their usages [7]. The edible oil as a feedstock for hydrocarbon biofuel, however, poses a threat to the food crisis and food vs fuel issues. Nonedible oils, such as Jatropha, Karanja, Mahua, Neem, algal, and waste cooking oil, are thus commonly used for the production of hydrocarbon biofuels. These feedstocks have no

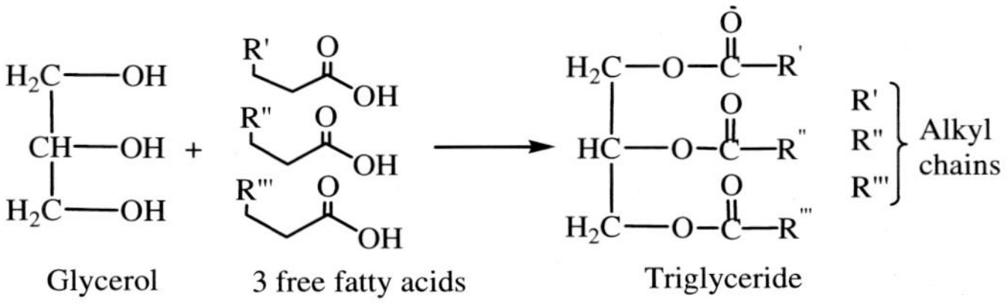

Fig. 4.1 Chemical structure of triglycerides.

Table 4.1 Fatty acids composition of vegetable oil, wt% [7,9–12].

	≤10	12:0	14:0	16:0	16:1	18:0	18:1	18:2	18:3	20:0	20:1	22:0	22:1	24:0
Coconut	14.5	47.9	18.4	8.4	0	1.65	5.7	1.4	0	0	00	0	0	0
Cottonseed	0	0	0	26.2	0	1.3	13.3	59.1	0	0	00	0	0	0
Corn	0	0	0	6.5	0.6	1.4	65.6	25.2	0.1	0.1	0.1	0	0.1	0.1
Canola	0	0	0	4.4	0	2.05	61.6	20.4	8.5	0.7	1.4	0.3	0.1	0
Groundnut	0	0	0	10.4	0	3.9	47.1	27.6	0	3	1.6	4.0	0.7	1.7
Olives	0	0	0	11.6	1	3.1	75.0	7.8	0.6	0.3	0	0.1	0	0.1
Palm	0	0.1	0.7	36.7	0.1	6.6	46.1	8.6	0.3	0.4	0.2	0.1	0	0.1
Peanut	0	0	0.1	8.0	0	1.8	53.3	28.4	0.3	0.9	2.4	3.0	0	1.8
Rapeseed	0	0	0	4.9	0	1.6	33.0	20.4	7.9	0	9.3	0	23	0
Safflower	0	0	0	7.9	0	1.9	12.6	77.5	0	0	0	0	0	0
Soybean	0	0	0	11.3	0.1	3.6	24.9	53.0	6.1	0.3	0.3	0	0.3	0.1
Sesame	0	0	0	11.2	0	5.2	41.9	40.1	0	0	0	0	0	0
Sunflower	0	0	0	6.2	0.1	3.7	25.2	63.1	0.2	0.3	0.2	0.7	0.1	0.2
Lard	0	0	2	27	4	11	44.0	11	0	0	0	0	0	0
Tallow	0.1	0.2	2.6	23.9	2.6	17.8	41.4	4.3	0.9	0.2	0.6	0.1	0.1	0
Jatropha	0	0	1.4	14.6	1.47	7.4	41.4	35.4	0.2	0	0.3	0	0	0
Waste oil	0	0	0.2	9.3	0.5	3.9	54.6	29.7	0.3	0	0	0	0	0
Microalgae	0	0	0.04	4.4	0	4.4	32.2	56.2	0	0	0.4	0	0	0

conflicts with the food and are preferable compared to edible oils. Micro and macroalgae-derived oils have emerged as the recent sources of triglycerides with tremendous potentials. The high expectations are due to their faster growth rate, higher productivity, and greater oil content than oil crops [8,9]. The compatibility with nonagricultural land and wastewater for their cultivation are other advantages of this feedstock. The high cost of algal oil is, however, the primary bottleneck.

4.1.2 Cracking vs hydrodeoxygenation of triglyceride

Catalytic cracking in the presence of zeolites (such as HZSM-5) and thermal cracking are the possible technologies to remove oxygen from triglycerides. These processes are carried out in the absence of air or hydrogen under atmospheric pressure and produce gasoline-range biofuels, known as green gasoline [13]. The thermal cracking of triglycerides is carried out at high temperatures in the range of 573–773 K. In these processes, the oxygen is mainly removed as CO_2 and CO by decarboxylation and decarbonylation reactions, respectively. The cracking is, however, the dominating reaction in catalytic cracking with the removal of oxygen in the form of CO_2, H_2O, and CO. As cracking reactions are severe, these technologies result in a large quantity of coke formation, huge loss of carbon as the gaseous product, and a low yield of green gasoline [14, 15].

In contrast, HDO is the most competent technology for the removal of oxygen from triglycerides. It is analogous to the existing hydroprocessing technology of petroleum refinery. It has the advantages of coprocessing of triglycerides with various crude oil fractions and use of existing refinery technology with their infrastructures. Contrary to transesterification reaction, this technology is insensitive to the presence of free fatty acids in the triglycerides. HDO occurs catalytically over supported metal catalysts at a temperature of 523–773 K and 25–50 bar hydrogen pressure. In this route, the deoxygenation of triglycerides generally occurs through decarbonylation and HDO reactions with the formation of CO and water, respectively. The cracking reactions are insignificant in the HDO process. The carbon loss in the form of the gaseous product is thus minimal in this process. This process produces green diesel (oxygen-free linear chain diesel-range hydrocarbons) in high yield. The fuel properties of green diesel are comparable with diesel and far better than biodiesel (Table 4.2).

4.2 Reaction mechanism

The removal of oxygen can be achieved by cleavage of the C—O bond or C—C bond. The selective cleavage of these bonds is governed by the reaction conditions and the types of catalyst used. To control the selectivity of these bond's cleavage, several supported metal (noble, transition, and bimetal) catalysts have been employed for the HDO of triglycerides. The deoxygenation of triglycerides in the presence of hydrogen involves multistep reactions over these supported metal catalysts, as shown in Fig. 4.2. In the first step,

Table 4.2 Fuel properties of green diesel, biodiesel, and diesel [16–21].

	Biodiesel	Diesel	Green diesel					
			Ecofining	NEXBTL	Karanja oil	Jatropha oil	Rapeseed oil	Sunflower oil
Yield, %	–	–		–	80	84	–	–
Specific gravity	0.88	0.84	0.78		0.857	0.81	0.79	0.81
Viscosity, cP	4–5	2–3	–	2.9–3.5	3.81	–	3.92	2.65
Sulfur, ppm	<1	<10	<1	<10	–	–	–	–
Lower heating value, MJ/kg	38	43	44	44	39.94	44	–	–
Flash point, K	–	–	–	–	411	351	398	322
Fire point, K	–	–	–	–	418	–	–	–
Cloud point, K	268–288	268	253–293	248–268	–	–	298–300	262
Pour point, K	–	–	–	–	282	276	298	255
Cetane number	50–60	40	70–90	84–99	–	–	–	65
Stability	Marginal	Good	Good	–	Good	–	–	–
Oxygen content, wt%	11	0	–	0	0.53	–	–	–
H/C	–	–	–	2.14	1.97	–	–	–
Chemical composition, wt%								
<C_{15}	–	–	–	–	12.8	13.7	–	–
C_{15}	–	–	–	–	12.9	11	–	–
C_{16}	–	–	–	–	5.2	6.6	5.8	–
C_{17}	–	–	–	–	42.3	51.0	24.9	–
C_{18}	–	–	–	–	4.3	17.7	58.1	–
>C_{18}	–	–	–	–	18.1	–	4.5	–

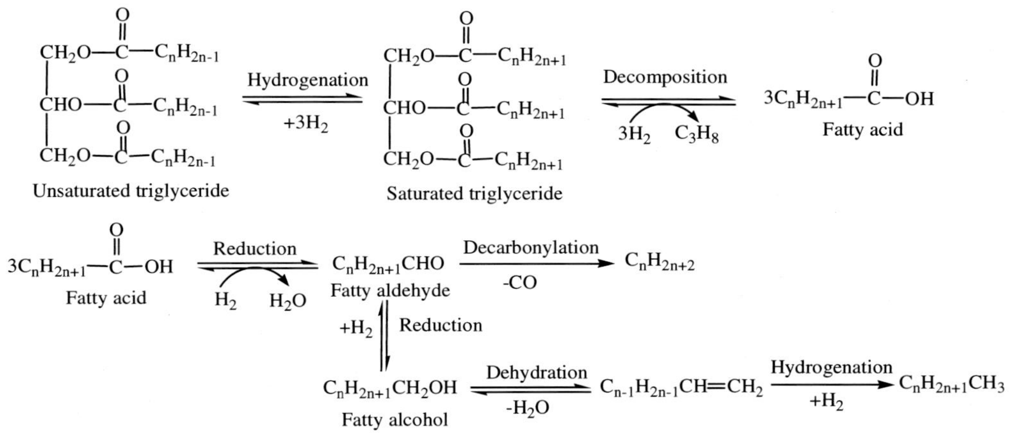

Production of green diesel   103

Fig. 4.2 Reaction mechanism for HDO of triglycerides [4, 18, 22–26].

the unsaturated fatty acid backbone of triglycerides undergoes the hydrogenation reaction [22]. The saturated triglycerides then decompose into fatty acids. In this step, propane is coproduced from the three-carbon glycerol backbone of the triglyceride by deoxygenation reaction. Both hydrogenation and triglyceride decomposition reactions occur over metal sites of the catalyst. Both of these reactions are quite fast as compared to subsequent deoxygenation of fatty acids [22]. A high concentration of fatty acids is thus observed during the initial stage of the reaction.

There is a possibility of deoxygenation of fatty acids through decarboxylation reaction. It is a dominating reaction in thermal cracking at elevated temperatures and occurs in absence of hydrogen. To elucidate this possibility, the reaction of stearic acid was performed using $Ni/\gamma-Al_2O_3$ catalyst at 623 K in the absence of hydrogen [4]. However, the conversion of stearic acid was negligible under high nitrogen (inert) pressure. This evidence eliminates the possibility of decarboxylation reaction under HDO reaction conditions. The removal of oxygen from fatty acids thus occurs primarily through reductive deoxygenation reactions, where fatty acids reduce to their respective fatty aldehydes by H_2 [4, 23]. This reaction occurs over metal or bifunctional reducible oxide (ZrO_2, TiO_2, and CeO_2) supported metal catalysts [27]. In the case of the metal catalyst, fatty acid first adsorbs via C—O and C=O bonds (Fig. 4.3). The hydrogen molecule separately chemisorbs on the metal sites, resulting in metal hydride formation. The hydride is then transferred to the carbon of the —COOH group. The OH group from fatty acid is thus eliminated as a water molecule and fatty acid is reduced to fatty aldehyde. In the case of reducible oxide supported metal catalysts, the fatty acid adsorbs on the oxygen vacant sites of the reducible oxide through the oxygen atom with the formation of carboxylate. Simultaneously, dissociative chemisorption of molecular hydrogen occurs on the metal sites of the catalyst. The carboxylate is then converted to ketene by the abstraction of α-hydrogen, followed by hydrogenation to aldehyde.

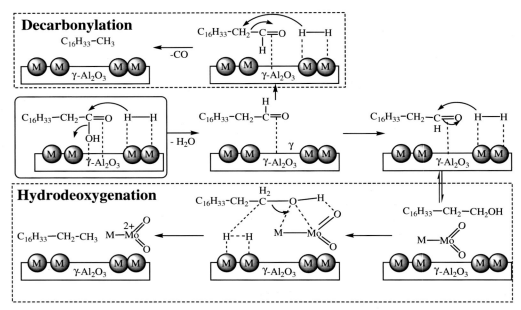

Fig. 4.3 Molecular mechanism for HDO of fatty acid over supported metal and bimetal catalysts [23].

The deoxygenation of fatty aldehyde then proceeds following two parallel reaction routes (Fig. 4.2) [23, 24]. In one of the routes, the oxygen from fatty aldehyde is eliminated following the decarbonylation reaction. It results in alkane which is one carbon less than the parent fatty acid. In another route, fatty aldehyde undergoes further reduction to fatty alcohol. The fatty alcohol then undertakes the dehydration reaction to the corresponding alkene, followed by the hydrogenation reaction, resulting in the formation of alkane. The carbon number of alkane in this route is the same as the parent fatty acid. The decarbonylation, hydrogenation, and reduction reaction occur on metal sites of the catalyst, while dehydration reaction occurs over the acidic sites of the catalysts [4]. For HDO of stearic acid, the octadecanol was identified as one of the intermediates over supported metal catalyst [4], while octadecanal was additionally observed over supported NiMo and CoMo catalysts [23]. The hydrogenation of olefins is a relatively fast reaction under prevailing reaction conditions and the olefin intermediate in the HDO route was thus not observed. To validate these reaction routes, the HDO reaction of stearic acid was performed using Ni catalyst supported on neutral (silica), mild acid (γ-alumina), and strong acid (HZSM-5) (Fig. 4.4) [4]. The reaction proceeds primarily through the decarbonylation route over Ni/SiO$_2$ and Ni/γ-Al$_2$O$_3$ catalysts with heptadecane as the primary alkane product. The reaction, however, follows both decarbonylation and HDO route over Ni/HZSM-5 catalyst with a substantial quantity of octadecane as the product. The conversion of stearic acid was also higher over Ni/HZSM-5 due to its bifunctionality.

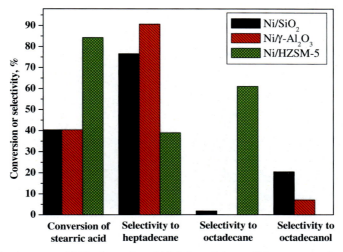

Fig. 4.4 Role of supports on HDO of stearic acid [4]. Reaction conditions: conversion at 6 h reaction time and selectivity at 40% conversion of stearic acid, 0.18 kmol/m³ stearic acid, 0.5 (w/v)% catalyst, 543 K, 8 bar initial H₂ pressure.

The deoxygenation of fatty alcohol, however, occurs on the $M^{2+}\cdot MoO_2$ (M=Ni/Co) sites of NiMo and CoMo catalysts [23]. The coordination of electron-deficient M^{2+} with Mo reduces electron density on Mo. The electronegative oxygen of —CH₂OH thus forms a bridge with Mo and M^{2+} (Fig. 4.3). The hydride then transfers to the carbon atom, thereby reducing fatty alcohol to an alkane. The high hydrogen pressure favors this route [28]. The water formed in this reaction reduces the catalytic activity and stability of catalysts. The decarbonylation reaction dominates over mono-metal (Pt, Pd, Ni, Co, etc.) catalyst, whereas HDO reaction occurs over bimetal (NiMo and CoMo) catalysts [4, 22–24, 29]. The reaction was further performed using the tristearate/tripalmitate mole ratio of 2:1 [22]. The stearic acid/palmitic acid and heptadecane/octadecane mole ratio remained 2:1 for a wide range of reaction conditions over Ni/γ-Al₂O₃ catalyst. This result indicates that the reaction routes are independent of fatty acid chain length. The methane was also formed via methanation reaction. The cracking reaction is also significant over acid catalysts at elevated reaction temperatures.

4.3 Catalysts

The design and selection of appropriate catalysts are crucial to achieve maximum deoxygenation of triglycerides/fatty acids/oxygenated intermediates, high yield of green diesel, and high selectivity to targeted hydrocarbons. The design of the catalysts in terms of metals and supports are based on their catalytic activity, time-on-stream stability, and desired fuel properties of green diesel. Fig. 4.5 illustrates a brief overview

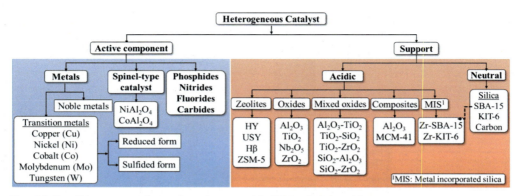

Fig. 4.5 Possible supported metal catalysts used in HDO of triglycerides.

of supported metal catalysts used in HDO of triglycerides. The catalyst properties, such as active metals, the nature of supports, acidity, porosity, and textural properties, are critical factors in the performance of these catalysts. The active metals and support also decide the reaction routes, as discussed in the reaction mechanism section. Zeolites (HZSM-5, HY, USY, SAPOs, and AlPOs) and metal oxides (Al_2O_3 and TiO_2) are widely used in petroleum refinery as either catalyst or support. The bulky structure of oxygenated intermediate compounds in HDO reaction also forced to develop supports with different textural properties and pore size, such as mesoporous materials. Therefore, this chapter presents a detailed discussion of the above mentioned factors for HDO of triglycerides. Generally, supported noble metal (Pd, Pt, Ru, and Rh) and transition monometal and bimetal (Ni, Co, Mo, CoMo, NiMo, NiW, and CoW) have been employed for HDO of triglycerides, as shown in Fig. 4.5. The transition metal and bimetal catalysts are used as either sulfided or reduced form.

4.3.1 Noble metal catalysts

The noble metal catalysts show excellent catalytic activity [25, 26, 30]. For example, Snare et al. studied HDO of stearic acid over carbon-supported Pd, Pt, Ru, Rh, and Ir catalysts [31]. Carbon-supported 5 wt% Pd showed 100% conversion with more than 98% selectivity to C_{17} alkane. Both decarboxylation and decarbonylation reaction was observed over supported noble metal catalysts. The decarboxylation reaction was dominant over the Pd/C catalyst, while the decarbonylation reaction was dominating over the Pt/C catalyst. For normalized metal content, HDO activity of stearic acid was in the descending order of Pd, Pt, Ni, Rh, Ir, Ru, and Os. However, the large-scale application of noble metals is not economical due to their high cost. The inexpensive transition metals were thus employed widely for HDO of triglycerides.

4.3.2 Sulfided transition metal catalysts

The sulfided transition metal catalysts were employed for HDO of triglycerides due to their similarity with hydroprocessing technology. For example, Kaluža et al. studied HDO of rapeseed oil in the fixed-bed reactor over γ-Al$_2$O$_3$ supported sulfided mono-metal (Co, Ni, and Mo) and bimetal (CoMo and NiMo) catalysts [32]. The catalytic activity was in the order of NiMo $\geq$ Mo $\sim$ CoMo $>$ Co $\sim$ Ni. The high catalytic activity of NiMo compared to its monometallic counterpart was due to the synergy between Ni and Mo. Hence, NiMo-S$_x$/γ-Al$_2$O$_3$ and CoMo-S$_x$/γ-Al$_2$O$_3$ catalysts are considered actively for HDO of triglycerides [33].

The support plays a beneficial role in the dispersion of metal and the formation of active sites. For example, coprocessing of soybean oil with refinery oil was studied over mesoporous SiO$_2$-Al$_2$O$_3$ and γ-Al$_2$O$_3$ supported sulfided NiW and NiMo catalysts [34]. The NiW/SiO$_2$-Al$_2$O$_3$ catalyst displayed the highest catalytic performance in terms of hydrocracking activity and selectivity toward green diesel and green kerosene. On the other side, sulfided NiMo/γ-Al$_2$O$_3$ was selective to HDO reaction and mainly produced green diesel due to its low cracking ability. The activity of alumina supported catalysts measured in terms of soybean oil conversion was in the following descending order: sulfided NiMo (92.9%) $>$ 4.29 wt% Pd (91.9%) $>$ sulfided CoMo (78.9%) $>$ 4.95 wt% Pt (50.8%) $>$ 3.06 wt% Ru (39.7%) [35]. The isomerization and cracking reactions were dominant over sulfided CoMo catalyst [35]. The sulfided Mo catalyst showed four times higher conversion of rapeseed oil compared to Ni and Co [32]. The incorporation of Zn in the Mo catalyst, however, reduced the catalytic activity [36].

In sulfided catalysts, the lower availability of active hydrogen leads to the olefin formation. The olefins participate in polymerization reaction through carbonium ion over the acidic sites, resulting in coke formation. This problem can be resolved by providing active hydrogen. In this regard, Sibi et al. brought up a novel approach for multifunctional catalyst design where Pt was incorporated inside the sodalite cage encapsulated in Ni and Mo supported on ZSM-5 [37]. Sodalite cage (2.8 Å) provides the access to molecular hydrogen (2.8 Å kinetic diameter) but prohibits the access of poisons (such as sulfur and nitrogen compounds and metals present in the feedstock) and coke precursors. A molecular modeling study confirmed the accessibility of active hydrogen (formed via dissociative adsorption of molecular hydrogen on Pt sites) to bulky molecules through the spillover effect. The almost complete conversion of triglycerides was observed over NiMo/Pt@SOD@HZSM-5 catalyst with six times higher kerosene yield than NiMo/HZSM-5 catalyst. It was due to the influence of spillover H$_2$ from Pt clusters inside sodalite cages. However, sulfided catalysts pose the problem of sulfur leaching-induced deactivation and product contamination by sulfur [34, 38, 39]. Therefore, sulfur-free catalysts attracted huge attention [4,8,22,39–42].

4.3.3 Reduced transition metal catalysts

The reduced transition metals are widely used in HDO of triglycerides to overcome the limitations of sulfided catalysts [8,18,26,40,43,44]. The catalytic activity of Ni was observed to be higher compared to Fe, Cu, and Co [13]. The lower catalytic activity of Co was due to the rapid catalyst deactivation and agglomeration of Co. The incorporation of Mo, W, Zr, and Ce promotes catalytic activity due to their multivalency. These promoters enhance the catalytic activity by activating oxygen-containing functional groups. The incorporation of Mo in Ni catalysts also improves the reducibility of metal oxides, resulting in increased active metal content [40]. Similar to a sulfided form, the reduced form of NiMo catalyst also exhibits higher catalytic activity than CoMo [32, 45]. The NiMo catalyst also showed better stability compared to CoMo during coprocessing of waste cooking oil with heavy atmospheric gas oil [46].

The Ni/Mo mole ratio plays an important role in the formation of various active species that influence both catalytic activity and reaction mechanism. Both Ni and NiMo alloy coexist in NiMo catalyst, but NiMo alloy is catalytically superior to Ni species [24]. The Ni catalyst thus showed lower catalytic activity compared to the NiMo catalyst for stearic acid HDO (Fig. 4.6). The NiMo catalyst further showed enhanced catalytic activity with the increase in Ni/Mo mole ratio due to the enhancement of NiMo alloy in the catalyst. Despite the lower conversion of stearic acid, Ni catalyst showed the highest catalytic activity for the conversion of intermediate oxygenated compounds (fatty alcohol and fatty aldehyde) to alkanes (Fig. 4.6). The conversion of intermediate

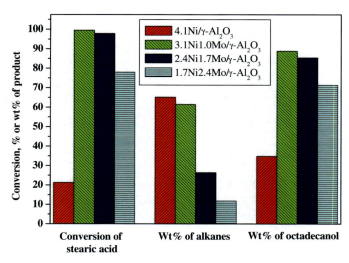

Fig. 4.6 Effect of Ni/Mo mole ratio for HDO of stearic acid [24]. Conditions: conversion at 2 h reaction time, wt% at 65% stearic acid conversion, 0.18 kmol/m^3 stearic acid, 543 K, 20 bars H$_2$, and 0.5 (w/v)% catalysts.

oxygenated compounds to alkanes was also enhanced with a rising amount of Ni in the NiMo catalysts. These results demonstrate that the reduction of fatty acid to fatty alcohol is faster over NiMo alloy than Ni species, while conversion of fatty alcohol to alkane is rapid over Ni species than NiMo alloy.

The calcination temperature during the synthesis of supported metal catalysts plays an important role in metal–support interaction that influences the metal dispersion. The metal dispersion was enhanced for CoMo, NiMo, and Ni catalyst and reduced for Co catalyst with increasing calcination temperature [23]. Hence, the activity of Ni, NiMo, and CoMo catalyst was increased with the rise in calcination temperatures, while it was decreased for Co catalyst during HDO of stearic acid. The optimum calcination temperature was reported to be 1073, 773, and 973 K for Ni, Co, and NiMo/CoMo catalysts, respectively. At these calcination temperatures, the highest conversion of stearic acid was observed over Co and CoMo catalysts followed by NiMo and Ni (Fig. 4.7). However, Ni and Co catalysts showed very high selectivity to alkanes, while CoMo and NiMo catalysts showed high selectivity to octadecanol. Moreover, heptadecane was the primary alkane over Ni catalyst, while octadecane was the major alkane over NiMo and CoMo catalyst. The significant quantity of both heptadecane and octadecane was, however, observed over Co catalyst. The nonsulfided NiMo and CoMo catalysts are also selective toward paraffin and isoparaffin with high deoxygenation activity which is important for improving cold flow properties of the green diesel [47]. However, the reduced form of metal catalysts favors methanation reaction at elevated temperatures [48].

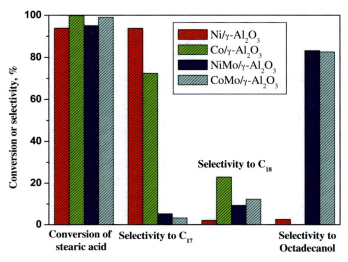

Fig. 4.7 Performance of Ni, Co, NiMo, and CoMo catalysts for HDO of stearic acid [23]. Conditions: conversion at 6 h, selectivity at around 100% conversion of stearic acid, 0.18 kmol/m^3 stearic acid, 0.5 (w/v)% catalysts, 20 bars hydrogen, and 543 K.

4.3.4 Nitride, phosphide, and carbide catalysts

The undesirable methanation and cracking reactions are significant over reduced metal catalysts. The catalyst development is thus needed to overcome these limitations. In this context, nitrides, carbides, and phosphides form of metal are promising catalysts for HDO reaction [8]. The catalytic activity of carbide and nitride catalysts depends on the presence of anionic vacancies (generates by thermal reduction process) and Brønsted acidity of support. The γ-Al_2O_3 supported Mo_2N catalyst exhibited high catalytic activity for HDO of oleic acid and canola oil with almost complete conversion to long-range hydrocarbons and stability up to 450 h [49]. While VN/γ-Al_2O_3 catalyst showed good deoxygenation activity by decarbonylation and decarboxylation pathways.

Souza Macedo et al. applied two different approaches to synthesize Mo_2C-based catalyst [50]. The first approach led to cubic α-MoC_{1-x}/CNF (carbon nanofiber) while the carbothermal reduction synthesis approach resulted in hexagonal β-Mo_2C/CNF. The higher activity of α-MoC_{1-x} attributed to either carbide particle size or active site density. Although higher site density gives better catalytic performance for several reactions but the lower site density of α-MoC_{1-x} (0.1906 Mo atom/Å) than β-Mo_2C (0.1402 Mo atom/Å) catalyst shows more space around the Mo atoms in α-MoC_{1-x}. Hence, it makes the Mo atoms more accessible for bulky reactant molecules in this reaction. The α-MoC_{1-x}/CNF catalyst thus favors rapid conversion of intermediates compared to β-Mo_2C/CNF during HDO of stearic acid. At 2 h reaction time, around 60% conversion of stearic acid was observed over α-MoC_{1-x}/CNF while it was 40% over β-Mo_2C/CNF. Hollak et al. synthesized W_2C and Mo_2C catalysts using a carbothermal approach using the same carbon support [51]. The HDO study of oleic acid demonstrated higher olefin-rich products over W_2C/CNF catalyst than Mo_2C/CNF.

The incorporation of phosphorus promotes catalytic activity due to enhanced metal dispersion. It reduces coke formation and helps to create new Brønsted and Lewis acid sites on support. Yang et al. studied HDO of methyl oleate over Ni_2P and Ni catalysts supported on SBA-15 [52]. It was observed that Ni_2P nanoparticles dispersed uniformly on SBA-15 support compared to only Ni metal. The activity of the Ni_2P phase was explained using the results of the extended X-ray absorption fine structure (EXAFS) study [53]. The presence of fourfold coordinated tetrahedral Ni sites [denoted as Ni(1)] and fivefold coordinated square pyramidal [denoted as Ni(2)] over Ni_2P crystallite surface are mainly responsible for catalytic activity. EXAFS line shape analysis indicates that both Ni(1) and Ni(2) sites are present on large crystallites of Ni_2P but as the crystallite size of Ni_2P increases the number of Ni(2) decreases. The Ni_2P crystallite rich in Ni(2) sites is superior in hydrogenation ability. Among various metal phosphides (Fe, Co, and Ni), silica-supported Ni_2P was found to be highly active than Fe_2P, CoP, and Co_2P [54].

Table 4.3 Performance of Ni-alumina composite catalyst [55].

	15 wt% Ni-alumina composite catalyst	15 wt% Ni/γ-Al$_2$O$_3$	15 wt% Ni/ mesoporous alumina
Surface area, m^2/g	213	154	150
Metallic surface area, m^2/g	7.15	6.76	3.92
Conversion of oxygenates at 6 h, %	46.7	41.6	40.0

Conditions: 613 K, 10 (w/v)% karanja oil, and 10 (w/w)% catalyst, 20 bars hydrogen.

4.3.5 Composite catalyst

The ordered mesoporous nickel-alumina composite catalysts displayed high surface area and strong metal-support interaction with enhanced metal dispersion compared to supported metal catalysts. HDO of karanja oil was studied using ordered mesoporous Ni-alumina composite catalysts [55]. This catalyst was prepared by the evaporation-induced self-assembly method and exhibited much higher catalytic activity than mesoporous alumina and γ-Al$_2$O$_3$ supported Ni catalysts (Table 4.3). The higher catalytic activity of the composite catalyst was owing to the existence of tetrahedral (or octahedral) coordinated Ni in the Al$_2$O$_3$ framework which was catalytically superior. Srifa et al. studied HDO of palm oil using NiAl$_2$O$_4$ spinel-type catalyst [56]. The highest conversion of triglycerides with a high yield of diesel-range hydrocarbons was achieved over NiAl$_2$O$_4$ spinel-type catalyst.

4.3.6 Role of acidity of support

The moderate and highly acidic supports, such as alumina, zirconia, ceria, titania, niobium, and zeolites, are high ionic potential materials. The alumina is used as support in steam reforming and hydrodesulfurization processes in the petroleum refinery. γ-Al$_2$O$_3$ is thus considered as promising support for HDO of triglycerides due to its outstanding mechanical and textural properties. Most importantly, the availability of hydroxyl groups on alumina is responsible for high metal dispersion. The acidity of support and its morphology also play a vital role in the HDO reaction. For instance, stearic acid HDO was studied over silica, γ-alumina, and HZSM-5 supported 10 wt% Ni catalysts [4]. The strongly acidic Ni/HZSM-5 catalyst showed greater catalytic activity than neutral SiO$_2$ and weakly acidic γ-Al$_2$O$_3$ supported catalyst (Fig. 4.4). The highest conversion of stearic acid for Ni/HZSM-5 was due to the presence of dual catalytic centers (metal and acid). The lower activity of Ni/SiO$_2$ catalyst compared to Ni/γ-Al$_2$O$_3$ was due to a weaker metal-support interaction with poor metal dispersion. The study was further extended to HDO of neat karanja oil using Ni-based catalyst in a semibatch reactor [18]. The HZSM-5 and 25 wt% Ni/HZSM-5 are highly acidic catalysts and hence, showed higher conversion of oxygenates and higher selectivity to lighter hydrocarbons

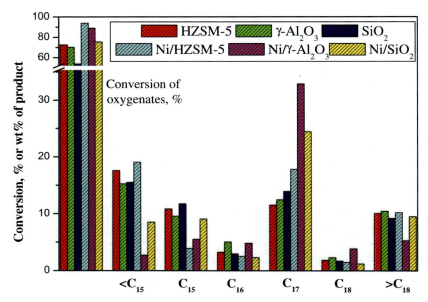

Fig. 4.8 Effect of supports on HDO of neat karanja oil [18]. Reaction conditions: 653 K, 35 bar hydrogen, 4 (w/w)% catalyst, 100 mL hydrogen/min, wt% of product at 55% conversion of oxygenates.

(lower than C_{15}) compared to SiO_2 and γ-Al_2O_3 supported 25 wt% Ni catalysts (Fig. 4.8). This result indicates a significant extent of catalytic cracking of oxygenated intermediates or hydrocarbons over HZSM-5 and Ni/HZSM-5 catalysts due to their high acidity.

The cracking reaction, however, reduces with increasing Si/Al ratio with the simultaneous drop in catalytic activity due to the decline in acidity. Lowering the Si/Al ratio increases the conversion due to the enhancement in hydrocracking and isomerization reaction [40, 57, 58]. The main issue of alumina is the susceptibility of water at elevated reaction conditions. It leads to a reduction of the surface area [7]. Activated carbon is thus employed as a support to overcome this issue. It was found as a suitable support owing to its hydrophobic nature and high surface area, pore volume (both mesopores and macropores), and thermal stability [7]. The high mesoporosity of activated carbon in Mo_2N/AC was reported to facilitate the diffusion of the reactants to the internal surface [7]. Moreover, activated carbon prevents the adsorption of water on the catalyst's active sites due to its hydrophobic nature [7].

For methyl palmitate HDO over 7 wt% Ni catalyst, the catalytic activity was highest for SiO_2, followed by γ-Al_2O_3, SAPO-11, H-ZSM-5, and HY [41]. γ-Al_2O_3 and SAPO-11 showed lower catalytic cracking owing to their medium acidity. The weakly acidic silicates supported catalysts are also widely used. Nava et al. performed olive oil upgradation via HDO over sulfided CoMo catalysts supported on different mesoporous silicates (SBA-15, SBA-16, DMS-1, and HMS) [59]. HMS supported catalyst was found to be least reactive than the rest silicate supported catalysts.

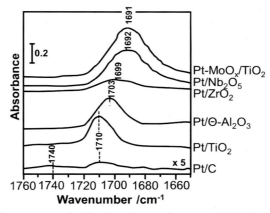

Fig. 4.9 IR spectra of the acetone adsorbed on the catalysts at 40°C [61].

4.3.7 Reducible oxides as support

ZrO_2, CeO_2, CeO_2-ZrO_2, and TiO_2 (both rutile and anatase) are reducible oxides with the ability to activate oxy-compounds [48]. They are thus widely examined supports for HDO reaction. For example, HDO of triglycerides was studied over TiO_2 supported Pt-MoO_x catalyst [60]. Almost complete conversion and a high yield of alkanes were observed at 48 h. The higher catalytic performance was observed due to the cooperative effect between Pt and Lewis acid sites of MoO_x/TiO_2. It plays a significant role in the dissociation of molecular hydrogen and weakening of coordinated ketones by the Mo^{n+} Lewis acid sites by adsorption of the C=O bond. The significance of Mo^{n+} Lewis acid sites on Pt-MoO_x/TiO_2 was confirmed by the IR adsorption study of carbonyl group-containing compound [61]. The IR spectrum band of acetone coordinated to the Mo^{n+} Lewis acid sites was observed at $1690/cm^{-1}$, which is lower than Pt/TiO_2 ($1709/cm^{-1}$) and Pt/Al_2O_3 ($1703/cm^{-1}$). The stretching band at the lower wavenumber ($1960/cm^{-1}$) indicates the strong adsorption of the C=O bond on strong Lewis acid sites, resulting in higher catalytic activity (Fig. 4.9).

4.3.8 Mesoporous support

The small pore size of zeolites limits the diffusion of bulky molecules. The diffusion constraints can be overcome by introducing mesoporous supports [62]. Typically, two different approaches are used for the synthesis of mesoporous materials: top-down or post-synthetic modification, and bottom-up or primary synthesis approach. The bottom-up approach is mostly used to synthesize mesoporous zeolites. The bottom-up approach involves the use of either hard or soft templates [63]. The hierarchical mesoporous zeolite further shows different morphology compared to the conventional ones. Verma et al. synthesized hierarchical mesoporous molecular sieve with tunable crystallinity, acidity, and porosity using the bottom-up approach [64]. It was employed as a support for NiMo

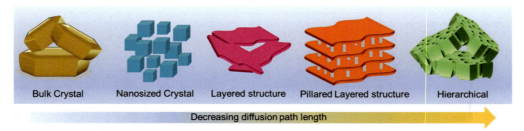

Fig. 4.10 Catalysts development for HDO of triglycerides.

catalyst in HDO of various triglycerides (jatropha oil and algal oil). The high acidity of high surface area semicrystalline (HSACS) is responsible for higher cracking and lower isomerization selectivity ($i/n=2$) compared to low surface area crystalline (LSAC) ($i/n=14$) at 673 K. The obtained green jet fuel met the basic requirements of conventional Jet A-1 fuel. MFI (n-HZSM-5) type zeolite not only provides easy diffusion of bulky molecules but also enhances metal dispersion (Fig. 4.10), resulting in high catalytic activity. For example, Schreiber et al. studied HDO of methyl stearate, tristearate, and algae oil over the self-pillared nanosheet of MFI supported Ni catalyst [65]. The Ni dispersion was higher for this catalyst which in turn led to higher catalytic activity than Ni/HZSM-5. Turnover frequency (TOF) of Ni/n-HZSM-5 was found to be 241 and 230 mol reactant (mol accessible Ni h)$^{-1}$ for HDO of tristearate and microalgae oil, respectively.

The isomerization of linear alkanes is important to improve the cold flow properties of green diesel. The Brønsted acid site favors isomerization reaction. For example, SAPO-11 and Al-SBA-15 supported NiMo catalysts exhibited greater isomerization activity than γ-Al$_2$O$_3$ for HDO of methyl stearate due to the presence of Brønsted acid sites [58]. SAPO-11 has an appropriate pore size and moderate acidity suitable for selective formation of mono-branched alkane with minimal cracking [58]. However, the cracking reaction was significant over NiMo/Al-SBA-15 due to strong acidity and large pore size [58]. γ-Al$_2$O$_3$ and SAPO-11 have optimum acidity for high catalytic activity with minimum cracking [42]. The Ni/SAPO-11 further showed high isomerization activity with excellent stability due to its optimum acidity, high Ni dispersion, and ease of diffusion of bulky molecules [41]. The acidic support further interacts strongly with metal, forming different active phases. For example, NiMo catalyst exhibited two NiMo phases (NiMoO$_4^-$ and MoO$_3$) with highly acidic mesoporous SAPO-11 (strong acidity: 610 μmol NH$_3$/g) compared to a single NiMo phase (Mo^{4+} polymeric octahedral) for mesoporous γ-Al$_2$O$_3$ (strong acidity: 190 μmol NH$_3$/g) [66]. The NiMoO$_4^-$ and MoO$_3$ phases showed higher decarboxylation activity, whereas Mo^{4+} polymeric octahedral phase showed higher HDO activity. HDO of rapeseed oil was studied over CoMo catalysts supported on MCM-41, Al$_2$O$_3$, and organized mesoporous alumina (OMA) [67]. OMA supported CoMo catalysts showed higher catalytic activity for deoxygenation of rapeseed oil compared MCM-41 and alumina supported CoMo catalysts. The high catalytic activity of OMA supported CoMo catalysts was due to their high surface area and better dispersion of active species.

Pt/SAPO-11 catalyst produces long-chain *n*-paraffins during the catalytic dewaxing process in the petroleum refinery. Hence, SAPO (silicoaluminophosphate) types of zeolite supported metal catalysts are used for HDO of triglycerides. For example, Verma et al. studied hydroprocessing of jatropha oil using sulfided NiMo and NiW catalysts supported on hierarchical mesoporous SAPO-11 at 648–723 K, 5–6 MPa hydrogen, and 1/h LHSV in a fixed-bed reactor [19]. The low hydrogen consumption was observed due to the formation of about 8% aromatics. The mild acidic support suppresses cracking reactions. For example, palm oil HDO was studied using Ni/mesoporous SAPO-11 as the catalyst [68]. The catalytic cracking was suppressed due to weak and medium acidity. The well-dispersion of Ni creates a balance between Ni and SAPO-11 functions, resulting in a 70% yield of green diesel with more than 80% isomerization selectivity.

4.3.9 Catalyst deactivation

The decline in catalytic activity with time-on-stream is one of the bottlenecks in the vegetable oil HDO. The identification of the exact cause of catalysts deactivation is thus crucial for the design of a suitable catalyst. The catalyst deactivation is broadly categorized into six types based on intrinsic mechanism: (i) poisoning, (ii) fouling (carbon deposition), (iii) thermal degradation (sintering), (iv) vapor compound formation accompanied by transport, (v) vapor-solid and/or solid-solid reactions, and (vi) attrition/crushing. Among them, the first three causes are well-known for catalyst deactivation in HDO of triglycerides. The catalyst deactivation is either temporary or permanent. Coke deposition and sulfur leaching from active sites are the temporary deactivations of catalysts, and this problem is resolved by reactivating the catalyst. While catalyst replacement is the only solution for permanent deactivation. Water, which is formed as a by-product during the HDO of triglycerides, is also responsible for catalyst deactivation (metal leaching or reduce the active metal dispersion or by changing the sulfide phase). Badawi et al. investigated the role of water on the stability of sulfided Mo and CoMo catalysts in HDO of phenolic compounds. DFT calculation showed that the stability of the MoS_2 phase toward water can be increased by incorporating Co at the edge of the MoS_2 phase [69].

4.4 Effect of process conditions

The process conditions, such as temperature and pressure, play a major role in the deoxygenation of fatty acids/triglycerides and product distribution. The elevated reaction temperature favors the higher degree of deoxygenation. For example, oxygenated intermediate compounds were significant at 583 K, but they were absent at 633 K for rapeseed oil HDO using NiMo catalyst [20]. The elevated reaction temperatures also favor cracking reaction, resulting in a higher yield of short-chain alkanes. For example, Kumar et al. studied HDO of stearic acid over alumina supported Ni catalysts at four different reaction temperatures, 533, 543, 553, and 563 K [4]. The selectivity to C_{15} and C_{16} alkanes (cracking products) was increased with increasing reaction temperature for a fixed conversion of stearic acid. For neat karanja oil HDO over Ni/γ-Al$_2$O$_3$ catalyst, thermal cracking was significant above 613 K with the formation of a large quantity of low molecular weight alkanes [18].

The hydrogen should be available in the liquid phase for hydrogenation and HDO reactions (Fig. 4.2). The hydrogen solubility in alkane increases with increasing temperature, pressure, and the number of carbon atoms present in the alkane solvent [70]. The high hydrogen pressure is thus needed to maintain sufficient hydrogen concentration in the liquid phase for the progress of the above reactions. The oxygenates conversion was thus improved by increasing the H_2 pressure for HDO of karanja oil with an insignificant effect on product distribution [55]. The complete deoxygenation was reported at 14 MPa hydrogen pressure, whereas a large quantity of unreacted fatty acids was observed at 7 MPa [71]. The ratio of C_{18}/C_{17} alkane was significantly lower at 0.7 MPa during HDO of rapeseed oil over organized-mesoporous-alumina (OMA) supported CoMo catalysts [67].

4.5 Coprocessing of triglycerides with the petroleum feedstock

Hydroprocessing is a well-established technology for the removal of sulfur, oxygen, nitrogen, and chlorine from crude oil distillates. However, the vegetable oils contain long linear-chain fatty acids and a larger quantity of oxygen compared to petroleum distillates. Therefore, HDO of neat vegetable oils at high LHSV may cause reactor plugging by wax formation [5]. The coprocessing of crude vegetable oils with petroleum distillates is thus a promising approach to overcome these problems. In this case, HDO of triglycerides and hydrodesulfurization/hydrodenitrification of gas oil occur simultaneously over hydrotreating catalyst (Fig. 4.11). Alumina supported sulfided NiMo, CoMo, and NiW catalysts were employed for coprocessing of various vegetable oils, such as jatropha [72], soybean [34, 73], sunflower oil [74, 75], waste cooking oil [46], and rapeseed oil [76], at 350–450°C and 50–100 bar H_2 pressure. It results in green diesel with about 66%–95% yield, as shown in Table 4.4. The sulfur removal efficiency is not affected much by the coprocessing of vegetable oil. For example, the HDS efficiency of NiMo catalyst was about 92% during coprocessing of cooking oil compared to 89% for pure petroleum distillate (Table 4.4) [46]. The addition of vegetable oil also improved the yield of diesel-range hydrocarbons and it was around 95% during coprocessing of cooking oil (Table 4.4). The selectivity to diesel-range hydrocarbons was decreased with increasing temperature. However, the sulfur removal efficiency was improved at the higher temperatures. The extent of reaction and product selectivity were evaluated for the competing deoxygenation and desulfurization reactions. Hydrocracking catalysts favored the

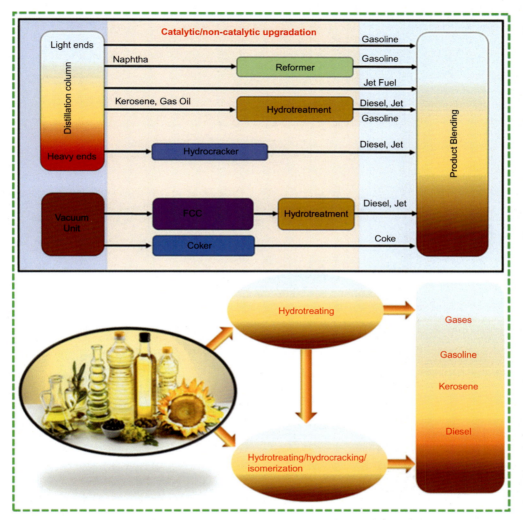

Fig. 4.11 Graphical representation of hydroprocessing and catalytic reaction for triglycerides.

maximum yield of middle distillates. The reaction also showed an increment in the selectivity to HDO products. However, HDO reaction competes with hydrodesulfurization reactions over active catalyst sites, resulting in a decrease in desulfurization of gas oil. However, Sinha et al. studied coprocessing of pretreated jatropha oil and reported an increase in desulfurization as compared to pure gas-oil [72]. Walendziewski et al. observed partial hydrocracking during HDO of rape oil and light gas oil mixture using commercial alumina supported NiMo catalysts, which resulted in a high yield of kerosene-range hydrocarbons (Table 4.4) [76].

Table 4.4 Co-processing of vegetable oils with petroleum distillates [46, 72–77].

Feed (space velocity)	Cooking		Soybean		Rapeseed		Sunflower	
	350°C (1/h)	370°C (1/h)	350°C (4/h)	370°C (1/h)	350°C (2/h)	370°C (2/h)	350°C (1/h)	370°C (1/h)
Naphtha (IBP–150)	0	1	0	0	0	0	0.09	2
Kerosene (150–250)	5	8	5	4.5	32.4	34.4	0.05	1
Diesel (250-FBP)	95	91	95	95.5	67.6	65.6	99.86	97
HDS (%)	92	96	89	NA	NA	NA	34	48
C_{17}/C_{18}	NA	NA	1.6	0.75	NA	NA	0.49	0.9
C_{15}/C_{16}	NA	NA	1.15	1	NA	NA	0.5	0.75
CO_2	NA	NA	NA	NA	NA	NA	0.08	0.66
CO	NA	NA	NA	NA	NA	NA	0.025	0.07

Berzergianni et al. studied the coprocessing of waste cooking oil with heavy atmospheric gas oil over NiMo and CoMo catalysts [46]. The extent of desulfurization was not affected during coprocessing over NiMo catalyst, while it was affected significantly over CoMo catalyst. Some studies, however, reported the enhancement of desulfurization during coprocessing of gas oil with cooking oil and sunflower oil [75]. The high desulfurization activity was observed for jatropha oil (96%) and soybean oil (89%), while it was only 34% for sunflower oil for coprocessing with gas oil. The coprocessing further improves the cetane number of diesel. For example, coprocessing of sunflower oil with gas oil at 693 K and 18 MPa showed the increase of cetane number by 1–7 units [21]. It was due to the high n-paraffin content in the products. The reaction temperature also influences the catalytic activity and extent of HDO and desulfurization reaction. The sulfur and oxygen removal was increased to 95% with increasing reaction temperature [77]. However, sulfur removal remained similar for both pure heavy atmospheric gas oil and its coprocessing with waste cooking oil [74].

4.6 Commercialization status

The use of low-cost and inedible oils, such as waste cooking oil, inedible oil, waste animal fats, and microalgae oils, are the key to make HDO technology economically viable. The commercial production of green diesel has been operating for more than a decade. For example, Neste Oil, a Finland based company, developed a stand-alone process called NEXBTL for producing green diesel from waste animal fat, rapeseed oil, and palm oil with an annual production capacity of about 3 million tons in Finland, Netherlands, and Singapore together [78]. NEXBTL technology combines the HDO of vegetable oils and isomerization of linear alkanes to branched alkanes for improving cold flow properties. NEXBTL-derived green diesel exhibits a high heating value (44 MJ/kg) and higher cetane number (84–99) compared to diesel and biodiesel (Table 4.2). UOP/Eni Ecofining developed a technology for the production of green diesel from fats, grease, and oils in collaboration with UOP Honeywell. It is a two-stage process (Fig. 4.12) [16]. In the first stage, oils/fats/grease undergo catalytic HDO in an adiabatic reactor where hydrogenation of unsaturated bonds and deoxygenation reactions occur simultaneously in the presence of hydrogen. The carbon dioxide, water, and volatile hydrocarbons are immediately separated from the product of the reactor to obtain diesel boiling-range alkanes. In the second stage, the diesel-range alkanes undertake hydro-isomerization to produce diesel-range branched alkanes with improved cold flow properties. The excess hydrogen is then separated from isomerized products and recycled to both HDO and isomerization reactors. Ecofining-derived green diesel displays a similar heating value (44 MJ/kg) and high cetane number (70–90) compared to diesel (Table 4.2). They have set up the production units with a capacity of 0.9 and 0.78 million tons located in the United States and Italy,

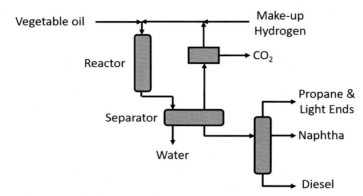

Fig. 4.12 Schematic representation of Ecofining process [16].

respectively [78]. Valero Energy Corporation in collaboration with Darling Ingredients Inc. is another manufacturer of renewable transportation fuels. They have set up a renewable diesel refinery, Diamond Alternative Energy LLC, with a capacity of 275 million gallons of renewable diesel per annum in the Louisiana state, United States. The tallow from recycled animal by-product and used cooking oil from restaurants are the major feedstocks in their technology. In this technology, the final product meets the ASTM International standard for diesel fuel (D-975).

4.7 Process design and economics

HDO of triglyceride can be envisioned by two possible approaches: direct HDO and two-step HDO. Recently, these processes were designed using the pinch analysis approach and the process economics were evaluated (Fig. 4.13) [79]. The first process involves hydrogenation of unsaturated bonds at 100 bar H_2 pressure and 573 K and HDO of vegetable oil at 623 K and 100 bar hydrogen pressure, followed by separation of excess hydrogen from green diesel. The propane is coproduced in the process, which is purified from CO_2 using an alkanolamine. The latter process involves hydrolysis of vegetable oil at 533 K and 56 bars, followed by HDO of crude fatty acids at 623 K and 100 bar hydrogen. In the downstream, green diesel is separated from excess hydrogen, water, CO_2, and a small amount of C_2-C_3 alkanes. The coproduct, glycerol is dehydrated by the combination of evaporation and vacuum distillation. The economic analysis showed that the latter process involved higher capital investment and utility consumption and lesser hydrogen consumption compared to the former process. The higher H_2 consumption in the direct HDO process was due to the removal of oxygen from glycerol backbone to form propane. The amount of coproduct in the latter

Production of green diesel

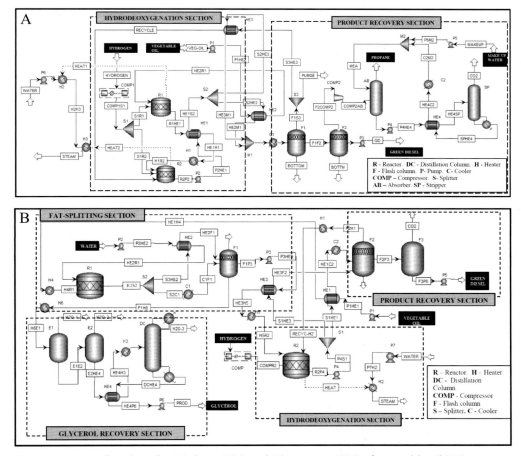

Fig. 4.13 Process flowsheet for (A) direct HDO and (B) two-step HDO of vegetable oil [79].

process was thus much more compared to the former process. The electricity consumption in direct HDO was slightly higher than the two-step process due to a large capacity compressor for pressurizing the large volume of H_2. The green diesel manufacturing cost was slightly lesser in the two-step process (0.798 $/kg) than direct HDO (0.84 $/kg) for an optimum plant capacity of 0.12 MMT karanja oil/annum. The minimum selling price of green diesel in a two-step process (1.21 $/kg) was, however, slightly more compared to direct HDO (1.186 $/kg) for an 8.5% rate of return. It was due to slightly higher capital investment in the two-step HDO process. The feedstock is the major factor contributing about 75% of the production cost. The indirect cost contributes only about 10% of the production cost. It indicates the simplicity of the process with less capital investment.

4.8 Conclusions

HDO of triglycerides is a promising catalytic conversion technology for the production of green diesel in hydrocarbon biorefinery. This chapter thus presents an overview of chemistry, catalysts, commercial status, and economics of HDO of triglycerides. The chapter further provides an insight into the reaction mechanism and catalytic performance of different active metals and supports. Decarbonylation and HDO are two main reaction routes for the removal of oxygen from triglycerides. While HDO reaction dominates over sulfided/reduced bimetallic (CoMo and NiMo), reducible oxides, and solid-acid catalysts, deoxygenation occurs mainly through decarbonylation reaction over Ni catalysts. Both of these reaction routes are, however, significant over Co catalyst. The composite catalyst improves catalytic activity due to the incorporation of metal in the framework of support. The cracking reaction is significant over acidic catalysts. Phosphided and nitrided catalysts exhibit a lower tendency of cracking and methanation reaction. The mesoporous support improves catalytic performance due to its high surface area. At the moment, Ecofining and NEXBTL are two notable technologies for HDO triglycerides. The cost of green diesel is comparable with the current diesel price. Green diesel has thus a tremendous potential for market penetration subjected to the availability of feedstock.

References

[1] IEA, World Energy Outlook 2015, IEA, Paris, 2015. https://www.iea.org/reports/world-energy-outlook-2015.

[2] S.K. Maity, Opportunities, recent trends and challenges of integrated biorefinery: part II, Renew. Sust. Energ. Rev. 43 (2015) 1446–1466, https://doi.org/10.1016/j.rser.2014.08.075.

[3] U.S. Energy Information Administration, Annual Energy Outlook 2012. https://www.eia.gov/outlooks/aeo/er/pdf/tbla2.pdf.

[4] P. Kumar, S.R. Yenumala, S.K. Maity, D. Shee, Kinetics of hydrodeoxygenation of stearic acid using supported nickel catalysts: effects of supports, Appl. Catal. A Gen. 471 (2014) 28–38, https://doi.org/10.1016/j.apcata.2013.11.021.

[5] T.V. Choudhary, C.B. Phillips, Renewable fuels via catalytic hydrodeoxygenation, Appl. Catal. A Gen. 397 (2011) 1–12, https://doi.org/10.1016/j.apcata.2011.02.025.

[6] A. Srivastava, R. Prasad, Triglycerides-based diesel fuels, Renew. Sust. Energ. Rev. 4 (2000) 111–133.

[7] S.L. Douvartzides, N.D. Charisiou, K.N. Papageridis, M.A. Goula, Green diesel: biomass feedstocks, production technologies, catalytic research, fuel properties and performance in compression ignition internal combustion engines, Energies 12 (2019), https://doi.org/10.3390/en12050809.

[8] M. Ameen, M.T. Azizan, S. Yusup, A. Ramli, M. Yasir, Catalytic hydrodeoxygenation of triglycerides: an approach to clean diesel fuel production, Renew. Sust. Energ. Rev. 80 (2017) 1072–1088, https://doi.org/10.1016/j.rser.2017.05.268.

[9] M.Y. Choo, L.E. Oi, P.L. Show, J.S. Chang, T.C. Ling, E.P. Ng, et al., Recent progress in catalytic conversion of microalgae oil to green hydrocarbon: a review, J. Taiwan Inst. Chem. Eng. 79 (2017) 116–124, https://doi.org/10.1016/j.jtice.2017.06.028.

[10] M.J. Ramos, C.M. Fernández, A. Casas, L. Rodríguez, Á. Pérez, Influence of fatty acid composition of raw materials on biodiesel properties, Bioresour. Technol. 100 (2009) 261–268, https://doi.org/10.1016/j.biortech.2008.06.039.

[11] A. Kumar Tiwari, A. Kumar, H. Raheman, Biodiesel production from jatropha oil (Jatropha curcas) with high free fatty acids: an optimized process, Biomass Bioenergy 31 (2007) 569–575, https://doi.org/10.1016/j.biombioe.2007.03.003.

[12] S.M. Sultan, N. Dikshit, U.J. Vaidya, Oil content and fatty acid composition of soybean (Glysine max L.) genotypes evaluated under rainfed conditions of Kashmir Himalayas in India, J. Appl. Nat. Sci. 7 (2015) 910–915, https://doi.org/10.31018/jans.v7i2.706.

[13] A. Srifa, N. Viriya-empikul, S. Assabumrungrat, K. Faungnawakij, Catalytic behaviors of Ni/γ-Al2O3 and Co/γ-Al2O3 during the hydrodeoxygenation of palm oil, Catal. Sci. Technol. 5 (2015) 3693–3705, https://doi.org/10.1039/C5CY00425J.

[14] S. Lestari, P. Mäki-Arvela, J. Beltramini, G.Q.M. Lu, D.Y. Murzin, Transforming triglycerides and fatty acids into biofuels, ChemSusChem 2 (2009) 1109–1119.

[15] P. Kumar, P. Kumar, P.V.C. Rao, N.V. Choudary, G. Sriganesh, Saw dust pyrolysis: effect of temperature and catalysts, Fuel 199 (2017) 339–345, https://doi.org/10.1016/j.fuel.2017.02.099.

[16] T.N. Kalnes, T. Marker, D.R. Shonnard, K.P. Koers, Green diesel production by hydrorefining renewable feedstocks, Biofuels Technol. (2008) 7–11.

[17] Neste, NExBTL® Renewable Diesel, Neste Corporation, 2009.

[18] S.R. Yenumala, S.K. Maity, D. Shee, Hydrodeoxygenation of karanja oil over supported nickel catalysts: influence of support and nickel loading, Cat. Sci. Technol. 6 (2016) 3156–3165, https://doi.org/10.1039/C5CY01470K.

[19] D. Verma, B.S. Rana, R. Kumar, M.G. Sibi, A.K. Sinha, Diesel and aviation kerosene with desired aromatics from hydroprocessing of jatropha oil over hydrogenation catalysts supported on hierarchical mesoporous SAPO-11, Appl. Catal. A Gen. 490 (2015) 108–116, https://doi.org/10.1016/j.apcata.2014.11.007.

[20] P. Šimáček, D. Kubička, G. Šebor, M. Pospíšil, Fuel properties of hydroprocessed rapeseed oil, Fuel 89 (2010) 611–615, https://doi.org/10.1016/j.fuel.2009.09.017.

[21] P. Šimáček, D. Kubička, I. Kubičková, F. Homola, M. Pospíšil, J. Chudoba, Premium quality renewable diesel fuel by hydroprocessing of sunflower oil, Fuel 90 (2011) 2473–2479, https://doi.org/10.1016/j.fuel.2011.03.013.

[22] S. Reddy, Y. Sunil, D. Shee, Reaction mechanism and kinetic modeling for the hydrodeoxygenation of triglycerides over alumina supported nickel catalyst, React. Kinet. Mech. Catal. 120 (2016) 109–128, https://doi.org/10.1007/s11144-016-1098-2.

[23] P. Kumar, S.K. Maity, D. Shee, Hydrodeoxygenation of stearic acid using Mo modified Ni and Co/alumina catalysts: effect of calcination temperature, Chem. Eng. Commun. 207 (2020) 1–16, https://doi.org/10.1080/00986445.2019.1630396.

[24] P. Kumar, S.K. Maity, D. Shee, Role of NiMo alloy and Ni species in the performance of NiMo/alumina catalysts for hydrodeoxygenation of stearic acid: a kinetic study, ACS Omega 4 (2019) 2833–2843, https://doi.org/10.1021/acsomega.8b03592.

[25] I. Simakova, O. Simakova, P. Mäki-Arvela, A. Simakov, M. Estrada, D.Y. Murzin, Deoxygenation of palmitic and stearic acid over supported Pd catalysts: effect of metal dispersion, Appl. Catal. A Gen. 355 (2009) 100–108, https://doi.org/10.1016/j.apcata.2008.12.001.

[26] I. Hachemi, D.Y. Murzin, Kinetic modeling of fatty acid methyl esters and triglycerides hydrodeoxygenation over nickel and palladium catalysts, Chem. Eng. J. 334 (2018) 2201–2207, https://doi.org/10.1016/j.cej.2017.11.153.

[27] B. Peng, X. Yuan, C. Zhao, J.A. Lercher, Stabilizing catalytic pathways via redundancy: selective reduction of microalgae oil to alkanes, J. Am. Chem. Soc. 134 (2012) 9400–9405, https://doi.org/10.1021/ja302436q.

[28] D. Kubička, M. Bejblová, J. Vlk, Conversion of vegetable oils into hydrocarbons over CoMo/MCM-41 catalysts, Top. Catal. 53 (2010) 168–178, https://doi.org/10.1007/s11244-009-9421-z.

[29] D. Kubička, L. Kaluža, Deoxygenation of vegetable oils over sulfided Ni, Mo and NiMo catalysts, Appl. Catal. A Gen. 372 (2010) 199–208, https://doi.org/10.1016/j.apcata.2009.10.034.

[30] J.G. Immer, M.J. Kelly, H.H. Lamb, Catalytic reaction pathways in liquid-phase deoxygenation of C18 free fatty acids, Appl. Catal. A Gen. 375 (2010) 134–139, https://doi.org/10.1016/j.apcata.2009.12.028.

[31] M. Snåre, I. Kubičková, P. Mäki-Arvela, K. Eränen, D.Y. Murzin, Heterogeneous catalytic deoxygenation of stearic acid for production of biodiesel, Ind. Eng. Chem. Res. 45 (2006) 5708–5715, https://doi.org/10.1021/ie060334i.

[32] L. Kaluža, D. Kubička, The comparison of Co, Ni, Mo, CoMo and NiMo sulfided catalysts in rapeseed oil hydrodeoxygenation, React. Kinet. Mech. Catal. 122 (2017) 333–341, https://doi.org/10.1007/s11144-017-1247-2.

[33] O.I. Şenol, T.R. Viljava, A.O.I. Krause, Effect of sulphiding agents on the hydrodeoxygenation of aliphatic esters on sulphided catalysts, Appl. Catal. A Gen. 326 (2007) 236–244, https://doi.org/10.1016/j.apcata.2007.04.022.

[34] R. Tiwari, B.S. Rana, R. Kumar, D. Verma, R. Kumar, R.K. Joshi, et al., Hydrotreating and hydrocracking catalysts for processing of waste soya-oil and refinery-oil mixtures, Catal. Commun. 12 (2011) 559–562, https://doi.org/10.1016/j.catcom.2010.12.008.

[35] B. Veriansyah, J.Y. Han, S.K. Kim, S.A. Hong, Y.J. Kim, J.S. Lim, et al., Production of renewable diesel by hydroprocessing of soybean oil: effect of catalysts, Fuel 94 (2012) 578–585, https://doi.org/10.1016/j.fuel.2011.10.057.

[36] X. Zhao, L. Wei, S. Cheng, E. Kadis, Y. Cao, E. Boakye, et al., Hydroprocessing of carinata oil for hydrocarbon biofuel over Mo-Zn/Al2O3, Appl. Catal. B Environ. 196 (2016) 41–49, https://doi.org/10.1016/j.apcatb.2016.05.020.

[37] M.G. Sibi, A. Rai, M. Anand, S.A. Farooqui, A.K. Sinha, Improved hydrogenation function of Pt@SOD incorporated inside sulfided NiMo hydrocracking catalyst, Catal. Sci. Technol. 6 (2016) 1850–1862, https://doi.org/10.1039/c5cy01243k.

[38] S. Harnos, G. Onyestyák, D. Kalló, Hydrocarbons from sunflower oil over partly reduced catalysts, React. Kinet. Mech. Catal. 106 (2012) 99–111, https://doi.org/10.1007/s11144-012-0424-6.

[39] A. Peeters, R. Ameloot, D.E.D.V. Vos, Carbon dioxide as a reversible amine-protecting agent in selective Michael additions and acylations, Green Chem. 15 (2013) 1550–1557, https://doi.org/10.1039/c3gc40568k.

[40] S. Chen, G. Zhou, C. Miao, Green and renewable bio-diesel produce from oil hydrodeoxygenation: strategies for catalyst development and mechanism, Renew. Sust. Energ. Rev. 101 (2019) 568–589, https://doi.org/10.1016/j.rser.2018.11.027.

[41] C. Kordulis, K. Bourikas, M. Gousi, E. Kordouli, A. Lycourghiotis, Development of nickel based catalysts for the transformation of natural triglycerides and related compounds into green diesel: a critical review, Appl. Catal. B Environ. 181 (2016) 156–196, https://doi.org/10.1016/j.apcatb.2015.07.042.

[42] H. Zuo, Q. Liu, T. Wang, L. Ma, Q. Zhang, Q. Zhang, Hydrodeoxygenation of methyl palmitate over supported Ni catalysts for diesel-like fuel production, Energy Fuel 26 (2012) 3747–3755, https://doi.org/10.1021/ef300063b.

[43] M. Toba, Y. Abe, H. Kuramochi, M. Osako, T. Mochizuki, Y. Yoshimura, Hydrodeoxygenation of waste vegetable oil over sulfide catalysts, Catal. Today 164 (2011) 533–537, https://doi.org/10.1016/j.cattod.2010.11.049.

[44] E. Kordouli, L. Sygellou, C. Kordulis, K. Bourikas, A. Lycourghiotis, Probing the synergistic ratio of the NiMo/γ-Al2O3 reduced catalysts for the transformation of natural triglycerides into green diesel, Appl. Catal. B Environ. 209 (2017) 12–22, https://doi.org/10.1016/j.apcatb.2017.02.045.

[45] O.I. Şenol, E.M. Ryymin, T.R. Viljava, A.O.I. Krause, Effect of hydrogen sulphide on the hydrodeoxygenation of aromatic and aliphatic oxygenates on sulphided catalysts, J. Mol. Catal. A Chem. 277 (2007) 107–112, https://doi.org/10.1016/j.molcata.2007.07.033.

[46] S. Bezergianni, A. Dimitriadis, G. Meletidis, Effectiveness of CoMo and NiMo catalysts on co-hydroprocessing of heavy atmospheric gas oil-waste cooking oil mixtures, Fuel 125 (2014) 129–136, https://doi.org/10.1016/j.fuel.2014.02.010.

[47] M. Krár, S. Kovács, D. Kalló, J. Hancsók, Fuel purpose hydrotreating of sunflower oil on CoMo/Al$_2$O$_3$ catalyst, Bioresour. Technol. 101 (2010) 9287–9293, https://doi.org/10.1016/j.biortech.2010.06.107.

[48] V.A. Yakovlev, S.A. Khromova, O.V. Sherstyuk, V.O. Dundich, D.Y. Ermakov, V.M. Novopashina, et al., Development of new catalytic systems for upgraded bio-fuels production from bio-crude-oil and biodiesel, Catal. Today 144 (2009) 362–366, https://doi.org/10.1016/j.cattod.2009.03.002.

[49] J. Monnier, H. Sulimma, A. Dalai, G. Caravaggio, Hydrodeoxygenation of oleic acid and canola oil over alumina-supported metal nitrides, Appl. Catal. A Gen. 382 (2010) 176–180, https://doi.org/10.1016/j.apcata.2010.04.035.

[50] L. Souza Macedo, R.R. Oliveira, T. van Haasterecht, V. Teixeira da Silva, H. Bitter, Influence of synthesis method on molybdenum carbide crystal structure and catalytic performance in stearic acid hydrodeoxygenation, Appl. Catal. B Environ. 241 (2019) 81–88, https://doi.org/10.1016/j.apcatb.2018.09.020.

[51] S.A.W. Hollak, R.W. Gosselink, D.S. Van Es, J.H. Bitter, Comparison of tungsten and molybdenum carbide catalysts for the hydrodeoxygenation of oleic acid, ACS Catal. 3 (2013) 2837–2844, https://doi.org/10.1021/cs400744y.

[52] Y. Yang, C. Ochoa-Hernández, V.A. de la Peña O'Shea, J.M. Coronado, D.P. Serrano, Ni2P/SBA-15 as a hydrodeoxygenation catalyst with enhanced selectivity for the conversion of methyl oleate into n-octadecane, ACS Catal. 2 (2012) 592–598.

[53] S.T. Oyama, Y.K. Lee, The active site of nickel phosphide catalysts for the hydrodesulfurization of 4,6-DMDBT, J. Catal. 258 (2008) 393–400, https://doi.org/10.1016/j.jcat.2008.06.023.

[54] S.T. Oyama, X. Wang, Y.K. Lee, W.J. Chun, Active phase of Ni2P/SiO2 in hydroprocessing reactions, J. Catal. 221 (2004) 263–273, https://doi.org/10.1016/S0021-9517(03)00017-4.

[55] S.R. Yenumala, P. Kumar, S.K. Maity, D. Shee, Hydrodeoxygenation of karanja oil using ordered mesoporous nickel-alumina composite catalysts, Catal. Today 348 (2020) 45–54, https://doi.org/10.1016/j.cattod.2019.08.040.

[56] A. Srifa, R. Kaewmeesri, C. Fang, V. Itthibenchapong, K. Faungnawakij, NiAl2O4 spinel-type catalysts for deoxygenation of palm oil to green diesel, Chem. Eng. J. 345 (2018) 107–113, https://doi.org/10.1016/j.cej.2018.03.118.

[57] B. Peng, Y. Yao, C. Zhao, J.A. Lercher, Towards quantitative conversion of microalgae oil to diesel-range alkanes with bifunctional catalysts, Angew. Chem. Int. Ed. 51 (2012) 2072–2075, https://doi.org/10.1002/anie.201106243.

[58] E.W. Qian, N. Chen, S. Gong, Role of support in deoxygenation and isomerization of methyl stearate over nickel–molybdenum catalysts, J. Mol. Catal. A Chem. 387 (2014) 76–85, https://doi.org/10.1016/j.molcata.2014.02.031.

[59] R. Nava, B. Pawelec, P. Castaño, M.C. Álvarez-Galván, C.V. Loricera, J.L.G. Fierro, Upgrading of bio-liquids on different mesoporous silica-supported CoMo catalysts, Appl. Catal. B Environ. 92 (2009) 154–167, https://doi.org/10.1016/j.apcatb.2009.07.014.

[60] K. Kon, T. Toyao, W. Onodera, S.M.A.H. Siddiki, K. Shimizu, Hydrodeoxygenation of fatty acids, triglycerides, and ketones to liquid alkanes by a Pt–MoOx/TiO2 catalyst, ChemSusChem 9 (2017) 2822–2827.

[61] Y. Nakamura, K. Kon, A.S. Touchy, K. Shimizu, W. Ueda, Selective synthesis of primary amines by reductive amination of ketones with Ammonia over supported Pt catalysts, ChemSusChem 7 (2015) 921–924.

[62] J. Přech, P. Pizarro, D.P. Serrano, J. Áejka, From 3D to 2D zeolite catalytic materials, Chem. Soc. Rev. 47 (2018) 8263–8306, https://doi.org/10.1039/c8cs00370j.

[63] K. Li, J. Valla, J. Garcia-Martinez, Realizing the commercial potential of hierarchical zeolites: new opportunities in catalytic cracking, ChemSusChem 6 (2014) 46–66.

[64] D. Verma, R. Kumar, B.S. Rana, A.K. Sinha, Aviation fuel production from lipids by a single-step route using hierarchical mesoporous zeolites, Energy Environ. Sci. 4 (2011) 1667–1671, https://doi.org/10.1039/c0ee00744g.

[65] M.W. Schreiber, D. Rodriguez-Niño, O.Y. Gutiérrez, J.A. Lercher, Hydrodeoxygenation of fatty acid esters catalyzed by Ni on nano-sized MFI type zeolites, Catal. Sci. Technol. 6 (2016) 7976–7984, https://doi.org/10.1039/c6cy01598k.

[66] H. Chen, Q. Wang, X. Zhang, L. Wang, Effect of support on the NiMo phase and its catalytic hydrodeoxygenation of triglycerides, Fuel 159 (2015) 430–435, https://doi.org/10.1016/j.fuel.2015.07.010.

[67] D. Kubic, P. Šimáček, N. Žilkova, Transformation of vegetable oils into hydrocarbons over mesoporous-alumina-supported CoMo catalysts, Top. Catal. 52 (2009) 161–168, https://doi.org/10.1007/s11244-008-9145-5.

[68] Q. Liu, H. Zuo, Q. Zhang, T. Wang, L. Ma, Hydrodeoxygenation of palm oil to hydrocarbon fuels over Ni/SAPO-11 catalysts, Chin. J. Catal. 35 (2014) 748–756, https://doi.org/10.1016/s1872-2067(12)60710-4.

[69] M. Badawi, J.F. Paul, S. Cristol, E. Payen, Y. Romero, F. Richard, et al., Effect of water on the stability of Mo and CoMo hydrodeoxygenation catalysts: a combined experimental and DFT study, J. Catal. 282 (2011) 155–164, https://doi.org/10.1016/j.jcat.2011.06.006.

[70] M. Safamirzaei, H. Modarress, Hydrogen solubility in heavy n-alkanes; modeling and prediction by artificial neural network, Fluid Phase Equilib. 310 (2011) 150–155, https://doi.org/10.1016/j.fluid.2011.08.004.

[71] G.N.R. da Filho, D. Brodzki, G. Djéga-Mariadassou, Formation of alkanes, alkylcycloalkanes and alkylbenzenes during the catalytic hydrocracking of vegetable oils, Fuel 72 (1993) 543–549.

[72] R. Kumar, B.S. Rana, R. Tiwari, D. Verma, R. Kumar, R.K. Joshi, et al., Hydroprocessing of jatropha oil and its mixtures with gas oil, Green Chem. 12 (2010) 2232–2239, https://doi.org/10.1039/c0gc00204f.

[73] B.S. Rana, R. Kumar, R. Tiwari, R. Kumar, R.K. Joshi, M.O. Garg, et al., Transportation fuels from co-processing of waste vegetable oil and gas oil mixtures, Biomass Bioenergy 56 (2013) 43–52, https://doi.org/10.1016/j.biombioe.2013.04.029.

[74] S. Bezergianni, A. Kalogianni, I.A. Vasalos, Hydrocracking of vacuum gas oil-vegetable oil mixtures for biofuels production, Bioresour. Technol. 100 (2009) 3036–3042, https://doi.org/10.1016/j.biortech.2009.01.018.

[75] G.W. Huber, P. O'Connor, A. Corma, Processing biomass in conventional oil refineries: production of high quality diesel by hydrotreating vegetable oils in heavy vacuum oil mixtures, Appl. Catal. A Gen. 329 (2007) 120–129, https://doi.org/10.1016/j.apcata.2007.07.002.

[76] J. Walendziewski, M. Stolarski, R. Łuzny, B. Klimek, Hydroprocesssing of light gas oil-rape oil mixtures, Fuel Process. Technol. 90 (2009) 686–691, https://doi.org/10.1016/j.fuproc.2008.12.006.

[77] S. Bezergianni, A. Dimitriadis, Temperature effect on co-hydroprocessing of heavy gas oil-waste cooking oil mixtures for hybrid diesel production, Fuel 103 (2013) 579–584, https://doi.org/10.1016/j.fuel.2012.08.006.

[78] A. Amin, Review of diesel production from renewable resources: catalysis, process kinetics and technologies, Ain Shams Eng. J. 10 (2019) 821–839, https://doi.org/10.1016/j.asej.2019.08.001.

[79] S. Mailaram, S.K. Maity, Techno-economic evaluation of two alternative processes for production of green diesel from karanja oil: a pinch analysis approach, J. Renewable Sustainable Energy 11 (2019) 025906, https://doi.org/10.1063/1.5078567.

CHAPTER 5

Advances in liquefaction for the production of hydrocarbon biofuels

Gabriel Fraga[a], Nuno Batalha[a], Adarsh Kumar[b,d], Thallada Bhaskar[b,d], Muxina Konarova[c], and Greg Perkins[a]

[a]School of Chemical Engineering, Faculty of Engineering, Architecture and Information Technology, The University of Queensland, Brisbane, QLD, Australia
[b]Material Resource Efficiency Division, CSIR-Indian Institute of Petroleum, Dehradun, Uttarakhand, India
[c]Australian Institute for Bioengineering and Nanotechnology (AIBN), The University of Queensland, Brisbane, QLD, Australia
[d]Academy of Scientific and Innovative Research (AcSIR), Ghaziabad, India

Contents

5.1 Introduction	128
5.2 Hydrothermal liquefaction	129
5.2.1 Process fundamentals	129
5.2.2 Reaction pathways	130
5.2.3 Process parameters	138
5.2.4 Effect of feedstock type	141
5.2.5 Catalyst selection	144
5.2.6 Biocrude properties and storage	146
5.2.7 Biocrude upgrading	147
5.2.8 Aqueous phase valorization	150
5.3 Liquefaction using organic solvents	152
5.3.1 Liquefaction in pure organic compounds	153
5.3.2 Implication of organic solvents for commercial application	155
5.4 Advances in commercialization	158
5.4.1 Hydrothermal liquefaction	158
5.4.2 Liquefaction using organic solvents	161
5.5 Process economics	165
5.5.1 Hydrothermal liquefaction	166
5.5.2 Liquefaction using organic solvents	167
5.6 Conclusions	168
References	169

Abbreviations

5-HMF	hydroxymethylfurfural
bbl	barrel
BP	boiling point
d.a.f.	dry ash-free
DP	degree of depolymerization
FCC	fluid catalytic cracking
HDN	hydrodenitrogenation

Hydrocarbon Biorefinery
https://doi.org/10.1016/B978-0-12-823306-1.00009-1

Copyright © 2022 Elsevier Inc.
All rights reserved.

HDO	hydrodeoxygenation
HDS	hydrodesulfurization
HHV	higher heating value
HTL	hydrothermal liquefaction
HVGO	heavy vacuum gas oil
LBET	Lobry de Bruyn-Alberda van Ekenstein
LCO	light cycle oil
LGE	liquid gasoline equivalent
LVGO	light vacuum gas oil
MFSP	minimum fuel selling price
PEG	polyethylene glycol
PERC	Pittsburgh Energy Research Center
PNNL	Pacific Northwest National Laboratory
TAGs	triacylglicerides
TCI	total capital investment
TEA	techno-economic analysis
THC	total unburnt hydrocarbon emissions
VGO	vacuum gas oil
WGS	water-gas-shift
WWTP	wastewater treatment plant

5.1 Introduction

The conversion of biomass and carbonaceous wastes through liquefaction using water (hydrothermal) or organic solvents is considered a promising thermochemical route for producing hydrocarbon biofuels [1]. Biomass liquefaction technology consists of the thermal depolymerization of the feedstock in the presence of a solvent, under mild temperatures (200–400°C) and high pressures (5–25 MPa), generating four streams: biocrude, aqueous phase, gas, and solid char [2]. The biocrude can be further upgraded through hydrotreatment into a mixture of hydrocarbons. This process is particularly suitable for wet feedstocks, such as algae and sewage sludge, avoiding the costly drying step, although even plastics have been shown to undergo depolymerization through liquefaction [3, 4]. The first discussions about using a solvent to produce oil from plant materials date back from the 1940s in the United States, motivated by concerns regarding possible oil scarcity [5]. Detailed studies on biomass liquefaction were conducted at the Pittsburgh Energy Research Center (PERC) in 1970 [6]. Wood conversion using wood recycled oil as solvent and Na_2CO_3 as a catalyst in the presence of a reducing gas (CO) achieved oil yields as high as 40 wt%. However, the process failed due to the inability to handle concentrated slurries, which led to poor economics. Since then, liquefaction technology has evolved, and the most promising approaches are moving to larger-scale plants. Hence, there is a significant interest in this technology to advance biorefineries. Fig. 5.1 shows a schematic diagram of the main streams involved in the liquefaction process.

This chapter builds on existing literature, with details of the chemistry behind biomass conversion through liquefaction, along with process parameters, biocrude upgrading to

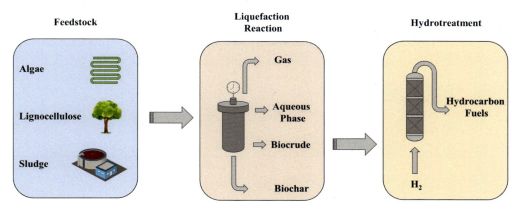

Fig. 5.1 Scheme of the liquefaction process.

liquid hydrocarbon biofuels, and aqueous phase valorization. An analysis of organic solvents as an alternative to water is presented, where the implications to commercialization are discussed. Finally, a summary of advances in commercialization elaborates the challenges and opportunities associated with scale-up, and the economic viability is analyzed where data is available.

5.2 Hydrothermal liquefaction
5.2.1 Process fundamentals

Hydrothermal liquefaction (HTL) is a technology capable of converting biomass into a biocrude, which can be further upgraded into transportation fuels, e.g., gasoline, jet fuel, and diesel. The process consists of the thermal depolymerization of the organic material in the presence of water under moderate temperatures (200–400°C) and high pressures (5–25 MPa) [2]. HTL offers the convenience to handle wet feedstocks, eliminating the energy and capital-intensive requirements of the feedstock drying, as opposed to fast pyrolysis or gasification technologies. As a conceptually simple process with high carbon efficiency, HTL has been applied to the valorization of several high-moisture content biological feedstock, such as algae [7] and sewage sludge [8, 9], as well as lignocellulosic biomass wastes [10].

The role of water in HTL is crucial to explain the fundamentals behind this process. While water does not react with organic substrates at ambient conditions of temperature and pressure, this is no longer the case under the process conditions. First of all, the water dielectric constant (ε), i.e., the relative permittivity, quickly decreases as the temperature increases. For instance, the dielectric constant of water at 225°C and 300°C is equivalent to that of methanol ($\varepsilon_{293.15,liq}=32.6$) and acetone ($\varepsilon_{293.15,liq}=21.0$) at room temperature, respectively. The reduction in water polarity is caused by a more uniform share between the electrons of oxygen and hydrogen, which

Table 5.1 Properties of water at different conditions of temperature and pressure.

	Ambient water	Subcritical water		Supercritical water	
Temperature (°C)	25	250	350	400	400
Pressure (MPa)	0.1	5	25	25	50
Density (g/cm^3)	1.00	0.80	0.60	0.17	0.58
Dielectric constant (F/m)	78.5	27.1	14.1	5.9	10.5
Ionic product (pK_w)	14.0	11.2	12.0	19.4	11.9
Heat capacity (kJ/kg K)	4.2	4.9	10.1	13.0	6.8
Dynamic viscosity (mPa/s)	0.89	0.11	0.06	0.03	0.07

Modified from S.S. Toor, L. Rosendahl, A. Rudolf, Hydrothermal liquefaction of biomass: a review of subcritical water technologies, Energy 36 (2011) 2328–2342. Copyright 2011 Elsevier.

reduces oxygen electronegativity [11]. This lower polarity facilitates the dissolution of hydrophobic compounds, such as lipids and hydrocarbons.

In addition to lowering water polarity, high temperatures significantly increase the dissociation rate of water, enabling it to act as an acid and basic catalyst, helping to promote feedstock conversion. The dramatic changes in water properties at high temperatures favor a set of organic chemical reactions, such as hydrolysis, dehydration, and polymerization reactions [12]. Under these hydrothermal conditions, water has been proven to depolymerize organic materials at rates quicker than those expected to be caused only by temperature increase [12]. Under conditions close to the critical point ($T_{C, H_2O} = 374°C$, $P_{C,H_2O} = 22 MPa$), the viscosity and density of water significantly reduce, while the ionization increases nearly three orders of magnitude compared to ambient conditions (Table 5.1). HTL processes typically operate under subcritical and supercritical conditions, near the water critical point [10, 13]. At conditions beyond the critical point, the hydrothermal gasification process takes over (Fig. 5.2).

5.2.2 Reaction pathways

The main objective of the HTL process is to liquefy biomass macromolecules and remove excess oxygen to obtain a biocrude with high energy content. Biomass is a general term for materials with different compositions of building blocks, such as carbohydrates, lignin, proteins, and lipids, in different ratios according to the species. Under hydrothermal conditions, biomass macromolecules are hydrolyzed into their basic units, which may further undergo conversion through other reactions, such as dehydration, decarboxylation, and polymerization. The fragments generated in this pool of reactions may repolymerize to form solids or recombine to form long-chain molecules according to the process conditions and chemical nature of the molecules. Appreciation of the mechanisms involved in these conversions is crucial to optimize biocrude yields and quality, based on the type of feedstock processed.

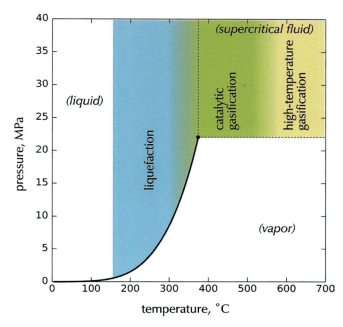

Fig. 5.2 Phase diagram for water. *(Reproduced with permission from A.A. Peterson, F. Vogel, R.P. Lachance, M. Froling, M.J. Antal, J.W. Tester, Thermochemical biofuel production in hydrothermal media: a review of sub- and supercritical water technologies, Energ. Environ. Sci. 1 (2008) 32–65. Copyright 2008 Royal Society of Chemistry.)*

5.2.2.1 Cellulose

Cellulose, a highly crystalline linear polymer of β-1,4 linked glucose units, is the main constituent of lignocellulosic biomass, representing 35–50 wt% depending on the feedstock [14]. Its crystallinity makes cellulose insoluble in water and resistant to enzyme attack under ambient conditions. Despite that, cellulose becomes highly soluble and readily hydrolyzed under hydrothermal conditions.

The cellulose degradation mechanism, under HTL conditions, starts with the hydrolysis of glycosidic linkages [15]. This reaction may occur homogeneously or heterogeneously (Fig. 5.3), depending on the degree of polymerization (DP) of the oligomers generated and the physical state of water (subcritical or supercritical) [16]. The fragments with low DP undergo degradation after dissolution in subcritical water, while larger fragments degrade heterogeneously. A significant difference, when the reaction proceeds heterogeneously, is the generation of pyrolysis products, which are not formed via the homogeneous mechanism.

The study of glucose as a cellulose monomer provides insights into the decomposition mechanism of cellulose. Dehydration is one of the main types of reactions that occur during HTL, as a result of both chemical and physical processes. Elimination of the residual

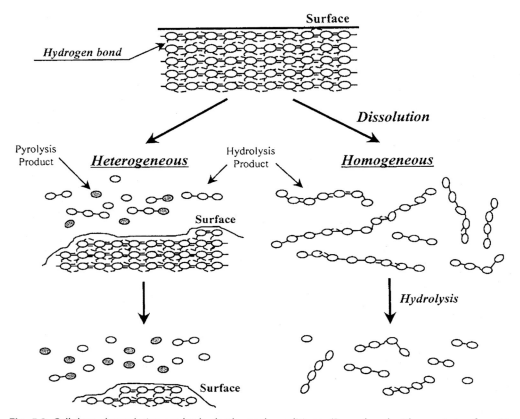

Fig. 5.3 Cellulose degradation under hydrothermal conditions. *(Reproduced with permission from M. Sasaki, Z. Fang, Y. Fukushima, T. Adschiri, K. Arai, Dissolution and hydrolysis of cellulose in subcritical and supercritical water, Ind. Eng. Chem. Res. 39 (2000) 2883–2890. Copyright 2000 American Chemical Society.)*

moisture of biomass due to an increase in hydrophobicity of the solid residue represents physical dehydration [17]. In contrast, chemical dehydration is characterized by the elimination of hydroxyl groups leading, for instance, to the production of levoglucosan. On the other hand, glucose may isomerize into fructose, which further dehydrates into hydroxymethylfurfural (5-HMF) [18]. This product may dehydrate once more into levulinic acid and formic acid, rearrange into trihydroxybenzene or polymerize into polyfuranic intermediates (Fig. 5.4). Retro-aldol condensation reactions lead to the formation of dihydroxyacetone, glyceraldehyde, and erythrose. Also, intermediate acids, such as lactic acid, acetic acid, and glycolic acid, may be formed by secondary reactions.

An important parameter affecting glucose degradation selectivity is the pH of the medium. In the strong alkaline pathway (initial pH > 11), mainly carboxylic acids, such as lactic acid, are formed, while in the weak alkaline pathway (initial pH < 11) higher

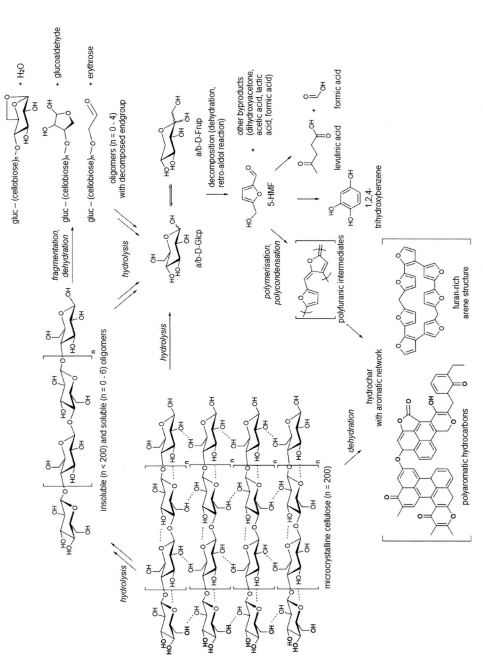

Fig. 5.4 Reaction pathways for cellulose dehydration during HTL. (Modified from M. Möller, F. Harnisch, U. Schröder, Hydrothermal liquefaction of cellulose in subcritical water—the role of crystallinity on the cellulose reactivity, RSC Adv. 3 (2013) 11035–11044. Copyright 2013 Royal Society of Chemistry.)

concentrations 5-HMF are found [19]. Carboxylic acids formed in weak alkaline solutions neutralize the medium and, when the pH becomes less than 7, the acidic pathway dominates. Therefore, pH is particularly relevant for the contribution of cellulose to HTL main product streams, since 5-HMF tends to form oil products, as opposed to carboxylic acids, which end up mostly in the water phase.

5.2.2.2 Hemicellulose

Hemicellulose is another major component of lignocellulose, representing 20–35 wt% of the biomass composition. Much like cellulose, hemicellulose is a carbohydrate-based polymer. However, hemicellulose is a less uniform heteropolymer, i.e., multiple carbohydrates, with an abundance of side groups, and a lower degree of crystallinity than cellulose. Due to this fact, hemicelluloses are more readily hydrolyzed than cellulose. While cellulose requires at least 250°C to undergo hydrolysis, hemicellulose is converted at temperatures above 180°C [20].

Hemicellulose is formed by several sugar units, including C_6, e.g., mannose, galactose, and glucose, and C_5, e.g., arabinose and xylose. Xylose is the main component of hemicellulose, representing up to 80% of its composition [21]. Therefore, this carbohydrate is frequently used for mechanistic studies on the transformation behavior of hemicellulose under HTL conditions. The degradation mechanism follows a similar trend to that of glucose, showing dehydration and retro-aldol pathways, but also the Lobry de Bruyn-Alberda van Ekenstein (LBET) transformation (Fig. 5.5) [22]. Degradation of xylose is highly dependent on water density, which is related to the operating pressure. For example, Aida et al. showed that increasing water density from 0.52 to 0.69 g/cm^3, at 400°C, by increasing pressure from 40 to 100 MPa, reduces the retro-aldol reaction rate of D-xylose [22]. Additionally, high water density shifts the equilibrium of the LBET reaction between D-xylose and D-xylulose, favoring D-xylulose production. D-xylulose is likely a primary product and, as a consequence, an intermediate for retro-aldol products and furfural.

5.2.2.3 Lignin

Lignin is the third major component of lignocellulosic feedstock, representing 10–25 wt% of the biomass composition. Lignin is a complex branched polymer made of hydroxyl and methoxyl substituted phenylpropane units. Its recalcitrance poses a challenge on hydrothermal conversion, since a significant amount of solid residue, i.e., char, may be generated from lignin due to repolymerization of reactive fragments [23]. Under hydrothermal conditions and above 200°C, lignin is depolymerized into phenolic oligomers and monomers by hydrolysis reactions (Fig. 5.6). These phenolic fragments can undergo secondary reactions of alkylation and condensation. Additionally, side-chain compounds lead to the formation of carboxylic acids, alcohols, ketones, and aldehydes (Fig. 5.6) [24].

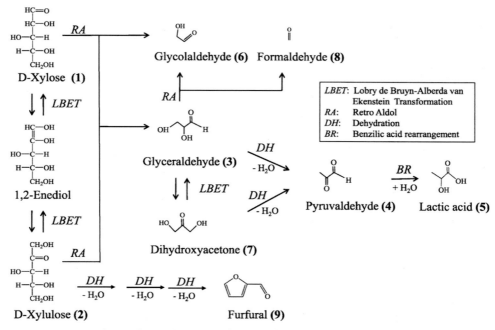

Fig. 5.5 Reaction pathways for D-xylose degradation under hydrothermal conditions. *(Reproduced with permission from T.M. Aida, N. Shiraishi, M. Kubo, M. Watanabe, R.L. Smith, Reaction kinetics of D-xylose in sub- and supercritical water, J. Supercrit. Fluids 55 (2010) 208–216. Copyright 2010 Elsevier.)*

5.2.2.4 Lipids and proteins

Lipids and proteins are particularly relevant for the hydrothermal conversion of algal biomass since they can represent 20–50 wt% [25] and 20–70 wt% [26] of the feedstock composition, respectively. Lipids consist of nonpolar triacylglycerides (TAGs), which are insoluble in water at ambient conditions. However, under hydrothermal conditions, the dielectric constant of water decreases, improving lipids miscibility. Consequently, lipids can undergo hydrolysis reactions, which leads to the formation of free fatty acids and glycerol (Fig. 5.7). With the increased concentration of fatty acids in the medium, the solubility of the lipids augments, accelerating the reaction rate [27]. Even though fatty acids are resistant to degradation, they may undergo decarboxylation, yielding hydrocarbons [28]. Like free fatty acids, glycerol may be converted, leading to the formation of allyl alcohol, acetaldehyde, and acrolein (Fig. 5.7) [29].

Proteins are polymeric derivatives of amino acids coupled together by peptide bonds, i.e., C—N bond between a carboxyl and an amine group. As such, they are responsible for most of the nitrogen content in biomass. Although more stable than glycosidic bonds, peptide bonds are hydrolyzed at temperatures above 230°C, followed by fast degradation of the resulting amino acids [13]. The heterogeneity of different amino

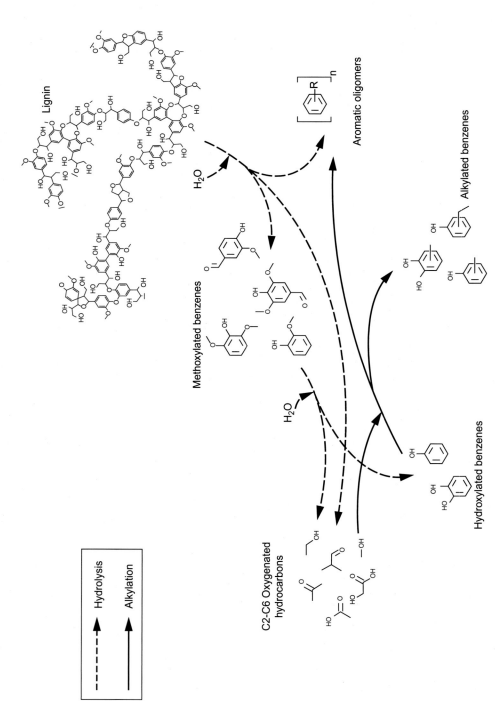

Fig. 5.6 Reaction pathways of the lignin under HTL conditions. (*Reproduced with permission from J. Barbier, N. Charon, N. Dupassieux, A. Loppinet-Serani, L. Mahe, J. Ponthus, M. Courtiade, A. Ducrozet, A.A. Quoineaud, F. Cansell, Hydrothermal conversion of lignin compounds. A detailed study of fragmentation and condensation reaction pathways, Biomass Bioenergy 46 (2012) 479–491. Copyright 2012 Elsevier.*)

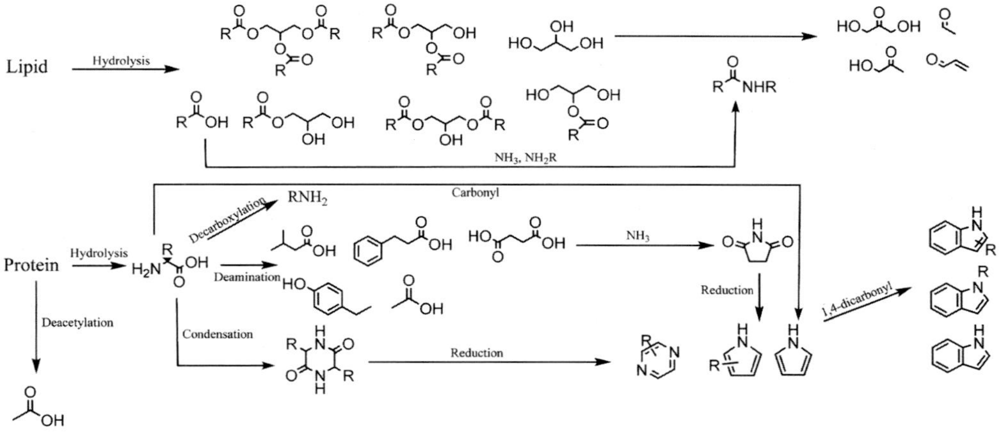

Fig. 5.7 Reaction pathways of lipids and proteins under HTL conditions. (*Modified from R.B. Madsen, R.Z.K. Bernberg, P. Biller, J. Becker, B.B. Iversen, M. Glasius, Hydrothermal co-liquefaction of biomasses—quantitative analysis of bio-crude and aqueous phase composition, Sustainable Energy Fuels 1 (2017) 789–805. Copyright 2017 Royal Society of Chemistry.*)

acids makes a detailed degradation mechanism a challenge. Still, all amino acids follow the same basic mechanism of (i) deamination, to produce ammonia and organic acids, and (ii) decarboxylation, forming carboxylic acids and amines [13]. Additionally, the presence of sugars in the medium promotes Maillard reactions, which yield nitrogen–containing cyclic organic compounds, as shown in Fig. 5.7.

5.2.2.5 Combined reaction pathway

The reaction network, extent, and occurrence, during HTL conversion of biomass, is ultimately controlled by the various component concentrations and process conditions. Additionally, it is important to understand that each component of biomass, i.e., lipids, proteins, cellulose, hemicellulose, and lignin, and their by-products does not react independently. Madsen et al. studied the product distribution of the coliquefaction of different biomasses [30]. In this study, the authors observed a modification in the product spectrum according to the liquefaction of combined biomasses. The effect was particularly important when high–protein content algae were liquified in the presence of lignocellulosic biomass. Ammonia and amines resulting from protein decomposition acted as shift bases and reacted with furans from lignocellulose, significantly increasing nitrogen content in the biocrude [30]. Similar conclusions were taken by Yang et al. for the hydrothermal conversion of soy protein mixed with cellulose and xylan [31].

In another study, Yang et al. investigated the synergistic and antagonistic effects between the different constituents of biomass by analyzing coliquefaction of different combinations of these and comparing the yields with the calculated mass averaged yield [32]. For instance, the interaction between lipids and both cellulose and hemicellulose exhibits a synergistic effect on biocrude yield through the reaction between free fatty acids and degradation intermediates, such as glyceraldehyde and furfural. The increase in ketone yield evidenced the synergistic effect of cellulose and lipids. The rather large amount of ketones identified in the products is expected to be generated through a sequence of hydrogenation of fatty acids from lipid hydrolysis, followed by alkylation of furfural and furans from cellulose hydrolysis. Additionally, short-chain carboxylic acids from hemicellulose hydrolysis showed advantages for lipid hydrolysis, generating a higher yield of free fatty acids. On the other hand, lignin did not show significant interaction with lipids, indicating that they follow independent behaviors in the HTL process. Lignin interaction with cellulose promoted an antagonistic effect on phenol yield due to the reaction between phenol and aldehydes, but it ultimately led to an increase in the biocrude yield.

Maillard reactions between carbohydrates (cellulose and hemicellulose) and protein were shown to give an antagonistic effect on the yield of typical decomposition products from carbohydrates such as aldehydes and furans. Since these compounds are highly reactive under reaction conditions, they are expected to react with degradation products from proteins, resulting in their absence in the biocrude derived from this analysis. In addition, the interaction between lipids and proteins showed a positive effect on amide and ester formation. The reaction pathway for the combined conversion of the different biomass components is shown in Fig. 5.8. This analysis elucidates some of the interactions among intermediate products under typical HTL conditions. Nevertheless, more mechanistic studies are required to understand the interactions when three or more of these components are present simultaneously.

5.2.3 Process parameters

The conversion of biomass and product selectivity in the HTL process relies on several process parameters, such as temperature, residence time, slurry concentration, pressure, and heating rate [33]. These parameters are discussed in this section, while the impact of feedstock type and catalyst selection is addressed in the following sections.

5.2.3.1 Temperature

Temperature plays a significant role in the biocrude yields [34]. The activation energy for bond cleavage reactions can be easily overcome at higher temperatures [35], while low temperatures are ineffective for cleaving the peptide bonds in protein derivatives [36]. Nevertheless, while excessive low temperatures ($<250°C$) suppress biocrude yields due to slow reaction rates, high reaction temperatures favor not only hydrolysis but also repolymerization of free radicals into char or decomposition of intermediates into gaseous products (Fig. 5.9). Additionally, high temperatures also favor secondary reactions and Boudard gas reactions ($CO_2 + C \rightleftharpoons 2CO$) improving gas formation. Hence, the optimal temperature for biocrude production relates to the balance between hydrolysis/fragmentation reactions and repolymerization/decomposition reactions [37].

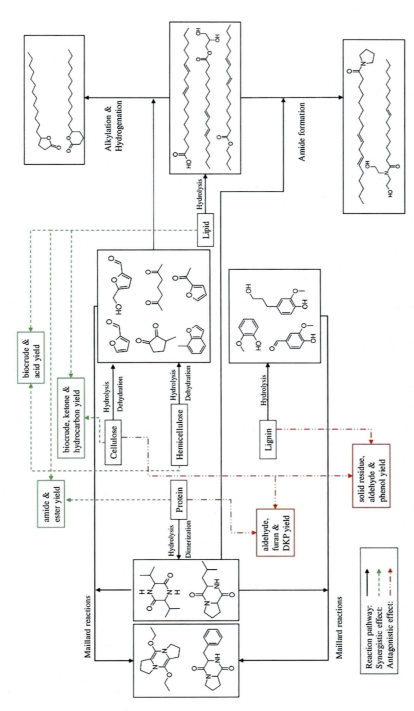

Fig. 5.8 Combined reaction network of lipids, proteins, cellulose, hemicellulose, and lignin under HTL conditions. *(Reproduced with permission from J. Yang, Q. He, H.B. Niu, K. Corscadden, T. Astatkie, Hydrothermal liquefaction of biomass model components for product yield prediction and reaction pathways exploration, Appl. Energy 228 (2018) 1618–1628. Copyright 2018 Elsevier.)*

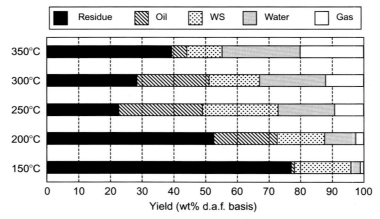

Fig. 5.9 Yield of products (dry ash-free—d.a.f) from HTL of eucalyptus over different temperatures. WS represents water-soluble material. *(Reproduced with permission from M. Sugano, H. Takagi, K. Hirano, K. Mashimo, Hydrothermal liquefaction of plantation biomass with two kinds of wastewater from paper industry, J. Mater. Sci. 43 (2008) 2476–2486. Copyright 2008 Springer.)*

5.2.3.2 Residence time

Residence time affects not only biocrude yield but also its composition. Since the reaction rate is somewhat fast, shorter residence times (<60 min) are preferred to avoid the decomposition of lighter products into gases [38]. Nevertheless, the long residence time may be advantageous for the decomposition of heavy compounds into lighter products. Generally, biocrude yields achieve a maximum value where decomposition reactions are dominant prior to decreasing for overly long residence times when condensation becomes notable (Table 5.2).

5.2.3.3 Slurry concentration

Another important process parameter is the ratio between biomass and water. When more water is present, the fragmented compounds are easily stabilized and solubilized

Table 5.2 Effect of residence time on yield of different streams of microalgae HTL at 330°C.

Residence time (min)	Mass yield (wt%)			
	Gas	Solid	Biocrude	Aqueous phase
15	20	18	35	24
30	5	21	60	14
45	9	11	62	18
60	12	16	55	17

Based on B.E. Eboibi, D.M. Lewis, P.J. Ashman, S. Chinnasamy, Effect of operating conditions on yield and quality of biocrude during hydrothermal liquefaction of halophytic microalga Tetraselmis sp, Bioresour. Technol. 170 (2014) 20–29.

[33]. On the other hand, at concentrated slurries, the interaction between biomass molecules and water becomes less important, and the process tends to behave more like pyrolysis. The heat and mass transfer are also hindered if a well-mixed suspension cannot be formed. This ratio is also crucial to guarantee the pumpability of the biomass slurry in commercial-scale operations.

5.2.3.4 Pressure

The main objective of pressure in the HTL process is to maintain a single-phase media since high energy input would be required for phase change [39]. Furthermore, solvent density increases with pressure, which promotes easier diffusion and enhances conversion. Conversely, the solvent cage effect slows down free-radical reactions at high pressure, due to intense collision frequency between the reactants and water, rather than with another reactant [40]. Upon supercritical conditions, the influence of pressure on water properties is minimal.

5.2.3.5 Heating rate

As opposed to the pyrolysis process, where the heating rate is critical for obtaining high biocrude yields, the effect of the heating rate in HTL is less important since the reactions take place in the liquid phase, where the solvent facilitates the dissolution of reactive intermediates [39]. Nevertheless, excessively slow heating rates promote secondary reactions that lead to char formation [33].

5.2.4 Effect of feedstock type

The type of feedstock influences HTL as the different components of biomass react differently. In this section, the effect of feedstock type is analyzed, with emphasis on biocrude yield and composition.

5.2.4.1 Wet feedstocks

HTL is particularly suitable for wet feedstocks, and, for this reason, significant research effort has been made toward the conversion of algae and sewage streams using this process [10]. Microalgae, in particular, are considered a promising energy source, which does not compete for agricultural land and displays a high growth rate. Another advantage of algae conversion is the possibility of nutrient recovery, e.g., N, and P, through reusing the aqueous phase for growing more algal biomass [7]. There are primarily two classes of algae considered: (1) low-lipid high-protein microalgae (e.g., *Nannochloropsis* sp.—50% protein, 14% lipid) and (2) high-lipid low-protein microalgae (e.g., *Chlorella* sp.—60% lipid, 9% protein) [41]. Processing these two types of feedstock has been shown to generate significantly different biocrude yields and product spectra [42]. Nevertheless, at high temperatures (375°C), the difference in biocrude yield tends to become less significant [43]. Despite all the positive impacts of using algae as a feedstock for HTL, drawbacks such as the presence of heteroatoms (mainly O and N) in the biocrude and, consequently, the need for expensive posttreatment has limited their application. Additionally, the cost of wastewater treatment would become excessively high if all the aqueous phase product needs to be treated. The yield of biocrude from microalgae can be predicted by the following equation [44]. The order of the highest yield per type of component follows lipids > proteins > carbohydrates.

$$\text{Biocrude yield (dry wt\%)} = 0.97(\pm 0.10) \times L + 0.42\,(\pm 0.0) \times P + 0.17\,(\pm 0.35) \times C \tag{5.1}$$

where L = lipids, P = proteins, and C = carbohydrates.

Another feedstock is sewage sludge, which can be processed through HTL to minimize environmental and hygienic risks [8]. Compared to landfilling and incineration, HTL has the potential to sterilize the sludge and generate a biocrude with 11-fold higher energy recovery as well as to reduce greenhouse gas emissions by 78%–85%. The dry basis composition includes lignin, ash (30–50 wt%), proteins (40 wt%), lipids (10–25 wt%), and carbohydrates (14 wt%) [45]. Similar to algae, the same process limitations have hindered a wide-spread application of HTL for this purpose.

5.2.4.2 Lignocellulosic biomass

Lignocellulosic biomass is often seen as a promising feedstock for the production of biofuels through HTL. The widespread availability, low price, and the fact lignocellulosic biomass is not used for food purposes represent its main advantages. Various types of wastes can be utilized as feedstock, such as agricultural and forestry wastes, due to their abundance and low price. Agricultural waste includes a variety of different types of biomass, such as sugarcane bagasse, wheat straw, and corn stover. Forestry residue includes both softwood and hardwood. Although these are all lignocellulosic feedstocks, they are expected to behave differently upon HTL conversion.

An analysis conducted for lignocellulosic feedstock showed that the highest biocrude yields per type of component follow the order: cellulose > hemicellulose > lignin, as shown in Fig. 5.10 [32]. This information can be used to develop coliquefaction processes, where a mixture of different biomass feedstocks is used to improve biocrude yields and properties.

5.2.4.3 Summary

Overall, the HTL process allows flexibility to process a large range of feedstocks, with particular advantages for wet feedstocks that otherwise would require the costly drying step (e.g., algae and sewage sludge). However, the feedstock composition significantly affects yields, and oily feedstocks tend to produce higher yields of biocrude. Table 5.3 shows the biocrude characteristics after HTL conversion of different feedstocks, and

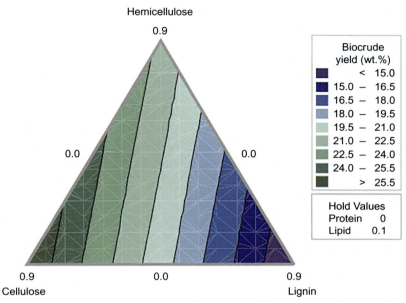

Fig. 5.10 Contour plot of biocrude yield versus the composition for lignocellulosic feedstock. *(Modified from J. Yang, Q. He, H.B. Niu, K. Corscadden, T. Astatkie, Hydrothermal liquefaction of biomass model components for product yield prediction and reaction pathways exploration, Appl. Energy 228 (2018) 1618–1628. Copyright 2018 Elsevier.)*

Table 5.3 Biocrude characteristics after HTL of different feedstocks.

Feedstock	C	H	O	N	S	HHV (MJ/kg)	Yield (%)	References
Spirulina algae	68.9	8.9	6.5	14.9	0.86	33.2	32.6	[46]
Swine manure	71.2	9.5	3.7	15.6	0.12	34.7	30.2	[46]
Anaerobic sludge	66.6	9.2	4.3	18.9	0.97	32.0	9.4	[46]
Coffee grounds	71.2	7.1	3.0	18.7	–	31.0	47.3	[47]
Aspen wood	75.2	8.2	15.8	0.5	0.3	34.3	27.5	[48]

Table 5.4 shows the main classes of compounds found in some types of biocrude. In algae feedstocks with high lipid content, the biocrude concentrates fatty acids. On the other hand, for lignocellulosic feedstocks, the presence of the phenolic corresponds to the larger percentage of the biocrude. Ketones are also present in more quantities in lignocelluloses due to the decomposition of carbohydrates.

Table 5.4 Chemical composition of biocrude from HTL of different feedstocks.

Feedstock	Biocrude (wt%)					
	Ketones	Nitrogenates	Phenolics	Carboxylic acids	Fatty acids	Alcohols
Miscanthus	12.8	0.3	43.7	19.8	20.0	3.5
Poplar	13.8	0.3	38.0	5.5	38.7	3.7
Willow	21.6	0.3	23.2	19.3	30.3	5.2
Spirulina	0.3	0.2	9.2	4.7	84.4	1.2

Based on R.B. Madsen, R.Z.K. Bernberg, P. Biller, J. Becker, B.B. Iversen, M. Glasius, Hydrothermal co-liquefaction of biomasses—quantitative analysis of bio-crude and aqueous phase composition, Sustainable Energy Fuels 1 (2017) 789–805.

5.2.5 Catalyst selection

5.2.5.1 *Homogeneous catalysts*

Even though not necessary for biomass conversion to take place, catalysts are key for HTL as they are capable of increasing oil yield and enable superior control over biocrude composition, improving its quality by removing heteroatoms, e.g., O, N, and S. Water-soluble homogeneous alkali salt catalysts, such as K_2CO_3, Na_2CO_3, KOH, and NaOH, are the most common catalysts [49] and are used in processes, such as Cat-HTR, developed by Licella [50–52], and Hydrofaction, from Steeper Energy [53]. However, homogeneous acids, such as phosphoric, acetic, formic, hydrochloric, and sulfuric acid, have also been used as catalysts [49, 54, 55]. Even though acid catalysts have shown promising results with high production of biocrude [49], alkaline catalysts demonstrate better performance in denitrification, deoxidation, and desulfurization reactions [56]. These reactions are essential for the improvement of the quality of the biocrude, and lower concentrations of heteroatoms. These factors directly translate to lower expenses for upgrading.

When present, homogeneous alkaline catalysts improve the C—C bond cleavage rate, enhance hydrolysis and promote dehydration and decarboxylation of feedstock [57]. Multiple studies have identified alkaline carbonates, in particular K_2CO_3, as the most promising catalysts for HTL [56, 58–60]. Additionally, potassium-based salts (e.g., K_2CO_3) were also reported to provide better biocrude yield than sodium equivalents (Na_2CO_3) [60].

Under reaction conditions, alkaline catalysts interact with the feedstock and water creating secondary promoters (Eqs. 5.2–5.5). Therefore, during the reaction, the conversion of the feedstock is promoted not only by the original homogeneous catalyst, but by a mixture of multiple alkaline salts, which include hydroxides, bicarbonates, carbonates, and formates (Eqs. 5.2–5.5) [57–59, 61]. These ions can also participate in the conversion of the feedstock, helping to promote dehydration, decarbonylation, decarboxylation, and hydroxylation reactions.

$$K_2CO_3 + H_2O \rightarrow KHCO_3 + KOH \tag{5.2}$$

$$KOH + CO \rightarrow HCOOK \tag{5.3}$$

$$HCOOK + H_2O \rightarrow KHCO_3 + H_2 \tag{5.4}$$

$$2KHCO_3 \rightarrow H_2O + K_2CO_3 + CO_2 \tag{5.5}$$

Although homogeneous catalysts can significantly improve the yield and quality of the biocrude [57–59], they cannot be separated from the HTL aqueous streams and must be disposed of, implying extra costs and environmental concerns. Additionally, there is a risk of salt precipitation because of the reduction in the dielectric constant as the temperature approaches the critical point of water. Operational problems, such as abrasion and clogging, could occur in these conditions. Nevertheless, due to pH increase, the presence of alkali salts inhibits the degradation of biomass monomers. Also, they promote the water-gas-shift (WGS) reaction, which may act as a reducing agent to improve the quality of the biocrude [13].

Despite the many benefits of homogeneous catalysts, which can lead to increases in biocrude yields of 5%–30%, the quality is still not suitable to produce large amounts of hydrocarbons [57]. Therefore, further upgrading is required for the production of transportation fuels compatible with current standards.

5.2.5.2 Heterogeneous catalysts

The use of heterogeneous catalysts could help to mitigate some of the challenges typically associated with their homogeneous alternatives [57]. Solid heterogeneous catalysts are not in the same phase as the reaction media. Therefore, they offer the possibility of being separated after reaction and reused multiple times. These catalysts can also be designed to be more resistant to the harsh operating conditions that typically destroy homogeneous catalyst materials [57]. Indeed, heterogeneous catalysts are usually preferred for industrial applications for these reasons.

Different types of catalysts, such as noble metals (Pt, Pd, Ru) [62, 63], oxides (Fe_2O_3) [64], sulfided transition metals from hydrodesulfurization (NiMo or CoMo/γ-Al_2O_3) [62, 63], and zeolites [62, 65, 66], can also improve biocrude yield from 35% to 55% when compared with the noncatalytic reaction [62]. The quality of biocrude, in terms of viscosity and heteroatoms concentration, was also improved [62].

Despite their potential, heterogeneous catalysts tend to suffer severe deactivation under HTL conditions. Traditional hydrogenation catalysts, such as supported Pd, Pt, Ru, or CoMo, NiMo, can undergo permanent deactivation by metal sintering, poisoning, and dissolution under reaction conditions [3]. Additionally, tar and coke formation on the catalysts also leads to deactivation, reducing the effectiveness of these catalysts [3]. Zeolites and aluminas, which are often used as support and active catalysts, are severely damaged under hydrothermal conditions [67, 68]. Furthermore, homogeneous alkali

5.2.6 Biocrude properties and storage

Biocrude is a dark, viscous liquid with a distinctive odor, insoluble in water, but miscible with organic solvents. It is a complex mixture of several hundred chemical compounds, with ketones, phenols, fatty acids, and carboxylic acids as the main components. Table 5.5 shows the physical characteristics of HTL-derived biocrude from wood and algae and the comparison with petroleum diesel. Typically, oxygen content in biocrude is between 10 and 20 wt%, and the heating value is in the range of 30–35 MJ/kg [10]. The high oxygen content compared to petroleum diesel, for instance, leads to a lower higher heating value (HHV) by around 20%. When algae are used as feedstock, the concentration of nitrogen in the biocrude is also of concern and must be reduced to prevent NOx emissions. Due to these problems, further upgrading is required to improve its properties.

The unstable nature of biocrude is also a concern. Undesirable side reactions may occur during storage, increasing viscosity. The chemical and physicochemical properties of biocrude limit its utilization as transportation fuel [73]. Palomino et al. studied the effect of processing parameters on the storage stability of biocrude from algal biomass, produced at 300°C and 350°C, over 60 days, at the storage temperatures of 4°C, 20°C, and 35°C [74]. Biocrude produced at a higher temperature (350°C) has lower initial viscosity than the one produced at a lower temperature (300°C) due to the more intense depolymerization of macromolecules that takes place at high temperatures. The viscosity of biocrudes stored at 35°C increased significantly over the period evaluated, as opposed to the samples stored at 4°C that were less affected; thus, characterizing the occurrence of esterification and polymerization reactions. This study draws attention to the fact that producing high-quality biocrude is pivotal for minimizing issues with storage.

Table 5.5 Properties of biocrude from different biomass sources and comparison with petroleum diesel.

	Aspen pine	Microalga *S. platensis*	Petroleum diesel
Density (kg/m^3, at 15°C)	1055	970	840
Viscosity (mm^2/s, 40°C)	3975	196	2.64
HHV (MJ/kg)	38	34	44
C (wt%)	79.4	73.7	86.1
H (wt%)	9.1	8.9	13.8
O (wt%)	10.9	10.2	0
S (wt%)	–	0.9	<0.005
N (wt%)	0.6	6.3	0
References	[70]	[71]	[72]

5.2.7 Biocrude upgrading

The most common method for upgrading biocrude is hydrotreatment. In this process, heteroatoms (O, N, and S) are eliminated by the addition of H_2 to the system at high pressure (2–20 MPa) and moderate temperatures (250–450°C), using a suitable catalyst [13]. Oil quality and stability are improved by the reduction of reactive compounds, such as organic acids and ketones, via hydrodeoxygenation (HDO). During HDO, oxygen is removed in the form of water upon hydrogenolysis, i.e., break of C—O bonds, hydrogenation, decarboxylation, and decarbonylation take place. Additionally, hydrocracking and thermal cracking promote the reduction of large molecules into smaller ones. In addition to oxygen elimination by HDO, reactions of hydrodenitrogenation (HDN) and hydrodesulfurization (HDS) may also occur during biocrude hydrotreatment, removing nitrogen and sulfur, respectively, in the form of NH_3 and H_2S. The main reactions observed in biocrude upgrading are represented below:

$$\textbf{Cracking}: R1 - CH_2 - CH_2 - CH_2 - CH_2 - R2 \rightarrow R1 - CH - CH_2 + H_2C = CH - R2 \tag{5.6}$$

$$\textbf{Decarbonylation}: R - HCO \rightarrow R - H + CO \tag{5.7}$$

$$\textbf{Decarboxylation}: R - COOH \rightarrow R - H + CO_2 \tag{5.8}$$

$$\textbf{Hydrocracking}: R1 - CH_2 - CH_2 - R2 + H_2 \rightarrow R1 - CH_3 + H_3C - R2 \tag{5.9}$$

$$\textbf{Hydrodeoxygenation}: R - OH + H_2 \rightarrow R - H + H_2O \tag{5.10}$$

$$\textbf{Hydrogenation}: RCH = CH_2 + H_2 \rightarrow RCH_2 - CH_3 \tag{5.11}$$

$$\textbf{Hydrodesulfurization}: R - SH + H_2 \rightarrow R - H + H_2S \tag{5.12}$$

$$\textbf{Hydrodenitrogenation}: R_3N + 3H_2 \rightarrow 3RH + NH_3 \tag{5.13}$$

Although the HTL process has the flexibility to handle different types of feedstock, the biocrude chemical composition varies accordingly, which poses challenges for developing an adequate upgrading process. An ideal process should remove all heteroatoms, reduce the average molecular weight of the oil while minimizing hydrogen consumption and excessive aromatic ring saturation. Besides, an important feature of the catalysts is the tolerance against water, present as residual moisture (5%–10%), and produced in situ. While for lignocellulosic biomass, the main objective is removing oxygen from the biocrude, when dealing with algae-derived feedstock, removing excess nitrogen is also relevant.

The commonly applied catalysts for hydrotreatment are commercial sulfided NiMo or CoMo on Al_2O_3 support and noble metals, e.g., Pt, Pd, or Ru [57]. In addition to their high cost, noble metals are usually not preferred since they are sensitive to impurities in the biocrude. Therefore, traditional sulfided catalysts utilized in HDS are the ones typically studied, despite the often need to add supplementary sulfur species to maintain the activity of these catalysts.

Reduction of oxygen content to less than 1 wt% was shown possible for batch mode hydrotreatment of wood-derived biocrude utilizing high metal loading NiMo/Al$_2$O$_3$ with 62.2% oil yield [75]. Characterization of chemical compounds in the hydrotreated biocrude showed substituted cycloalkanes, aliphatic hydrocarbons (C$_7$-C$_{24}$), and aromatic hydrocarbons. In continuous operation for the same type of biocrude and catalyst, temperature and residence time showed to have a strong influence on conversion and oxygen elimination, while increasing hydrogen pressure had less impact [70]. An increase in temperature from 320°C to 370°C could reduce the oxygen content from 4.7 to 2.2 wt%. Nevertheless, a single-stage hydrogenation step is generally not sufficient for the full upgrading of biocrude, and a second hydrotreatment step may be required. Frequently, two stages are employed for this type of process, where the first stage operates at mild temperature (ca. 275°C) and focuses on removing most of the oxygen and stabilizing the biocrude to prevent secondary reactions. The second operates at a higher temperature (>400°C) to complete the removal of heteroatoms [56]. This improved performance is due to the high rate of coke formation at harsh conditions, which ultimately leads to rapid catalyst deactivation and, potentially, reactor plugging [76].

For algae-derived feedstock, removing excess nitrogen during hydrotreatment is extremely relevant. However, the complete removal of nitrogen by hydrodenitrogenation (HDN) remains a challenge. Due to thermodynamical limitations, the hydrogenolysis of the aliphatic C—N bond is more challenging than C—O or C—S removal [77]. HDN mechanisms involve nucleophilic substitution, elimination, and saturation of intermediates, as illustrated in Fig. 5.11 by the typical model compounds, such as pyridine/piperidine, quinoline/tetrahydroquinoline, and indole/indoline. In the upgrading of algal biocrude, the catalyst must be active for both HDO and HDN, which remains a difficult task as full nitrogen removal is often not achieved. Although most hydrogenation catalysts can remove nitrogen to some extent, the amount left in the upgraded oil is still in the range of 1–4 wt% [77]. In ASTM and EN standards, there is no regulation for nitrogen content. However, the US DOE-funded studies have a target of less than 0.05% nitrogen content in the algal upgraded fuel [78].

The work conducted by the Pacific Northwest National Laboratory (PNNL) achieved a particularly significant reduction in nitrogen content of algal biocrude by continuous hydrotreatment using a commercial sulfided CoMo/F-Al$_2$O$_3$ [78]. Upgraded oil yields of 85% were obtained, and nitrogen reduction from 5.5 to 0.3 wt%, at 400°C and 100 bar H$_2$. However, due to the low sulfur content of algal biocrude, using sulfided catalysts requires the addition of an external sulfur source, which is not desired because of the consequent formation of hydrogen sulfide (H$_2$S). Also, they are not as selective toward C—N bond cleavage as noble metals. Due to the high cost and deactivation of noble metals, other approaches involve doping molybdenum carbides with noble metals [79]. In this case, although the activity toward HDN reactions can be improved when

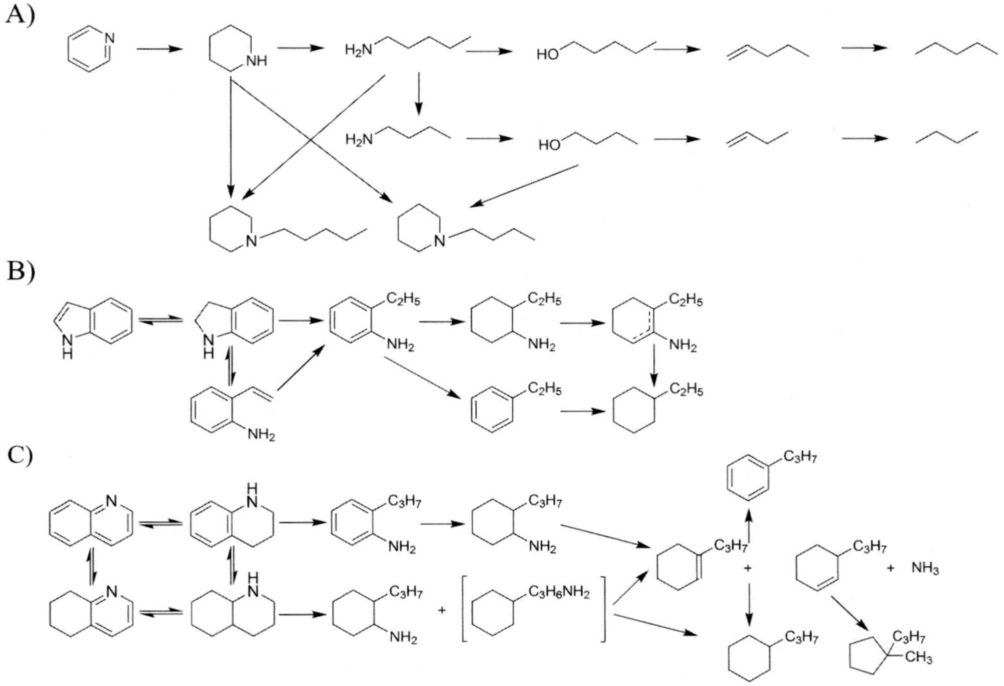

Fig. 5.11 Pathways for hydrodenitrogenation (HDN) of model nitrogenated compounds: (A) pyridine, (B) quinolone, and (C) indole. *(Reproduced with permission from C.Y. Yang, R. Li, C. Cui, S.P. Liu, Q. Qiu, Y.G. Ding, Y.X. Wu, B. Zhang, Catalytic hydroprocessing of microalgae-derived biofuels: a review, Green Chem. 18 (2016) 3684–3699. Copyright 2016 Royal Chemistry Society.)*

compared to sulfided CoMo and NiMo, a balance between optimum HDN and HDO performances requires further optimization.

As an example of the challenges involved in HDO and HDN of biocrude obtained from different feedstock, a comparison study was conducted for Spirulina, *Miscanthus*, and sewage sludge using a $NiMo/Al_2O_3$ catalyst, as shown in Table 5.6 [80]. Mild hydrotreating conditions (350°C and 80 bar H_2) are sufficient to fully deoxygenate biocrude from sewage sludge and Spirulina algae, while full deoxygenation of biocrude from

Table 5.6 Degree of deoxygenation and denitrogenation for different feedstocks at 80 bar H_2 [80].

Temperature	350°C		400°C	
Feedstock	de-O (%)	de-N (%)	de-O (%)	de-N (%)
Spirulina	100	48	100	47
Miscanthus	78	50	94	9
Sewage sludge	100	36	100	77

150 Hydrocarbon biorefinery

Miscanthus was not achieved even at harsh conditions (400°C and 80 bar H_2). The final product from sewage sludge and algae primarily consisted of paraffins, which are suitable for the diesel range. At the same time, *Miscanthus* led to more aromatics and naphthenes, which are suitable for gasoline blends. However, complete denitrogenation was still proven challenging.

5.2.8 Aqueous phase valorization

One of the major technical and commercial challenges for scaling-up HTL technologies is dealing with the aqueous phase generated [81]. With most of the attention placed on biocrude as the main product, until recently, the understanding of the aqueous phase had not seen equal emphasis. The same parameters affecting biocrude yield and composition also affect the aqueous phase, such as temperature and the biochemical composition of the feedstock [82]. Interestingly, the aqueous phase contains a high concentration of organic compounds, as can be seen in Table 5.7, which may enable different valorization strategies. Despite the low concentration of organic derivatives in the aqueous phase ($\sim$ 10 wt %), the carbon yield should not be neglected since between 20% and 50% of the organics from the feedstock are transferred into the aqueous phase [83]. The major compounds in the aqueous phase are small organic acids (e.g., acetic acid, glutaric acid, and pentanoic acid), cyclic oxygenates (e.g., cyclopentanone, butyrolactone, and valerolactone), oxygenated aromatics (benzoic acid, cresol, and phenol), nitrogenates (e.g., pyrazine and derivatives), and fatty acids (e.g., palmitic acid, stearic acid, and linoleic acid).

Several approaches have been evaluated for the valorization of the aqueous phase, including the separation of valuable chemicals, anaerobic digestion, hydrothermal gasification, use for biomass cultivation, and recycle back for HTL [81]. Due to the large concentration of organic compounds present in the aqueous phase, direct separation through chromatography and/or membranes is a potential route [84]. Nevertheless, these methods yet represent bench-scale experiments and would possibly lead to high operating costs on a commercial scale because the complex composition of the aqueous phase makes it challenging to extract selected compounds with high purity. One sensible approach is to collect the nutrients (N and P) as struvite, which possesses high quality as fertilizer [85]. Subsequently, anaerobic digestion can be employed as a simple and effective process to generate methane from the organics left in the liquid [86]. The challenge with anaerobic digestion of the HTL aqueous phase is its complex composition, which is abundant in nitrogenated organics, furfurals, and inorganics, typically not found in other wastewater streams [87, 88]. Alternatively, hydrothermal gasification at temperatures higher than 400°C in the presence of suitable catalysts can generate basic molecules, such as hydrogen, methane, carbon monoxide, and carbon dioxide [89]. This technology has the potential to convert most of the organic compounds into gaseous products, even though it requires high energy input.

Table 5.7 Chemical composition of the aqueous phase from HTL of different feedstocks [30].

Feedstock	(mg/L)							
	Small organic acids	Fatty acids	Oxygenated aromatics	Cyclic oxygenates	Nitrogenates	Total organic carbon (TOC)	Total nitrogen (TN)	pH
Miscanthus	7552	41	393	1683	155	20,845	231	6.5
Poplar	7831	109	345	1580	78	21,095	87	5.8
Willow	8063	127	98	1703	74	20,800	83	5.7
Spirulina	6720	257	239	341	796	26,715	10,150	9.5

A different approach is to utilize the aqueous phase from HTL to cultivate biomass rather than using synthetic growth media due to the presence of organic matter [90]. The primary challenge is the presence of inhibitors that could lead to excessively low growth rates [91]. Therefore, a pretreatment step may be required to decrease the concentration of inhibitors, with techniques, such as partial oxidation with ozone and H_2O_2 or adsorption with activated carbon and zeolite [92, 93]. Despite this fact, it has been shown that algae could grow to similar concentrations when compared to synthetic media [94]. This could potentially lead to a circular process, where the aqueous phase from algal biocrude is utilized for growing more algae. Similarly, other microorganisms can also grow at different concentrations of the HTL aqueous phase and, potentially, produce other valuable products [95, 96]. Further research in this field and improvements through metabolic engineering may enhance the yields and resistance of microorganisms to grow in this media.

Finally, one of the technically simplest solutions is to recycle the aqueous phase in the process [97]. Since the biomass to water ratio is typically about 1:4, recycling the aqueous phase contributes to both reduce the demand for make-up water and the expenditure with wastewater treatment. The recycle can then be utilized as a dilution agent for preparing a pumpable slurry in continuous processes [98]. This is likely to be the primary choice for commercial application of the first HTL technologies since it is easy to scale up. This strategy requires purge from time to time to avoid the build-up of recalcitrant compounds. Long-term experimental results have shown that recycling can improve biocrude yields and increase the organic matter concentration in the aqueous phase, which also facilitates its valorization on downstream processing [98, 99]. Consequently, there is a lower carbon discharge to the aqueous phase. The higher biocrude yield may allow to reduce operational temperature, which would also generate energy savings in commercial operations [97]. This improvement in yields may be attributed to the saturation of the aqueous phase with organics and the re-polymerization of nitrogenous compounds from the aqueous phase (proteins) back to the biocrude.

5.3 Liquefaction using organic solvents

As opposed to water, other solvents can be utilized for biomass liquefaction. The solvent plays an important role in the solvation of the fragments generated during chain scission, thus having a direct influence on product distribution [100]. Alcohols, ketones, and hydrocarbons are the most common options to replace water as a solvent, and due to their lower critical point, milder operating conditions can be implemented. Additionally, they have lower dielectric constants than water, as seen in Table 5.8. Consequently, they have a better affinity to high-molecular-weight apolar molecules from celluloses, hemicelluloses, and lignin than water. Biswas et al. compared *Sargassum tenerrimum* (macroalgae) liquefaction in water with methanol and ethanol as solvents [101]. They observed that, under the same conditions, the biocrude yield increased from 16 to 23 wt% when

Table 5.8 Properties of organic solvents utilized for liquefaction.

	T_C (°C)	P_C (MPa)	ρ_C (g/cm³)	Dielectric constant (ε) at 20°C
Water	374	22.1	0.322	78.5
Methanol	240	8.1	0.273	32.6
Ethanol	241	6.3	0.276	24.6
2-Propanol	235	4.8	0.271	18.3
Acetone	235	4.8	0.273	21.0
Glycerol	577	7.5	0.368	42.5
Ethylene glycol	447	8.2	0.325	37.7
Toluene	320	4.1	0.317	2.38

using alcohols as solvents. Besides, the analysis of the liquid products shows higher selectivity toward esters with alcoholic solvents, which could indicate the incorporation of the solvent into the biocrude. Another study, by Wang et al. also observed higher conversion with solvents, such as ethanol and acetone, when compared to water [102]. In this case, ethanol favored phenolic derivatives in the biocrude, while acetone favored ketones.

5.3.1 Liquefaction in pure organic compounds

From all organic solvents, alcohols, such as ethanol or methanol, are the most common alternatives to water in biomass liquefaction. Alcohol solvents were proven to improve the degradation of lignin and dissolution of lignin fragments [103]. Hence, improving the conversion of this polymer into biocrude. Ethanol has been shown to result in high biocrude yield (44%) for liquefaction of giant fennel compared to 2-butanol (39%), 2-propanol (38%), and methanol (36%) [104]. Similar performance was also observed for algae and cellulose [105, 106]. The replacement of alcohols for acetone resulted in a higher yield (47%) for both algae and cellulose. In terms of biocrude composition, it was found that methanol resulted in more hydrocarbon formation, while ethanol and acetone favor esters and ketones, respectively [106]. It was due to the different free radical species formed by each one of these solvents. Also, several comparative studies have shown that mixing water and alcohols can improve the conversion of lignocellulosic feedstock [103, 107–110]. The synergy between water and ethanol leads to improved properties of hydrolysis/depolymerization of cellulose and hemicellulose, together with lignin dissolution, preventing re-polymerization [103].

The product selectivity of wood liquefaction in multiple organic solvents is shown in Fig. 5.12. In general, the order of solvent preference to achieve high biocrude yields is oxygenated > aromatics > paraffinic [111]. Still, protic solvents, such as phenol, were found to lead to slow biomass conversion rates, needing longer reaction times [100]. Thus, optimized solvents can be prepared, aiming to improve interaction with biomass, by blending a major aromatic solvent with minor amounts of oxygenated compounds.

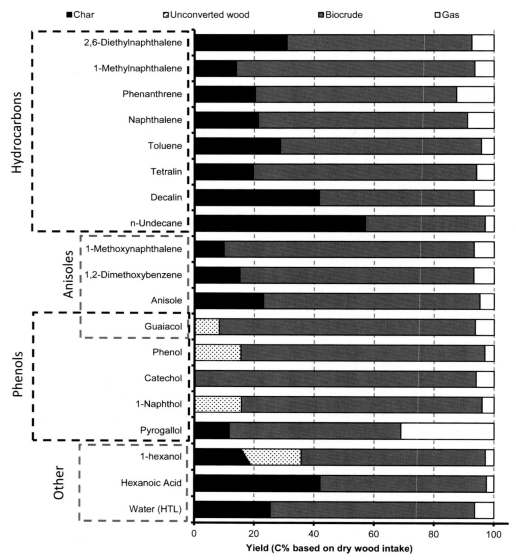

Fig. 5.12 Wood liquefaction yields (C basis) over different solvents, without catalyst (310°C and 30 min). *(Reproduced with permission from M. Castellví Barnés, J. Oltvoort, S.R.A. Kersten, J.-P. Lange, Wood liquefaction: role of solvent, Ind. Eng. Chem. Res. 56 (2017) 635–644. Copyright 2017 American Chemical Society.)*

The impact of the interaction between solvent and biomass can be analyzed by the Hildebrand (δ) and the Hansen (Ra) solubility parameters. The Hildebrand (δ) solubility parameter is defined as the square root of the cohesive energy density, which represents the degree of van der Waals forces holding the molecules of the liquid together, as follows [112]:

$$\delta = \sqrt{\frac{\Delta H_v - RT}{\nu_m}} \tag{5.14}$$

where ΔH_v is the enthalpy of vaporization, R is the ideal gas constant, T is the temperature, and ν_m is the molar volume of the pure solvent.

Compounds with comparable Hildebrand values are expected to be miscible. However, this parameter does not take into consideration polar interactions and hydrogen bonds. Therefore, Hansen proposed a three-dimensional solubility parameter (Ha) for polar molecules by dividing the Hildebrand parameter into three parts: dispersion (δD), polarity (δP), and H bonding (δH) [113]. The correspondence between the Hildebrand parameter and the partial Hansen solubility is described by the following equation:

$$\delta = \sqrt{\delta D^2 + \delta P^2 + \delta H^2} \tag{5.15}$$

The Hasan parameter is called "solubility radius" (Ra) between a solute and a solvent. A lower Ra means higher interaction. Ra is calculated by the following equation:

$$Ra = \sqrt{4 \times (\delta D_2 - \delta D_1)^2 + (\delta P_2 - \delta P_1)^2 + (\delta H_2 - \delta H_1)^2} \tag{5.16}$$

Therefore, minimum char yields or maximum biocrude yields occur when the Hildebrand parameter value is similar to that of individual biomass components or when a small Hansen (Ra) distance is observed with respect to biomass components (Fig. 5.13). These solubility parameters may also justify why adding water or phenolics to the medium improves the biocrude yield. A more polar solvent increases the Hildebrand parameter and reduces the Ra distance, resulting in improved interaction with the biomass components.

5.3.2 Implication of organic solvents for commercial application

While different solvents have been evaluated in multiple research studies, the recovery of the solvent from the biocrude is often overlooked. For instance, polyaromatic hydrocarbons were shown to have superior recovery (>95%) when compared to oxygenated solvents (<90%) [100]. Still, the cost of organic solvents (e.g., alcohols or

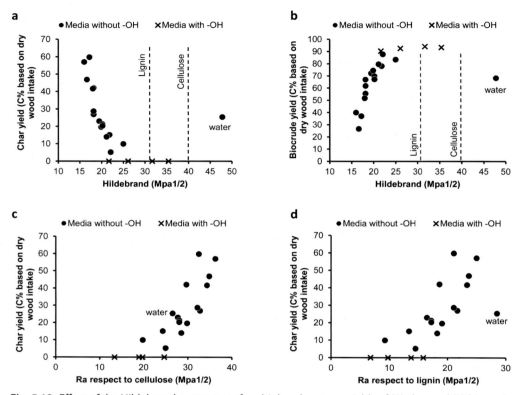

Fig. 5.13 Effect of the Hildebrand parameter of multiple solvents on yields of (A) char and (B) biocrude and correlation between char yield and *Ra* of the solvent with respect to (C) cellulose and (D) lignin. (Reproduced with permission from M. Castellví Barnés, J. Oltvoort, S.R.A. Kersten, J.-P. Lange, Wood liquefaction: role of solvent, Ind. Eng. Chem. Res. 56 (2017) 635–644. Copyright 2017 American Chemical Society.)

hydrocarbons) can be in the thousands of dollars per tonne, and therefore, the use of organic solvents instead of water poses a disadvantage in terms of the cost required to make up the solvent losses [114]. Thus, a commercial-scale liquefaction process requires solvent recovery of at least 99% to keep profitability [115]. This is particularly relevant for the category of hydrogen-donor solvents, which are capable of hydrogenating without the need to provide molecular hydrogen. The presence of hydrogen reduces the oxygen content of the biocrude while avoiding re-polymerization reactions between the reaction intermediates [116]. For instance, tetralin is a known hydrogen-donor solvent since the 1980s, capable of improving biocrude deoxygenation and yield in liquefaction processes [6]. Despite the many advantages, after dehydrogenation, hydrogen donor solvents, such as tetralin, require external hydrogenation to restore the original solvent to maintain high recovery.

The high cost of pure organic solvents and their difficult recovery increased the interest in using refinery streams, such as VGO (vacuum gas oil) and LCO (light cycle oil), as liquefaction solvents to achieve profitability at a commercial scale. In terms of integration with an existing refinery, VGO seems to be the best option since it is normally fed to the Fluid Catalytic Cracking (FCC) unit, which is less sensitive to water or oxygen. Besides their relatively low cost, these refinery streams are advantageous because of their higher boiling point. Currently, one of the bottlenecks of liquefaction processes toward scale-up is the high pressure required because of engineering limitations in biomass feeding [117]. Hence, efficiently converting biomass under mild conditions of pressure means a significant improvement for the commercial application of liquefaction technologies [115]. Nonetheless, the temperature range for the process (200–400°C) restricts the number of candidates. Within these limits, the solvent should have a boiling point of at least 170°C. Those solvents could be refinery streams such as LCO (BP = 195–400°C) [118], VGO (BP = 320–580°C) [119], tetralin (BP = 207°C) [120], and polyalcohols, such as polyethylene glycol (PEG-400) or glycerol (BP = 290°C) [121].

Finally, when using hydrocarbon solvents, algae streams are not as attractive as for HTL, due to the need for dry feedstock. Therefore, the lignocellulosic feedstock is often the preferred choice, making liquefaction in organic solvents less flexible than HTL. The biocrude obtained from liquefaction using organic solvents is somewhat similar to the one from HTL, and also requires further upgrading to be utilized in the transportation fuel sector. Table 5.9 shows the properties of different biocrude streams produced from wood liquefaction using hydrocarbons as solvents at different process conditions. Properties are within the range of those presented in Table 5.5 for biocrude in HTL.

The biocrude from organic solvent liquefaction must be upgraded via hydrotreatment, focused on hydrodeoxygenation (HDO), which can be achieved in one or more steps [123]. In general, the first HDO step allows the stabilization of biocrude by the

Table 5.9 Comparison of biocrude from different processes.

	Solvent		
Properties	**VGO**	**VGO**	**Hydrotreated LCO and naphthalene-depleted aromatic**
Total acid number (mg KOH/g)	101	32	51
Density (kg/m^3)	1097	1120	–
Viscosity (mm^2/s, 40°C)	2.3	–	17.7
HHV (MJ/kg)	–	30.7	38.1
C (wt%)	51.1	75.2	–
H (wt%)	6.2	7.2	–
O (wt%)	42.6	17.6	23.2
References	[122]	[117]	[118]

conversion of ketones and carboxylic acids, while the subsequent hydrogenation stages focus on complete oxygen removal. Several studies evaluated hydrotreatment of the liquefaction biocrude using sulfided $CoMo/Al_2O_3$ [123–125]. The operating conditions are typically harsh: 400°C and 120 bar of H_2, but almost full oxygen elimination was shown possible (<1.2 wt% oxygen), with low catalyst deactivation. For another liquefaction process, a bench study of hydrotreatment in two steps using a sulfided $NiMo/Al_2O_3$ catalyst at 177°C (Step 1) and 302°C (Step 2), under 155 bar over more than 1-month operation was possible without plugging and the oxygen target of less than 1 wt% was achieved [126]. In this case, catalyst deactivation was observed through the increase in oxygen concentration on the treated oil, requiring to increase in the temperature in the second reactor to 310°C to keep oxygen content within limits. However, based on the results and the constitution of the solvent (heavy aromatics mainly), the analysis of the biocrude oil suggested that hydrocracking might be the suitable upgrading process in this case.

5.4 Advances in commercialization

5.4.1 Hydrothermal liquefaction

5.4.1.1 Licella's Cat-HTR

Although HTL has potential as a biomass conversion technology to hydrocarbon biofuels, there have been few successful attempts of commercial-scale operation. One of them is led by Licella, an Australian company behind the Cat-HTR process, where water and organic material are pressurized to more than 220 bar between 350°C and 420°C, reaching supercritical conditions [51]. In the Cat-HTR process, the reaction mixture moves through a continuous flow reactor for several minutes before being cooled down below 200°C and depressurized back to atmospheric pressure (Fig. 5.14). The reactor effluent is separated into three fractions: gas, solid, and liquid. Although Licella patents mention several catalysts, homogeneous base catalysts (e.g., NaOH and KOH) are the preferred choices [50–52]. The addition of a second catalyst, such as HCOONa, acts as a reducing agent to favor hydrolysis and hydrogen transfer with oxygen elimination and consequent biocrude stabilization. Licella's patents also mention the possibility of utilizing a mixture of water and alcohol, e.g., methanol or ethanol, between 10 and 30 wt% of the solvent [50, 52]. The use of a cosolvent reduces the critical temperature and vapor pressure of the solvent mixture, besides promoting the alcoholysis of biomass, improving biocrude yield.

As presented in Fig. 5.14, biomass is combined with water and fed to a ball mill to reduce its particle size (<1 mm). Next, the slurry is pressurized and heated to around 50°C below the supercritical point of the solvent. Another stream, consisting of the solvent (water + alcohol), is heated to above the supercritical point of the mixture using a furnace. These two streams are combined in an injector where the resulting mixture is in

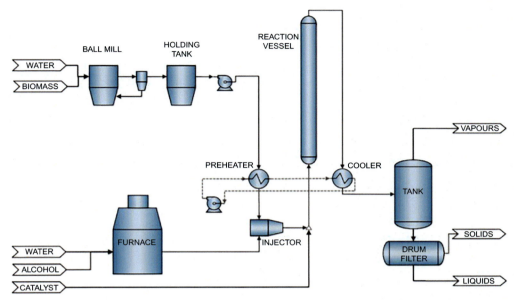

Fig. 5.14 Process flow diagram of the Licella CAT-HTR process. *(Reproduced with permission from G. Perkins, N. Batalha, A. Kumar, T. Bhaskar, M. Konarova, Recent advances in liquefaction technologies for production of liquid hydrocarbon fuels from biomass and carbonaceous wastes, Renew. Sustain. Energy Rev. 115 (2019) 109400. Copyright 2019 Elsevier.)*

its subcritical state, followed by mixing with catalyst and feeding into the reaction vessel. The reaction residence time is kept in the range of minutes (~30 min) before cooling, filtering the solids, and recovering the gases. The desired organic oil fraction can be obtained by density difference or further distillation, while the aqueous phase is treated and recycled back to the beginning of the process.

Licella's pilot plant in Somersby, Australia, can process up to 7500 tonnes per annum of organic material into biocrude [127]. The oxygen content of the biocrude produced from lignocellulosic biomass is within the range of 10–20 wt%, with a heating value between 30 and 38 MJ/kg [127]. Tests of blending as much as 20% of the company's raw biocrude with regular diesel and using in diesel engines showed similar brake power, whereas total unburnt hydrocarbon emissions (THC) increased by 13% and particulate emissions were lower than diesel [127]. Licella has established a joint venture with a major Canadian pulp and paper company, Canfor Pulp, to integrate the Cat-HTR technology into pulp mills, with the first project a 500,000 bbl/year plant in Canada.

5.4.1.2 Steeper Energy Hydrofaction

Another commercial effort toward commercial-scale HTL is being conducted by Steeper Energy, a Danish-Canadian venture that is advancing its Hydrofaction

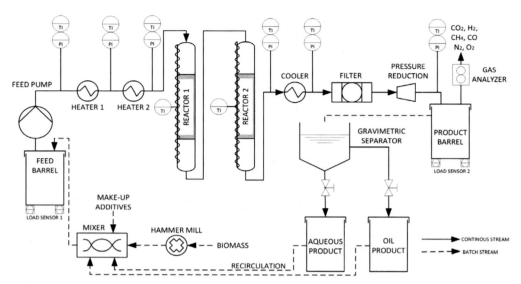

Fig. 5.15 Process flow diagram of Steeper Energy Hydrofaction process. *(Reproduced with permission from C.U. Jensen, J.K. Rodriguez Guerrero, S. Karatzos, G. Olofsson, S.B. Iversen, Fundamentals of Hydrofaction™: renewable crude oil from woody biomass, Biomass Conservs. Biorefin. 7 (2017) 495–509. Copyright 2017 Springer.)*

technology [53]. Similar to Licella, the process operates above the critical point of water (390–420°C and 300–350 bar), using the homogeneous alkali salt, K_2CO_3 to control the pH of the medium (>8) and achieve the desired reaction conditions. The catalyst helps to promote hydrolysis, depolymerization, decarboxylation, and WGS reactions. The aqueous phase containing water-soluble organics and the homogeneous catalyst recirculates back to the first step to assist in the preparation of the slurry. A purge of 5%–15% is required to regulate the accumulation of trace components, such as chloride. In addition, a portion of the biocrude is recycled to the preconditioning step to preheat the slurry and promote the stabilization of reactive intermediates in the reaction. Steeper Energy owns a pilot plant at the Aalborg University in Denmark, with the capacity 30 kg/h of slurry from wood particles (<2 mm particle size) (Fig. 5.15). A two-stage hydrotreatment study of biocrude from the Hydrofaction process using sulfided catalysts showed oxygen decrease from 10.9 to 1.8 wt%, resulting in 52% diesel fraction, 12% gasoline, 20% light vacuum gas oil (LVGO), 11% heavy vacuum gas oil (HVGO), and 5% residue (BP > 550°C) [70]. A demonstration plant in partnership with Silva Green Fuel, in Norway, with a production capacity of 4000 L/day and budgeted at US$60 M is predicted for 2021.

5.4.2 Liquefaction using organic solvents
5.4.2.1 High-pressure liquefaction
5.4.2.1.1 Iowa State University and Chevron

Iowa State University, together with Chevron, evaluated a continuous noncatalytic liquefaction process using a mixture of LCO and naphthalene-depleted heavy aromatic as solvent at 400°C and 43 bar [118]. The unit had a processing capacity of 1 kg/h biomass, with online solids removal and biocrude separation using acetone. Up to 55.3 wt% biocrude yields (solvent-free, dry biomass basis) with less than 6% oxygen content was obtained from loblolly pine. Fig. 5.16 shows the schematic process flow diagram of this

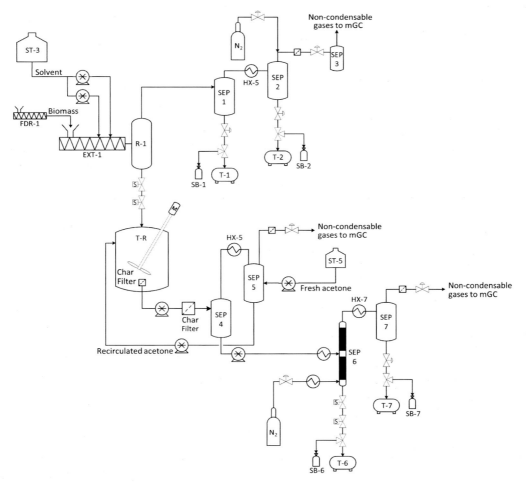

Fig. 5.16 Process flow diagram for liquefaction pilot plant using hydrocarbons at Iowa State University. *(Reproduced with permission from M.R. Haverly, T.C. Schulz, L.E. Whitmer, A.J. Friend, J.M. Funkhouser, R.G. Smith, M.K. Young, R.C. Brown, Continuous solvent liquefaction of biomass in a hydrocarbon solvent, Fuel 211 (2018) 291–300. Copyright 2018 Elsevier.)*

process. Biomass is fed into a pressurized reaction by a twin-screw extruder, where part of the reaction occurs. The reactor vessel allows residence time and gas/liquid separation. The solids settle in the bottom of the vessel and are separated by filtration, aided by acetone. The biocrude is separated into different fractions, and a cut near the boiling point range of the hydrocarbon solvent is recirculated to make the slurry. Up to 93% of the initial solvent is recovered during this process. For the upgrading step, continuous lab-scale studies were conducted by Chevron using commercial $NiMo/\gamma\text{-}Al_2O_3$ catalyst in two reactors in parallel [126].

5.4.2.2 Low-pressure liquefaction

Alternatively, the liquefaction process can operate under near atmospheric pressure, in a so-called low-pressure liquefaction process. Two distinct steps are required. First of all, biomass is mixed with the organic solvent, commonly designated as a carrier, under near atmospheric pressure at temperatures generally below 200°C. This preconditioning step is required to remove residual moisture, to improve the diffusion of the solvent within the biomass, and, sometimes, to promote particle size reduction (Fig. 5.17). The preconditioning stage also forms a homogeneous slurry to be fed to the liquefaction reactor. As the system operates under low pressure, the continuous separation between the reaction products, i.e., aqueous phase, apolar organic phase, and noncondensable gases, is possible by evaporation, followed by condensation and decanting (Fig. 5.17). The organic phase has physical properties similar to transportation diesel, whereas the aqueous phase includes ketones, alcohols, and acids [122]. The solvent is recycled back to the preconditioning step after the separation of the remaining solid char from the slurry. Solvent make-up is required as the solvent is partially converted into products under reaction

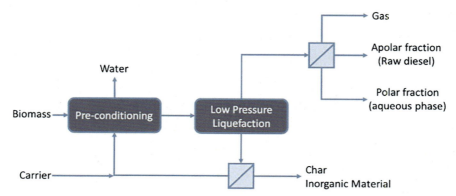

Fig. 5.17 Block flow diagram for low-pressure liquefaction process. *(Reproduced with permission from G. Perkins, N. Batalha, A. Kumar, T. Bhaskar, M. Konarova, Recent advances in liquefaction technologies for production of liquid hydrocarbon fuels from biomass and carbonaceous wastes, Renew. Sustain. Energy Rev. 115 (2019) 109400. Copyright 2019 Elsevier.)*

conditions. Important features of this process include efficient and rapid liquid–phase heat transfer due to the presence of the solvent, rapid separation of the products by evaporation from the reaction media, and solvolysis chemistry. These features place low-pressure liquefaction process in between traditional approaches, such as fast pyrolysis and hydrothermal liquefaction. Similar to pyrolysis, the reaction takes place at atmospheric pressure with low residence time, resulting in a low re-polymerization rate of the reactive intermediates [119]. The presence of a solvent, which characterizes it as liquefaction, affects not only the chemistry of solvolysis reactions but also has the advantage of improving the rate of heat transfer.

5.4.2.2.1 BDI—BioEnergy International bioCRACK process

One of the few commercial attempts of noncatalytic low-pressure liquefaction is the bio-CRACK process (BDI-BioEnergy International, GmbH) in Austria. The technology uses a refinery stream (VGO) as a solvent for the liquefaction of lignocellulosic feedstock under atmospheric pressure (Fig. 5.18). A pilot plant with 2.4 tonnes/day throughput of feedstock is integrated with the OMV Refinery at Schwechat, Austria, as shown in Fig. 5.19 [128]. The reaction takes place at 350–400°C, with the solvent/biomass ratio in the range of 3–6. The vapors that leave the reaction vessel are condensed and separated into aqueous and oil phases. The final product, called bioCRACK oil, is obtained after distillation of the biocrude, and the remaining residue is recycled to be used as solvent again. The outlet streams are processed in the refinery. The solvent is upgraded in the FCC unit, and the bioCRACK oil is further hydrogenated to allow full oxygen removal and to meet fuel specifications. The noncondensable gases are combusted to provide energy for the plant, while the pyrolysis oil and char are not processed by the refinery. Since there is a mix of biogenic and nonbiogenic carbon feedstock concomitantly processed, ^{14}C analysis is required to determine the actual biomass conversion. The overall mass balance demonstrated the distribution of biogenous carbon in four product streams (Fig. 5.19): bio-CRACK oil (22%—apolar fraction), pyrolysis oil (20%—polar fraction), char (37%), solvent VGO after the process (16%), and gas (6%). The solvent, VGO, is also distributed between the different fractions of products, except the polar stream pyrolysis oil [119]. Indeed, within the diesel product, 93.7% (carbon basis) is derived from VGO [119]. The boiling point range of VGO (320–550°C) and the formation of a heterogeneous azeotrope with water are responsible for the presence of nonbiogenic carbon in the bio-CRACK oil [129]. Still, process integration with a refinery makes solvent recovery less relevant for process economics than for other liquefaction processes, which is a significant advantage. Still, VGO conversion into bioCRACK oil limits liquefaction temperature to a maximum of 400°C.

After the reaction, the VGO stream contains 2% of biogenic carbon, around 0.5% of oxygen, and is richer in heavier compounds due to evaporation during the bioCRACK process. Hence, the overall quality of the processed VGO is inferior to pure VGO [130].

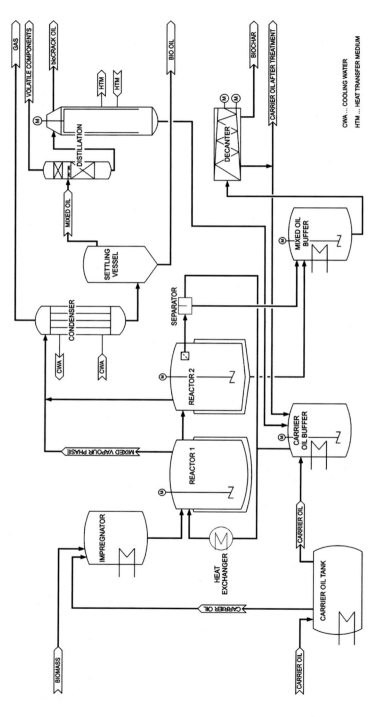

Fig. 5.18 Process flow diagram of the bioCRACK process. (From N. Schwaiger, D.C. Elliott, J. Ritzberger, H. Wang, P. Pucher, M. Siebenhofer, Hydrocarbon liquid production via the bioCRACK process and catalytic hydroprocessing of the product oil, Green Chem. 17 (2015) 2487–2494. Published by The Royal Society of Chemistry.)

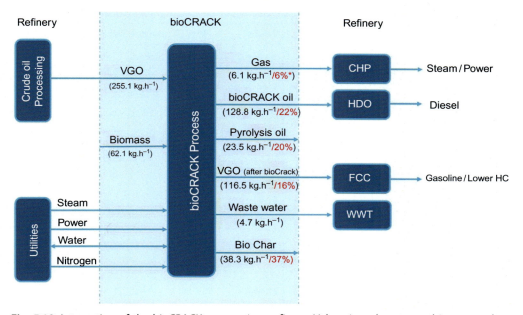

Fig. 5.19 Integration of the bioCRACK process in a refinery. Values in *red* represent biomass carbon distribution (^{14}C) between reaction outlet streams for the conversion of spruce wood at 375°C. *(Reproduced with permission from G. Perkins, N. Batalha, A. Kumar, T. Bhaskar, M. Konarova, Recent advances in liquefaction technologies for production of liquid hydrocarbon fuels from biomass and carbonaceous wastes, Renew. Sustain. Energy Rev. 115 (2019) 109400. Copyright 2019 Elsevier.)*

It was shown experimentally that FCC processing of the VGO effluent from bioCRACK resulted in lower gasoline and LCO yields than pure VGO. Additionally, trial attempts were made to convert the pyrolysis oil fraction into fuel. It can be upgraded through hydrotreatment, and the resulting pyrolysis oil is more stable and with lower polarity; therefore, it can be combined with the bioCRACK oil, leading to a total carbon yield of 32% in the diesel fraction [131]. Even though introducing biogenic feedstock in the infrastructure of a refinery may be seen as an attraction, oxygenated compounds can have severe impacts on the overall operation and need further and detailed consideration to evaluate this integration in a large-scale operation.

5.5 Process economics

Several studies on the economics of liquefaction processes have been published in the literature. Despite the promising aspect of biomass liquefaction, robust technologies must be competitive with the existing petroleum-based industry to speed up commercial implementation. In this section, techno-economic analyses (TEA) of hydrocarbon biofuel from different liquefaction processes are compared. Four TEA reports are

Table 5.10 Overview of published techno-economic studies for biomass liquefaction [132].

Feedstock	Microalgae	Wood	WWTP sludge	Wood
Location	United States	Europe	United States	United States
Configuration	Centralized	Centralized	Distributed	Centralized
Feed scale (tonnes per day, dry ash free)	2000.00	992.00	95.84	2000.00
Feed cost (US$/tonne, dry)	66.00	82.61	0.00	66.00
Process (type)	Catalytic HTL	Catalytic HTL	Thermal HTL	Thermal liquefaction with a hydrocarbon solvent
Biocrude yield (wt%)	36.0	37.5	43.6	–
Total capital investment (TCI) (US$ million)	583.96	193.20	512.40	848.70
Fuel products (ML/year)	258.38	117.02	143.01	266.11
Overall liquid fuel yield (wt%)	28.02	26.21	33.10	30.27
MFSP (US$/LGE)	0.72	0.97	0.92	0.84
References	[133]	[134]	[135]	[126]

summarized in Table 5.10, including information about the yield of biocrude and liquid fuel, the total capital investment (TCI), and the minimum fuel selling price (MFSP) [126, 132]. The study by Tzanetis et al. considered gasoline, diesel, jet, and heavy fuel as fuel products [134], while the remaining studies only considered gasoline and diesel. The study by Snowden-Swan et al. considered 10 distributed plants of sludge waste feeding into a centralized upgrading facility [135].

The current MFSP estimated for liquefaction to liquid fuel products ranges between $0.72 and $0.97 per liter gasoline-equivalent (LGE), which is around 2–3 times more than fossil fuel-based hydrocarbons [132]. Even over different studies, sensitivity analysis shows that the biocrude yield is the critical parameter for the fuel production cost, which emphasizes the need to achieve optimum performance in the liquefaction process. Also, feedstock price and hydrogen consumption in the upgrading step are other key parameters that affect the MFSP. A study by Pedersen et al. evaluated lignocellulosic biomass coliquefaction with glycerol and concluded that this is unlikely to be a feasible scenario due to the glycerol price [136]. This indicates that using solvents other than water will possibly reduce the viability of biomass liquefaction technologies. The process of economics becomes highly dependent on the solvent price, which in turn leads to low–profit margins.

5.5.1 Hydrothermal liquefaction

Fig. 5.20 shows the capital cost distribution for the HTL process in the TEA studies previously mentioned, indicating the units that impact the most on the product cost.

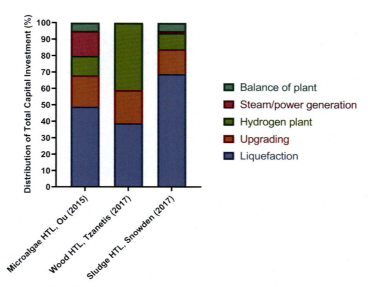

Fig. 5.20 Distribution of total capital investment from TEA reports of HTL processes [132].

For centralized plants, the cost of the liquefaction section represents around 40%–50% of the capital investment, while for distributed, it increases to almost 70%. Although hydrotreatment consumes significant amounts of high-pressure hydrogen, the upgrading section costs only, on average, about 10%–20% of the total installed capital.

One possibility to improve the overall efficiency of HTL technologies is to upgrade the aqueous phase [137]. A study conducted by the PNNL evaluated strategies for the valorization of the aqueous phase from the HTL of lignocellulosic biomass [83]. The base case considered was to produce methane (CH_4) through anaerobic digestion of the aqueous phase. Two valorization processes were analyzed. Case 1 focused on single-step ketonization of acids to ketones and ketone condensation into branched C_4 and C_5 olefins on a $Zn_xZr_yO_z$ catalyst (450°C, 1 atm). Case 2 focused on a two-step process of ketonization in a $La_xZr_yO_z$ (300°C, 97 atm), followed by steam reforming with cobalt (Co) as the active metal. Case 1 can reduce the biofuel cost by 13% when compared to the base case through coproduct credits. Despite Case 2 not offering cost reduction, producing H_2 by reforming lowers the carbon impact of the process. Therefore, upgrading the aqueous phase can be explored for commercial applications of HTL. It was estimated that including an aqueous-phase upgrading step increases capital investment by US$80–100M.

5.5.2 Liquefaction using organic solvents

While abundant literature is available for the economic potential of HTL, the information available for liquefaction using organic solvents is fairly limited. A study conducted at

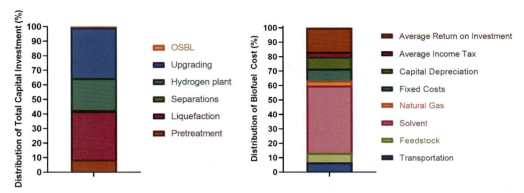

Fig. 5.21 Distribution of total capital investment and biofuel cost for a liquefaction process using hydrocarbon solvents developed at Iowa State University and Chevron [126].

Iowa State University in partnership with Chevron evaluated the feasibility of a pinewood liquefaction process using a mixture of refinery streams as solvents without the addition of any catalyst or reducing gas [126]. While the MFSP is within the same range as for HTL processes (Table 5.10), the TCI for the plant of US$849 M exceeds by at least 1.5 times the highest value for an HTL plant. Although the biocrude yield variability is also crucial for this type of liquefaction, sensitivity analysis shows that the biofuel cost is dominated by the solvent price rather than the feedstock price. Therefore, as the solvent is a crude oil fraction, the MFSP is strongly attached to the crude oil price. Liquefaction and upgrading are the costliest operations involved, as shown in Fig. 5.21. Additionally, a short study conducted by Shell for pinewood liquefaction using a similar mixture of refinery streams as solvent estimated the biocrude price between $55 and 70/bbl (energy-equivalent to crude oil) and TCI at US$222 M, without considering the upgrading step [138]. However, more research is required to analyze the economic potential of liquefaction using hydrocarbon solvents to confirm the assumptions made, given the technology readiness level.

5.6 Conclusions

In this chapter, biomass liquefaction toward the production of hydrocarbon biofuels has been discussed. Liquefaction technologies offer a relatively simple pathway to produce high yield biocrude, where the reaction takes place in the liquid phase under moderate temperatures. Flexibility toward dry and wet feedstocks is a significant advantage of liquefaction technologies. The prevailing technology involves using water as the solvent and reaction media, i.e., hydrothermal liquefaction(HTL). However, new alternatives are emerging that use hydrocarbon solvents, such as refinery streams like VGO. Although the production of renewable hydrocarbon biofuels from biomass is advancing, the complex structure of biomass, with several types of compounds, creates significant technical and commercial challenges. The biocrude upgrading needs to be adjusted for achieving proper deoxygenation and denitrogenation.

Challenges related to pumping slurries safely at high pressure need to be overcome as evident from the development of at least three pilot-scale processes (Licella, Steeper Energy, and Iowa State University). Additionally, atmospheric pressure liquefaction using refinery streams was proven feasible in a pilot scale (bioCRACK). The refinery integration remains a major risk, due to the recirculation of oxygenated streams, which needs to be deeply evaluated. Nevertheless, there are indications that these technologies are moving toward demonstration-scale plants in the short term.

In terms of economic potential, the capital cost related to the reactors and heat exchangers in the liquefaction step remains a challenge for scale-up given the uncertainties around catalyst performance, heat and mass transfer, and the impact of recycling streams. Handling waste feedstock is also critical for enabling these technologies to reach a commercial scale. However, the presence of contaminants, such as sulfur, chlorine, and heavy metals, accelerates catalyst deactivation and causes poisoning. The performance of hydrotreatment catalysts over long periods has not been demonstrated yet, and contaminants in wastes will bring challenges for demonstration and commercial scale. Hence, clean streams, such as woody biomass, are more likely to be used in initial commercial-scale plants. As the cost of biofuels from liquefaction technologies remains more than two times that of fossil fuels, other improvements are sought. For instance, upgrading the aqueous phase to energy or hydrogen generation seems a sensible approach, although large-scale testing is still required at this stage. Government policies, subsidies, and incentives will likely play a major role in the commercial deployment of these technologies.

Finally, a deeper understanding of the process fundamentals is still required to fill gaps in the knowledge. In particular, more fundamental studies are necessary to understand the intermediates and mechanistic pathways of this process, to improve deoxygenation and improve carbon efficiency in all technologies. Such analyses could help optimize reaction conditions and determine synergistic effects between reaction intermediates, feedstocks, and solvents.

References

[1] D.M. Alonso, J.Q. Bond, J.A. Dumesic, Catalytic conversion of biomass to biofuels, Green Chem. 12 (2010) 1493–1513.

[2] A. Dimitriadis, S. Bezergianni, Hydrothermal liquefaction of various biomass and waste feedstocks for biocrude production: a state of the art review, Renew. Sustain. Energy Rev. 68 (2017) 113–125.

[3] D. López Barreiro, W. Prins, F. Ronsse, W. Brilman, Hydrothermal liquefaction (HTL) of microalgae for biofuel production: state of the art review and future prospects, Biomass Bioenergy 53 (2013) 113–127.

[4] W.-T. Chen, K. Jin, N.-H. Linda Wang, Use of supercritical water for the liquefaction of polypropylene into oil, ACS Sustain. Chem. Eng. 7 (2019) 3749–3758.

[5] E. Berl, Production of oil from plant material, Science 99 (1944) 309–312.

[6] N.P. Vasilakos, D.M. Austgen, Hydrogen-donor solvents in biomass liquefaction, Ind. Eng. Chem. Process. Des. Dev. 24 (1985) 304–311.

[7] D.C. Elliott, T.R. Hart, A.J. Schmidt, G.G. Neuenschwander, L.J. Rotness, M.V. Olarte, A.H. Zacher, K.O. Albrecht, R.T. Hallen, J.E. Holladay, Process development for hydrothermal liquefaction of algae feedstocks in a continuous-flow reactor, Algal Res. 2 (2013) 445–454.

[8] W.T. Chen, M.A. Haque, T. Lu, A. Aierzhati, G. Reimonn, A perspective on hydrothermal processing of sewage sludge, Curr. Opin. Environ. Sci. Health 14 (2020) 63–73.

[9] L. Qian, S. Wang, P.E. Savage, Hydrothermal liquefaction of sewage sludge under isothermal and fast conditions, Bioresour. Technol. 232 (2017) 27–34.

[10] A.R.K. Gollakota, N. Kishore, S. Gu, A review on hydrothermal liquefaction of biomass, Renew. Sustain. Energy Rev. 81 (2018) 1378–1392.

[11] P.E. Savage, Organic chemical reactions in supercritical water, Chem. Rev. 99 (1999) 603–622.

[12] M. Siskin, A.R. Katritzky, Reactivity of organic compounds in hot water: geochemical and technological implications, Science 254 (1991) 231–237.

[13] S.S. Toor, L. Rosendahl, A. Rudolf, Hydrothermal liquefaction of biomass: a review of subcritical water technologies, Energy 36 (2011) 2328–2342.

[14] F.H. Isikgor, C.R. Becer, Lignocellulosic biomass: a sustainable platform for the production of biobased chemicals and polymers, Polym. Chem. 6 (2015) 4497–4559.

[15] O. Bobleter, Hydrothermal degradation of polymers derived from plants, Prog. Polym. Sci. 19 (1994) 797–841.

[16] M. Sasaki, Z. Fang, Y. Fukushima, T. Adschiri, K. Arai, Dissolution and hydrolysis of cellulose in subcritical and supercritical water, Ind. Eng. Chem. Res. 39 (2000) 2883–2890.

[17] T.C. Acharjee, C.J. Coronella, V.R. Vasquez, Effect of thermal pretreatment on equilibrium moisture content of lignocellulosic biomass, Bioresour. Technol. 102 (2011) 4849–4854.

[18] M. Möller, F. Harnisch, U. Schröder, Hydrothermal liquefaction of cellulose in subcritical water—the role of crystallinity on the cellulose reactivity, RSC Adv. 3 (2013) 11035–11044.

[19] S. Yin, A.K. Mehrotra, Z. Tan, Alkaline hydrothermal conversion of cellulose to bio-oil: influence of alkalinity on reaction pathway change, Bioresour. Technol. 102 (2011) 6605–6610.

[20] M. Deniel, G. Haarlemmer, A. Roubaud, E. Weiss-Hortala, J. Fages, Energy valorisation of food processing residues and model compounds by hydrothermal liquefaction, Renew. Sustain. Energy Rev. 54 (2016) 1632–1652.

[21] C. Schädel, A. Blöchl, A. Richter, G. Hoch, Quantification and monosaccharide composition of hemicelluloses from different plant functional types, Plant Physiol. Biochem. 48 (2010) 1–8.

[22] T.M. Aida, N. Shiraishi, M. Kubo, M. Watanabe, R.L. Smith, Reaction kinetics of D-xylose in sub- and supercritical water, J. Supercrit. Fluids 55 (2010) 208–216.

[23] T. Wahyudiono, M. Kanetake, M. Sasaki, Goto, decomposition of a lignin model compound under hydrothermal conditions, Chem. Eng. Technol. 30 (2007) 1113–1122.

[24] J. Barbier, N. Charon, N. Dupassieux, A. Loppinet-Serani, L. Mahe, J. Ponthus, M. Courtiade, A. Ducrozet, A.A. Quoineaud, F. Cansell, Hydrothermal conversion of lignin compounds. A detailed study of fragmentation and condensation reaction pathways, Biomass Bioenergy 46 (2012) 479–491.

[25] A. Molino, A. Iovine, P. Casella, S. Mehariya, S. Chianese, A. Cerbone, J. Rimauro, D. Musmarra, Microalgae characterization for consolidated and new application in human food, animal feed and nutraceuticals, Int. J. Environ. Res. Public Health 15 (2018) 2436.

[26] Y. Zhang, W.T. Chen, Chapter 5—Hydrothermal liquefaction of protein-containing feedstocks, in: L. Rosendahl (Ed.), Direct Thermochemical Liquefaction for Energy Applications, Woodhead Publishing, Oxford, UK, 2018, pp. 127–168.

[27] J.W. King, R.L. Holliday, G.R. List, Hydrolysis of soybean oil, Green Chem. 1 (1999) 261–264.

[28] M. Watanabe, T. Iida, H. Inomata, Decomposition of a long chain saturated fatty acid with some additives in hot compressed water, Energ. Conver. Manage. 47 (2006) 3344–3350.

[29] L. Qadariyah, Mahfud, Sumarno, S. Machmudah, Wahyudiono, M. Sasaki, M. Goto, Degradation of glycerol using hydrothermal process, Bioresour. Technol. 102 (2011) 9267–9271.

[30] R.B. Madsen, R.Z.K. Bernberg, P. Biller, J. Becker, B.B. Iversen, M. Glasius, Hydrothermal co-liquefaction of biomasses—quantitative analysis of bio-crude and aqueous phase composition, Sustainable Energy Fuels 1 (2017) 789–805.

[31] J. Yang, Q. He, K. Corscadden, H. Niu, J. Lin, T. Astatkie, Advanced models for the prediction of product yield in hydrothermal liquefaction via a mixture design of biomass model components coupled with process variables, Appl. Energy 233–234 (2019) 906–915.

[32] J. Yang, Q. He, H.B. Niu, K. Corscadden, T. Astatkie, Hydrothermal liquefaction of biomass model components for product yield prediction and reaction pathways exploration, Appl. Energy 228 (2018) 1618–1628.

[33] J. Akhtar, N.A.S. Amin, A review on process conditions for optimum bio-oil yield in hydrothermal liquefaction of biomass, Renew. Sustain. Energy Rev. 15 (2011) 1615–1624.

[34] U. Jena, K.C. Das, J.R. Kastner, Effect of operating conditions of thermochemical liquefaction on biocrude production from Spirulina platensis, Bioresour. Technol. 102 (2011) 6221–6229.

[35] R. Shakya, J. Whelen, S. Adhikari, R. Mahadevan, S. Neupane, Effect of temperature and Na2CO3 catalyst on hydrothermal liquefaction of algae, Algal Res. 12 (2015) 80–90.

[36] T.M. Brown, P. Duan, P.E. Savage, Hydrothermal liquefaction and gasification of Nannochloropsis sp, Energy Fuel 24 (2010) 3639–3646.

[37] J. Remon, J. Randall, V.L. Budarin, J.H. Clark, Production of bio-fuels and chemicals by microwave-assisted, catalytic, hydrothermal liquefaction (MAC-HTL) of a mixture of pine and spruce biomass, Green Chem. 21 (2019) 284–299.

[38] A. Funke, F. Ziegler, Hydrothermal carbonization of biomass: a summary and discussion of chemical mechanisms for process engineering, Biofuels Bioprod. Biorefin. 4 (2010) 160–177.

[39] M.K. Jindal, M.K. Jha, Hydrothermal liquefaction of wood: a critical review, Rev. Chem. Eng. 32 (2016) 459.

[40] A. Kruse, E. Dinjus, Hot compressed water as reaction medium and reactant: properties and synthesis reactions, J. Supercrit. Fluids 39 (2007) 362–380.

[41] H. Li, Z. Liu, Y. Zhang, B. Li, H. Lu, N. Duan, M. Liu, Z. Zhu, B. Si, Conversion efficiency and oil quality of low-lipid high-protein and high-lipid low-protein microalgae via hydrothermal liquefaction, Bioresour. Technol. 154 (2014) 322–329.

[42] F. Cheng, Z. Cui, L. Chen, J. Jarvis, N. Paz, T. Schaub, N. Nirmalakhandan, C.E. Brewer, Hydrothermal liquefaction of high- and low-lipid algae: bio-crude oil chemistry, Appl. Energy 206 (2017) 278–292.

[43] D. López Barreiro, C. Zamalloa, N. Boon, W. Vyverman, F. Ronsse, W. Brilman, W. Prins, Influence of strain-specific parameters on hydrothermal liquefaction of microalgae, Bioresour. Technol. 146 (2013) 463–471.

[44] S. Leow, J.R. Witter, D.R. Vardon, B.K. Sharma, J.S. Guest, T.J. Strathmann, Prediction of microalgae hydrothermal liquefaction products from feedstock biochemical composition, Green Chem. 17 (2015) 3584–3599.

[45] D. Xu, G. Lin, L. Liu, Y. Wang, Z. Jing, S. Wang, Comprehensive evaluation on product characteristics of fast hydrothermal liquefaction of sewage sludge at different temperatures, Energy 159 (2018) 686–695.

[46] D.R. Vardon, B.K. Sharma, J. Scott, G. Yu, Z. Wang, L. Schideman, Y. Zhang, T.J. Strathmann, Chemical properties of biocrude oil from the hydrothermal liquefaction of Spirulina algae, swine manure, and digested anaerobic sludge, Bioresour. Technol. 102 (2011) 8295–8303.

[47] L. Yang, L. Nazari, Z. Yuan, K. Corscadden, C. Xu, Q. He, Hydrothermal liquefaction of spent coffee grounds in water medium for bio-oil production, Biomass Bioenergy 86 (2016) 191–198.

[48] T.H. Pedersen, I.F. Grigoras, J. Hoffmann, S.S. Toor, I.M. Daraban, C.U. Jensen, S.B. Iversen, R.B. Madsen, M. Glasius, K.R. Arturi, R.P. Nielsen, E.G. Søgaard, L.A. Rosendahl, Continuous hydrothermal co-liquefaction of aspen wood and glycerol with water phase recirculation, Appl. Energy 162 (2016) 1034–1041.

[49] Y. Xue, H. Chen, W. Zhao, C. Yang, P. Ma, S. Han, A review on the operating conditions of producing bio-oil from hydrothermal liquefaction of biomass, Int. J. Energy Res. 40 (2016) 865–877.

[50] L. Humphreys, Bio-oil production method, US Patent 9,005,312 B2, April 14, 2015.

[51] L.J. Humphreys, Process and apparatus for converting organic matter into a product, US Patent 8,579,996 B2, November 12, 2013.

[52] T. Maschmeyer, L.J. Humphreys, Methods for biofuel production, US Patent 9,944,858 B2, April 17, 2018.

[53] C.U. Jensen, J.K. Rodriguez Guerrero, S. Karatzos, G. Olofsson, S.B. Iversen, Fundamentals of Hydrofaction™: renewable crude oil from woody biomass, Biomass Conservs. Biorefin. 7 (2017) 495–509.

[54] A.B. Ross, P. Biller, M.L. Kubacki, H. Li, A. Lea-Langton, J.M. Jones, Hydrothermal processing of microalgae using alkali and organic acids, Fuel 89 (2010) 2234–2243.

[55] Y. Chen, Y. Wu, R. Ding, P. Zhang, J. Liu, M. Yang, P. Zhang, Catalytic hydrothermal liquefaction of D. tertiolecta for the production of bio-oil over different acid/base catalysts, AICHE J. 61 (2015) 1118–1128.

[56] D. Xu, G. Lin, S. Guo, S. Wang, Y. Guo, Z. Jing, Catalytic hydrothermal liquefaction of algae and upgrading of biocrude: a critical review, Renew. Sustain. Energy Rev. 97 (2018) 103–118.

[57] A. Galadima, O. Muraza, Hydrothermal liquefaction of algae and bio-oil upgrading into liquid fuels: role of heterogeneous catalysts, Renew. Sustain. Energy Rev. 81 (2018) 1037–1048.

[58] J. Akhtar, S.K. Kuang, N.S. Amin, Liquefaction of empty palm fruit bunch (EPFB) in alkaline hot compressed water, Renew. Energy 35 (2010) 1220–1227.

[59] Z. Zhu, S.S. Toor, L. Rosendahl, D. Yu, G. Chen, Influence of alkali catalyst on product yield and properties via hydrothermal liquefaction of barley straw, Energy 80 (2015) 284–292.

[60] S. Karagöz, T. Bhaskar, A. Muto, Y. Sakata, T. Oshiki, T. Kishimoto, Low-temperature catalytic hydrothermal treatment of wood biomass: analysis of liquid products, Chem. Eng. J. 108 (2005) 127–137.

[61] Y.Y. Lin, C. Zhang, M.C. Zhang, J. Zhang, Deoxygenation of bio-oil during pyrolysis of biomass in the presence of CaO in a fluidized-bed reactor, Energy Fuel 24 (2010) 5686–5695.

[62] P.G. Duan, P.E. Savage, Hydrothermal liquefaction of a microalga with heterogeneous catalysts, Ind. Eng. Chem. Res. 50 (2011) 52–61.

[63] P. Biller, R. Riley, A.B. Ross, Catalytic hydrothermal processing of microalgae: decomposition and upgrading of lipids, Bioresour. Technol. 102 (2011) 4841–4848.

[64] J. Bian, Q. Zhang, P. Zhang, L. Feng, C. Li, Supported Fe2O3 nanoparticles for catalytic upgrading of microalgae hydrothermal liquefaction derived bio-oil, Catal. Today 293–294 (2017) 159–166.

[65] B. Zhang, Q. Lin, Q. Zhang, K. Wu, W. Pu, M. Yang, Y. Wu, Catalytic hydrothermal liquefaction of Euglena sp. microalgae over zeolite catalysts for the production of bio-oil, RSC Adv. 7 (2017) 8944–8951.

[66] W. Yang, Z. Wang, S. Song, H. Chen, X. Wang, J. Cheng, R. Sun, J. Han, Understanding catalytic mechanisms of HZSM-5 in hydrothermal liquefaction of algae through model components: glucose and glutamic acid, Biomass Bioenergy 130 (2019) 105356.

[67] S. Prodinger, M.A. Derewinski, Recent progress to understand and improve zeolite stability in the aqueous medium, Pet. Chem. 60 (2020) 420–436.

[68] C.H. Chang, R. Gopalan, Y.S. Lin, A comparative study on thermal and hydrothermal stability of alumina, titania and zirconia membranes, J. Membr. Sci. 91 (1994) 27–45.

[69] Z.T. Bi, J. Zhang, E. Peterson, Z.Y. Zhu, C.J. Xia, Y.N. Liang, T. Wiltowski, Biocrude from pre-treated sorghum bagasse through catalytic hydrothermal liquefaction, Fuel 188 (2017) 112–120.

[70] P. Haghighat, A. Montanez, G.R. Aguilera, J.K. Rodriguez Guerrero, S. Karatzos, M.A. Clarke, W. McCaffrey, Hydrotreating of Hydrofaction™ biocrude in the presence of presulfided commercial catalysts, Sustainable Energy Fuels 3 (2019) 744–759.

[71] U. Jena, K.C. Das, Comparative evaluation of thermochemical liquefaction and pyrolysis for bio-oil production from microalgae, Energy Fuel 25 (2011) 5472–5482.

[72] F.M. Hossain, J. Kosinkova, R.J. Brown, Z. Ristovski, B. Hankamer, E. Stephens, T.J. Rainey, Experimental investigations of physical and chemical properties for microalgae HTL bio-crude using a large batch reactor, Energies 10 (2017) 467.

[73] D.C. Hietala, C.M. Godwin, B.J. Cardinale, P.E. Savage, The independent and coupled effects of feedstock characteristics and reaction conditions on biocrude production by hydrothermal liquefaction, Appl. Energy 235 (2019) 714–728.

[74] A. Palomino, R.D. Godoy-Silva, S. Raikova, C.J. Chuck, The storage stability of biocrude obtained by the hydrothermal liquefaction of microalgae, Renew. Energy 145 (2020) 1720–1729.

[75] J.L. Yu, P. Biller, A. Mamahkel, M. Klemmer, J. Becker, M. Glasius, B.B. Iversen, Catalytic hydrotreatment of bio-crude produced from the hydrothermal liquefaction of aspen wood: a catalyst screening and parameter optimization study, Sustainable Energy Fuels 1 (2017) 832–841.

[76] M.S. Haider, D. Castello, L.A. Rosendahl, Two-stage catalytic hydrotreatment of highly nitrogenous biocrude from continuous hydrothermal liquefaction: a rational design of the stabilization stage, Biomass Bioenergy 139 (2020) 105658.

[77] C.Y. Yang, R. Li, C. Cui, S.P. Liu, Q. Qiu, Y.G. Ding, Y.X. Wu, B. Zhang, Catalytic hydroprocessing of microalgae-derived biofuels: a review, Green Chem. 18 (2016) 3684–3699.

[78] S.B. Jones, Y. Zhu, L.J. Snowden-Swan, D. Anderson, R.T. Hallen, A.J. Schmidt, K.O. Albrecht, D.-C. Elliott, Whole Algae Hydrothermal Liquefaction: 2014 State of Technology, Pacific Northwest National Lab. (PNNL), Richland, WA, United States, 2014.

[79] M. Lewandowski, A. Szymanska-Kolasa, P. Da Costa, C. Sayag, Catalytic performances of platinum doped molybdenum carbide for simultaneous hydrodenitrogenation and hydrodesulfurization, Catal. Today 119 (2007) 31–34.

[80] D. Castello, M.S. Haider, L.A. Rosendahl, Catalytic upgrading of hydrothermal liquefaction biocrudes: different challenges for different feedstocks, Renew. Energy 141 (2019) 420–430.

[81] J. Watson, T. Wang, B. Si, W.-T. Chen, A. Aierzhati, Y. Zhang, Valorization of hydrothermal liquefaction aqueous phase: pathways towards commercial viability, Prog. Energy Combust. Sci. 77 (2020) 100819.

[82] D.C. Hietala, C.M. Godwin, B.J. Cardinale, P.E. Savage, The individual and synergistic impacts of feedstock characteristics and reaction conditions on the aqueous co-product from hydrothermal liquefaction, Algal Res. 42 (2019) 101568.

[83] K.O. Albrecht, R.A. Dagle, D.T. Howe, J.A. Lopez-Ruiz, S.D. Davidson, B. Maddi, A.R. Cooper, E.A. Panisko, Final Report for the Project Characterization and Valorization of Aqueous Phases Derived From Liquefaction and Upgrading of Bio-Oils, in, United States, PNNL-27848, Pacific Northwest National Laboratory, Richland, WA, 2018.

[84] M. Usman, H. Chen, K. Chen, S. Ren, J.H. Clark, J. Fan, G. Luo, S. Zhang, Characterization and utilization of aqueous products from hydrothermal conversion of biomass for bio-oil and hydro-char production: a review, Green Chem. 21 (2019) 1553–1572.

[85] S.R. Shanmugam, S. Adhikari, R. Shakya, Nutrient removal and energy production from aqueous phase of bio-oil generated via hydrothermal liquefaction of algae, Bioresour. Technol. 230 (2017) 43–48.

[86] R. Posmanik, R.A. Labatut, A.H. Kim, J.G. Usack, J.W. Tester, L.T. Angenent, Coupling hydrothermal liquefaction and anaerobic digestion for energy valorization from model biomass feedstocks, Bioresour. Technol. 233 (2017) 134–143.

[87] B. Si, J. Li, Z. Zhu, M. Shen, J. Lu, N. Duan, Y. Zhang, Q. Liao, Y. Huang, Z. Liu, Inhibitors degradation and microbial response during continuous anaerobic conversion of hydrothermal liquefaction wastewater, Sci. Total Environ. 630 (2018) 1124–1132.

[88] M. Zheng, L.C. Schideman, G. Tommaso, W.-T. Chen, Y. Zhou, K. Nair, W. Qian, Y. Zhang, K. Wang, Anaerobic digestion of wastewater generated from the hydrothermal liquefaction of Spirulina: toxicity assessment and minimization, Energ. Conver. Manage. 141 (2017) 420–428.

[89] J. Watson, B. Si, H. Li, Z. Liu, Y. Zhang, Influence of catalysts on hydrogen production from wastewater generated from the HTL of human feces via catalytic hydrothermal gasification, Int. J. Hydrogen Energy 42 (2017) 20503–20511.

[90] L. Garcia Alba, C. Torri, D. Fabbri, S.R.A. Kersten, D.W.F. Brilman, Microalgae growth on the aqueous phase from hydrothermal liquefaction of the same microalgae, Chem. Eng. J. 228 (2013) 214–223.

[91] L. Leng, J. Li, Z. Wen, W. Zhou, Use of microalgae to recycle nutrients in aqueous phase derived from hydrothermal liquefaction process, Bioresour. Technol. 256 (2018) 529–542.

[92] L. Yang, B. Si, X. Tan, H. Chu, X. Zhou, Y. Zhang, Y. Zhang, F. Zhao, Integrated anaerobic digestion and algae cultivation for energy recovery and nutrient supply from post-hydrothermal liquefaction wastewater, Bioresour. Technol. 266 (2018) 349–356.

[93] D. Quispe-Arpasi, R. de Souza, M. Stablein, Z. Liu, N. Duan, H. Lu, Y. Zhang, A.L.D. Oliveira, R. Ribeiro, G. Tommaso, Anaerobic and photocatalytic treatments of post-hydrothermal liquefaction wastewater using H2O2, Bioresour. Technol. Rep. 3 (2018) 247–255.

[94] P. Biller, A.B. Ross, S.C. Skill, A. Lea-Langton, B. Balasundaram, C. Hall, R. Riley, C.A. Llewellyn, Nutrient recycling of aqueous phase for microalgae cultivation from the hydrothermal liquefaction process, Algal Res. 1 (2012) 70–76.

[95] M. Nelson, L. Zhu, A. Thiel, Y. Wu, M. Guan, J. Minty, H.Y. Wang, X.N. Lin, Microbial utilization of aqueous co-products from hydrothermal liquefaction of microalgae Nannochloropsis oculata, Bioresour. Technol. 136 (2013) 522–528.

[96] Y. He, X. Li, X. Xue, M.S. Swita, A.J. Schmidt, B. Yang, Biological conversion of the aqueous wastes from hydrothermal liquefaction of algae and pine wood by *Rhodococci*, Bioresour. Technol. 224 (2017) 457–464.

[97] E.A. Ramos-Tercero, A. Bertucco, D.W.F. Brilman, Process water recycle in hydrothermal liquefaction of microalgae to enhance bio-oil yield, Energy Fuel 29 (2015) 2422–2430.

[98] M. Klemmer, R.B. Madsen, K. Houlberg, A.J. Mørup, P.S. Christensen, J. Becker, M. Glasius, B.B. Iversen, Effect of aqueous phase recycling in continuous hydrothermal liquefaction, Ind. Eng. Chem. Res. 55 (2016) 12317–12325.

[99] P. Biller, R.B. Madsen, M. Klemmer, J. Becker, B.B. Iversen, M. Glasius, Effect of hydrothermal liquefaction aqueous phase recycling on bio-crude yields and composition, Bioresour. Technol. 220 (2016) 190–199.

[100] M. Castellví Barnés, J. Oltvoort, S.R.A. Kersten, J.-P. Lange, Wood liquefaction: role of solvent, Ind. Eng. Chem. Res. 56 (2017) 635–644.

[101] B. Biswas, A. Arun Kumar, Y. Bisht, R. Singh, J. Kumar, T. Bhaskar, Effects of temperature and solvent on hydrothermal liquefaction of *Sargassum tenerrimum* algae, Bioresour. Technol. 242 (2017) 344–350.

[102] Y. Wang, H. Wang, H. Lin, Y. Zheng, J. Zhao, A. Pelletier, K. Li, Effects of solvents and catalysts in liquefaction of pinewood sawdust for the production of bio-oils, Biomass Bioenergy 59 (2013) 158–167.

[103] S.H. Feng, R.F. Wei, M. Leitch, C.C. Xu, Comparative study on lignocellulose liquefaction in water, ethanol, and water/ethanol mixture: roles of ethanol and water, Energy 155 (2018) 234–241.

[104] T. Aysu, M.M. Kucuk, A. Demirbas, Optimization of process variables for supercritical liquefaction of giant fennel, RSC Adv. 4 (2014) 55912–55923.

[105] H. Durak, T. Aysu, Thermochemical liquefaction of algae for bio-oil production in supercritical acetone/ethanol/isopropanol, J. Supercrit. Fluids 111 (2016) 179–198.

[106] X. Wang, X.A. Xie, J. Sun, W. Liao, Effects of liquefaction parameters of cellulose in supercritical solvents of methanol, ethanol and acetone on products yield and compositions, Bioresour. Technol. 275 (2019) 123–129.

[107] H.M. Liu, Cypress liquefaction in a water/methanol mixture: effect of solvent ratio on products distribution and characterization of products, Ind. Eng. Chem. Res. 52 (2013) 12523–12529.

[108] S. Cheng, I. D'cruz, M. Wang, M. Leitch, C. Xu, Highly efficient liquefaction of woody biomass in hot-compressed alcohol–water co-solvents, Energy Fuel 24 (2010) 4659–4667.

[109] Y. Liu, X.-Z. Yuan, H.-J. Huang, X.-L. Wang, H. Wang, G.-M. Zeng, Thermochemical liquefaction of rice husk for bio-oil production in mixed solvent (ethanol–water), Fuel Process. Technol. 112 (2013) 93–99.

[110] H. Ahmed Baloch, S. Nizamuddin, M.T.H. Siddiqui, N.M. Mubarak, D.K. Dumbre, M.P. Srinivasan, G.J. Griffin, Sub-supercritical liquefaction of sugarcane bagasse for production of bio-oil and char: effect of two solvents, J. Environ. Chem. Eng. 6 (2018) 6589–6601.

[111] G. van Rossum, W. Zhao, M. Castellvi Barnes, J.P. Lange, S.R. Kersten, Liquefaction of lignocellulosic biomass: solvent, process parameter, and recycle oil screening, ChemSusChem 7 (2014) 253–259.

[112] J.H. Hildebrand, Solubility of Non-Electrolytes, Reinhold Publishing Corporation, New York, 1936.

[113] C.M. Hansen, in: C. Hansen (Ed.), Hansen Solubility Parameters A User's Handbook, second ed., Taylor & Francis, Boca Raton, 2007.

[114] A.J.J. Straathof, A. Bampouli, Potential of commodity chemicals to become bio-based according to maximum yields and petrochemical prices, Biofuels Bioprod. Biorefin. 11 (2017) 798–810.

[115] J.P. Lange, Lignocellulose liquefaction to biocrude: a tutorial review, ChemSusChem 11 (2018) 997–1014.

[116] K.M. Isa, T.A.T. Abdullah, U.F.M. Ali, Hydrogen donor solvents in liquefaction of biomass: a review, Renew. Sustain. Energy Rev. 81 (2018) 1259–1268.

[117] S. Kumar, J.P. Lange, G. Van Rossum, S.R. Kersten, Liquefaction of lignocellulose in fluid catalytic cracker feed: a process concept study, ChemSusChem 8 (2015) 4086–4094.

[118] M.R. Haverly, T.C. Schulz, L.E. Whitmer, A.J. Friend, J.M. Funkhouser, R.G. Smith, M.K. Young, R.C. Brown, Continuous solvent liquefaction of biomass in a hydrocarbon solvent, Fuel 211 (2018) 291–300.

[119] K. Treusch, J. Ritzberger, N. Schwaiger, P. Pucher, M. Siebenhofer, Diesel production from lignocellulosic feed: the bioCRACK process, R. Soc. Open Sci. 4 (2017) 171122.

[120] M. Grilc, B. Likozar, J. Levec, Simultaneous liquefaction and hydrodeoxygenation of lignocellulosic biomass over NiMo/Al2O3, Pd/Al2O3, and zeolite Y catalysts in hydrogen donor solvents, ChemCatChem 8 (2016) 180–191.

[121] Z.X. Lu, L.W. Fan, Z.G. Wu, H. Zhang, Y.Q. Liao, D.Y. Zheng, S.Q. Wang, Efficient liquefaction of woody biomass in polyhydric alcohol with acidic ionic liquid as a green catalyst, Biomass Bioenergy 81 (2015) 154–161.

[122] N. Schwaiger, D.C. Elliott, J. Ritzberger, H. Wang, P. Pucher, M. Siebenhofer, Hydrocarbon liquid production via the bioCRACK process and catalytic hydroprocessing of the product oil, Green Chem. 17 (2015) 2487–2494.

[123] K. Treusch, A.M. Mauerhofer, N. Schwaiger, P. Pucher, S. Muller, D. Painer, H. Hofbauer, M. Siebenhofer, Hydrocarbon production by continuous hydrodeoxygenation of liquid phase pyrolysis oil with biogenous hydrogen rich synthesis gas, React. Chem. Eng. 4 (2019) 1195–1207.

[124] K. Treusch, N. Schwaiger, K. Schlackl, R. Nagl, A. Rollett, M. Schadler, B. Hammerschlag, J. Ausserleitner, A. Huber, P. Pucher, M. Siebenhofer, High-throughput continuous hydrodeoxygenation of liquid phase pyrolysis oil, React. Chem. Eng. 3 (2018) 258–266.

[125] K. Treusch, N. Schwaiger, K. Schlackl, R. Nagl, P. Pucher, M. Siebenhofer, Temperature dependence of single step hydrodeoxygenation of liquid phase pyrolysis oil, Front. Chem. 6 (2018) 297.

[126] R. Brown, L. Whitmer, R. Smith, M. Wright, M. Young, Liquefaction of Forest Biomass to "Drop-In" Hydrocarbon Biofuels, Iowa State University, Ames, IA, United States, 2017. 58 p.

[127] M.N. Nabi, M.M. Rahman, M.A. Islam, F.M. Hossain, P. Brooks, W.N. Rowlands, J. Tulloch, Z.D. Ristovski, R.J. Brown, Fuel characterisation, engine performance, combustion and exhaust emissions with a new renewable Licella biofuel, Energ. Conver. Manage. 96 (2015) 588–598.

[128] J. Ritzberger, P. Pucher, N. Schwaiger, M. Siebenhofer, The BioCRACK process—a refinery integrated biomass-to-liquid concept to produce diesel from biogenic feedstock, Chem. Eng. Trans. 39 (2014) 1189.

[129] N. Schwaiger, R. Feiner, K. Zahel, A. Pieber, V. Witek, P. Pucher, E. Ahn, P. Wilhelm, B. Chernev, H. Schrottner, M. Siebenhofer, Liquid and solid products from liquid-phase pyrolysis of softwood, Bioenergy Res. 4 (2011) 294–302.

[130] M. Berchtold, J. Fimberger, A. Reichhold, P. Pucher, Upgrading of heat carrier oil derived from liquid-phase pyrolysis via fluid catalytic cracking, Fuel Process. Technol. 142 (2016) 92–99.

[131] K. Treusch, A. Huber, S. Reiter, M. Lukasch, B. Hammerschlag, J. Ausserleitner, D. Painer, P. Pucher, M. Siebenhofer, N. Schwaiger, Refinery integration of lignocellulose for automotive fuel production via the bioCRACK process and two-step co-hydrotreating of liquid phase pyrolysis oil and heavy gas oil, React. Chem. Eng. 5 (2020) 519–530.

[132] G. Perkins, N. Batalha, A. Kumar, T. Bhaskar, M. Konarova, Recent advances in liquefaction technologies for production of liquid hydrocarbon fuels from biomass and carbonaceous wastes, Renew. Sustain. Energy Rev. 115 (2019) 109400.

[133] L.W. Ou, R. Thilakaratne, R.C. Brown, M.M. Wright, Techno-economic analysis of transportation fuels from defatted microalgae via hydrothermal liquefaction and hydroprocessing, Biomass Bioenergy 72 (2015) 45–54.

[134] K.F. Tzanetis, J.A. Posada, A. Ramirez, Analysis of biomass hydrothermal liquefaction and biocrude-oil upgrading for renewable jet fuel production: the impact of reaction conditions on production costs and GHG emissions performance, Renew. Energy 113 (2017) 1388–1398.

[135] L.J. Snowden-Swan, Y. Zhu, S.B. Jones, D.C. Elliott, A.J. Schmidt, R.T. Hallen, J.M. Billing, T.R. Hart, S.P. Fox, G.D. Maupin, Hydrothermal Liquefaction and Upgrading of Municipal Wastewater Treatment Plant Sludge: A Preliminary Techno-Economic Analysis, Rev.1, Pacific Northwest National Lab. (PNNL), Richland, WA, United States, 2016. 40 p.

[136] T.H. Pedersen, N.H. Hansen, O.M. Perez, D.E.V. Cabezas, L.A. Rosendahl, Renewable hydrocarbon fuels from hydrothermal liquefaction: a techno-economic analysis, Biofuels Bioprod. Biorefin. 12 (2018) 213–223.

[137] D.C. Elliott, G.G. Neuenschwander, T.R. Hart, L.J. Rotness, A.H. Zacher, K.A. Fjare, B.C. Dunn, S.L. McDonald, G. Dassor, Hydrothermal Liquefaction of Agricultural and Biorefinery Residues Final Report—CRADA #PNNL/277, Pacific Northwest National Laboratory (PNNL), Richland, WA, United States, 2010.

[138] S. Kumar, A. Segins, J.P. Lange, G. Van Rossum, S.R.A. Kersten, Liquefaction of lignocellulose in light cycle oil: a process concept study, ACS Sustain. Chem. Eng. 4 (2016) 3087–3094.

CHAPTER 6

Advances in the conversion of methanol to gasoline

Jyoti Prasad Chakraborty[a], Satyansh Singh[a], and Sunil K. Maity[b]

[a]Department of Chemical Engineering and Technology, Indian Institute of Technology (Banaras Hindu University), Varanasi, Uttar Pradesh, India
[b]Department of Chemical Engineering, Indian Institute of Technology Hyderabad, Kandi, Sangareddy, Telangana, India

Contents

6.1 Introduction		178
6.2 Production of methanol		179
	6.2.1 Production of methanol from coal	180
	6.2.2 Conversion of natural gas to methanol	181
	6.2.3 Conversion of COG to methanol	182
	6.2.4 Production of methanol from biomass	182
	6.2.5 Production of methanol from carbon dioxide (CO_2)	183
6.3 Methanol to gasoline		184
	6.3.1 Reaction mechanism	184
	6.3.2 Shape-selectivity of zeolites	187
	6.3.3 Effect of Si/Al ratio in zeolite	189
	6.3.4 Effect of size of crystal and surface properties	189
	6.3.5 Effect of modification of catalyst surface	191
	6.3.6 Effect of process parameters	191
	6.3.7 Effect of reactor configuration	193
6.4 Industrial development		194
6.5 Conclusions		196
References		197

Abbreviations

COG	coke oven gas
DME	dimethyl ether
FTIR	Fourier-transform infrared spectroscopy
MON	motor octane number
MTG	methanol-to-gasoline
NMR	nuclear magnetic resonance
RON	research octane number
RWGSR	reverse water-gas shift reaction
TOS	time-on-stream
TPD	temperature programmed desorption
WHSV	weight hourly space velocity

Hydrocarbon Biorefinery
https://doi.org/10.1016/B978-0-12-823306-1.00008-X

Copyright © 2022 Elsevier Inc.
All rights reserved.

6.1 Introduction

With the rapid depletion of fossil-based fuels and the rise in energy demand, various alternative energy sources, such as bio-hydrogen, biodiesel, and bioethanol, are considered actively to replace fossil-based fuels [1]. The increase in oil price and concerns of global warming, air pollution, etc., make the use of fossil-based fuels not permissible in the near future [2]. The rapid increase in per capita energy demand is due to a sharp increase in the world's population and industrial growth in many developing nations [3]. The world's total oil consumption rose by 13% from 2008 to around 4662 Mtoe (million tons of oil equivalent) in 2018, and it will continue to grow due to the rapid growth of emerging economies [4]. The oil production throughout the world in 2018 was 96 million barrels per day [5]. Besides, the consumption of gasoline in the world was increased from 114,000 barrels per day in 2008 to 408,000 barrels per day in 2018, an increase of 254% [5]. Generally, gasoline is produced from crude oil by distillation, fluid catalytic cracking, and alkylation processes in petroleum refineries. Sometimes gasoline is also produced from coal-based sources [6].

With such a huge demand for fuels, gasoline production from alternate sources is quite imperative for balancing the demand and supply chain. In this regard, the methanol-to-gasoline (MTG) process is a promising route for gasoline production [5]. At present, methanol is produced from fossil-based sources, such as coal and natural gas. Methanol may also be obtained from renewable carbon sources, such as biomass, making it an attractive route for the production of renewable gasoline [7]. The methanol might be used directly in combustion engines or blended with gasoline. However, the direct application of methanol as engine fuel requires substantial changes in the design of the engine due to some drawbacks associated with it, such as high volatility, low calorific value, high water miscibility, corrosiveness, and toxicity [5]. Hence, methanol is generally converted to gasoline by the MTG process [7]. Gasoline obtained in this process is quite similar to the current gasoline with identical calorific value and fuel mileage. The MTG process was first introduced by Mobil in 1970 by using synthetic zeolite, known as ZSM-5. The optimum process parameters for 80% selectivity of gasoline were 400°C and 15–20 atm pressure. The remaining 20% product was mainly LPG [8]. The zeolite catalysts generally show high catalytic activity with good selectivity and stability. The zeolite-based catalysts are thus widely used in catalytic cracking in petroleum refineries for gasoline production with higher selectivity and enhanced octane number [5]. The zeolite-based catalysts are thus commonly used in the MTG process, and they play a crucial role. This chapter summarizes different processes for the production of methanol from various feedstocks. Afterward, this chapter discusses the MTG process using various zeolite-based catalysts. The chapter also provides an overview of the reaction mechanism, the impact of the characteristic feature of zeolites and process variables on the quality and yield of product, and a brief description of the commercial processes.

6.2 Production of methanol

The technology related to the production of methanol experienced rapid development over the last few decades. Methanol is now produced almost entirely from syngas (mixtures of carbon monoxide, carbon dioxide, and hydrogen). The syngas is obtained from various feedstocks such as coal, natural gas, and naphtha, as well as renewable feedstock, such as biomass [9, 10]. The gasification of coal/biomass and steam reforming of naphtha/natural gas are generally used for the production of syngas. In 1926, BASF first developed a commercial technology for the production of methanol from syngas using chromium and manganese oxide-based catalyst. This process, however, involves extremely high pressure (50–220 bar) and high temperature (450°C). The recent commercial processes are operated at moderate temperature (200–300°C) and lower pressure (50–100 bar) using alumina supported copper and zinc oxide catalysts with various promoters [11]. The promoters enhanced the dispersion of copper and reduced particle size, thereby improving the catalytic activity. These catalysts are, however, sensitive to sulfur poisoning. Hence, the desulfurization of feedstock is needed before the methanol synthesis. These technologies include Imperial Chemical Industries (ICI), Lurgi, Topsoe, Mitsubishi, etc. These processes are mainly distinguished by their reactor configuration, types of feedstock, and heat energy integration. A general process flow diagram for the production of methanol is shown in Fig. 6.1.

$$CO + 2H_2 \leftrightarrow CH_3OH \quad \Delta H_{298K} = -90.77 \, kJ/mol \tag{6.1}$$

$$CO_2 + 3H_2 \leftrightarrow CH_3OH + H_2O \quad \Delta H_{298K} = -49.16 \, kJ/mol \tag{6.2}$$

$$CO_2 + H_2 \leftrightarrow CO + H_2O \quad \Delta H_{298K} = +41:21 \, kJ/mol \tag{6.3}$$

The methanol is formed by either hydrogenation of CO or CO_2 on Cu sites of the catalyst, as shown above [12, 13]. However, the exact mechanism of methanol formation from syngas is still not clear. These reactions of Eqs. (6.1), (6.2) are highly exothermic, while reaction (6.3) is endothermic. Reaction (6.3) is also called reverse water gas shift reaction (RWGSR). It is generally well accepted that CO is first converted to CO_2 by WGSR in this process (reverse of Eq. 6.3) [14]. The CO_2 is then hydrogenated on the catalyst's surface to produce methanol (Eq. 6.2). The methanol synthesis is equilibrium limited reactions with incomplete CO conversion per pass (around 20%). The unreacted syngas is separated from the products and recycled into the methanol production reactor with an overall conversion of more than 99%. Methanol synthesis is performed using slight excess hydrogen with the stoichiometric ratio (Eq. 6.4) of slightly more than 2. Slight excess hydrogen is maintained to reduce the mole fraction of water that generally blocks the catalytic sites.

$$\text{Stoichiometric ratio} = (H_2 - CO_2)/(CO + CO_2) \tag{6.4}$$

where H_2, CO, and CO_2 are expressed as vol.%.

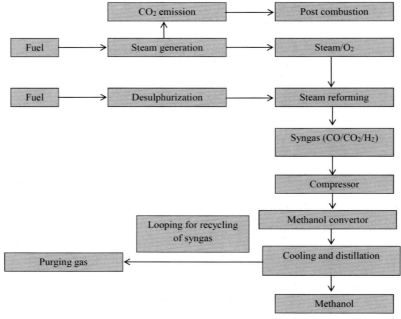

Fig. 6.1 A general flowchart for the production of methanol. *(Based on F.G. Üçtuğ, S. Ağralı, Y. Arıkan, E. Avcıoğlu, Deciding between carbon trading and carbon capture and sequestration: an optimisation-based case study for methanol synthesis from syngas, J. Environ. Manage. 132 (2014) 1–8).*

The reaction is more than 99% selective to methanol. The small quantities of higher alcohols, dimethyl ether (DME), esters, ketones, and hydrocarbons are formed as by-products in this reaction. These by-products are formed due to the presence of impurities in the catalyst that promote various reactions, such as Fischer-Tropsch synthesis (iron, cobalt, and nickel), higher alcohols (alkali), and DME (acidic alumina) [14]. The crude methanol containing water and by-products is purified using two distillation columns. The volatile compounds (H_2, $CO/CO_2/CH_4$, and DME) are removed in the first distillation column, while water and higher alcohols are separated in the second column.

6.2.1 Production of methanol from coal

The process starts with the gasification of coal, where partial combustion occurs at high temperatures (around 1000°C) in the limited supply of oxidizing agents, such as air or oxygen. The syngas produced from gasification is further processed to make it suitable for methanol production. The Purisol process coupled with Claus sulfur recovery is used for the processing of syngas. Further, the WGSR is used to maintain the ratio of carbon monoxide to hydrogen. A process flow diagram for the production of methanol from

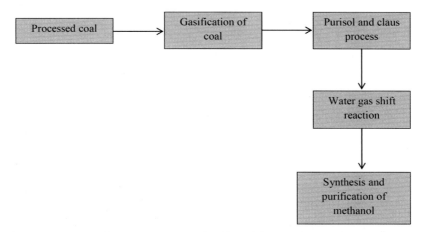

Fig. 6.2 A general flowchart for the production of methanol from coal. *(Based on F.G. Üçtuğ, S. Ağralı, Y. Arıkan, E. Avcıoğlu, Deciding between carbon trading and carbon capture and sequestration: an optimisation-based case study for methanol synthesis from syngas, J. Environ. Manage. 132 (2014) 1–8).*

coal is depicted in Fig. 6.2. Chen et al. suggested a simulated model for the production of methanol from coal [15]. Based on their model, the methanol production unit housed various process units for air separation, coal gasification, cleaning of gaseous products, methanol synthesis unit, and methanol purification column.

6.2.2 Conversion of natural gas to methanol

The natural gas and naphtha are generally steam reformed at high temperatures (700–800°C) in the presence of alumina supported nickel catalyst. The syngas is then converted to methanol, as described before. The capital investment and operating cost are lesser for natural gas compared to heavy feedstock, such as coal. Natural gas is thus the preferred feedstock for methanol production from economic perspectives. However, coal is abundant in some countries, such as China and India, and preferred feedstock for this process. Methanol may also be produced from natural gas directly by selective oxidation. Verma et al. studied the homogeneous reaction kinetics for methane to methanol conversion at a lower temperature range [16]. They concluded that 100°C reaction temperature with longer retention time, methane to oxygen ratio of 0.005/0.995, and 5 ppm concentration of OH species could be considered as the optimum condition for methane to methanol conversion. Okazaki et al. considered the mixture of methane and water vapor as feedstock for methanol production directly [17]. They used a thin glass tube reactor for ultrashort pulse barrier discharge to study nonequilibrium plasma chemical reaction under ambient pressure. They also reported that the addition of rare gases, such as krypton or argon, may enhance the production of methanol. This reaction is, however,

6.2.3 Conversion of COG to methanol

Coke oven gas (COG) may be considered as a by-product of the coking plant. It mainly consists of hydrogen (around 55%–60%), nitrogen (around 3%–5%), carbon monoxide (around 5%–8%), and methane (around 23%–27%), as well as a small fraction of NH_3, H_2S, and other hydrocarbons [18]. Methanol may be obtained from COG via the production of syngas using steam or dry reforming process using activated carbon as a catalyst [1]. Bermúdez et al. concluded that the presence of a small amount of hydrocarbon in COG would facilitate syngas production, resulting in higher methanol production [19]. Bermúdez et al. recommended methanol production from CO_2 reforming of COG [20].

6.2.4 Production of methanol from biomass

All organic matter may be employed as a feedstock for methanol production [21]. Among the various resources used for methanol production, biomass has been proven to be the most cost-effective resource [22, 23]. Meanwhile, native biomass has certain inherent drawbacks, such as higher moisture content and higher H/C and O/C ratios that decrease the process efficiency [24]. Thus, to tackle the challenges with native biomass, prior pretreatment processes, such as torrefaction, hydrothermal liquefaction, carbonization, etc., may be employed [25]. The processing stages adopted for the methanol production from biomass are similar to the production of methanol from coal and natural gas, which consists of mainly three steps, namely, production of syngas, methanol synthesis, and finally methanol purification [26]. A plant for the methanol production using biomass as a raw material consists of two units: (1) gasification unit for syngas production, and (2) methanol synthesis unit. A general process flow diagram for the production of methanol from biomass is presented in Fig. 6.3. For biomass gasification, the solid feedstock is first dried to reduce the moisture content to less than 15%–20%, ground to obtain uniform particle size, and then sent to the gasifier together with air or oxygen. Gasification is a mild endothermic process. The biomass is thus partially combusted in situ to supplement the heat energy needed for gasification. The formation of oxygenated tar is one of the major challenges in biomass gasification. Clausen et al. studied the techno-economic aspects of methanol production plants based on biomass gasification and electrolysis of water [27]. They summarized that electrolysis of water and auto thermal reforming of natural gas significantly reduced the cost of syngas production. Shabangu et al. implemented the techno-economic study of methanol production based on the application of

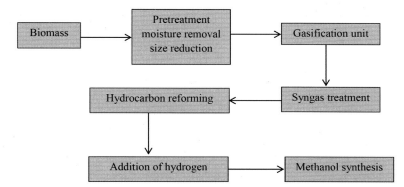

Fig. 6.3 A general flowchart for the production of methanol using biomass. *(Based on X. Zhen, Y. Wang, An overview of methanol as an internal combustion engine fuel, Renew. Sustain. Energy Rev. 52 (2015) 477–493.)*

biochar [28]. They concluded that biochar application as a soil amendment agent might facilitate the overall economy of methanol production.

6.2.5 Production of methanol from carbon dioxide (CO_2)

For methanol production from hydrogen, a proper carbon source is needed [29]. The utilization of CO_2 as a sustainable carbon source is thus an important strategy for methanol production by catalytic CO_2 hydrogenation, thereby mitigating CO_2 emission [30, 31]. Similar to methanol synthesis from syngas, copper-based catalysts are commonly used for direct hydrogenation of CO_2 to methanol [32]. Lurgi AG, jointly with Süd-Chemie, developed a highly active Cu-ZnO-based catalyst for methanol production by CO_2 hydrogenation at about 260°C reaction temperature. This catalyst showed excellent selectivity to methanol [33]. The deactivation of this catalyst is quite similar to the commercial catalyst for methanol synthesis from syngas. The commercial methanol synthesis catalyst from syngas developed by Süd-Chemie, however, showed a slower reaction rate for CO_2 hydrogenation compared to CO. The methanol productivity from CO_2 is 3–10 times lower than CO. In contrast, the methanol selectivity was higher for CO_2 than CO. Further, the Cu-ZnO-based catalyst was found to deactivate progressively at higher CO_2 partial pressure. The catalyst deactivation was reported to be due to the sintering of Cu and ZnO in the presence of water, formed during CO_2 hydrogenation reaction [33]. However, when the reaction is performed using the mixture of CO and CO_2, the water reacts with CO by WGSR to form CO_2, thereby reducing the catalyst deactivation. The methanol synthesis using RWGSR of CO_2, followed by CO hydrogenation, is thus an alternative approach. This process involves RWGSR over $ZnAl_2O_4$ catalyst, followed by removal of water and then methanol production using alumina supported Cu-ZnO-based methanol synthesis catalyst [32].

6.3 Methanol to gasoline

In the methanol to gasoline (MTG) process, methanol is first converted to DME as a result of the dehydration reaction. The established equilibrium mixture contains methanol, DME, and water. The light olefins are then formed from the methanol-DME equilibrium mixture. These light olefins subsequently are converted into higher olefins, paraffins, and aromatics as a result of polycondensation and alkylation reactions [34]. The overall reactions involved in MTG are shown in Scheme 6.1. The MTG reaction is generally carried out using zeolite type of catalysts, especially ZSM-5. ZSM-5 belongs to a primary associate of the material of the pentasil family [35]. ZSM-5 has unique properties that enable it to act as a catalyst in many processes, such as the MTG process, olefins, aromatics, paraffin, etc. [5]. However, the activity of native ZSM-5 catalyst decreases rapidly as a result of coke deposition on the active sites because of the simple microstructure and long path for the diffusion of reactant molecules. The coke formation on the catalyst surface is supported by the strong acidic sites of the catalyst [36]. Thus, the improvement of physicochemical characteristics of ZSM-5 is of prime importance to increase its catalytic activity. The characteristics of native ZSM-5 catalysts are enhanced by altering the morphology, aluminum content, and crystal size and introducing the other suitable elements in the framework of the ZSM-5 catalyst [37–39]. Table 6.1 summarizes the several altered ZSM-5 catalysts used for the methanol conversion to various hydrocarbons/chemicals.

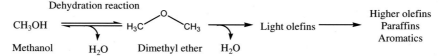

Scheme 6.1 Overall reactions involved in MTG.

6.3.1 Reaction mechanism

MTG reaction mechanism is quite complex, and it is a subject of debate for quite a long time, especially how the first C—C bond forms from methanol having a single carbon atom and what are the initial C—C species. Several reaction mechanisms have been proposed in the past to explain the formation of the initial C—C bond from methanol or DME during the induction period, such as carbene-carbenoid and oxonium-ylide [48]. These mechanisms are known as the direct mechanisms. In these mechanisms, the reaction proceeds slowly during the initial stage of the reaction, known as the induction period. In this period, sufficient hydrocarbon species (reaction centers) are generated for subsequent autocatalysis. In the direct mechanisms, it is generally well accepted that methoxy species is first formed during the induction period by the adsorption of methanol on the Brønsted acid sites of the catalyst. This species has carbene/ylide

Table 6.1 Effects of ZSM-5 and modified ZSM-5 on the MTG process.

Catalyst	Reaction conditions	Methanol conversion (%)	Yield/selectivity (%)	Refs.
H–ZSM-5/CaCO$_3$	400°C, catalyst loading $=0.5$ g, WHSV (MeOH) $=28$/h	75.02	100	[40]
H–ZSM-5/Na$_2$CO$_3$		90.93	100	
H–ZSM-5/NaOH		87.45	100	
CuO/NH$_4$–ZSM-5 (3%)	400°C	99.6	99.6	[41]
CuO/NH$_4$–ZSM-5 (5%)		99.7	99.7	
CuO/NH$_4$–ZSM-5 (7%)		99.9	99.9	
CuO/NH$_4$–ZSM-5 (9%)		99.0	99.0	
H–Al–TPABr	400°C, WHSV $=1.12$/h	70	65	[42]
H–Al–HDA		73	58	
H–Al–PA		48	62	
H–FFAE–TPABr		99	52	
H–FFAE–HDA		98	59	
H–FFAE–PA		97	60	
H–ZSM-5	400°C, WHSV $=1.12$/h	99.9	30.81	[43]
HMZ-0C-5T		99.8	19.51	
HMZ-5C-0T		99.7	14.11	
HMZ-0C-2T		99.9	26.02	
HMZ-1C-1T		99.7	19.48	
HMZ-0.5C-1.5T		99.8	21.26	
9 wt% CuO/HZSM-5	673 K, ratio of catalyst to feed $=0.129$	5	32.28	[44]
7 wt% CuO/HZSM-5		97	42.18	
5 wt% CuO/HZSM-5		95.7	39.55	
3 wt% CuO/HZSM-5		92.2	37.7	
H–ZSM-5	550°C, WHSV $=1$/h	99.9	30.8	[45]
ATZ-0R		99.8	25	
ATZ-0.2R		99.7	19.9	
ATZ-0.4R		99.8	15.4	
ATZ-0.6R		99.7	19.1	

Continued

Table 6.1 Effects of ZSM-5 and modified ZSM-5 on the MTG process—cont'd

Catalyst	Reaction conditions	Methanol conversion (%)	Yield/selectivity (%)	Refs.
Zn/ZSM-5	550°C	100	22.90	[46]
Zn–Mg–P/ZSM-5		99.90	88.45	
AT–ZSM-5		100	23.41	
Zn/AT–ZSM-5		100	22.82	
Zn–Mg–P/AT–ZSM-5		99.95	84.80	
Zn/HZ5/0.3AT	436.85°C, WHSV = 3.2/h	100	99.4	[47]
HZ5/0.3AT		100	99.3	
HZ5/0.1AT		100	99.2	
HZ5		97.5	99.8	

characteristics and is formed due to the polarization of the C—H bond of the methoxy group by adjacent oxygen (Fig. 6.4) [49]. The methoxy species then initiate the formation of hydrocarbon pool species. Various mechanisms have been proposed for the formation of hydrocarbon pool compounds from methoxy species and methanol/DME. Most notables are Koch carbonylation, methane-Al/oxonium, methane-formaldehyde, methoxymethyl cation, and methyleneoxy mechanism [49]. Following a short induction period for the formation of hydrocarbon pool species, the steady-state formation of hydrocarbons continues under the autocatalytic mode. This concept is known as the hydrocarbon pool mechanism. The basic concept of this mechanism is that it involves the reaction of methanol or DME with hydrocarbon pool species trapped inside the cavities of the catalyst. The hydrocarbon pool species is larger than the pore size of the zeolite, and hence, these species cannot move out from the cavities. These hydrocarbon pool species react with methanol or DME to form an intermediate compound (Fig. 6.4). The olefin is then formed from this intermediate compound with the regeneration of hydrocarbon pool species. The steady-state production of olefins continues following this cycle. Various hydrocarbon pool species are proposed in the literature, including polyalkylated aromatics, large alkylated olefins, and carbenium ions (Fig. 6.4). The olefins further undergo reaction with methanol by chain growth mechanism, followed by cracking and hydrogen transfer, to produce a mixture of paraffins and aromatics (Fig. 6.4).

6.3.2 Shape-selectivity of zeolites

HSAPO-34 is a small-pore three-dimensional silico-aluminophosphate material. It is composed of eight-membered ring cages (c. 1 nm) that are interconnected through smaller windows opening (c. 38 nm). On the other hand, HZSM-5 is the medium-pore zeolite with two intersecting channels (10-membered rings). Although HSAPO-34 can easily accommodate fairly large molecules, they cannot diffuse out of its cages or windows due to the smaller channel dimension. The products in the MTG process are thus limited to lighter hydrocarbons over HSAPO-34, such as ethylene and propylene with a smaller amount of butanes and traces of linear alkanes [50]. However, the pore structure of HZSM-5 allows cyclization and hydride transfer reactions with the formation of alkylbenzenes and isoalkanes, such as isobutane and isopentane [50]. The molecular size of durene (1,2,4,5-tetramethyl benzene) is quite close to the pore size of HZSM-5. A significant quantity of durene is thus produced over the HZSM-5 catalyst. The channel dimension of H-ferrierite is slightly smaller than HZSM-5. Therefore, cyclization and hydride transfer reactions cannot proceed to a large extent, thereby restricting the products to olefins (mainly butenes and pentenes) [50]. On the other hand, the large-pore mordenite (12-membered rings) can easily accommodate polymethyl benzenes. The polymethyl benzenes products are thus observed primarily over mordenite [48].

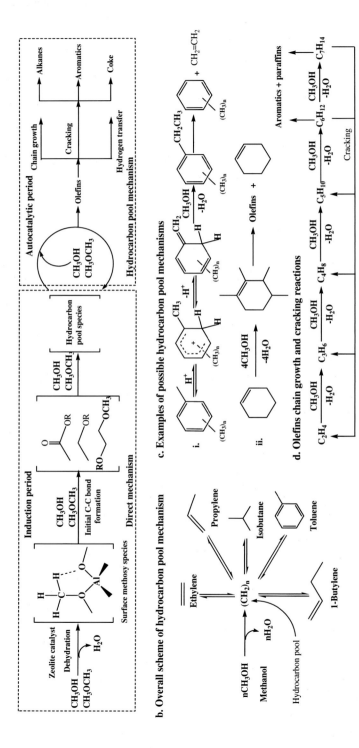

Fig. 6.4 A general reaction mechanism for the conversion of methanol to hydrocarbons.

6.3.3 Effect of Si/Al ratio in zeolite

The formation of coke on zeolite catalysts due to strong acid sites is of prime concern as it has two types of detrimental impact on the process: (i) it prevents the access of the reactant molecules to the active acid sites and (ii) the competitive removal of active acid sites [51]. The ZSM-5 catalysts having a large number of active acid sites can cause the overreaction of reactants resulting in a large amount of coke formation and blockage of reaction pathways [5]. Various analytical techniques, such as FTIR, NH_3-TPD, and NMR, have been used for qualitative and quantitative analysis of active acid centers.

The Si/Al ratios of catalysts have a significant impact on Brønsted/Lewis acidity ratio, as reported in published literature. The Brønsted/Lewis acidity ratio has a direct influence on the distribution and strength of active acid sites, which ultimately affect the selectivity of different products [52–54]. Benito et al. obtained a relation concerning the Si/Al ratios in the ZSM-5 catalyst and the characteristics of the primary product during the MTG process [52]. They reported that lighter products, such as ethane, propene, and butane, are formed initially. However, with the increase in Si/Al ratio, consequently increase in Brønsted/Lewis acidity ratio, the overall active acid sites decreases resulting in an increase in a higher fraction of heavier alkenes in products. Gao et al. performed the MTG process by varying Si/Al ratio from 20 to 60 in zeolite (ZSM-5) catalyst [54]. As the Si/Al ratio increases, the total acidity of catalysts decreases, while the strong active site to weak active site ratio increases. They also found that with the increase in Si/Al ratio from 20 to 60, the life span of catalysts increased by 87%. The fraction of C_4, C_5, and aromatics hydrocarbons increased, while the fraction of C_9^+ hydrocarbons decreased. The impact of the Si/Al ratios on the MTG process has been presented in Table 6.2.

The octane number of olefins/branched paraffins and aromatic hydrocarbons is comparatively high. Thus gasoline having a higher fraction of olefins/branched paraffins and aromatic have high octane number with better fuel properties [57]. Thus, the higher Si/Al ratio in ZSM-5 catalysts increases the life span of catalysts as well as increases the quality of gasoline by increasing the fraction of olefins and aromatic compounds.

6.3.4 Effect of size of crystal and surface properties

The crystal size of catalysts plays a prominent role in defining the performance, the selectivity of products, and the life span of the catalyst. Csicsery et al. reported that larger size zeolite catalysts are more shape-selective [58]. However, it causes faster deactivation because of the longer diffusion path for the catalytic reaction. In another study, it was observed that the larger crystal size of the catalyst facilitates a lower life span of the catalyst due to the faster dealkylation reaction of polymethylbenzene, hydrogen transfer reaction, and breakdown of nonaromatic compounds in the range of C_5-C_9 [59]. Takamitsu et al. examined the impact of the crystal size of the ZSM-5 catalyst on the selectivity of propylene during the conversion of methanol to propylene [60]. Their results showed that

Table 6.2 Effect of Si/Al ratio on the MTG process.

Zeolite type	Si/Al ratio	S_{BET} (m^2/g)	NH_3-TPD (mmol/g)	Brønsted/ Lewis	Strong/ weak	Catalyst life span (h)	C_9^+ Selectivity (%)[a]	Refs.
ZSM-5	24	420	0.50	2.86	–	–	–	[52]
	42	440	0.27	3.40	–	–	–	
	78	365	0.19	3.63	–	–	–	
	154	395	0.10	4.42	–	–	–	
ZSM-5	34	379	0.73	–	0.39	–	18.8	[55]
	58	373	0.49	–	0.34	–	25.3	
	69	372	0.39	–	0.10	–	24.3	
	80	377	0.31	–	0.09	–	23.4	
ZSM-5	27	419	0.36	–	–	18	30	[56]
ZSM-11	39	372	0.37	–	–	43	35	
	46	313	0.35	–	–	53	44	
ZSM-5	20	380	3.28	–	1.47	40	53.6	[54]
	34	325	1.69	–	1.60	51	52.1	
	43	316	1.65	–	1.62	71	46.0	
	58	321	1.34	–	1.68	75	44.0	

[a]Aromatic compounds are included.

the smaller crystal size of the catalyst has lower propylene selectivity. The smaller crystal size of the catalyst results in undesirable reactions due to the large surface area of the catalyst (Table 6.3).

6.3.5 Effect of modification of catalyst surface

The surface alteration of zeolite-based catalysts could be done by insertion of different elements into the framework, ion exchange, reactions of functional groups present on the outer surface of catalysts, dealumination, etc. [5]. Depending on the processes used for the surface modification, the properties of catalysts changed accordingly. Zaidi and Pant examined the impact of loading of copper oxide on the properties of HZSM-5 catalyst and subsequently on the selectivity of hydrocarbons from the MTG process [64]. They reported that with the increase in copper oxide content over HZSM-5 catalyst, the selectivity of heavy hydrocarbons increased markedly. However, the rate of deposition of coke over the catalyst increased with the increase in copper oxide content. Bjorgen et al. investigated the desilication effect on H-ZSM-5 catalyst with the treatment of catalyst with NaOH solution by examining the product selectivity and life span of the catalyst [56]. They reported that after the treatment of the catalyst with NaOH, the conversion increased by 3.3 times, while gasoline selectivity increased by 1.7 times.

6.3.6 Effect of process parameters

Temperature plays a crucial role in the MTG process. Generally, the MTG process occurs at 400°C [34]. Since the MTG reaction is exothermic, different levels of conversion could be observed for different temperature patterns. Also, the activity of catalysts is significantly governed by the temperature. The activity of catalysts increased linearly with temperature, and the least activity of catalysts is observed at the lowest reaction temperature [65]. The influence of process parameters (temperature, feed intake, space velocity, and pressure) on selectivity and yield of the MTG process has been summarized in Table 6.4. Pressure has a significant impact on the MTG process. It is a very important parameter for achieving maximum activity of catalyst and selectivity of gasoline. Lower pressure favors the formation of a large number of hydrocarbons. However, higher pressure favors a large amount of C_1-C_5 range hydrocarbons [71, 72]. Many studies are available in the published literature on the production of hydrocarbons from methanol in a fixed-bed reactor [66, 68, 70]. They observed that higher pressure favors the formation of more aromatic compounds at the cost of higher coke formation on the catalyst surface [73]. More aromatics mean gasoline with a higher octane number. However, the catalyst should be regenerated frequently.

The WHSV is a crucial parameter in MTG. The product selectivity is highly driven by WHSV during the MTG process over zeolite-based catalysts. If the WHSV is increased after a degree of conversion of more than 99%, the distribution of hydrocarbon

Table 6.3 Effect of crystal size of zeolite catalysts on product selectivity.

Zeolite	Average crystal size (µm)	Si/Al ratio	S_{BET} (m²/g)	Average pore volume (cm³/g)	NH_3-TPD (mmol/g)	C_6^+ selectivity (%)	Catalyst life span (h)	Reaction conditions	Refs.
HZSM-5	0.10	70	387	0.53	1.02	22	58	500°C, 1 bar,	[61]
	0.25	70	372	0.40	0.89	24	42	5 h TOS,	
	0.50	73	347	0.23	0.82	28	30	1.58/h WHSV	
Zn-ZSM-5	0.25	39	350	0.21	0.56	61	–	390°C, 5 bar,	[62]
	0.50	37	337	0.23	0.57	55	–	12.5 h TOS,	
	1.00	36	341	0.26	0.57	52	–	3.2/h WHSV	
	2.00	36	327	0.53	0.57	48	–		
HZSM-5	0.22	25	357	0.35	0.64	55	100	420°C, 1 bar,	[59]
	1.59	25	239	0.23	0.57	52	40	4 h TOS, 2.1/h	
	2.46	25	283	0.15	0.65	45	20	WHSV	
ZSM-5	0.05–0.10	19	345	–	0.78	24	–	350°C, 5 min	[63]
	2.00	43	453	–	0.26	20	–	TOS, 1.8/h	
	5.00	25	320	–	1.00	19	–	WHSV	
	15.00–20.00	53	305	–	0.31	17	–		

Advances in the conversion of methanol to gasoline 193

Table 6.4 Effect of temperature on the MTG process.

Catalysts	Process parameters	Conversion of methanol (%)	Yield/ selectivity	Refs.
H-UZM-12	350°C, 1 atm, catalyst (0.1 g), 0.67/h, 1–12 h	60	50	[66]
ZSM-5	340°C, 1 MPa, catalyst (1.4 g), 2/h, 5–100 h	93.1	18.50	[67]
ZSM-5	360°C, 1 MPa, catalyst (1.4 g), 2/h, 5–100 h	96.1	23.70	[67]
ZSM-5	380°C, 1 MPa, catalyst (1.4 g), 2/h, 5–100 h	98.3	27.60	[67]
ZSM-5	400°C, 1 MPa, catalyst (1.4 g), 2/h, 5–100 h	99.2	41.20	[67]
ZSM-5	420°C, 1 MPa, catalyst (1.4 g), 2/h, 5–100 h	99.5	42.30	[67]
HZSM-5	380°C, 1 MPa, catalyst (0.9 g), 781/h, 6 h	97	45	[68]
HZSM-5	350°C, 1.1 MPa, catalyst (0.7 g), 1.2/h, 24 h	100	59	[69]
HZSM-5	380°C, 1.0 MPa, catalyst (1.4 g), 2/h, 1 h	98	28	[70]
HZSM-5	310°C, 0.13 MPa, catalyst (0.06 g), 7/h, 20 min	5	23	[71]
HZSM-5	330°C, 0.13 MPa, catalyst (0.06 g), 7/h, 20 min	52	39	[71]
HZSM-5	390°C, 0.13 MPa, catalyst (0.06 g), 7/h, 20 min	91	38	[71]
HZSM-5	460°C, 1.0 atm, catalyst (2 g), 1/h, 12 h	87	95	[72]

products changed significantly [74]. The composition of feed can affect the quality of the product during gasoline production from methanol. The research showed that the addition of diluents to the feed could increase the selectivity of alkenes in the product stream [69, 71, 72]. The addition of water with methanol in the feed stream can increase the activity of catalysts. Also, adding water to the reactant stream can facilitate alkenes production and decrease the production of aromatic and paraffinic compounds [75].

6.3.7 Effect of reactor configuration

The effect of reactor configuration on the yield of products from the MTG process has been summarized in Table 6.5 [74, 76]. It can be observed that the fluidized-bed reactor gives a higher yield of gasoline as compared to the fixed-bed reactor. The fluidized-bed

194 Hydrocarbon biorefinery

Table 6.5 Effect of reactor configuration on hydrocarbon yield during MTG process [74, 76].

	Fluidized-bed reactor	**Fixed-bed reactor**
Hydrocarbons (wt%)		
Light hydrocarbons	4.3	1.3
Propylene	4.3	0.2
Propane	4.4	4.6
Normal butane	2.0	2.7
Isobutane	11	8.8
Butylenes	5.8	1.1
C_5^+ gasoline (without alkylate)	68.2	81.3
C_5^+ gasoline (with alkylate)	91.2	83.9
Octane numbers of gasoline		
RON	95	93
MON	85	83

reactor can save 10% of total energy consumed compared to the fixed-bed reactor considering the quality and yield of products obtained from the MTG process [76].

6.4 Industrial development

In the 1970s, the research group in ExxonMobil was working to produce high-octane gasoline by introducing a methyl or carbene group of methanol into butane over HZSM-5 catalyst [50]. They observed gasoline-range liquid product containing alkanes and aromatics with a small amount of gaseous product. Surprisingly, the butane remained completely unreacted, and the entire product was formed from methanol. This work led to an accidental discovery of the MTG conversion process in 1975. The methanol is produced from natural gas by employing the steam reforming process. In this process, the equilibrium amount of DME, methanol, and water is first formed from methanol over amorphous alumina catalyst at high pressure (2–50 bar) in a fixed-bed reactor (Fig. 6.5) [77]. The methanol conversion to DME is an exothermic reaction and releases about 416.9 kcal heat energy per kg of methanol. This step is equilibrium limited and releases only about 15%–20% of the overall heat of the reaction. Thereafter, the outlet stream from the DME reactor is mixed with recycled gas and passed to another fixed-bed reactor (MTG reactor) operated at 20 bar under the adiabatic condition with 350°C and 410°C inlet and outlet temperature, respectively. In the MTG reactor, the equilibrium mixture of methanol, DME, and water is converted to liquid hydrocarbon biofuels over shape-selective ZSM-5 catalysts. For catalyst regeneration by coke burn-off, parallel MTG reactor configuration is used for uninterrupted operation. The MTG reactor is operated under adiabatic conditions, and the temperature is controlled by diluting the feed stream

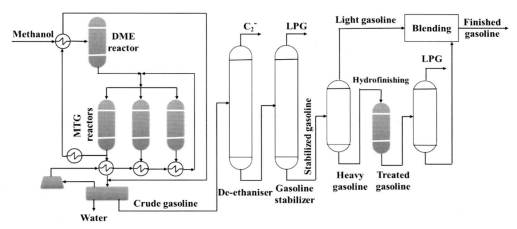

Fig. 6.5 ExxonMobil MTG process.

with recycled gas (volatile hydrocarbons, H_2, and CO/CO_2). In this reactor, DME is first dehydrated to light olefins, which are then converted to higher olefins by oligomerization reaction [78]. The higher olefins are converted into paraffins, naphthenes, and aromatics by the combination of reactions. The hydrocarbons formed in this reaction are limited up to about C_{11} due to the shape selectivity of the zeolite. The selectivity to liquid fuels is about 75% at 100% methanol conversion with more than 90% carbon conversion to gasoline [77].

MTG reactor outlet stream contains mainly gasoline-range hydrocarbons and water with a small amount of hydrogen, CO/CO_2, and C_1-C_4 volatile hydrocarbons. This stream is sent to a separator where light hydrocarbons and water are separated from crude gasoline. The light hydrocarbons are compressed and recycled to the MTG reactor. The crude gasoline stream is sent to de-ethanizer to separate C_1-C_2 hydrocarbons. The bottom stream from de-ethanizer is then sent to the stabilizer column, where LPG is obtained as the overhead product. The stabilized gasoline is further separated into light and heavy gasoline. Excessive 1,2,4,5-tetramethylbenzene (durene) formation is the major drawback in the MTG process. More than 4% durene content in gasoline is undesirable, especially under cold climates due to its high melting point (79°C). The durene content in heavy gasoline is thus reduced by the hydrofinishing step. In this step, dealkylation reaction takes place with the formation of a small amount of LPG. MTG process mainly produces gasoline-range hydrocarbon biofuel with only a small amount of LPG. This hydrocarbon biofuel is composed of paraffins, olefins, naphthenes, and aromatics with no sulfur, a small amount of benzene within the acceptable limit, and high octane number (Table 6.6). Table 6.6 shows the fuel properties and composition of gasoline obtained from the MTG process [78]. The properties of gasoline from the MTG process are quite similar to conventional gasoline.

Table 6.6 Properties of gasoline obtained from the MTG process.

Properties	Gasoline from MTG[a]	Conventional gasoline
Octane, RON	92	90–100
Octane, MON	82	81–90
Octane (RON+MON)/2	87	86–94
Reid vapor pressure, psi	9	8–15
T (50) F	201	–
T (90) F	320	–
Paraffins, vol%	53	–
Olefins, vol%	12	–
Naphthenes, vol%	9	–
Arimatics, vol%	26	–
Benzene, vol%	0.3	–
Sulfur	Nil	–

RON, research octane number; MON, motor octane number.
[a]Data adapted from ExxonMobil [78].

Primus Green Energy at Hillsborough, NJ set up a plant of 100,000 gal of gasoline production [77]. Natural gas and municipal solid waste was used as a feedstock for the production of syngas, followed by gasoline production using syngas. Similarly, the MTG plant has been operated from 1986 to 1996 in New Zealand. This plant was based on low-cost natural gas, which was finally shut down due to the high cost of natural gas. Sundrop fuels proposed an MTG plant based on the gasification of waste residue from the forest.

6.5 Conclusions

MTG process is a promising route for the production of renewable and clean gasoline-range hydrocarbon biofuel with lesser nitrogen and sulfur content. This process involves the production of syngas from various carbon sources, conversion of syngas to methanol, and production of gasoline from methanol. This chapter provides a brief overview of the production of methanol from the various feedstock, such as coal, biomass, CO_2, etc., followed by recent developments of the MTG process. Zeolite-based catalysts are most widely used in the MTG process. Although zeolite-based catalysts are prone to coking and produce a diverse range of hydrocarbons, the product selectivity and coking of catalysts can be controlled by modification of catalyst with the transition metal, alkali earth metal, and soil metal. Zeolites are shape-selective catalysts, and product distribution depends on their pore structure. Small-pore catalysts produce lighter hydrocarbons, while polyalkylated aromatics dominate over large-pore zeolites. MTG reaction proceeds through two different stages: induction period and autocatalytic mode. The initial C—C bonds and hydrocarbon pool species are formed during the short induction period.

The olefins are then formed from methanol/DME under the autocatalytic mode through hydrocarbon pool species. These olefins are then transformed to higher olefins and aromatics via the chain-growth mechanism. ExxonMobil MTG process mainly produces gasoline-range hydrocarbon biofuel with only a small amount of LPG. The gasoline obtained in this process is composed of paraffins, olefins, naphthenes, and aromatics with a small amount of benzene within the acceptable limit. However, the formation of a vast range of hydrocarbons along with the coking of catalysts is the major challenge associated with the MTG process.

References

[1] X. Zhen, Y. Wang, An overview of methanol as an internal combustion engine fuel, Renew. Sustain. Energy Rev. 52 (2015) 477–493.

[2] R.J. Nichols, The methanol story: a sustainable fuel for the future, J. Sci. Ind. Res. 62 (2003) 97–105.

[3] S. Singh, J.P. Chakraborty, M.K. Mondal, Optimization of process parameters for torrefaction of Acacia nilotica using response surface methodology and characteristics of torrefied biomass as upgraded fuel, Energy 186 (2019) 115865.

[4] M. Finley, The oil market to 2030—implications for investment and policy, Econ. Energy Environ. Policy 1 (2012) 25–36.

[5] E. Kianfar, S. Hajimirzaee, S. Mousavian, A.S. Mehr, Zeolite-based catalysts for methanol to gasoline process: a review, Microchem. J. 156 (2020) 104822.

[6] J.G. Speight, The Refinery of the Future, Elsevier, USA, 2010.

[7] C.D. Chang, J.C. Kuo, W.H. Lang, S.M. Jacob, J.J. Wise, A.J. Silvestri, Process studies on the conversion of methanol to gasoline, Ind. Eng. Chem. Process. Des. Dev. 17 (1978) 255–260.

[8] J. Cejka, A. Corma, S. Zones, Zeolites and Catalysis: Synthesis, Reactions and Applications, Wiley-VCH, Weinheim, 2010.

[9] L.A. Pellegrini, G. Soave, S. Gamba, S. Langè, Economic analysis of a combined energy–methanol production plant, Appl. Energy 88 (2011) 4891–4897.

[10] J. Bermúdez, B. Fidalgo, A. Arenillas, J. Menéndez, CO_2 reforming of coke oven gas over a Ni/γ-Al_2O_3 catalyst to produce syngas for methanol synthesis, Fuel 94 (2012) 197–203.

[11] J. Ott, V. Gronemann, F. Pontzen, E. Fiedler, G. Grossmann, D.B. Kersebohm, G. Weiss, C. Witte, Methanol, in: Ullmann's Encyclopedia of Industrial Chemistry, Wiley-VCH, 2000.

[12] F.G. Üçtuğ, S. Ağralı, Y. Arıkan, E. Avcıoğlu, Deciding between carbon trading and carbon capture and sequestration: an optimisation-based case study for methanol synthesis from syngas, J. Environ. Manage. 132 (2014) 1–8.

[13] G.A. Olah, Beyond oil and gas: the methanol economy, Angew. Chem. Int. Ed. 44 (2005) 2636–2639.

[14] O. Jörg, V. Gronemann, F. Pontzen, E. Fiedler, G. Grossmann, D.B. Kersebohm, G. Weiss, C. Witte, Methanol, Ullmann's Encyclopedia of Industrial Chemistry, Wiley-VCH Verlag GmbH & Co. KGaA, Weinheim, Germany, 2012.

[15] P.-C. Chen, H.-M. Chiu, Y.-P. Chyou, C.-S. Yu, Processes simulation study of coal to methanol based on gasification technology, World Acad. Sci. Eng. Technol. 41 (2010) 988–996.

[16] S. Verma, To study the direct transformation of methane into methanol in the lower temperature range, Energ. Conver. Manage. 43 (2002) 1999–2008.

[17] K. Okazaki, T. Kishida, K. Ogawa, T. Nozaki, Direct conversion from methane to methanol for high efficiency energy system with exergy regeneration, Energ. Conver. Manage. 43 (2002) 1459–1468.

[18] J.M. Bermúdez, B. Fidalgo, A. Arenillas, J. Menéndez, Dry reforming of coke oven gases over activated carbon to produce syngas for methanol synthesis, Fuel 89 (2010) 2897–2902.

[19] J. Bermúdez, A. Arenillas, J. Menéndez, Equilibrium prediction of CO2 reforming of coke oven gas: suitability for methanol production, Chem. Eng. Sci. 82 (2012) 95–103.

[20] J. Bermúdez, A. Arenillas, J. Menéndez, Syngas from CO_2 reforming of coke oven gas: synergetic effect of activated carbon/Ni–γ-Al_2O_3 catalyst, Int. J. Hydrogen Energy 36 (2011) 13361–13368.

[21] M. Specht, A. Bandi, The Methanol-Cycle—Sustainable Supply of Liquid Fuels, Centre for Solar Energy and Hydrogen Research (ZSW), Stuttgart, Germany, 1999.

[22] S. Leduc, J. Lundgren, O. Franklin, E. Dotzauer, Location of a biomass based methanol production plant: a dynamic problem in northern Sweden, Appl. Energy 87 (2010) 68–75.

[23] N. Shamsul, S.K. Kamarudin, N.A. Rahman, N.T. Kofli, An overview on the production of bio-methanol as potential renewable energy, Renew. Sustain. Energy Rev. 33 (2014) 578–588.

[24] S. Singh, J.P. Chakraborty, M.K. Mondal, Pyrolysis of torrefied biomass: optimization of process parameters using response surface methodology, characterization, and comparison of properties of pyrolysis oil from raw biomass, J. Clean. Prod. 272 (2020) 122517.

[25] S. Singh, J.P. Chakraborty, M.K. Mondal, Torrefaction of Acacia nilotica: oxygen distribution and carbon densification mechanism based on in-depth analyses of solid, liquid, and gaseous products, Energy Fuel 34 (2020) 12586–12597.

[26] N. Ouellette, H.-H. Rogner, D. Scott, Hydrogen from remote excess hydroelectricity. Part II: hydrogen peroxide or biomethanol, Int. J. Hydrogen Energy 20 (1995) 873–880.

[27] L.R. Clausen, N. Houbak, B. Elmegaard, Technoeconomic analysis of a methanol plant based on gasification of biomass and electrolysis of water, Energy 35 (2010) 2338–2347.

[28] S. Shabangu, D. Woolf, E.M. Fisher, L.T. Angenent, J. Lehmann, Techno-economic assessment of biomass slow pyrolysis into different biochar and methanol concepts, Fuel 117 (2014) 742–748.

[29] P.G. Cifre, O. Badr, Renewable hydrogen utilisation for the production of methanol, Energ. Conver. Manage. 48 (2007) 519–527.

[30] A. Boretti, Renewable hydrogen to recycle CO_2 to methanol, Int. J. Hydrogen Energy 38 (2013) 1806–1812.

[31] I. Ganesh, Conversion of carbon dioxide into methanol–a potential liquid fuel: fundamental challenges and opportunities (a review), Renew. Sustain. Energy Rev. 31 (2014) 221–257.

[32] M.D. Porosoff, B. Yan, J.G. Chen, Catalytic reduction of CO_2 by H_2 for synthesis of CO, methanol and hydrocarbons: challenges and opportunities, Energ. Environ. Sci. 9 (2016) 62–73.

[33] A. Goeppert, M. Czaun, J.-P. Jones, G.S. Prakash, G.A. Olah, Recycling of carbon dioxide to methanol and derived products–closing the loop, Chem. Soc. Rev. 43 (2014) 7995–8048.

[34] L. Zhang, C. Xu, P. Champagne, Overview of recent advances in thermo-chemical conversion of biomass, Energ. Conver. Manage. 51 (2010) 969–982.

[35] G.A. Jablonski, L. Sand, J. Gard, Synthesis and identification of ZSM-5ZSM-11 pentasil intergrowth structures, Zeolites 6 (1986) 396–402.

[36] Y. Ji, H. Yang, W. Yan, Strategies to enhance the catalytic performance of ZSM-5 zeolite in hydrocarbon cracking: a review, Catalysts 7 (2017) 367.

[37] C.S. Triantafillidis, A.G. Vlessidis, L. Nalbandian, N.P. Evmiridis, Effect of the degree and type of the dealumination method on the structural, compositional and acidic characteristics of H–ZSM-5 zeolites, Microporous Mesoporous Mater. 47 (2001) 369–388.

[38] L. Shirazi, E. Jamshidi, M. Ghasemi, The effect of Si/Al ratio of ZSM-5 zeolite on its morphology, acidity and crystal size, Cryst. Res. Technol. 43 (2008) 1300–1306.

[39] T. Armaroli, L. Simon, M. Digne, T. Montanari, M. Bevilacqua, V. Valtchev, J. Patarin, G. Busca, Effects of crystal size and Si/Al ratio on the surface properties of H–ZSM-5 zeolites, Appl. Catal. A. Gen. 306 (2006) 78–84.

[40] S. Fathi, M. Sohrabi, C. Falamaki, Improvement of HZSM-5 performance by alkaline treatments: comparative catalytic study in the MTG reactions, Fuel 116 (2014) 529–537.

[41] E. Kianfar, M. Salimi, V. Pirouzfar, B. Koohestani, Synthesis of modified catalyst and stabilization of CuO/NH4-ZSM-5 for conversion of methanol to gasoline, Int. J. Appl. Ceram. Technol. 15 (2018) 734–741.

[42] R.N. Missengue, P. Losch, N.M. Musyoka, B. Louis, P. Pale, L.F. Petrik, Conversion of south African coal fly ash into high-purity ZSM-5 zeolite without additional source of silica or alumina and its application as a methanol-to-olefins catalyst, Catalysts 8 (2018) 124.

[43] J. Ahmadpour, M. Taghizadeh, Catalytic conversion of methanol to propylene over high-silica meso-porous ZSM-5 zeolites prepared by different combinations of mesogenous templates, J. Nat. Gas Sci. Eng. 23 (2015) 184–194.

[44] H. Zaidi, K. Pant, Catalytic activity of copper oxide impregnated HZSM-5 in methanol conversion to liquid hydrocarbons, Can. J. Chem. Eng. 83 (2005) 970–977.

[45] J. Ahmadpour, M. Taghizadeh, Selective production of propylene from methanol over high-silica mesoporous ZSM-5 zeolites treated with NaOH and NaOH/tetrapropylammonium hydroxide, C. R. Chim. 18 (2015) 834–847.

[46] J. Zhang, X. Zhu, S. Zhang, M. Cheng, M. Yu, G. Wang, C. Li, Selective production of para-xylene and light olefins from methanol over the mesostructured Zn–Mg–P/ZSM-5 catalyst, Cat. Sci. Technol. 9 (2019) 316–326.

[47] P.-H. Chao, S.-T. Tsai, S.-L. Chang, I. Wang, T.-C. Tsai, Hexane isomerization over hierarchical Pt/MFI zeolite, Top. Catal. 53 (2010) 231–237.

[48] C.D. Chang, The methanol-to-hydrocarbons reaction: a mechanistic perspective, in: Shape-Selective Catalysis, American Chemical Society, Washington, 1999, pp. 96–114.

[49] I. Yarulina, A.D. Chowdhury, F. Meirer, B.M. Weckhuysen, J. Gascon, Recent trends and fundamental insights in the methanol-to-hydrocarbons process, Nat. Catal. 1 (2018) 398–411.

[50] J.F. Haw, W. Song, D.M. Marcus, J.B. Nicholas, The mechanism of methanol to hydrocarbon catalysis, Acc. Chem. Res. 36 (2003) 317–326.

[51] D.M. Bibby, R.F. Howe, G.D. McLellan, Coke formation in high-silica zeolites, Appl. Catal. A. Gen. 93 (1992) 1–34.

[52] P.L. Benito, A.G. Gayubo, A.T. Aguayo, M. Olazar, J. Bilbao, Effect of Si/Al ratio and of acidity of H-ZSM5 zeolites on the primary products of methanol to gasoline conversion, J. Chem. Technol. Biotechnol. 66 (1996) 183–191.

[53] P. Li, W. Zhang, X. Han, X. Bao, Conversion of methanol to hydrocarbons over phosphorus-modified ZSM-5/ZSM-11 intergrowth zeolites, Catal. Lett. 134 (2010) 124–130.

[54] Y. Gao, B. Zheng, G. Wu, F. Ma, C. Liu, Effect of the Si/Al ratio on the performance of hierarchical ZSM-5 zeolites for methanol aromatization, RSC Adv. 6 (2016) 83581–83588.

[55] R. Wei, C. Li, C. Yang, H. Shan, Effects of ammonium exchange and Si/Al ratio on the conversion of methanol to propylene over a novel and large partical size ZSM-5, J. Nat. Gas Chem. 20 (2011) 261–265.

[56] M. Bjørgen, F. Joensen, M.S. Holm, U. Olsbye, K.-P. Lillerud, S. Svelle, Methanol to gasoline over zeolite H-ZSM-5: improved catalyst performance by treatment with NaOH, Appl. Catal. A. Gen. 345 (2008) 43–50.

[57] B.M. Abu-Zied, W. Schwieger, A. Unger, Nitrous oxide decomposition over transition metal exchanged ZSM-5 zeolites prepared by the solid-state ion-exchange method, Appl. Catal. Environ. 84 (2008) 277–288.

[58] S.M. Csicsery, Shape-selective catalysis in zeolites, Zeolites 4 (1984) 202–213.

[59] Y. Zhang, Y. Qu, J. Wang, Effect of crystal size on the catalytic performance of HZSM-5 zeolite in the methanol to aromatics reaction, Pet. Sci. Technol. 36 (2018) 898–903.

[60] Y. Takamitsu, K. Yamamoto, S. Yoshida, H. Ogawa, T. Sano, Effect of crystal size and surface modification of ZSM-5 zeolites on conversion of ethanol to propylene, J. Porous. Mater. 21 (2014) 433–440.

[61] J. Meng, J. Park, D. Tilotta, S. Park, The effect of torrefaction on the chemistry of fast-pyrolysis bio-oil, Bioresour. Technol. 111 (2012) 439–446.

[62] X. Niu, J. Gao, K. Wang, Q. Miao, M. Dong, G. Wang, W. Fan, Z. Qin, J. Wang, Influence of crystal size on the catalytic performance of H-ZSM-5 and Zn/H-ZSM-5 in the conversion of methanol to aromatics, Fuel Process. Technol. 157 (2017) 99–107.

[63] F.L. Bleken, S. Chavan, U. Olsbye, M. Boltz, F. Ocampo, B. Louis, Conversion of methanol into light olefins over ZSM-5 zeolite: strategy to enhance propene selectivity, Appl. Catal. A. Gen. 447–448 (2012) 178–185.

[64] H.A. Zaidi, K.K. Pant, Transformation of methanol to gasoline range hydrocarbons using HZSM-5 catalysts impregnated with copper oxide, Korean J. Chem. Eng. 22 (2005) 353–357.

[65] G. Echevskii, K. Ione, G. Nosyreva, G. Litvak, Effect of the temperature regime of methanol conversion to hydrocarbons on coking of zeolite catalysts and their regeneration, Appl. Catal. 43 (1988) 85–89.

[66] J.H. Lee, M.B. Park, J.K. Lee, H.-K. Min, M.K. Song, S.B. Hong, Synthesis and characterization of ERI-type UZM-12 zeolites and their methanol-to-olefin performance, J. Am. Chem. Soc. 132 (2010) 12971–12982.

[67] Z. Di, C. Yang, X. Jiao, J. Li, J. Wu, D. Zhang, A ZSM-5/MCM-48 based catalyst for methanol to gasoline conversion, Fuel 104 (2013) 878–881.

[68] F.L. Bleken, K. Barbera, F. Bonino, U. Olsbye, K.P. Lillerud, S. Bordiga, P. Beato, T.V. Janssens, S. Svelle, Catalyst deactivation by coke formation in microporous and desilicated zeolite H-ZSM-5 during the conversion of methanol to hydrocarbons, J. Catal. 307 (2013) 62–73.

[69] Z. Wan, W. Wu, W. Chen, H. Yang, D. Zhang, Direct synthesis of hierarchical ZSM-5 zeolite and its performance in catalyzing methanol to gasoline conversion, Ind. Eng. Chem. Res. 53 (2014) 19471–19478.

[70] M. Bjørgen, F. Joensen, K.-P. Lillerud, U. Olsbye, S. Svelle, The mechanisms of ethene and propene formation from methanol over high silica H-ZSM-5 and H-beta, Catal. Today 142 (2009) 90–97.

[71] M. Bjørgen, S. Svelle, F. Joensen, J. Nerlov, S. Kolboe, F. Bonino, L. Palumbo, S. Bordiga, U. Olsbye, Conversion of methanol to hydrocarbons over zeolite H-ZSM-5: on the origin of the olefinic species, J. Catal. 249 (2007) 195–207.

[72] F.B. Shareh, M. Kazemeini, M. Asadi, M. Fattahi, Metal promoted mordenite catalyst for methanol conversion into light olefins, Pet. Sci. Technol. 32 (2014) 1349–1356.

[73] R. Comelli, N. Fígoli, Effect of pressure on the transformation of methanol into hydrocarbons on an amorphous silica—alumina, Appl. Catal. 73 (1991) 185–194.

[74] M. Stöcker, Methanol-to-hydrocarbons: catalytic materials and their behavior, Microporous Mesoporous Mater. 29 (1999) 3–48.

[75] E. Kianfar, M. Salimi, A review on the production of light olefins from hydrocarbons cracking and methanol conversion, in: Advances in Chemistry Research, vol. 59, Nova Science Publishers, Inc., New York, NY, 2020.

[76] S.A. Tabak, S. Yurchak, Conversion of methanol over ZSM-5 to fuels and chemicals, Catal. Today 6 (1990) 307–327.

[77] A.M. Brownstein, Renewable Motor Fuels: The Past, the Present and the Uncertain Future, Butterworth-Heinemann, USA, 2014.

[78] . https://www.exxonmobilchemical.com/en/exxonmobil-chemical/newsroom. (Accessed February 4, 2021).

SECTION 2

Biological and biochemical conversion processes

CHAPTER 7

Biomass pretreatment technologies

Ayaz Ali Shah[a,b], Tahir Hussain Seehar[a,b], Kamaldeep Sharma[a], and Saqib Sohail Toor[a]

[a]Department of Energy Technology, Aalborg University, Aalborg, Denmark
[b]Department of Energy & Environment Engineering, Dawood University of Engineering & Technology, Karachi, Sindh, Pakistan

Contents

7.1 Introduction	203
7.2 Lignocellulosic feedstock composition and pretreatment	204
7.3 Pretreatment techniques for lignocellulosic feedstock	207
7.3.1 Physical pretreatment	207
7.3.2 Physicochemical pretreatment	208
7.3.3 Chemical pretreatment	209
7.3.4 Biological pretreatment	210
7.4 Sewage sludge composition and pretreatment	211
7.5 Pretreatment techniques for the sewage sludge	213
7.5.1 Mechanical treatment	213
7.5.2 Thermal treatment	217
7.5.3 Chemical treatment	218
7.5.4 Biological pretreatment	221
7.6 Challenges and future perspectives	222
7.7 Conclusions	223
References	223

7.1 Introduction

Pretreatment is an important phase to modify the structure of the biomass feedstock to make it more accessible during any biomass conversion process. The pretreatment step is a key bottleneck in biomass processing either for the biochemical or thermochemical process for the production of biofuels and other valuable chemicals. For energy generation, waste materials and purpose-grown energy crops are the two main sources of biomass [1,2]. It is hard to describe the prominent pretreatment method due to its dependency on some factors such as feedstock type and the desired products obtained from it. The huge range of biomass categories prevents the use of just a single type of pretreatment method for various feedstocks. A particularly efficient method for one specific feedstock might not translate to an efficient process for another type of feedstock even though some pretreatment methods show apparent advantages. Based on biomass

---

Hydrocarbon Biorefinery
https://doi.org/10.1016/B978-0-12-823306-1.00014-5

Copyright © 2022 Elsevier Inc.
All rights reserved.

characteristics, the pretreatment method is a tailor-made process that should be precisely selected and planned for every individual biomass [3,4].

Due to the production of various products, the selection of the optimum pretreatment process conditions mainly depends on the objectives of biomass pretreatment. Besides this, the selection of a specific pretreatment method should be based on its economic and environmental impacts too rather than only on its potential yield. To overcome biomass recalcitrance, the pretreatment techniques are essentially required as generally categorized into physical, chemical, and biological approaches [5].

Many pretreatment methods have been explored. However, broad research is still needed for the growth of new and more effective techniques for yielding promising results. It is problematic to assess and relate pretreatment technologies due to the involvement of processing cost and capital investment. However, mass balance analysis can be used to confirm the effectiveness of a pretreatment process with any specified feedstock [6]. For designing an industrial-based biomass process, an economic assessment of the pretreatment methods should be prepared to define their feasibility [7].

This chapter presents an overlook of pretreatment technologies used for the feedstocks, especially lignocellulosics and sewage sludge, which include physical, mechanical, chemical, and biological treatments as demonstrated in Fig. 7.1 for energy enhancement in the form of bio-crude and biogas by thermochemical and biological processes. The technical issues and challenges encountered during pretreatment processes, along with the future research needs, are also discussed.

7.2 Lignocellulosic feedstock composition and pretreatment

Lignocellulosic material is a highly rich renewable energy resource in the form of forest and agricultural biomass throughout the world. Lignocellulose biomass such as different kinds of straws, i.e., rice, wheat, and barley straw, sugarcane bagasse, etc., are mostly used for bioenergy production. The composition of lignocellulosic feedstock comprises of 40%–50% cellulose, 25%–30% hemicellulose, and 7%–20% lignin [8].

Cellulose is the main and valuable component of the lignocellulosic biomass that stores a huge quantity of energy preserved by the photosynthesis process. By nature, water-insoluble fibrous and linear polymer material contain thousands of glucose monomers. From the structure point of view, cellulose consists of a crystalline structure which is a kind of barrier for cellulose to degrade or decompose within the process. One of the purposes of pretreatment is also to reduce the crystalline structure of cellulose fragments [9]. Hemicellulose is heterogeneous and mostly comprises xylose, galactose, mannose, and arabinose which require enzymes to decompose hemicellulose fractions. The degradation of hemicellulose produces acetic acid and monomeric sugars [10]. Lignin is an aromatic polymer composed of primarily three phenolic components, namely

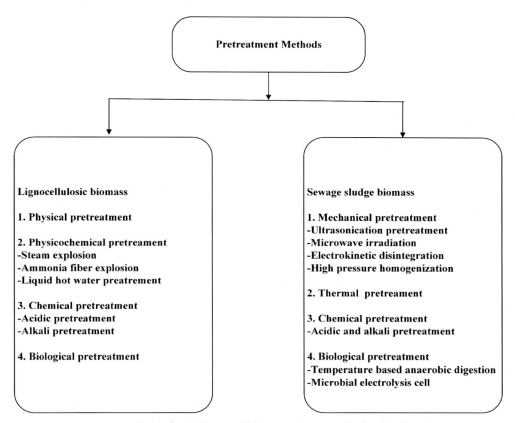

Fig. 7.1 Pretreatment methods for the lignocellulosic and sewage sludge feedstock.

coniferyl alcohol, *p*-coumaryl alcohol, and sinapyl alcohol. In the biomass structure, lignin is mainly available in the grouping with cellulose and hemicellulose. To obtain the energy from the carbohydrates, it is important to break down or reduce the lignin fragments to have easy access to carbohydrates via pretreatment techniques. Fig. 7.2 demonstrates the complex structure dynamics of carbohydrates and lignin. Cellulose chains fill the structural biological materials often known as fibers. Those microfibers are linked together by hemicellulose and different polymers of sugars, pectin, etc. which are covered by the lignin content. The concentration of carbohydrates and lignin content in the cell wall layers mainly depends upon the composition of the lignocellulosic biomass. Among the wall layers, the higher concentration of the cellulose fractions is present in the second portion of the secondary wall. However, the higher availability of the lignin content can be found in the lamella portion of the plant [8].

For the transformation of biomass to biofuels and chemicals, the pretreatment is considered as a subprocess. However, for the production of ethanol from a lignocellulosic

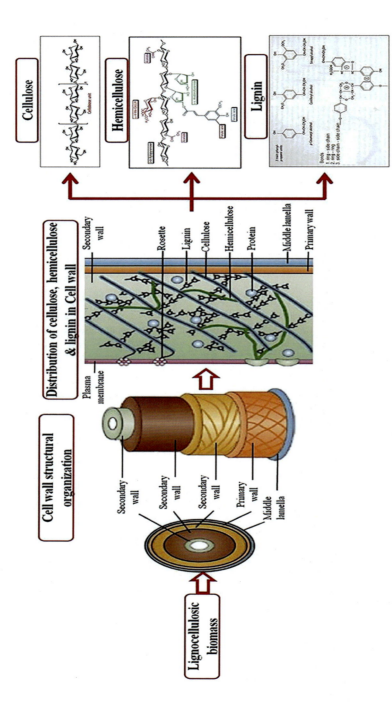

Fig. 7.2 The internal structure of the lignocellulosic biomass reflecting carbohydrates and lignin content [8]. *(From V. Menon, M. Rao, Trends in bioconversion of lignocellulose: biofuels, platform chemicals & biorefinery concept, Prog. Energy Combust. Sci. 38 (4) (2012) 522–550. https://doi.org/10.1016/j.pecs.2012.02.002.)*

source, the main objectives covers the pretreatment are to maintain the particle size, bulk density, and to reduce the ash contents [11]. Additionally, the lignocellulosic feedstock requires pretreatment for the structural modification of cellulose and lignin present in the biomass and to utilize the plant fibers easily. The selection of the pretreatment method mainly depends upon the composition of the biomass like the concentration of carbohydrates, lignin contents, and process requirements, for example, to enhance the sugars and to avoid the reduction of the carbohydrates.

7.3 Pretreatment techniques for lignocellulosic feedstock

The following are the major pretreatments for the lignocellulosic feedstock.
1. Physical pretreatment
2. Physicochemical pretreatment
3. Chemical pretreatment
4. Biological pretreatment

7.3.1 Physical pretreatment

Physical pretreatment is considered as a prerequisite treatment before further application of the lignocellulosic biomass. Particle size reduction of the biomass is the basic goal of physical treatment before processing. The surface area increases significantly by reducing the particle size, crystallinity and by breaking the polymeric chains [12]. The fine particle size of the biomass is easy to utilize for the digestion process. This type of pretreatment is environmentally friendly as there is no exposure to toxic chemicals [13]. The process conditions like pressure, temperature, retention time, and feedstock type could significantly affect the outcomes of the physical pretreatments [14]. Physical treatment can also be termed as a mechanical treatment, which includes milling, grinding, and chipping. Zhu et al. experienced that for the pretreatment of a woody biomass, disk milling efficiently enhances the cellulose hydrolysis as compared to hammer milling [15]. By the milling process, the particle size of the biomass can be reduced by up to 0.2mm. Researchers reported that pretreatment via milling resulted in enhanced bioethanol and biogas yields. At the laboratory scale, Seehar et al. used a cyclone mill for reducing the particle size for the hydrothermal liquefaction of the wheat straw. The physical treatment has a few drawbacks, especially power consumption [16]. At the industrial or plant scale, physical pretreatment increases energy utilization, which leads to the escalation of electricity demand. It was experienced that the cost of power consumption during the milling and grinding operation is more than the price of biomass used for the process [17]. However, few researchers have reported that physical pretreatment, i.e., milling and grinding, can significantly lower the energy consumption if it is allied with chemical pretreatment [15,18]. It was reported that agricultural feedstock needs less energy as compared to hardwoods for pretreatment during the production of ethanol [19].

7.3.2 Physicochemical pretreatment

The grouping of physical and chemical treatment is well known as physicochemical treatment. As discussed earlier, the combined treatment is cost-effective and reduces energy consumption. Biomass digestibility can also be enhanced by combined treatment. This type of pretreatment comprises a wide range of pretreatment technologies like Steam pretreatment, Ammonia Fiber Explosion, liquid hot water pretreatment.

7.3.2.1 Steam explosion/auto-hydrolysis

Steam explosion is also well known as an auto-hydrolysis technique. In this technique, the lignocellulosic biomass is introduced to high-pressure steam under the temperature and pressure range 160–260°C and 0.69–4.83 MPa, respectively. Steam pretreatment aims to decompose the biomass material to enhance the hemicellulose hydrolysis in which the quick release of pressure is mandatory. The process involves the transformation of lignin content and degradation of the hemicellulose by increasing the temperature and results in the promotion of hydrolysis. In some cases, acids, especially acetic acid, are employed to increase the steam explosion pretreatment efficiency. During the steam pretreatment, the limited concentration of lignin content is reduced due to the depolymerization reactions [20]. Sulfuric acid or carbon dioxide can also be utilized as catalysts to decrease the time and temperature for the improvement in the hydrolysis process and enhances the removal efficiency of hemicellulose. By using H_2SO_4 and SO_2 [21] explored the effects of impregnation of softwood by steam explosion at 195–215°C, in which a higher yield of glucose was obtained with the addition of SO_2 and a lower yield with H_2SO_4.

7.3.2.2 Ammonia fiber explosion

Ammonia fiber explosion is another type of physicochemical pretreatment. It is mostly used for the pretreatment of wheat straw and different herbaceous crops, etc. Ammonia fiber explosion is well known for increasing the fermentation rate of the various lignocellulosic biomass types. Ammonia in liquid form is used to treat the lignocellulosic materials under the high temperature and pressure conditions. Generally, the dosage rate of ammonia is 1–2 kg of ammonia/kg of feedstock at 30 min residence time. It was also experienced that, for high lignin content biomass like aspen wood, the ammonia fiber explosion is not a suitable pretreatment method from an efficiency point of view [5].

7.3.2.3 Liquid hot water pretreatment

Liquid hot water pretreatment mainly utilizes hot water instead of steam at temperatures in the range 170–230°C. In this process, lignin is removed and hemicellulose undergoes the hydrolysis reaction. In the liquid hot water pretreatment process, no chemicals or catalysts are needed; however, the pH value is maintained to enhance the process efficiency [22]. Hongdan et al. conducted a study on the optimization of the process variables by using a liquid hot water pretreatment method for bagasse and successfully obtained 90% glucose recovery at 180°C, 30 min [23]. A catalyst plays a significant role during the liquid hot water pretreatment of the lignocellulosic biomass. It was found that the addition of an alkali catalyst (NaOH) improves the glucose yield during the liquid hot water pretreatment of rice straw as compared to without an alkali catalyst [24]. One of the drawbacks of liquid hot water pretreatment is the huge consumption of water during the process that may be a barrier to making it effective for being promoted on a commercial scale.

7.3.3 Chemical pretreatment

Chemical pretreatment for the lignocellulosic materials introduces the mixing of different chemical reagents to biomass. By the utilization of chemicals, the basic aim of this treatment is to disintegrate the crystallinity and polymerization of the biomass structure. However, the formation of toxic materials and loss of carbohydrate polymers are the drawbacks of this treatment [25]. There are two main techniques that are widely used as chemical pretreatment methods, namely alkali and acid pretreatment.

7.3.3.1 Alkali pretreatment

Alkali pretreatment of the lignocellulosic biomass is favorable due to processing at lower pressure and temperature as compared to other pretreatment techniques that mainly depend on the concentration of lignin contents [2]. However, the retention time for this treatment method is in hours or days. Among alkali agents, calcium hydroxide and sodium hydroxide are considered suitable candidates regarding the efficiency and economic perspective. Additionally, sodium, potassium, and calcium metals are also widely used for the alkali pretreatments. The reactions in the alkali mediums break down the ester linkage between the carbohydrate fractions and lignin through which solubilization occurs. Furthermore, alkali treatment also affects the structure of the lignocellulosic biomass by polymerization and crystallinity decreases the growth of cellulose [26]. Shah et al. used the corncob residue for the production of biogas and reported that pretreatment with alkali agents enhances the digestion process and removes the lignin content efficiently and produces more biogas than that produced without alkali treatment [27]. Additionally, Sakuragi et al. explored the effects of alkali pretreatment by using ammonia for hardwood species. They experienced that wood samples containing low lignin and a high concentration of xylan efficiently promote enzymatic hydrolysis [28]. In conclusion, the alkali pretreatment method is efficient to eliminate the lignin content and to expose the utilization of carbohydrates to further processing. However, a significant concentration of different alkaline solutions like sodium hydroxide, etc., leads to higher corrosion

7.3.3.2 Acid pretreatment

Acid pretreatment of the lignocellulosic biomass is mainly dependent on the utilization of acid materials to improve the enzymatic hydrolysis. Several studies are conducted in which organic and inorganic acids were used for the pretreatment techniques [29,30]. Acidic ions are mainly responsible for the breaking of cellulose and hemicellulose chains into sugar fragments. Among the acids, sulfuric acid is the one most widely adopted to treat the lignocellulosic feedstocks. For the acid pretreatment, highly concentrated acids or dilute acids can be used depending on the process requirements. Sahoo et al. concluded that dilute acid pretreatment is more favorable than alkali pretreatment for the enzymatic hydrolysis of wild rice grass [31]. One of the disadvantages of this treatment is the recovery of toxic acids, which requires maintenance or further operation to reduce the impact of acids. The inorganic acids like sulfuric acid, hydrochloric acid, nitric acid are mostly used as acid pretreatment agents. Due to toxic and corrosive nature, acid pretreatment decreases the pH level of the feedstock in the acidic range, which is not favorable for efficient hydrolysis for the HTL processing particular for the bio-crude production. These acids need to recover to make thermochemical processes like HTL and pyrolysis processes efficient and more sustainable.

7.3.4 Biological pretreatment

Biological pretreatment is also referred to as a microorganism pretreatment, which does not require abundant energy and water like other pretreatment techniques. It is used to reduce the concentration of lignin content and to hydrolyze the cellulose. The lignocellulosic biomass has a recalcitrant nature with complex composition and structures. In this type of treatment, mostly microorganisms such as fungi (white, brown, soft rot fungi) and/or bacteria are used to produce enzymes that can decompose the lignin and hemicellulose concentration available in the feedstock [32]. Fungi like Pteurotus spp., Ischnoderma benzoinum are appropriate for such methods as they are efficient to degrade carbohydrates and lignin. These fungi degrade the lignin content in the presence of carbon and nitrogen source. Among all fungi species, the white fungi such as *Pycnoporus cinnabarinus*, *Pleurotus ostreatus* are considered as a better option. Generally, antimicrobial substances can also be removed by biological pretreatment and that is not only limited to the reduction of lignin. Tian et al. also reported that white, brown, and soft root fungi can be utilized for reducing the lignin from wheat straw [25]. Depending on the process conditions, the high retention time is one of the drawbacks of biological treatment as it takes approximately 10–14 days to complete the process [1]. The other demerit of this pretreatment is the reduced concentration of carbohydrates, which was taken by the

Biomass pretreatment technologies   **211**

Table 7.1 The main observations experienced during the pretreatment techniques for lignocellulosic biomass.

Pretreatment	Observation	Reference
Physical pretreatment	For woody biomass, physical treatment via chipper disks (rpm: 40–1000) with motor power of 2500 hp. or 1864 kW was applied and efficiently decreased the biomass particles size in range between 6 mm to 5 cm for the energy enhancement.	[33]
Steam explosion	By applying steam explosion method, changes were observed in the cellulose structure and activation of lignin content that can produce new chemical bonds. The treated lignin then can be utilized as a binder to the lignocellulosic biomass to obtain the fiberboard.	[34]
Ammonia fiber explosion (AFEX)	Ammonia and water loading, temperature, and retention time can effect on the overall process economy for ethanol production and can be changed for the optimization of the AFEX pretreatment.	[35]
Liquid hot water	During liquid hot water treatment in temperature range 200–230°C, about 40%–60% of the total biomass is dissolved. For carbohydrates, 4%–22% of the cellulose and almost all of the hemicellulose is removed that can be recovered as monomer sugars. 35%–60% of the lignin can also be removed.	[2]
Alkaline treatment	For the production of bio-crude via hydrothermal liquefaction, alkaline pretreatment at 180°C by using NaOH is preferable for the pumpability perspective of woody biomass.	[36]
Acid treatment	During hydrolysis of lignocellulosic biomass, sulfuric, maleic, and oxalic, acids were investigated and concluded that maleic and oxalic dicarboxylic acids are more efficient to degrade the hemicelluloses fraction compared to sulfuric acid.	[37]
Biological pretreatment	The biological treatment via microbial agents with dosage 0.01% (w/w) with 15-day hydraulic retention time for biogas production, total biogas increased by 33.07%, specifically 75.57% more methane yield, compared with the untreated corn straw sample.	[38]

microorganisms. The main observations from some lignocellulosic pretreatment studies are listed in Table 7.1.

7.4 Sewage sludge composition and pretreatment

Sewage sludge is a heterogeneous mixture that comprises of (i) high water content 75%–98%, (ii) nontoxic organic compounds with 48%–55% of the volatile matter with heating values of 11–23 MJ/kg, (iii) nontoxic inorganic compounds of silicon, aluminum, calcium, zinc, iron, etc., (iv) inorganic toxic materials like chromium, nickel, mercury, and arsenic, originated mainly from corroded sewers with industrial wastes, (v) organic pollutants like dioxins and polycyclic aromatic hydrocarbons, (vi) Nitrogen and Phosphorus-containing compounds originated from proteins, sugars, and fatty acids, and (vii) biological pollutants such as microorganisms and pathogens [39,40]. The percentage of these constituents varies depending on the location and origin and the treatment process applied by the wastewater treatment plant [41]. The existence of a variety of the undesired nonorganic matter in the sewage sludge makes pretreatment an indispensable aspect for the valorization of the organic potential of the sewage sludge.

For energy recovery, the sewage sludge has been processed via two main processes, biological and thermochemical. The biological process includes anaerobic digestion, which is a low-cost biological method used to convert organic wastes with higher water content to biogas (CH_4 and CO_2) [42]. However, the reaction time for anaerobic digestion is long from 5 to 7 weeks with a lower conversion efficiency of the organic matter with a maximum around 40%–70% [43]. Previously, sewage sludge anaerobic digestion-based studies have indicated that hydrolysis is a rate-limiting step due to the existence of Extracellular polymeric substances (EPS), which lead to lower organic degradation and unsatisfactory methane production [41]. To speed up the hydrolysis reaction and increase methane production, several pretreatment options for sewage sludge are adopted nowadays [43,44].

In the scenario of thermochemical processes, hydrothermal liquefaction (HTL) (270–400°C) [45] and pyrolysis (350–600°C) [46] are the two main processes which are widely used for bio-oil production from sewage sludge. The presence of high inorganic compounds like metal oxides CaO, MgO, ZnO in the sewage sludge affects the overall processability of HTL and pyrolysis by reducing the bio-crude yield and increasing the char formation due to inefficient degradation or decomposition of the organic matter [47]. The adverse impact of inorganics on the bio-oil yield establishes a necessary requirement for the pretreatment process for the removal of inorganic elements via leaching [48] or elimination of char from the chamber during the pyrolytic reaction. This extracted solids to char can be further used in combustion for heat generation, or an economic alternate as a catalyst for the pyrolytic or HTL process [49].

Nowadays, the gasification of sludge is also getting attention for the hydrogen gas production at temperature ranges 800–1000°C [50]. The main challenges are the ash-related inorganic constituents and sludge composition (heavy metals, moisture, nitrogen, and sulfur contents). A substantial amount of ash content has several negative impacts on the operations of the gasifier, specifically agglomeration, sintering, formation of clinker, which cause frequent shutoffs and require maintenance of the reactor; moreover, low ash fusion disturbs the fuel flow, which diminishes the heat transfer potential and quality of the gas [51]. Hence researchers used dolomite, Ca-, Fe-, and Ni-based catalysts as a gasifier substrate with sewage sludge to overcome the ash–associated complications and maximize the hydrogen yield [52].

7.5 Pretreatment techniques for the sewage sludge

Four different approaches have been described in this section which includes mechanical, thermal, chemical, biological treatments. The description of some key findings of pretreatment of sewage sludge studies is given in Table 7.2.

7.5.1 Mechanical treatment

7.5.1.1 Ultrasonic pretreatment

Ultrasonication is based on the phenomenon of sludge disintegration. Ultrasound waves develop compression and rarefaction while transmitting through the medium in Fig. 7.3A. During this process, microbubbles are formed and then fall down within a few microseconds after attaining a threshold size. The sudden and violent changes to conditions at 4700°C with pressure 500 bar produce strong hydromechanical shear forces and some active radicals such as OH^- and H^+ [65]. The low-frequency ultrasound technique (20–40 kHz) was firstly adopted at a laboratory scale in the late 1960s, but nowadays it has been frequently used at the pilot and large scale for the disintegration of sludge [66].

Kapusta et al. treated sewage sludge through ultrasonication with the power of 300 W and frequency of 24 kHz and subsequently liquefied sewage sludge at subcritical conditions for the HTL process. The positive effect of ultrasonication over bio-crude yield was observed and the maximum increase in bio-crude yield and energy recovery was noticed at 320°C [60]. Martín et al. applied ultrasonication treatment on sewage sludge for anaerobic digestion and reported that ultrasonication pretreatment increased methane yield by around 95% [56]. Besides this, ultrasound treatment has also been used for attaining dewaterability, as Feng et al. [67] found that the optimal energy 800 kJ/kg total solids for the maximum degree dewaterability from the sludge with an optimal concentration of extra-cellular polymeric (EPS) of 400–500 mg/L with distribution of particles of size 80–90 μm in diameter.

7.5.1.2 Microwave irradiation

Microwave irradiation is a heating technology in which microwave irradiations work under an electromagnetic spectrum in wavelengths from 1 mm to 1 m within the frequency range of 0.3–300 GHz. From an industrial perspective, a shorter frequency from 900 MHz to 2450 MHz is mostly adopted. Microwave irradiation breaks the sludge cells in two ways as shown in Fig. 7.3B. The first way is the thermal effect via the rotation of dipoles under the vibration of electromagnetic fields that break up bacterial cells and the second way is the thermal effect which is produced by changing the dipole alignment of

Table 7.2 Main observations experienced from some pretreatment technique studies of the sewage sludge.

Process	Reactor	Pretreatment method	Key findings	Reference
Anaerobic digestion	Full-scale-semicontinuous and HRT of 15 days at 35°C	Acid treatment, HCL for obtaining lower values of pH around 6–1	Acid treatment reduced the hydraulic retention time (HRT) almost same methane yield was found at 13 and 21 days at an optimal dosage of acid at pH of 2.	[53]
Anaerobic digestion	Continuous reactor, HRT of 20 days (35°C)	Alkaline treatment of KOH under pH from 10 to 12	Almost 70% increase in biogas yield whereas 36% increase in COD removal was observed.	[54]
Anaerobic digestion	The batch reactor, HRT of 19–21 days at 37°C	Thermal treatment at moderate and high temperatures 80°C and 130–170, respectively.	Volatile fraction increased from 2% to 29%. Whereas at higher temperatures the volatile fractions increased from to 2% to 17% at 130°C and 44% 170°C.	[55]
Anaerobic digestion	Batch reactor, HRT 10 days at 35°C	Mechanical treatment via Sonication at the power of 150 W at 0–60 min	Improved methane yield from around 64%–95% with optimal time of sonication of 45 min.	[56]
Anaerobic digestion	10 semicontinuous reactors, HRT of 20 days at a temperature of 55°C and 35°C for thermophilic and mesophilic conditions, respectively.	Physiochemical treatment-microwave radiations with a frequency of 2.45 GHz for 6 min at 1250 W.	50% increase in the removal of volatile solids, whereas 63%–207% increment of biogas production was noticed.	[57]
HTL	Batch reactors 10 mL at 350°C and 400°C	Acid pretreatment with 5% solution of citric acid with sewage sludge and citric acid solution ratio of 1:10 for 4 h at 30°C	40% of the ash removal was achieved, whereas almost 38% of fat was lost during the pretreatment leaching.	[45,48]
HTL	Batch reactor of 1 L at HTL temperature of 300°C	Acid pretreatment-inorganic acids (HCl, HNO_3, and H_2SO_4) and organic acids (HCOOH, CH_3COOH, and HOOCC-OOH) with the concentration of 0.3–0.6 mol/L for 2 h at 25°C	Inorganic acids reduced the ash content, while organic acids increased the ash content. The bio-crude yield trend from highest to lowest is $HCl > HNO_3 > H_2SO_4 > HCOOH > CH_3COOH > HOOCCOOH$.	[58]

HTL	Batch reactor of 20 mL at HTL temperature of 340°C with RT of 20 min	Lignocellulosics biomass were used as filter cakes with via mechanical pretreatment particularly for increasing the dry matter content of sewage sludge	The energy recovery was increased from 67% to 75% via Lignocellulosics biomass filter cake The DM increased to 25% as compared to 5% of the untreated sewage sludge.	[59]
HTL	Temp. 280, 300, 320, 340, and 360°C	Ultrasonication-nominal power of 300 W and frequency of 24 kHz under the energy input of 4500 kJ/kg of solid sludge.	The ultrasonic pretreatment increased the bio-crude yield. The highest bio-crude yield (19%) was obtained at the temperature of 340°C.	[60]
HTL	Batch reactor 250 mL at 350°C	Microwave-frequency 2450 MHz for 20 min at the power from 180 to 900 W	Microwave treatment influenced the bio-crude yield from 202 and 10%. The highest bio-crude yield was measured at 360°C with HHV of 26.29 MJ/kg.	[61]
Pyrolysis	Continuous screw reactor at 500–800°C for 6–46 min, N_2 4–100 g/min mass	Mechanical treatment-drying	Gas and char yields were increased with an increasing temperature while bio-oil increased constantly up to 700°C then decreased rapidly at an extreme temperature of 800°C.	[62]
Pyrolysis	Microwave reactor with the power of 0.7–1 kW with (graphite, SiC, and residue absorbent)	Microwave treatment	SiC and graphite produced higher bio-crude and biogas yield. From various other absorbents, the yields are bio-crude (2%–3.5%), biogas (10%–14%) and bio-char (78%–84%)	[63]
Gasification	Three stage gasifier (650°C), fluidized bed (815°C) and tar cracking (815°C). Catalyst-activated carbon	Catalyst-activated carbon	Activated carbon increased the syngas yield by 12%. tar removal efficiently and carbon conversion increased by 26% and 10%, respectively.	[64]
Gasification	Fluidized bed gasifier (800°C). Catalyst-Dolomite	Catalyst-Dolomite	The usage of dolomite improved the tar removal efficiency to 71%. 20%–36% increase in H_2 production was noticed.	[52]

polar molecules. This phenomenon causes the breaking of hydrogen bonds and kills the microorganisms at lower temperatures [68].

Chen et al. processed sewage sludge through HTL after treating by microwave radiations and noticed that microwave treatment (180–900 W) had a noticeable positive impact on the bio-crude yield between 2.2 and 10.1 wt% [61]. The highest bio-crude yield (35 wt%) was obtained at 360 W with an HHV of 26.29 MJ/kg. Appels et al. [69] studied the effect of microwave pretreatment for sludge solubilization at a pilot-scale anaerobic digestion plant and found that microwave pretreatment improved the solubilization of organic matter with a 50% increase in biogas production. Besides the enhancement of methane recovery, microwave pretreatment can also be used for the destruction of the pathogens of sewage sludge. Hong et al. [70] used the microwave technique (2450 MHz) for the anaerobic digester feed and reported the removal of fecal coliforms about ≥2.66 log with irradiated sludge.

7.5.1.3 Electrokinetic disintegration

Electrokinetic disintegration operates at the high-voltage electric fields. The high-voltage field creates charges, which trigger the disintegration process by the sudden disruption of cellular membranes and hard sludge flocs. This process offers easy accessibility for the fermentation of bacteria as shown in Fig. 7.3C [41]. Choi et al. applied the electrokinetic disintegration technique at 19 kV, with a frequency of 110 Hz for 1.5 sec on activated sludge and reported an increase in the degree of solubilization of chemical oxygen demand (SCOD/TCOD) and methane production by 4.5 and 2.5 times, respectively. Here, SCOD refers to Solubilized Chemical oxygen demand (SCOD), whereas TCOD stands for Total Chemical oxygen demand [64]. Electrokinetic disintegration is newly established pretreatment technology for the sludge and it has been extensively implemented in the industrial sector. Rittmann et al. [71] has treated 63% of the waste sludge, which resulted in a 40% increase in biogas production via electrokinetic pretreatment.

7.5.1.4 High-pressure homogenization

High-pressure homogenization is based on the principle of high-pressure gradient turbulence. The strong shearing forces are produced due to depressurization of highly compressed sludge suspensions at the pressure of 900 bar (Fig. 7.3D) [41]. This process breaks cell membrane, sludge flocs, and releases the intracellular substances. In this way the high-pressure homogenization technique helps in sludge disintegration and enhancing the efficiency of biodegradation performance. Zhang et al. studied the impact of homogenization pressure from 20 to 80 MPa with four homogenization cycles on the degree of solubilization [72]. It was reported that the increasing homogenization pressure and number cycles from 20 to 80 MPa and 1–4, respectively, were desirable for sludge solubilization.

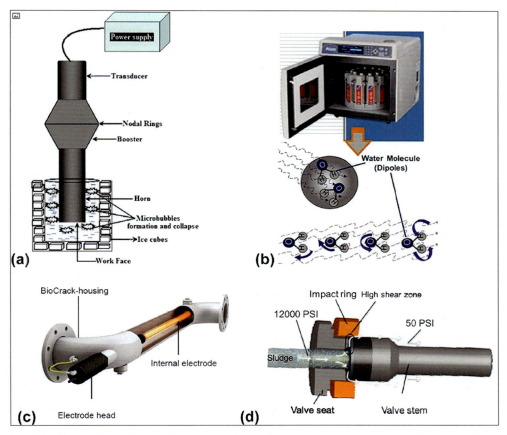

Fig. 7.3 (A) Configurations of ultrasonication, (B) microwave irradiation, (C) electrokinetic disintegration and (D) high-pressure homogenizer. *(From G. Zhen, X. Lu, H. Kato, Y. Zhao, Y. Li, Overview of pretreatment strategies for enhancing sewage sludge disintegration and subsequent anaerobic digestion: current advances, full-scale application and future perspectives. Renew. Sustain. Energy Rev. 69 (559) (2017) 577. https://doi.org/10.1016/j.rser.2016.11.187.)*

7.5.2 Thermal treatment

Thermal hydrolysis pretreatment technology is widely used on a commercial scale, particularly to enhance sludge dewaterability. The efficiency of the thermal hydrolysis process strictly depends on the temperature and treatment time [73]. Zhen et al. reported the conclusions of several studies related to thermal pretreatment in their review article and summarized that increasing temperature improved sludge solubilization and thermal hydrolysis influenced more on the biodegradability of carbohydrates and proteins than on lipids [41]. The optimal temperature for thermal hydrolysis was 170°C from 30 to 60 min, which reduced the retention time by 5 days and increased the biogas production.

Carrere and coworkers performed thermal hydrolysis from 60 to 210°C on six different sludges before anaerobic digestion and observed that solubility increased up to 190°C. However, a further increase in temperature decreased the biodegradability due to the creation of recalcitrant compounds via Maillard reactions like melanoidins, which are hard to degrade and slow down the rate of process of biodegradation [74]. The major advantages of thermal treatment include improved dewaterability, odor removal, and destruction of the pathogens [75].

7.5.3 Chemical treatment

7.5.3.1 Acidic and alkali pretreatment

Acidic and alkali pretreatments have been promising pretreatments for biomass solubilization due to the several merits (e.g., a simple device, low cost, easy to operate, removal of inorganics especially for thermochemical processing, and high methane conversion efficiency) of anaerobic digestion [44]. Furthermore, the detailed information about the pros and cons of the other pretreatment techniques is listed in Table 7.3. The addition of acid or alkali solutions bypasses the prerequisite of high temperatures and thus can be operated at moderate temperatures. For acid treatment, HCl, H_2SO_4, H_3PO_4, and HNO_3 are the notable acids, whereas alkali solutions such as KOH, NaOH, CaO Mg(OH)$_2$, Ca(OH)$_2$, including ammonia are widely used for the pretreatment process [76]. The usefulness of acidic or alkali treatment varies with the type and chemical composition of the sewage sludge, reaction time, and amount of acids or alkali used. The literature supports that in comparison to nitrogenous feedstocks, the application of acids is favorable for the lignocellulosic biomass. Hydrolysis is the key reaction involved in the acid treatment of the hemicellulose, which liberates the sugars and oligomers from the cell wall leading to improvement of the enzymatic digestibility [77]. Another disadvantage of acidic treatment is that it increases toxicity and corrosivity due to the extremely low values of pH, which give rise the need for the manufacturing of reactors with special materials. Recently, in a hydrothermal study, the pretreatment effect of citric acid (5% solution) on the secondary sewage sludge via the leaching process was investigated [45,48]. It was reported that citric acid not only removed 40% of the inorganic contents but also reduced almost 38% of the lipids. In another study, three inorganic acids and three organic acids were utilized to find their impact on the sludge properties for the bio-crude production via HTL [58]. It was reported that the inorganic acids reduced the ash content, while the organic acids increased the ash content. The bio-crude yield trend from highest to lowest was $HCl > HNO_3 > H_2SO_4 > HCOOH > CH_3COOH > HOOCCOOH$.

Alkali treatment is more adequate for lignin decomposition. For sewage sludges, from an anaerobic digestion perspective, alkali methods have been found to be very efficient methods before putting the sewage sludge into the reaction digesters. Alkali solutions provide buffer conditions for the system and process stability with specific methanogenic

Biomass pretreatment technologies 219

Table 7.3 Advantages and disadvantages of the pretreatment techniques.

Pretreatment method	Advantages	Disadvantages
Physical pretreatment Milling and grinding (Lignocellulosics)	Effective for particle size reduction. Help to maintain the homogenous mixture for pumpability. Decreases the crystallinity structure of cellulose.	Power consumption. High equipment cost.
Physicochemical pretreatment	Cost effective	Generation of compounds like phenolic compounds that cause hindrances to microorganisms.
Steam explosion (Lignocellulosics)	Less environmental impact	
Ammonia Fiber Explosion (Lignocellulosics)	Efficiently reduce hemicellulose concentration	Not suitable for lignin-rich biomass.
Liquid hot water (LHW) (Lignocellulosics)	Energy efficient	Energy-intensive process.
	Do not produce the compounds that cause hindrances to downstream processes. No chemical or catalyst is required. No toxic materials are formed.	A large amount of H_2O is required.
Chemical pretreatment	Performed at ambient conditions, no high temperature, pressure is required.	Need lot of time to complete (hours or days).
Alkaline treatment		Recovery of the added alkalis is a challenge.
(Lignocellulosics/ sewage sludge)	Suitable for low lignin-biomass.	High cost
Acid treatment (Lignocellulosics/ sewage sludge)	Changes lignin structure Hydrolyzes hemicellulose to sugars.	Cause corrosive Required maintenance due to toxic elements
Ozonation	Removal of ash to significant level (HTL and pyrolysis).	High energy required
(Sewage sludge)	Higher bio-crude yield with less char formation. Easy to integrate. Removal of pathogens. Increase in methane yield (anaerobic digestion). Flexible to operate	Mineralization potential of cellulose matter.

Continued

Table 7.3 Advantages and disadvantages of the pretreatment techniques—cont'd

Pretreatment method	Advantages	Disadvantages
Mechanical pretreatment Ultrasonication (Sewage sludge)	High methane yield (anaerobic digestion) low operational cost Easy maintenance.	High-energy demand and unsuitable for lignocellulosic biomass.
Microwave irradiation (Sewage sludge)	Fast and uniform heating easy to control	High-energy demand and scale up issues at larger plant.
Thermal treatment	Improved dewater-ability	High capital cost, potential for the ammonia generation.
(Sewage sludge)	Increase methane yield (anaerobic digestion) Higher bio-crude yield with less char formation (pyrolysis). Odor removal, volume reduction.	
Biological treatment	Low-energy demand.	Limitation in data parameters
Temperature-phased anaerobic digestion (Sewage sludge)	Higher solids degradation.	Optimization, energy analysis in needed for scaling up.
Microbial electrolysis cell (Sewage sludge)	Increase methane yield (anaerobic digestion) Sterilization of pathogens	pH issues, maintenance of electrode materials. High-energy demand.
	Increase methane yield (anaerobic digestion) Purification of biogas Higher process efficiency	

activity. For the anaerobic digestion process, among all alkali solutions, NaOH is the most efficient in terms of sludge solubilization and enhancing biogas production in the order of NaOH >KOH> $Mg(OH)_2$ > $Ca(OH)_2$ [78].

Nowadays alkali pretreatment has been allied with other sludge disintegration techniques such as microwave, thermal, ultrasound, and electrolysis, to enhance biodegradability and maximize the methane recovery. For HTL and pyrolysis processes, alkali compounds are specially used as a reaction catalyst to accelerate the rate of hydrolysis of high carbohydrate-containing sewage [45,48,79].

7.5.3.2 Ozonation

Ozonation is the most widely adopted peroxidation process, especially for the anaerobic digestion of sewage sludge. It can cleave the zoogloea structure and has been effectively used in the solubilization of the sludge. The efficiency of solubilization is directly dependent on the amount of applied ozone dosage[80]. The studies based on the kinetic reactions reveal that the increasing dosage of ozone at higher temperatures had not any significant impact on the degree of solubilization [81]. For sludge solubilization, the optimum dosage of the ozone varies in the range of 0.05–0.5 g O_3/g-TSS, depending on the composition and type of the sludge and pretreatment method employed [82]. Chu et al. carried out microbubble ozonation and observed that the efficiency of sludge solubilization increased from 15 and 30% to 25–40% at ozone doses ranging from 0.06 to 0.16 g O_3/g-TSS, respectively [83].

7.5.4 Biological pretreatment

Biological pretreatment is often used for the processability of the sewage sludge for anaerobic digestion. The temperature-phased anaerobic digestion treatment and microbial electrolysis cell has been commonly used in biological treatments for sludge processability. There are many other biological treatments like enzymatic hydrolysis, predigestion, as well as the use of fungi or bio-surfactants, which are not reported in this chapter.

7.5.4.1 Temperature-phased anaerobic digestion pretreatment (TPAD)

Temperature-phased anaerobic digestion combined with a thermophilic pretreatment unit before mesophilic anaerobic digestion is one of the leading approaches in the perspective of anaerobic digestion. This combination of the temperatures in TPAD stimulates the feedstock by the processes of hydrolysis and acidogenesis in thermophilic range; as a result, higher rate of acetogenesis and methanogenesis occur in the subsequent mesophilic stage [84]. Ge et al. used the thermophilic pretreatment parameters (retention time, pH, and temperature) to observe the effect on the overall degradability in TPAD and reported that methane production was up to 300 mL/g VS with the hydraulic retention time of 14 days at 35°C after thermophilic pretreatment at 1–2 days under the pH of 6–7 at 65°C [85].

7.5.4.2 Microbial electrolysis cell

Pretreatment based on the microbial electrolysis cell is an emerging technique nowadays for methane production through the process of electro-methanogenesis. In microbial electrolysis cell, the organic matter is consumed by exoelectrogenic bacteria in an anaerobic environment by transmitting electrons toward the anode and protons into the solution. Through the external circuit, the electrochemically active microorganisms accept electrons transferring to the cathode or another way to use cathodic H_2 as electron carrier to drive the formation of methane gas [86]. Liu et al. applied the microbial electrolysis technique for the pretreatment of sludge and the rate of methane production was enhanced to 91.8 g CH_4/m^3 reactor/day in microbial electrolysis anaerobic digestion

7.6 Challenges and future perspectives

Sewage sludge and lignocellulosic biomass feedstocks are important sources of renewable energy. Comparatively, ash contents of sludge are higher than that of lignocellulosic feedstocks due to high inorganics, which importantly requires pretreatments to increase its valorization to renewable biofuels. The pretreatment of lignocellulosic feedstocks is also required to increase the accessibility of sugar degradation by decreasing the crystalline characteristics of cellulose, increasing the available surface area, and removing the lignin. The biomass pretreatment conditions have shown a significant impact on the biomass processing and production of biofuels and biogas but also posed different challenges depending on the type of feedstock and pretreatment conditions that need to be addressed [43]. For instance, the alkali pretreatment of lignocellulosic feedstock brings many undesirable processes into existence, such as an increase of substrate pH during the digestion process and salt buildup on neutralization of pretreated products, which may significantly inhibit the methanogenesis due to the imbalance of ammonium ions and ammonia. Furthermore, the formation of salts may present additional challenges with disposal [25]. Another inherent challenge associated with the alkali pretreatment is the destruction of lignin instead of separation through which the lignin can be used for different applications. The bottleneck related to acid pretreatment of lignocellulosics is the presence of a significant quantity of klason lignin and pseudo-lignin as a component of most lignocelluloses. These lignins are incapable of being dissolved in acidic solutions used during the pretreatment under acidic conditions. Furthermore, the formation of lignin-like material (pseudo-lignin) is not favorable for the enzymatic hydrolysis process. Therefore the destruction of soluble lignins and some untreated forms of insoluble lignin is one of the main concerns of the acid pretreatment method. The production of pseudo-lignin by a combination of sugar molecules (e.g., carbohydrates) and other degradation products (aldehydes, phenols, furans, and other insoluble products) obtained during the thermal decomposition of polymeric carbohydrates is detrimental to the process that triggers irreversible cellulose loss [88]. Other challenges correlated with the pretreatment using concentrated acid comprise acid recovery, corrosion of instruments, and neutralization of waste acid when it is not recovered, which can be addressed using either organic acids (e.g., carbonic acid) or weak acids (e.g., acetic acid) in the pretreatment step.

Most of the methods known for the pretreatment of sludge are based on anaerobic digestion, employed for production of biogas. However, the main challenges related to the gasification of sewage sludge are high ash contents owing to the presence of high inorganic elements, sludge composition (heavy metals and water (80–90 wt%)), and tar

Biomass pretreatment technologies **223**

minimization. Furthermore, the high nitrogen and oxygen contents in sewage sludge, considerably higher than the lignocellulosic biomass, bring additional challenges to sludge gasification owing to the high possibilities of environment pollution. Additionally, the presence of the huge amount of heteroatoms and inorganics in sludge posed the utmost challenges in the continuous operation of a fluidized-bed gasifier. However, biological pretreatment processes are effective for sewage sludge; scaling-up is of utmost importance to make the biological methods efficiently applicable for the industrial operations on a large scale [41].

Nonetheless, the abovementioned challenges are being handled via several interventions, particularly the combination of various chemical processes with biological pretreatment methods to obtain high sugar yields, lesser use of expensive solvents, recycling of feedstock components from the aqueous phase, milder process conditions, and improvements in environmental and economic sustainability.

Consideration of anaerobic digestion as an efficient method for sewage sludge pretreatment should be sought in the future as an overall sustainable waste management perspective. The recovery of phosphorus from sewage sludge is also an important step for the further development of sewage treatment. For the future perspectives, the practical availability of pretreatment methods aiming at the alteration of the lignocellulosic feedstock structure to make the cellulose more available for biofuel production as well as removing heteroatoms and inorganic contents (e.g., metals) from sewage sludge can be beneficial from the economic, environmental, and energetic point of view.

7.7 Conclusions

This chapter has summarized pretreatment techniques adopted during thermochemical or biological process for the lignocellulosics and sewage sludge. It has been concluded that pretreatment techniques can significantly increase the energy production in the form of bio-oil, biogas, or bioethanol. However, the effectiveness of these pretreatment techniques depends upon the biomass structure, quantity of feedstocks, and operating conditions. Secondly, the cost applied to the pretreatment methods and their operational methods pose some of the hindrances in the practical approach on a larger scale.

References

[1] V.B. Agbor, N. Cicek, R. Sparling, A. Berlin, D.B. Levin, Biomass pretreatment: fundamentals toward application, Biotechnol. Adv. 29 (6) (2011) 675–685, https://doi.org/10.1016/j.biotechadv.2011.05.005.

[2] N. Mosier, C. Wyman, B. Dale, E. Richard, Y.Y. Lee, M. Holtzapple, M. Ladisch, Features of promising technologies for pretreatment of lignocellulosic biomass, Bioresour. Technol. 96 (6) (2005) 673–686, https://doi.org/10.1016/j.biortech.2004.06.025.

[3] G. Brodeur, E. Yau, K. Badal, C. John, K.B. Ramachandran, S. Ramakrishnan, Chemical and physicochemical pretreatment of lignocellulosic biomass: a review, Enzyme Res. 2011 (1) (2011), https://doi.org/10.4061/2011/787532.

[4] A.K. Kumar, S. Sharma, Recent updates on different methods of pretreatment of lignocellulosic feedstocks: a review, Bioresour. Bioprocess. 4 (1) (2017), https://doi.org/10.1186/s40643-017-0137-9.

[5] P. Kumar, D.M. Barrett, M.J. Delwiche, P. Stroeve, Methods for pretreatment of lignocellulosic biomass for efficient hydrolysis and biofuel production, Ind. Eng. Chem. Res. 48 (8) (2009) 3713–3729, https://doi.org/10.1021/ie801542g.

[6] R. Wooley, M. Ruth, J. Sheehan, H. Majdeski, A. Galvez, Lignocellulosic Biomass to Ethanol Process Design and Economics Utilizing Co-Current Dilute Acid Prehydrolysis and Enzymatic Hydrolysis Current and Futuristic Scenarios Lignocellulosic Biomass to Ethanol Process Design and Economics Utilizing Co-Current D, 1999. Contract, no. July: 132. NREL/TP-510-32438.

[7] T. Eggeman, R.T. Elander, Process and economic analysis of pretreatment technologies, Bioresource Technol. 96 (18 Spec. Iss) (2005) 2019–2025, https://doi.org/10.1016/j.biortech.2005.01.017.

[8] V. Menon, M. Rao, Trends in bioconversion of lignocellulose: biofuels, platform chemicals & biorefinery concept, Prog. Energy Combust. Sci. 38 (4) (2012) 522–550, https://doi.org/10.1016/j.pecs.2012.02.002.

[9] A.O. Wagner, N. Lackner, M. Mutschlechner, E.M. Prem, R. Markt, P. Illmer, Biological pretreatment strategies for second-generation lignocellulosic resources to enhance biogas production, Energies 11 (7) (2018), https://doi.org/10.3390/en11071797.

[10] C. Sánchez, Lignocellulosic residues: biodegradation and bioconversion by fungi, Biotechnol. Adv. 27 (2) (2009) 185–194, https://doi.org/10.1016/j.biotechadv.2008.11.001.

[11] Summary, Extended Executive, Biofuel technologies, Biofuel Technol. (November) (2013), https://doi.org/10.1007/978-3-642-34519-7.

[12] J. Baruah, B.K. Nath, R. Sharma, S. Kumar, R.C. Deka, D.C. Baruah, E. Kalita, Recent trends in the pretreatment of lignocellulosic biomass for value-added products, Front. Energy Res. 6 (Dec) (2018) 1–19, https://doi.org/10.3389/fenrg.2018.00141.

[13] E. Shirkavand, S. Baroutian, D.J. Gapes, B.R. Young, Combination of fungal and physicochemical processes for lignocellulosic biomass pretreatment – a review, Renew. Sust. Energ. Rev. 54 (2016) 217–234, https://doi.org/10.1016/j.rser.2015.10.003.

[14] W.C. Tu, J.P. Hallett, Recent advances in the pretreatment of lignocellulosic biomass, Curr. Opin. Green Sustain. Chem. 20 (2019) 11–17, https://doi.org/10.1016/j.cogsc.2019.07.004.

[15] J.Y. Zhu, G.S. Wang, X.J. Pan, R. Gleisner, Specific surface to evaluate the efficiencies of milling and pretreatment of wood for enzymatic saccharification, Chem. Eng. Sci. 64 (3) (2009) 474–485, https://doi.org/10.1016/j.ces.2008.09.026.

[16] T.H. Seehar, S.S. Toor, A.A. Shah, T.H. Pedersen, L.A. Rosendahl, Biocrude production from wheat straw at sub and supercritical hydrothermal liquefaction, Energies 13 (12) (2020) 3114, https://doi.org/10.3390/en13123114.

[17] A.T.W.M. Hendriks, G. Zeeman, Pretreatments to enhance the digestibility of lignocellulosic biomass, Bioresour. Technol. 100 (1) (2009) 10–18, https://doi.org/10.1016/j.biortech.2008.05.027.

[18] J.Y. Zhu, X.J. Pan, Woody biomass pretreatment for cellulosic ethanol production: technology and energy consumption evaluation, Bioresour. Technol. 101 (13) (2010) 4992–5002, https://doi.org/10.1016/j.biortech.2009.11.007.

[19] L. Cadoche, G.D. López, Assessment of size reduction as a preliminary step in the production of ethanol from lignocellulosic wastes, Biol. Wastes 30 (2) (1989) 153–157, https://doi.org/10.1016/0269-7483(89)90069-4.

[20] J. Li, G. Henriksson, G. Gellerstedt, Lignin depolymerization/repolymerization and its critical role for delignification of aspen wood by steam explosion, Bioresour. Technol. 98 (16) (2007) 3061–3068, https://doi.org/10.1016/j.biortech.2006.10.018.

[21] Z. Wang, G. Wu, L.J. Jönsson, Effects of impregnation of softwood with sulfuric acid and sulfur dioxide on chemical and physical characteristics, enzymatic digestibility, and fermentability, Bioresour. Technol. 247 (September 2017) (2018) 200–208, https://doi.org/10.1016/j.biortech.2017.09.081.

[22] H.Q. Li, W. Jiang, J.X. Jia, X. Jian, PH pre-corrected liquid hot water pretreatment on corn Stover with high hemicellulose recovery and low inhibitors formation, Bioresour. Technol. 153 (2014) 292–299, https://doi.org/10.1016/j.biortech.2013.11.089.

[23] Z. Hongdan, S. Xu, S. Wu, Enhancement of enzymatic saccharification of sugarcane bagasse by liquid hot water pretreatment, Bioresour. Technol. 143 (2013) 391–396, https://doi.org/10.1016/j.biortech.2013.05.103.

[24] S. Imman, J. Arnthong, V. Burapatana, V. Champreda, N. Laosiripojana, Influence of alkaline catalyst addition on compressed liquid hot water pretreatment of Rice straw, Chem. Eng. J. 278 (2015) 85–91, https://doi.org/10.1016/j.cej.2014.12.032.

[25] S.Q. Tian, R.Y. Zhao, Z.C. Chen, Review of the pretreatment and bioconversion of lignocellulosic biomass from wheat straw materials, Renew. Sustain. Energy Rev. 91 (June 2017) (2018) 483–489, https://doi.org/10.1016/j.rser.2018.03.113.

[26] S. Behera, A. Richa, N. Nandhagopal, S. Kumar, Importance of chemical pretreatment for bioconversion of lignocellulosic biomass, Renew. Sust. Energ. Rev. 36 (2014) 91–106, https://doi.org/10.1016/j.rser.2014.04.047.

[27] T.A. Shah, R. Tabassum, Enhancing biogas production from lime soaked corn cob residue, Int. J. Renew. Energy Res. 8 (2) (2018) 761–766.

[28] K. Sakuragi, K. Igarashi, M. Samejima, Application of ammonia pretreatment to enable enzymatic hydrolysis of hardwood biomass, Polym. Degrad. Stab. 148 (December 2017) (2018) 19–25, https://doi.org/10.1016/j.polymdegradstab.2017.12.008.

[29] H. Du, C. Liu, Y. Zhang, G. Yu, C. Si, B. Li, Preparation and characterization of functional cellulose nanofibrils via formic acid hydrolysis pretreatment and the followed high-pressure homogenization, Ind. Crop. Prod. 94 (2016) 736–745, https://doi.org/10.1016/j.indcrop.2016.09.059.

[30] M.A. Kärcher, Y. Iqbal, I. Lewandowski, T. Senn, Comparing the performance of Miscanthus x Giganteus and wheat straw biomass in sulfuric acid based pretreatment, Bioresour. Technol. 180 (2015) 360–364, https://doi.org/10.1016/j.biortech.2014.12.107.

[31] D. Sahoo, S.B. Ummalyma, A.K. Okram, A. Pandey, M. Sankar, R.K. Sukumaran, Effect of dilute acid pretreatment of wild Rice grass (*Zizania latifolia*) from Loktak Lake for enzymatic hydrolysis, Bioresour. Technol. 253 (November 2017) (2018) 252–255, https://doi.org/10.1016/j.biortech.2018.01.048.

[32] Y. Demirel, Biofuels. Comprehensive Energy Systems, Vol. 1–5, 2018, https://doi.org/10.1016/B978-0-12-809597-3.00125-5.

[33] L.J. Naimi, S. Sokhansanj, S. Mani, M. Hoque, T. Bi, A.R. Womac, S. Narayan, The Canadian society for bioengineering cost and performance of woody biomass size reduction for energy production, in: CSBE/SCGAB 2006 Annual Conference, 2006.

[34] J. Zandersons, J. Gravitis, A. Zhurinsh, A. Kokorevics, U. Kallavus, C.K. Suzuki, Carbon materials obtained from self-binding sugar cane bagasse and deciduous wood residues plastics, Biomass Bioenergy 26 (4) (2004) 345–360, https://doi.org/10.1016/S0961-9534(03)00126-0.

[35] B. Bals, C. Wedding, V. Balan, E. Sendich, B. Dale, Evaluating the impact of ammonia fiber expansion (AFEX) pretreatment conditions on the cost of ethanol production, Bioresour. Technol. 102 (2) (2011) 1277–1283, https://doi.org/10.1016/j.biortech.2010.08.058.

[36] I.M. Dãrãban, L.A. Rosendahl, T.H. Pedersen, S.B. Iversen, Pretreatment methods to obtain pumpable high solid loading wood-water slurries for continuous hydrothermal liquefaction systems, Biomass Bioenergy 81 (2015) 437–443, https://doi.org/10.1016/j.biombioe.2015.07.004.

[37] J.W. Lee, T.W. Jeffries, Efficiencies of acid catalysts in the hydrolysis of lignocellulosic biomass over a range of combined severity factors, Bioresour. Technol. 102 (10) (2011) 5884–5890, https://doi.org/10.1016/j.biortech.2011.02.048.

[38] W. Zhong, Z. Zhang, Y. Luo, S. Sun, W. Qiao, M. Xiao, Effect of biological pretreatments in enhancing corn straw biogas production, Bioresour. Technol. 102 (24) (2011) 11177–11182, https://doi.org/10.1016/j.biortech.2011.09.077.

[39] B.M. Cieślik, J. Namieśnik, P. Konieczka, Review of sewage sludge management: standards, regulations and analytical methods, J. Clean. Prod. 90 (March) (2015) 1–15, https://doi.org/10.1016/j.jclepro.2014.11.031.

[40] K. Fijalkowski, A. Rorat, A. Grobelak, M.J. Kacprzak, The presence of contaminations in sewage sludge – the current situation, J. Environ. Manag. 203 (2017) 1126–1136, https://doi.org/10.1016/j.jenvman.2017.05.068.

[41] G. Zhen, X. Lu, H. Kato, Y. Zhao, Y.Y. Li, Overview of pretreatment strategies for enhancing sewage sludge disintegration and subsequent anaerobic digestion: current advances, full-scale application and future perspectives, Renew. Sust. Energ. Rev. 69 (March 2016) (2017) 559–577, https://doi.org/10.1016/j.rser.2016.11.187.

[42] A. Raheem, V.S. Sikarwar, J. He, W. Dastyar, D.D. Dionysiou, W. Wang, M. Zhao, Opportunities and challenges in sustainable treatment and resource reuse of sewage sludge: a review, Chem. Eng. J. 337 (October 2017) (2018) 616–641, https://doi.org/10.1016/j.cej.2017.12.149.

[43] J. Oladejo, K. Shi, X. Luo, G. Yang, W. Tao, A review of sludge-to-energy recovery methods, Energies 12 (1) (2019) 1–38, https://doi.org/10.3390/en12010060.

[44] G. Zhen, X. Lu, Y.Y. Li, Y. Zhao, Combined electrical-alkali pretreatment to increase the anaerobic hydrolysis rate of waste activated sludge during anaerobic digestion, Appl. Energy 128 (2014) 93–102, https://doi.org/10.1016/j.apenergy.2014.04.062.

[45] A.A. Shah, S.S. Toor, F. Conti, A.H. Nielsen, L.A. Rosendahl, Hydrothermal liquefaction of high ash containing sewage sludge at sub and supercritical conditions, Biomass Bioenergy 135 (2020), https://doi.org/10.1016/j.biombioe.2020.105504.

[46] K. Malins, V. Kampars, J. Brinks, I. Neibolte, R. Murnieks, R. Kampare, Bio-oil from thermochemical hydro-liquefaction of wet sewage sludge, Bioresour. Technol. 187 (2015) 23–29, https://doi.org/10.1016/j.biortech.2015.03.093.

[47] J. Jin, Y. Li, J. Zhang, S. Wu, Y. Cao, P. Liang, J. Zhang, et al., Influence of pyrolysis temperature on properties and environmental safety of heavy metals in biochars derived from municipal sewage sludge, J. Hazard. Mater. 320 (2016) 417–426, https://doi.org/10.1016/j.jhazmat.2016.08.050.

[48] A.A. Shah, S.S. Toor, T.H. Seehar, R.S. Nielsen, A.H. Nielsen, T.H. Pedersen, L.A. Rosendahl, Biocrude production through aqueous phase recycling of hydrothermal liquefaction of sewage sludge, Energies 13 (2) (2020) 493, https://doi.org/10.3390/en13020493.

[49] H. Chen, D. Chen, L. Hong, Influences of activation agent impregnated sewage sludge pyrolysis on emission characteristics of volatile combustion and De-NOx performance of activated char, Appl. Energy 156 (2015) 767–775, https://doi.org/10.1016/j.apenergy.2015.05.098.

[50] A. Kelessidis, A.S. Stasinakis, Comparative study of the methods used for treatment and final disposal of sewage sludge in European countries, Waste Manag. 32 (6) (2012) 1186–1195, https://doi.org/10.1016/j.wasman.2012.01.012.

[51] M. Seggiani, M. Puccini, G. Raggio, S. Vitolo, Effect of sewage sludge content on gas quality and solid residues produced by cogasification in an updraft gasifier, Waste Manag. 32 (10) (2012) 1826–1834, https://doi.org/10.1016/j.wasman.2012.04.018.

[52] E. Roche, J.M. de Andrés, A. Narros, M.E. Rodríguez, Air and air-steam gasification of sewage sludge. The influence of dolomite and throughput in tar production and composition, Fuel 115 (2014) 54–61, https://doi.org/10.1016/j.fuel.2013.07.003.

[53] D.C. Devlin, S.R.R. Esteves, R.M. Dinsdale, A.J. Guwy, The effect of acid pretreatment on the anaerobic digestion and dewatering of waste activated sludge, Bioresour. Technol. 102 (5) (2011) 4076–4082, https://doi.org/10.1016/j.biortech.2010.12.043.

[54] A. Valo, H. Carrère, J.P. Delgenès, Thermal, chemical and thermo-chemical pre-treatment of waste activated sludge for anaerobic digestion, J. Chem. Technol. Biotechnol. 79 (11) (2004) 1197–1203, https://doi.org/10.1002/jctb.1106.

[55] H.B. Nielsen, A. Thygesen, A.B. Thomsen, J.E. Schmidt, Anaerobic digestion of waste activated sludge-comparison of thermal pretreatments with thermal inter-stage treatments, J. Chem. Technol. Biotechnol. 86 (2) (2011) 238–245, https://doi.org/10.1002/jctb.2509.

[56] M.Á. Martín, I. González, A. Serrano, J.Á. Siles, Evaluation of the improvement of sonication pretreatment in the anaerobic digestion of sewage sludge, J. Environ. Manag. 147 (2015) 330–337, https://doi.org/10.1016/j.jenvman.2014.09.022.

[57] N.M.G. Coelho, R.L. Droste, K.J. Kennedy, Evaluation of continuous mesophilic, thermophilic and temperature phased anaerobic digestion of microwaved activated sludge, Water Res. 45 (9) (2011) 2822–2834, https://doi.org/10.1016/j.watres.2011.02.032.

[58] R. Liu, W. Tian, S. Kong, Y. Meng, H. Wang, J. Zhang, Effects of inorganic and organic acid pretreatments on the hydrothermal liquefaction of municipal secondary sludge, Energy Convers. Manag. 174 (August) (2018) 661–667, https://doi.org/10.1016/j.enconman.2018.08.058.

[59] P. Biller, I. Johannsen, J.S. dos Passos, L.D.M. Ottosen, Primary sewage sludge filtration using biomass filter aids and subsequent hydrothermal co-liquefaction, Water Res. 130 (2018) 58–68, https://doi.org/10.1016/j.watres.2017.11.048.

[60] K. Kapusta, Effect of ultrasound pretreatment of municipal sewage sludge on characteristics of bio-oil from hydrothermal liquefaction process, Waste Manag. 78 (2018) 183–190, https://doi.org/10.1016/j.wasman.2018.05.043.

[61] G. Chen, M. Hu, D. Guiyue, S. Tian, Z. He, B. Liu, W. Ma, Hydrothermal liquefaction of sewage sludge by microwave pretreatment, Energy Fuels 34 (2) (2020) 1145–1152, https://doi.org/10.1021/acs.energyfuels.9b02155.

[62] N. Gao, C. Quan, B. Liu, Z. Li, C. Wu, A. Li, Continuous pyrolysis of sewage sludge in a screw-feeding reactor: products characterization and ecological risk assessment of heavy metals, Energy Fuels 31 (5) (2017) 5063–5072, https://doi.org/10.1021/acs.energyfuels.6b03112.

[63] R. Ma, S. Sun, H. Geng, F. Lin, P. Zhang, X. Zhang, Study on the characteristics of microwave pyrolysis of high-ash sludge, including the products, yields, and energy recovery efficiencies, Energy 144 (3688) (2018) 515–525, https://doi.org/10.1016/j.energy.2017.12.085.

[64] J.N. Choi, Enhanced anaerobic gas production of waste activated sludge pretreated by pulse power technique, Bioresour. Technol. 3 (2) (2006) 198–213, https://doi.org/10.1016/j.biortech.2005.02.023.

[65] C.P. Chu, D.J. Lee, B.V. Chang, C.S. You, J.H. Tay, 'Weak' ultrasonic pre-treatment on anaerobic digestion of flocculated activated biosolids, Water Res. 36 (11) (2002) 2681–2688, https://doi.org/10.1016/S0043-1354(01)00515-2.

[66] F. Hogan, S. Mormede, P. Clark, M. Crane, Ultrasonic sludge treatment for enhanced anaerobic digestion, Water Sci. Technol. 50 (9) (2004) 25–32, https://doi.org/10.2166/wst.2004.0526.

[67] X. Feng, J. Deng, H. Lei, T. Bai, Q. Fan, Z. Li, Dewaterability of waste activated sludge with ultrasound conditioning, Bioresour. Technol. 100 (3) (2009) 1074–1081, https://doi.org/10.1016/j.biortech.2008.07.055.

[68] I. Toreci, K.J. Kennedy, R.L. Droste, Evaluation of continuous mesophilic anaerobic sludge digestion after high temperature microwave pretreatment, Water Res. 43 (5) (2009) 1273–1284, https://doi.org/10.1016/j.watres.2008.12.022.

[69] L. Appels, S. Houtmeyers, J. Degrève, J. Van Impe, R. Dewil, Influence of microwave pre-treatment on sludge solubilization and pilot scale semi-continuous anaerobic digestion, Bioresour. Technol. 128 (2013) 598–603, https://doi.org/10.1016/j.biortech.2012.11.007.

[70] S.M. Hong, J.K. Park, N. Teeradej, Y.O. Lee, Y.K. Cho, C.H. Park, Pretreatment of sludge with microwaves for pathogen destruction and improved anaerobic digestion performance, Water Environ. Res. 78 (1) (2006) 76–83, https://doi.org/10.2175/106143005x84549.

[71] B.E. Rittmann, H.S. Lee, H. Zhang, J. Alder, J.E. Banaszak, R. Lopez, Full-scale application of focused-pulsed pre-treatment for improving biosolids digestion and conversion to methane, Water Sci. Technol. 58 (10) (2008) 1895–1901, https://doi.org/10.2166/wst.2008.547.

[72] Y. Zhang, P. Zhang, J. Guo, W. Ma, W. Fang, B. Ma, X. Xiangzhe, Sewage sludge solubilization by high-pressure homogenization, Water Sci. Technol. 67 (11) (2013) 2399–2405, https://doi.org/10.2166/wst.2013.141.

[73] E. Neyens, J. Baeyens, A review of thermal sludge pre-treatment processes to improve dewaterability, J. Hazard. Mater. 98 (1–3) (2003) 51–67, https://doi.org/10.1016/S0304-3894(02)00320-5.

[74] H. Carrère, C. Bougrier, D. Castets, J.P. Delgenès, Impact of initial biodegradability on sludge anaerobic digestion enhancement by thermal pretreatment, J. Environ. Sci. Health Part A 43 (13 Spec. Iss) (2008) 1551–1555, https://doi.org/10.1080/10934520802293735.

[75] H. Carrère, C. Dumas, A. Battimelli, D.J. Batstone, J.P. Delgenès, J.P. Steyer, I. Ferrer, Pretreatment methods to improve sludge anaerobic degradability: a review, J. Hazard. Mater. 183 (1–3) (2010) 1–15, https://doi.org/10.1016/j.jhazmat.2010.06.129.

[76] L. Appels, J. Baeyens, J. Degrève, R. Dewil, Principles and potential of the anaerobic digestion of waste-activated sludge, Prog. Energy Combust. Sci. 34 (6) (2008) 755–781, https://doi.org/10.1016/j.pecs.2008.06.002.

[77] E. Neyens, J. Baeyens, M. Weemaes, B. De Heyder, Hot acid hydrolysis as a potential treatment of thickened sewage sludge, J. Hazard. Mater. 98 (1–3) (2003) 275–293, https://doi.org/10.1016/S0304-3894(03)00002-5.

[78] J. Kim, C. Park, T.H. Kim, M. Lee, S. Kim, S.W. Kim, J. Lee, Effects of various pretreatments for enhanced anaerobic digestion with waste activated sludge, J. Biosci. Bioeng. 95 (3) (2003) 271–275, https://doi.org/10.1263/jbb.95.271.

[79] S.Y. Yokoyama, A. Suzuki, M. Murakami, T. Ogi, K. Koguchi, E. Nakamura, Liquid fuel production from sewage sludge by catalytic conversion using sodium carbonate, Fuel 66 (8) (1987) 1150–1155, https://doi.org/10.1016/0016-2361(87)90315-2.

[80] C. Bougrier, A. Battimelli, J.P. Delgenes, H. Carrere, Combined ozone pretreatment and anaerobic digestion for the reduction of biological sludge production in wastewater treatment, Ozone Sci. Eng. 29 (3) (2007) 201–206, https://doi.org/10.1080/01919510701296754.

[81] G. Manterola, I. Uriarte, L. Sancho, The effect of operational parameters of the process of sludge ozonation on the Solubilisation of organic and nitrogenous compounds, Water Res. 42 (12) (2008) 3191–3197, https://doi.org/10.1016/j.watres.2008.03.014.

[82] A. Salihu, M.Z. Alam, Pretreatment methods of organic wastes for biogas production, J. Appl. Sci. 16 (3) (2016) 124–137, https://doi.org/10.3923/jas.2016.124.137.

[83] L.B. Chu, S.T. Yan, X.H. Xing, A.F. Yu, X.L. Sun, B. Jurcik, Enhanced sludge solubilization by microbubble ozonation, Chemosphere 72 (2) (2008) 205–212, https://doi.org/10.1016/j.chemosphere.2008.01.054.

[84] W. Lv, F.L. Schanbacher, Y. Zhongtang, Putting microbes to work in sequence: recent advances in temperature-phased anaerobic digestion processes, Bioresour. Technol. 101 (24) (2010) 9409–9414, https://doi.org/10.1016/j.biortech.2010.07.100.

[85] H. Ge, P.D. Jensen, D.J. Batstone, Increased temperature in the thermophilic stage in temperature phased anaerobic digestion (TPAD) improves degradability of waste activated sludge, J. Hazard. Mater. 187 (1–3) (2011) 355–361, https://doi.org/10.1016/j.jhazmat.2011.01.032.

[86] S. Cheng, D. Xing, D.F. Call, B.E. Logan, Direct biological conversion of electrical current into methane by electromethanogenesis, Environ. Sci. Technol. 43 (10) (2009) 3953–3958, https://doi.org/10.1021/es803531g.

[87] W. Liu, W. Cai, Z. Guo, L. Wang, C. Yang, C. Varrone, A. Wang, Microbial electrolysis contribution to anaerobic digestion of waste activated sludge, leading to accelerated methane production, Renew. Energy 91 (2016) 334–339, https://doi.org/10.1016/j.renene.2016.01.082.

[88] S.D. Shinde, X. Meng, R. Kumar, A.J. Ragauskas, Recent advances in understanding the pseudo-lignin formation in a lignocellulosic biorefinery, Green Chem. 20 (10) (2018) 2192–2205, https://doi.org/10.1039/c8gc00353j.

CHAPTER 8

Generation of hydrocarbons using microorganisms: Recent advances

Bhabatush Biswas, Muthusivaramapandian Muthuraj, and Tridib Kumar Bhowmick
Department of Bioengineering, National Institute of Technology Agartala, Agartala, Tripura, India

Contents

8.1 Introduction	230
8.2 Biochemistry of hydrocarbon synthesis in microbes	231
8.3 Diverse microbial systems for hydrocarbon generation	236
8.3.1 Bacteria	237
8.3.2 Algae	240
8.3.3 Fungus	243
8.4 Biosynthesis of hydrocarbons	244
8.4.1 Alkenes	244
8.4.2 Alkanes	245
8.4.3 Alcohols	246
8.4.4 Esters	247
8.4.5 Aromatic hydrocarbons/polycyclic aromatic hydrocarbon/nitrogen-containing heterocyclic hydrocarbons	247
8.4.6 Lipids/phospholipid/phenolic lipids/neutral lipids	248
8.5 Conclusions and future perspectives	249
Acknowledgment	249
References	249

Abbreviations

AARs	acyl–ACP reductases
ACCase	acetyl-CoA carboxylase
ACL	ATP-citrate lyase
ACP	acyl carrier protein
AdhE	aldehyde/alcohol dehydrogenase
ADOs	aldehyde deformylating oxygenases
AMP	adenosine mono phosphate
AOR	aldehyde:ferredoxin oxidoreductase
CAR	carboxylic acid reductase
DAGAT	diacylglycerol acyltransferase
DHAP	dihydroxyacetone phosphate
ENR	enoyl-ACP reductase
ER	endoplasmic reticulum

Hydrocarbon Biorefinery
https://doi.org/10.1016/B978-0-12-823306-1.00012-1

Copyright © 2022 Elsevier Inc.
All rights reserved.

FAS	fatty acids synthetase
FAT	fatty acyl-ACP thioesterase
G3PDH	gycerol-3-phosphate dehydrogenase
GPAT	glycerol-3-phosphate acyltransferase
HD	3-hydroxyacylACP dehydratase
KAR	3-ketoacyl-ACP reductase
KAS	3-ketoacyl-ACP synthase
LPAAT	lyso-phosphatidic acid acyltransferase
LPAT	lyso-phosphatidylcholine acyltransferase
MAT	malonyl-CoA:ACP transacylase
MDH	malate dehydrogenase
ME	malic enzyme
MVA	mevalonate
PDH	pyruvate dehydrogenase complex
TAG	triacylglycerols

8.1 Introduction

Energy demand is sharply rising due to worldwide demographic expansion and industrialization. The international energy agency estimated that the population of the world is set to increase up to 9.1 billion by 2040 and the energy demand is set to grow by 30% [1]. Conventional fossil fuels are the primary sources of energy exploited at higher rates to meet the growing energy demands. Among all others, hydrocarbon fossil fuels account for more than 80% of the energy requirement of modern society. The utilization of conventional fossil fuels to meet energy demands remains the major cause of global warming and climate change attributed to the high levels of greenhouse gas emissions [1]. The rising demand for fuel, concerns for the environment, and reduction in usage of traditional fuels have attracted immense research interests into alternative, renewable, and clean energy resources, such as biofuels (biodiesel, bioethanol, and biobutanol), bio-crude, or bio-oils. In general, bio-crude or bio-oils are refined further to obtain hydrocarbons of our interest whereas biofuels are directly used in combination with the conventional fossil fuels at different ratios for energy generation.

Bio-crude or bio-oils are usually derived from processing the biomass via microbial processes or through physicochemical conversions. Among them, microbial processes are recognized as a sustainable source of renewable energy with lesser environmental impacts. Microorganisms can produce different types of hydrocarbon (alkanes, alkenes, alcohols, esters, acids, aromatic hydrocarbons, phospholipids, arenes, isoprenoids, etc.) compounds within a very short period. These compounds are synthesized by diverse microbes, implying that the mechanisms of synthesis are ancient and diversified [1]. Microorganisms yield hydrocarbons, such as alkanes and alkenes, by utilizing carbon from different sources through metabolic pathways [2]. For instance, a soil-isolated fungi

Penicillium cyclopium is known to accumulate a two carbon atom-containing hydrocarbon molecule ethylene [3]. Different cyanobacterial strains are well known for the accumulation of hydrocarbon of varying chain length from C15 to C19 with saturations and unsaturations or methylhexadecanes and methylheptadecanes [4]. Similarly, eukaryotic strains of algae, especially, brown algae, red algae, and *Dunaliella salina* are reported to accumulate *n*-pentadecane, *n*-heptadecane, 6-methyl hexadecane, 4-methyl octadecane, etc. Optimization of biosynthetic pathways of the engineered microorganism has also been of particular interest in the production of hydrocarbons and in a way to reduce the environmental stress of pollution and large dependency on the nonrenewable sources of derived oils [5]. The structure-based engineering approach has been taken to produce propane through whole-cell biotransformations [4]. More recently, researchers have exhibited enzymatic synthesis of ethane, propane, and other hydrocarbons [6]. Surprisingly, the nitrogenase enzyme, which reduces nitrogen gas to ammonia in nature, can also be used for producing gaseous alkanes by reacting with carbon monoxide. Specifically, the vanadium-containing nitrogenase enzyme has been shown to catalyze this reaction to synthesize hydrocarbon chains [7,8]. The molybdenum or vanadium-containing nitrogenase enzyme has been constructed in the laboratories to catalyze the reactions of hydrocarbon production [9]. These reactions are now being studied exhaustively. The bacterial strain *Vibrio furnissii* was also reported to produce C16–C28 alkanes in large quantities [10].

In the present scenario, establishing a sustainable bioprocess for commercial-scale production of hydrocarbons from microbial resources remains a challenge. The primary bottleneck lies in the development of a cellular factory with the inherent capability to directly accumulate high levels of hydrocarbon by utilizing either waste resources and atmospheric greenhouse gases, or agricultural residues. The secondary bottleneck is developing a process with economic feasibility and sustainability. This particular study aims to analyze and understand in detail different microbial cellular factories that can produce hydrocarbons.

8.2 Biochemistry of hydrocarbon synthesis in microbes

Hydrocarbons are synthesized in a wide range of microbes, which are often found in freshwater, marine, and terrestrial environments. The pathway of the synthesis of hydrocarbon is being encoded in the sequence of microbial species including other engineered microbes. Research has shown the amount of hydrocarbon (alkanes) produced by marine cyanobacterial species is approximately 2–540 pg/(mL day), which translates into an annual global ocean yield of around 308–771 million tons of production [11]. The major hydrocarbon synthesis in the microbes is usually through the fatty acid biosynthesis pathway, polyketide biosynthesis, and isoprenoid biosynthesis pathway. Besides, sterol biosynthesis pathways are involved in the synthesis of complex hydrocarbon or sterol rings whereas several enzymes are available which form alcohols. Acetyl–CoA is one of the primary precursor metabolites that lead to the branching pathways involved in almost all hydrocarbon biosynthesis pathways.

In the case of fatty acid biosynthesis-based hydrocarbon generation, the pathway commences with the conversion of acetyl-CoA into malonyl-CoA as the prime step. Further, a sequence of condensation, reduction, dehydration, and reduction of the malonyl-CoA acyl carrier protein (ACP) complex occurs, which results in the formation of fatty acyl chains [12]. The malonyl-CoA formed from acetyl-CoA acts as the sole carbon donor for the growing fatty acyl chain and with every one molecule of malonyl-CoA two molecules of hydrocarbon are added to the growing chain. The elongation of the growing fatty acyl chain-ACP complex is terminated by ACP-thioesterase or acyltransferases resulting in saturated fatty acids whereas the unsaturation is incorporated by desaturases. The free fatty acids formed from the fatty acid biosynthetic pathway usually undergo further two-step conversions to synthesize alkanes whereas a one-step conversion to synthesize alkenes involving different enzymes exists in microbes. The free fatty acids are reduced at their carboxylic terminal to form fatty-aldehydes in the first step catalyzed by fatty-aldehyde dehydrogenase, followed by conversion into alkanes in the second step by aldehyde deformylating oxygenases or long-chain aldehyde decarbonylase enzymes. In alkene biosynthesis, the free fatty acids are directly converted into 1-alkene via catalysis by fatty acid decarboxylase, which forms terminal olefins (OleT). In certain cases, the unsaturated fatty acid aldehydes are also converted into alkenes via decarbonylation. On the other hand, in the case of polyketide biosynthesis there exist different modules of polyketide synthase enzyme with different functional and catalytic domains that cordially operate together in initiation, elongation, and termination of polyketide chains from malonyl-CoA [13]. The starter units of the polyketide synthase complex vary significantly based on the organisms and results in a wide variety of hydrocarbon derivatives.

The mevalonate (MVA) pathway is a well-known metabolic pathway that involves the production of isoprenoids from the precursory metabolite acetyl–CoA present in the cytosol. The pathway utilizes mevalonate generated from acetyl-CoA to synthesize isopentenyl-diphosphate, which further generates isoprene rings. An alternate pathway to mevalonate involves methylerythritol-4-phosphate generated from 3-phosphoglycerate and phosphoenolpyruvate that synthesizes dimethylallyl-pyrophosphate and isopentenyl-pyrophosphate which act as a primer and extending units of isoprene rings, respectively [14]. Liu et al. reported construction of a plasmid for the MVA pathway to incorporate in the chromosomes of *E. coli* for the synthesis of isoprene hydrocarbon [15]. Thus, the synthesis of all hydrocarbons requires acetyl-CoA as the precursor metabolite and therefore, there are several mechanisms to replenish the acetyl-CoA used up in different metabolic reactions leading to hydrocarbon biosynthesis. The fatty alcohols or large carbon chain alcohols (C6–C18) are produced from the fatty acid biosynthesis pathway, whereas the short-chain alcohols are synthesized from acetyl-CoA or pyruvate or tricarboxylic acid intermediates. The production of long-chain fatty alcohols is usually concomitant with the release of low levels of organic acids such as acetate, formate, lactate, pyruvate, etc. Blocking the competing pathways leading toward organic acid biosynthesis resulted in a significant improvement in long-chain fatty alcohols.

Accumulation of lipids in yeast cells has been studied widely by several researchers [16,17]. The metabolic pathways (glycolysis and TCA cycle) involved in the yeast strains are interlinked to the fatty acid biosynthesis via pyruvate-malate translocase systems and citrate-malate translocase system. It was identified that due to the presence of citrate lyase enzyme in the cells, a significant amount of lipid is being synthesized in the yeast cells. This ATP-dependent enzyme produces acetyl-CoA in the cytosol, which is a precursor for generating several fatty acid hydrocarbons [18]. The property of utilizing various carbon sources (lignocellulosic biomass, agro-industrial residues, and wastewater) by the fungi has been exploited to generate lipid hydrocarbons. Researchers have identified that the fungi have certain advantages over algae, such as rapid growth period, light independence, etc. for scaling up the process. More specifically, fungal species that belong to the oleaginous microorganisms follow de novo and *ex novo* pathways to the biosynthesis of lipids. The de novo biosynthesis occurs in the cytosol of the fungus by a fatty acids synthetase (FAS) complex [19]. The fatty acid components of lipids are synthesized from the acetyl-CoA derived from the other central metabolic pathways like glycolysis and TCA cycle (Fig. 8.1). The citrate molecules present in the cells are cleaved to produce

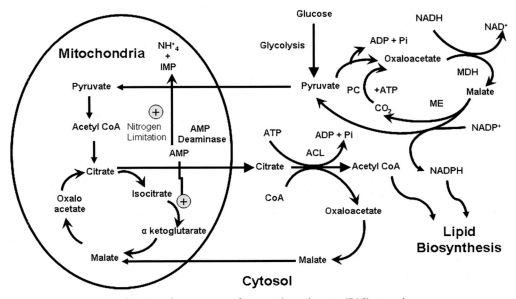

Fig. 8.1 Biosynthesis of lipid in fungus using fatty acid synthetase (FAS) complex.

acetyl-CoA. The pyruvate molecules generated from the glycolysis pathway are converted to acetyl-CoA in the mitochondria of the cells. The citrate produced from the TCA cycle is exported out to the cytosols of fungi by malate/citrate antiport transporters (Fig. 8.1). In the cytosol, citrate is cleaved by citrate lyase and generates acetyl-CoA and oxaloacetate [18,19].

The glycolytic pathway occurring in the cytosol leads to the formation of pyruvate. The pyruvate molecules are exported to the mitochondria. For the synthesis of fatty acid, a constant supply of acetyl-CoA is required. Acetyl-CoA is derived from pyruvate. In the mitochondria, acetyl-CoA undergoes the citric acid pathway. For more lipid accumulation in the cells, fermentation media is nitrogen deficient during the ongoing citrate pathway. AMP-deaminase supplies ammonium to the nitrogen-starved fungal cells. As a result, mitochondrial AMP concentration decreases, which leads to the lowering of activity of isocitrate dehydrogenase. The citric acid cycle is thus blocked at the level of isocitrate in the cell. Accumulation of citrate and its equilibrium with isocitrate is attained through aconitase activity. Excess citrate is exported out of the mitochondria via the malate/citrate antiport. The cytosolic ATP-citrate lyase (ACL) cleaves citrate to give oxaloacetate and acetyl-CoA. Acetyl-CoA carboxylases (ACC) catalyze the reaction to convert acetyl-CoA into malonyl-CoA. The acetyl-CoA unit is condensed with bicarbonate. NADPH is required for β-ketoacyl-ACP reductase (KR) and enoyl-ACP reductase (EAR) functions of FAS in the cytosol. For the elongations of the acyl chains, two molecules of NADPH are required. NADPH is derived from the pentose phosphate pathway and the trans-hydrogenase cycle, which transforms NADH into NADPH through the activity of the enzymes: pyruvate carboxylase (PC), malate dehydrogenase (MDH), and malic enzyme (ME). The final products synthesized from FAS are myristic or palmitic acids.

Algae are another class of microorganisms where researchers have employed different strategies to increase the lipid hydrocarbon accumulation in the cells. Researchers have leveraged the metabolic pathways to produce various kinds of hydrocarbons from algae (Fig. 8.2) [20].

Free fatty acid molecules are formed as a by-product of reaction mechanisms in the chloroplasts of algae. The conversion of acetyl-coenzyme A (CoA) to malonyl-CoA is catalyzed by acetyl-CoA carboxylase (ACCase) enzyme. This is an important step toward the biosynthesis of lipid in algae. A second important step is an increase in the expression of 3-ketoacyl-acyl carrier protein synthase (KAS) which catalyzes the conversion of malonyl-CoA into ketoacyl ACP. Ketoacyl ACP undergoes further reaction to form free fatty acids. This is exported out to the TAG assembly of the endoplasmic reticulum. The genes of the TAG assembly: glycerol-3-phosphate acyltransferase, lysophosphatidic acid acyltransferase, and diacylglycerol acyltransferase overexpress to generate lipid molecules. ACCase: acetyl-CoA carboxylase; ACP: acyl carrier protein; CoA: coenzyme A; DAGAT: diacylglycerol acyltransferase; DHAP: dihydroxyacetone phosphate; ENR:

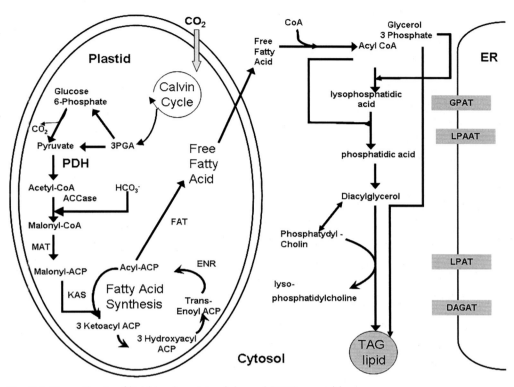

Fig. 8.2 Biosynthesis of lipid in algae triacylglycerol (TAG) assembly.

enoyl-ACP reductase; FAT: fatty acyl-ACP thioesterase; G3PDH: gycerol-3-phosphate dehydrogenase; GPAT: glycerol-3-phosphate acyltransferase; HD: 3-hydroxyacylACP dehydratase; KAR: 3-ketoacyl-ACP reductase; KAS: 3-ketoacyl-ACP synthase; LPAAT: lysophosphatidic acid acyltransferase; LPAT: lysophosphatidylcholine acyltransferase; MAT: malonyl-CoA:ACP transacylase; PDH: pyruvate dehydrogenase complex; TAG: triacylglycerols. In cyanobacterial strains, the hydrocarbons are also produced intrinsically from the fatty-acyl chain, which is converted to fatty aldehydes by fatty-acyl-ACP reductase and further synthesis of alkanes via the action of aldehyde decarbonylase on the fatty aldehydes.

In a different study, medium-chain (C9–C13) alkanes and alkenes were synthesized by two *E. coli* strains in the gas phase of the culture media. These strains of bacteria were engineered with a photo-enzyme, fatty acid photodecarboxylase (FAP), and thioesterase, which converts free fatty acid molecules into alkanes or alkenes by catalyzing the decarboxylation of free fatty acids using photons (Fig. 8.3). To produce the hydrocarbons from *E. coli* strains, these were transformed with genes derived from *Chlorella* and *Umbellularia* species. The genes specifically encode FAP and thioesterase enzymes [21].

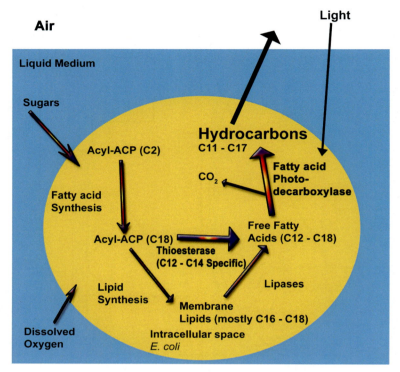

Fig. 8.3 Photoproduction of hydrocarbons by two *E. coli* strains, one strain transformed with a gene encoding a fatty acid FAP from *Chlorella* and another strain transformed with a gene encoding FAP and a specific thioesterase from *Umbellularia*.

8.3 Diverse microbial systems for hydrocarbon generation

Brassica napus L. (canola) from the family *Cruciferae* is the third largest oilseed crop on the planet that is predominantly utilized for hydrocarbon biosynthesis [22]. Biodiesel based on canola oil has an advantage in oil production as it contains approximately 40%–45% oil, which is comparatively more than other oilseeds, including soybeans (about 18%–20%) [23]. Plant-based biofuels are under research in the present scenario for the production of biofuel and hydrocarbons. However, there are numerous microbial strains ranging from simple bacteria to complex algal systems that are known to have immense potential in producing significant levels of different hydrocarbons. Among various microbial systems use of cyanobacterial, and eukaryotic algal strains have gained immense attention attributed to their intrinsic photosynthetic nature utilizing atmospheric inorganic CO_2 as the major carbon source and associated scope in carbon neutrality.

8.3.1 Bacteria

Both archea and eubacteria are well known for their hydrocarbon production, such as alkanes and alkenes. Table 8.1 represents the various bacterial sources that can synthesize different hydrocarbons. Soil bacteria *Clostridium tyrobutyricum* had drawn the attention of the scientists for producing butanol. n-Butanol was produced naturally by these *Clostridium* species through the ABE (acetone-butanol-ethanol) fermentation of sugars and carbon monoxide gases [27,30]. Some of the prominent *Clostridium* species, which are able to produce ethanol, include *Clostridium autoethanogenum*, *Clostridium ljungdahlii*, *Clostridium ragsdalei*, and *Clostridium carboxidivorans* [26]. Liew et al. had identified a bacterial species *Macrococcus caseolyticus*, which is capable of biosynthesizing terminal alkenes from fermentable sugars by harnessing a P450 fatty acid decarboxylase. Biosynthesis of furfuryl alcohol and lactic acid was achieved using a glucose coupled biphasic system in *Bacillus coagulans* species. Bu et al. had studied the one-pot biosynthesis of furfuryl alcohol, which is an important reduction product from biomass-derived furfural. Under mild reaction conditions, *Bacillus coagulans*, *Methanococcus deltae*, and *Escherichia coli* are able to transform furfural to furfuryl alcohol [31]. Passarini et al. had isolated bacteria strains (*Arthrobacter* and *Pseudoalteromonas*) from extremely cold environments (Antarctica) capable of producing hydrocarbons (long-chain alkanes between C_{11} and C_{35}) with potential for use as biofuels. Two of the identified bacteria *Arthrobacter livingstonensis* and *Pseudoalteromonas arctica* were able to produce the hydrocarbon undecane (CH_3-$(CH_2)_9$-CH_3). Production of 7- and 8-methylheptadecane hydrocarbon was reported in some cyanobacteria [33]. In the report, it had been mentioned that a halotolerant strain (*Vibrio furnissiiis*) is able to produce 15–24 long-chain intra- and extracellular hydrocarbons. A number of species capable of producing undecane collected from Antarctica are *Arthrobacter livingstonensis*, *Arthrobacter psychrochitiniphilus*, and *Pseudoalteromonas arctica*. These species were identified as phylogenetically related to the species *Arthrobacter livingstonensis* and *Pseudoalteromonas arctica*, which are capable of producing hydrocarbons like long-chain mono alkenes (C29 olefinic hydrocarbons) [40]. Bacterial strains *Arthrobacter nitroguajacolius*, isolated from Shule River soil, are capable of producing short-chain hydrocarbons [39].

Micrococcus luteus and *Arthrobacter aurescens* are the Gram-positive bacteria and *Stenotrophomonas maltophilia*, *Micrococcus*, and *Arthrobacter* strains are the Gram-negative bacteria, that synthesize long-chain olefinic hydrocarbons. Long-chain monoalkenes like cis-3,25-dimethyl-13-heptacosene were synthesized by *Arthrobacter* species [39]. There are methane-producing bacteria called methanogens, are strict anaerobes and are found in aquatic sediments, anaerobic sewage digesters, and a variety of other oxygen-free environments. These methanogens participate in the decomposition of organic matter and produce carbon dioxide and methane as two major by-products in the gaseous form [38]. Surger et al. claimed in a report for the first time, olefin synthesis in the genera of *Pseudarthrobacter*, *Paenarthrobacter*, *Glutamicibacter*, *Clavibacter*, *Rothia*, *Dermacoccus*, *Kytococcus*, *Curtobacterium*, and *Microbacterium* [37].

Table 8.1 Hydrocarbons produced from bacterial sources.

Hydrocarbon	Microorganism	Production level	Application	Carbon chain length	Reference
1-Propanol	Escherichia coli	7 g/L	Pharmaceutical solvents, textile industries, drugs, antiseptic solutions, cosmetics, and dyes	C3	[24]
Alcohols	Escherichia coli	210.1–449.2 mg/L	De-foamers, solubility retarders, consistency giving factors, and detergents	C12–C14	[5]
Alcohols	Escherichia coli	101.5 mg/L	Surfactants, natural oil feedstocks, and lubricants	C16–C18	[5]
Alkanes	Escherichia coli	81.8–101.7 mg/L	Biofuels	C13–C44	[25]
Bisabolene	Escherichia coli	1.15 g/L	Precursor of commercially valuable products, applications in biofuels, bioplastics, cosmetic industries, nutraceutical, and pharmaceutical	C15	[26]
Butanol	Escherichia coli	6.9 g/L	Biofuel	C4	[27]
Cinnamyl Alcohol	Escherichia coli	37.4 mM	Cancer drugs, cosmetic industry, flavoring agents, etc.		[28]
Enduracidin	Streptomyces fungicidicus	1.01 g/L	Biochemicals	C107	[29]
Ethanol	Escherichia coli	31 g/L	Biofuel	C2	[24]
Ethanol	Clostridium autoacetobutylicum	0.28 g/L	Biofuel	C2	[30]
Furfuryl alcohol	Bacillus coagulans	264 mM	Pharmaceuticals, pesticides, and rubber chemicals	C4–C5	[31]
Geraniol	Escherichia coli	2 g/L	Fragrant oils, essential oils, pharmaceuticals, specialty and commodity chemicals, and biofuels	C10	[32]
Isoprene	Escherichia coli	60 g/L	Springs, tyres, adhesives, and rubber-bands	C5	[12]
Lactic acid	Bacillus coagulans	64.2 g/L	Preserving agent	C3	[31]
Limonene	Escherichia coli	605 mg/L	Food, cosmetics, pharmaceuticals, biomaterials, and biofuels	C10	[26]

Long-chain alkanes	*Escherichia coli*	1.31 g/L	Plastics, lacquers, detergents, and fuels as starting materials	C15, C17	[33]
Methane	Engineered methanotrophic bacteria	–	Biochemicals	C1	[34]
Nonane	*E. coli* strain DH5α	0.62 μg/g	Aviation fuel	C9	[35]
Odd chain terminal alkenes	*Macrococcus caseolyticus*	17.78 mg/L	Polymers, lubricants, surfactants, and coatings	C11–C17	[36]
Pentadecaheptaene	*Escherichia coli*	140 mg/L	Biofuel	C15	[37]
Propane	*Escherichia coli*	32 mg/L	Water heating, cooking, and fuel	C3	[38]
Sabinene	*Escherichia coli*	2.7 g/L	Antibacterial and antiseptic agents in oil preparations	C10	[31]
Short-chain alkane	*Escherichia coli*	1.3–4.3 mg/L	Chemical industry	C3, C4, C5	[39]
Tridecane	*E. coli* strain DH5α	249 μg/g	Aviation fuel	C13	[35]
Undecane	*E. coli* strain DH5α	5.2 μg/g	Aviation fuel	C11	[35]
Undecane	*Arthrobacter livingstonensis*	1.39 mg/L	Biofuel	C11	[40]
Undecane	*Pseudoalteromonas arctica*	1.81 mg/L	Biofuel	C11	[40]
Vanillyl alcohol	*Escherichia coli*	499.36 mg/L	Flavoring agent in foods and beverages	C8	[41]

Lee et al. had engineered a gammaproteobacterial methanotroph (*M. alcaliphilum*) for the production of hydrocarbons at an industrial scale [36]. Boock et al. had shown engineered supercritical CO_2-tolerant strain (*Bacillus megaterium*), previously isolated from the CO_2 field, and had produced branched alcohols that have potential use as biofuels [42]. Zhang et al. has reported high yielding of cinnamyl alcohol from cinnamic acid using the recombinant *E. coli* strain BLCS, with the features of coexpressing carboxylic acid reductase (CAR) from *Nocardia iowensis* (*NiCAR*) and phosphopantetheine transferase (SFP) from *Bacillus subtilis* (*BsSFP*). CAR and SFP catalyze the reduction of cinnamic acid to cinnamic aldehyde. Cinnamaldehyde is further converted to cinnamyl alcohol [28]. In the study carried out by Cao et al., it was shown that optimization of the metabolic flux in recombinant *E. coli* strains is able to produce a high amount of alkanes [25]. In the study, an engineered *E. coli* strain was used for the synthesis of ethanol and propanol [43] and medium–chain length alkanes for the production of hydrocarbon-based fuel [35]. *E. coli* is also used for long-chain alkane production spatially organizing the enzymes of biosynthetic pathways [44]. Microbes use central metabolic pathways (glycolysis, pentose phosphate pathway, and tricarboxylic acid cycle) for generating energy from the carbon source. Manipulation of central metabolism and glycerol catabolism in the engineered *E. coli* cells had produced n-butanol from glycerol systematically [27]. Engineered *E. coli* had been optimized for the high-level and selective production of C—12/14 and C—16/18 alcohols through the fatty alcohol biosynthesis pathway [5]. Engineered *E. coli* was used to produce isoprenoids and geraniols through the isoprenoid alcohol pathway from a single nonrelated carbon source (glycerol) [32]. Another hydrocarbon vanillyl alcohol was synthesized in the *E. coli* through a de novo pathway [41].

8.3.2 Algae

The "World Oil Outlook 2020" is the report of the Organization of the Petroleum Exporting Countries (OPEC), which has reported that global primary energy demand is set to increase from 289 million barrels oil equivalent per day (mboe/d) in 2019 to about 361 mboe/d in 2045. This represents an average growth of 0.9% per annum. On the other hand, specific to oil-based products global demand is set to increase in the long-term from around 91 mboe/d in 2019 to 99.5 mboe/d in 2045. In the near future, it is assumed that increased biomass will be considered for the production of energy by the European policy [45]. Algae are considered to have significant scope toward the formation of carbon-neutral fuels because these microorganisms can uptake carbon in high amounts and have a high potential to recycle CO_2 emissions [46]. Currently, algae have gained immense importance as they are the potential source for the third-generation biofuels [47]. Cyanobacteria (the prokaryotic algal strains) species such as *Synechococcus elongatus* are well known for their increased accumulation of alkanes, terminal alkenes (i.e., 1-alkenes), and internal alkenes, which are ideal alternatives to the conventional fossil fuels [48]. Lea-Smith et al. reported the production of alkanes by two marine cyanobacterial species *Prochlorococcus* and *Synechococcus* [11]. They have pointed out that the amount of alkanes, being produced by the *Prochlorococcus*, are low nutrient-dependent and found in open ocean areas between Earth's 40°N and 40°S latitudes and the *Synechococcus* sp. are found up to 200 m depths of the ocean. In their report, it had been mentioned that *Prochlorococcus* and *Synechococcus* species produce and accumulate hydrocarbons mostly in the range of C15 and C17 alkanes [11]. *Spirulina platensis* is also known for the synthesis of n-hexane. Duongbia et al. reported that extraction of hydrocarbons from *Spirulina* was done by acid hydrolysis of the cell which generates hydrocarbons (hexadecane, heptadecane, palmitic acid, etc.) of different chain lengths, useful as alternative biofuel for automobile fuels [49]. Algal biomass is processed in different ways to extract hydrocarbons, namely fractionation, hydrothermal liquefaction, etc. [46]. A class of cyanobacterial species, *Oscillatoria*, is being investigated for synthesizing biofuels from its biomass through a pyrolysis process [50]. Most popular hydrocarbons, such as benzene, benzaldehydes, caprolactam, furans, guaiacol (mequinol), oximes, phenols, styrene, toluene, and xylene, were extracted from *Oscillatoria* species. Moreover, the bio-charcoal obtained from the pyrolysis of *Oscillatoria* biomass was found useful for water remediation purposes [50].

Chlorella is considered a promising resource for green diesel production through the hydrodeoxygenation process [51]. Moulin et al. had shown recovery of hydrocarbons by a novel mechanism, in its pure form without any mixture of fatty alcohols and aldehydes from the microalgal strain *Chlorella variabilis* [21]. Some of the algae strains used for the hydrocarbon production including biofuels are listed in Table 8.2.

Red algae species *Rhodymenia pseudopalmata* characterized for lipid hydrocarbon had shown an exceptional predominance of phospholipids, followed by glycolipids and neutral lipids [53]. Fixed-bed pyrolysis processing of a marine microalga *Nannochloropsis gaditana* under varied temperatures (400°C, 500°C, and 600°C) was done and different alkanes and alkenes, such as heptadecene, octadecene, pentadecane, tetradecane, etc. were obtained in different quantities in the liquid products. The lists of promising hydrocarbons produced by the marine algae are mentioned in Table 8.2. In another research Mass et al. had reported the extraction of monocyclic triterpenes compound from diatom species *Rhizosolenia setigera* [54]. Sugumar et al. had studied and evaluated the transesterification method for the production of biodiesel from isolated microalgae species *D. salina* from the Rameshwaram area of the Indian coast [55,56]. Sorigu et al. had reported the production of alkanes and alkene from microalgae species *Chlamydomonas reinhardtii* and *Chlorella variabilis*. It was reported that the length of the carbon chain found in the extracted alkanes and alkene was C15–C17. The study also identified one more strain *Nannochloropsis* sp., which can produce n-heptadecene [57].

Table 8.2 List of algae strains used for the production of hydrocarbons.

Hydrocarbon	Microorganism	Production levels	Application	Carbon chain length	Reference
Heptadecene	*Chlamydomonas reinhardtii*	0.04%–0.1%	Tools, personal care, arts_crafts, and toys	C15–C17	[52]
Heptadecene	*Chlorella*	0.04%–0.1%	Tools, personal_care, arts_crafts, and toys	C15–C17	[52]
Methyl 8, 11, 14 eicosatrienoate	*Rhodymenia pseudopalmata*	0.8 mg/g dry weight	Food products, personal care, and nutraceuticals	C20	[53]
Heptadecane	*Rhodymenia pseudopalmata*	2.5 mg/g dry weight	Food products, personal care, and nutraceuticals	C17	[53]
2-Pentadecanone, 6,10,14-trimethyl	*Rhodymenia pseudopalmata*	0.9 mg/g dry weight	Food products, personal care, and nutraceuticals	C15	[53]
Lipids	*Chlorella vulgaris*	41.95 g/m^2	Biofuels	–	[54]
Bioethanol	*Acanthophora spicifera*	240 mg/g	Biofuels	C2	[55]
Bioethanol	*Dictyopteris australis*	190 mg/g	Biofuels	C2	[55]

8.3.3 Fungus

The production of alcoholic beverages from baker's yeast is well established and is economically valuable for the wine industries. Production of alcohols in the range of C12–C18 from the engineered *Saccharomyces cerevisiae* cells is the next target for the researchers [58]. Engineered *S. cerevisiae* are under research focus to produce a high yield of isoamyl alcohol, isobutanol, and 2-phenylethanol from the yeast cell biomass [59]. Research had shown that *S. cerevisiae* is responsible for the production of long-chain alcohol and wax ester biosynthesis [52]. Similarly, another yeast strain, *Pichia pastoris* has also been investigated for the synthesis of isobutanol and isobutyl acetate, which are the resources for the production of economically efficient high energy renewable fuels. The approach has been adopted to produce the higher alcohols and their corresponding acetate esters by exploiting the amino acid biosynthetic pathway of the yeast strain (*Pichia pastoris*) [60]. Another interesting approach is the manipulation of cytosolic acyl-CoA and acyl-ACP (acyl carrier protein) pathways and harnessing lipogenic pathways in yeast *Yarrowia lipolytica*, for the sustainable production of hydrocarbon-based fuels and oleochemicals from carbohydrate resources [61]. Hydrocarbons such as alkenes, isobutene, are synthesized by the yeast *Rhodotorula minuta*. The yeast strain *Rhodotorula minuta* synthesize isobutene from branched-chain amino acids through the decarboxylation of the intermediate isovalerate molecule [62]. Different oil-producing yeast and molds are set out in Table 8.3.

Table 8.3 Oil producing yeast and molds.

Hydrocarbon	Microorganism	Production levels	Carbon chain length	Reference
Lipids	*Yarrowia lipolytica*	1.85 g/L h	–	[63]
Lipids	*Rhodotorula toruloides*	0.19 g/g	C18	[64]
Isoprene	*Saccharomyces cerevisiae*	2.5 g/L	C5	[12]
Limonene	*Yarrowia lipolytica*	24 mg/L	C10	[15]
Farnesene	*Saccharomyces cerevisiae*	130 g/L	C15	[65]
Lipids	*Cryptococcus* species	22%–63% w/w	–	[66,67]
Lipids	*Lipomyces* species	60%–70% w/w	–	[66,67]
Lipids	*Schwanniomyces occidentalis*	23% w/w	–	[18]
Lipids	*Trichosporon* species	40%–65% w/w	–	[67]
Lipids	*Hansenula ciferri*	22% w/w	–	[18]
Isoamyl alcohol	*Saccharomyces cerevisiae*	532–578 mg/L	C4–C5	[53]

8.4 Biosynthesis of hydrocarbons

Methanogenic bacteria have been used for a long time in the past to produce methane from the biomass that isused as cooking or heating fuel by biological conversion. However, currently, research thrust has been given to create platforms to produce different hydrocarbons utilizing sugars, lignocellulosic biomass, cellulosic biomass, and captured CO_2 in the cells through the photosynthetic process. The different hydrocarbons produced in the microbial biosynthesis process are discussed in detail below.

8.4.1 Alkenes

Alkenes are a class of hydrocarbon containing at least one double bond between two carbon atoms. The biological productions of alkenes are of great interest for their use as drop-in fuel [21]. The phytoplanktonic diatom strain *Rhizosolenia setigera* was investigated for the synthesis of straight-chain, highly branched-chain, and monocyclic alkene molecules. Reports have shown that certain straight-chain alkene molecules are synthesized in diatom species via the mevalonate (MVA) pathway. This route is also used by the diatoms for the biosynthesis of sterols and highly branched-chain alkene molecules [54]. In another study, a novel P450 fatty acid decarboxylase enzyme from the bacterial strain (*Macrococcus caseolyticus*) was used for the synthesis of terminal alkene (or 1–alkene) [36]. The study report had shown the direct biosynthesis of medium- and long-chain terminal hydrocarbon in engineered *E. coli* cells. Certain bacterial genera naturally produce olefins (unsaturated aliphatic hydrocarbons), which are an alternative and sustainable source of biofuels. Surger et al. reported that the presence of *oleABCD* genes in several Gram-positive bacteria (*Micrococcus*, *Kocuria*, *Arthrobacter*, and *Brevibacterium*) enables the production of olefin molecules via the ole pathway [37]. Alkenes of branch length C17 (n-heptadecene) have been found in cell pellet and liquid cultures of *Chlamydomonas reinhardtii* (Chlorophyceae) [57]. Long-chain alkenes have been identified in Gram-positive bacteria *Micrococcus* species, *Stenotrophomonas maltophilia*, and *Arthrobacter* species [68]. The alkene molecules with a carbon chain of C23–C31 are produced by *Micrococcus*. Cyclic hydrocarbons such as alkyl-benzenes are reported to have been produced by archaea species of the genera *Thermoplasma* and *Sulfolobus*. The report has shown that novel cycloheptane ring structures are produced in the bacteria of the genus *Alicyclobacillus* [63]. Hydrocarbon, sterane, consisting of four fused alicyclic rings, are reported to be biosynthesized by some methanotrophic proteobacteria [64]. Another class of hydrocarbons is the hopanoids, which consists of multiple, nonaromatic rings. These compounds structurally resemble cholesterol and are found in cyanobacteria, *Streptomyces* sp., and *Zymomonas mobilis*. Another alkene and isoprene were produced from engineered *Escherichia coli* by manipulating its mevalonate pathway [15]. Engineered microbial synthesis of gaseous isoprene molecules at 37°C using *Escherichia coli* and other bacteria is an interesting area of research. The synthesis of isoprene molecules in nature is observed widely.

Approximately, 500 million tons of isoprene molecules are produced by terrestrial plants, marine plankton, and bacteria annually [65]. Many bacteria produce isoprene, but only a few numbers of strains have been analyzed in the laboratory. It has been observed that *Bacillus subtilis* strains are able to produce isoprene molecules by a biosynthetic pathway named methylerythritol phosphate pathway (MEP) [14]. The isoprenoid compounds have carbon chain lengths with large variations, ranging from C10 to C110 [66]. The report has shown that an isoprenoid molecule (geraniol) was synthesized (2 g/L) by exploiting these native metabolic pathways in *E. coli* cells [32]. Such isoprenoids find their application in the pharmaceutical and chemical industries. Research has shown that more than 50,000 isoprenoid hydrocarbons are synthesized in eukaryotes as well as prokaryotes [67].

8.4.2 Alkanes

As terminal alkanes are chemically similar to the hydrocarbons found in petroleum, their demand for biofuels is increasing steadily. The US Department of Energy's "top 10 + 4" biochemicals from biomass include furfural (FAL), which finds its applications with a considerable economic interest in various useful chemical intermediates to produce flame-resistant composites, pharmaceuticals, pesticides, rubber chemicals, rocket fuels, and sealants [31]. Biologically, terminal alkanes are produced from fatty acyl-ACPs by acyl-ACP reductases (AARs) and aldehyde deformylating oxygenases (ADOs). One of the major concerns in n-alkane biosynthesis is the low catalytic turnover rates of ADOs. In a study, biosynthesis of n-alkane was enhanced in engineered *E. coli* cell using a chimeric ADO-AAR fusion protein and zinc finger protein-guided ADO/AAR assembly fixed upon DNA scaffolds to control their stoichiometric ratios and spatial arrangements in the biosynthesis pathway [43]. Long-chain alkanes are considered as good for hydrocarbon-based fuels due to their high energy density, hydrophobic property, and especially compatibility with modern internal combustion engines [25]. A metabolic engineering approach was adopted to enhance (36-fold) production of alkane molecules in *E. coli* cells in a bioreactor under fed-batch cultivation conditions [44]. In another study, other alkanes like tridecane, nonane, and undecane were also successfully produced from engineered *E. coli* cells [35]. Methane is another alkane that is natural gas or biogas synthesized by using microbial biocatalysis with methanase [34]. Biosynthesis of alkanes by microorganisms offers a sustainable and green supplement to traditional fossil fuels. The fatty aldehydes and other intermediate compounds play a critical role in microbial alkanes production [25]. *Arthrobacter nitroguajacolius* strain isolated from Shule river soil in Gansu province, China, was studied for the short-chain production of alkanes. Mass spectrometry and gas chromatography study of the synthesized alkanes had revealed the length of the chain in the range of C16–C22 hydrocarbons [39].

8.4.3 Alcohols

Alcohols are the most-sought-after hydrocarbons derived from microbes of all classifications. In several studies, it was found that alcohols such as ethanol, propanol, and isoamyl alcohol were produced by engineered microbes [5,24,28,59]. Fatty alcohols of hydrocarbon chain lengths C8-C18 obtained from engineered microbes are composed of a nonpolar, lipophilic carbon chain and a polar, hydrophilic hydroxyl group. These hydrocarbons find their application in a multitude of uses ranging from consistency giving factors, defoamers, detergents, lubricant additives, surfactants, and solubility retarders. Synthesis of such fatty alcohols is carried out in engineered *E. coli*. This is of particular interest to reduce the increasing pressure on the environment causing pollution and oils derived from nonrenewable sources [5]. In a study, synthesis of 1-propanol and ethanol was shown in engineered *Escherichia coli*, grown under anaerobic conditions [24]. The report had shown that engineered cells are able to produce nearly 7 g/L of 1-propanol and 31 g/L of ethanol in glycerol media while culturing the cells in an anaerobic fed-batch system. In a different study, *Clostridium tyrobutyricum* cells were engineered for butanol production by inactivating the acetate kinase (*ack*) gene or the phosphate butyryltransferase (*ptb*) gene and introducing the aldehyde/alcohol dehydrogenase (*adhE2*) gene. Batch fermentation of the engineered *Clostridium* cells led the butanol production level to 26.2 g/L [30]. *Clostridium acetobutylicum* utilizes a heterologous CoA-dependent pathway for the production of n-butanol [27]. Another study had reported that *Clostridium autoethanogenum* strain is able to produce ethanol in the gas fermentation process, where inactivation of two important genes aldehyde/alcohol dehydrogenase (AdhE), and aldehyde:ferredoxin oxidoreductase (AOR) in *Clostridium autoethanogenum* cell had increased ethanol production (up to 180%) [26]. *Saccharomyces cerevisiae* is well known for the production of ethanol, and also synthesizes higher alcohols, as shown in a study [59]. Higher alcohols of four or five carbon atoms medium-chained aliphatic compounds branched with aromatic side residues were produced by the *Saccharomyces cerevisiae* strain. Isobutanol is considered an advanced biofuel, which is synthesized in *P. pastoris* by exploiting the amino acid biosynthetic pathway and diverting the amino acid intermediates to the 2-keto acid degradation pathway for isobutanol production [60]. There are carbon compounds of chain length (C4–C5) branched with aromatic rings considered to have the characteristic of higher alcohols. These are used in the fermentation industry for the production of beverages. Interestingly, there are yeast strains that are responsible for the production of these higher alcohols as a result of the Ehrlich metabolic pathway. One example of this higher alcohol is isoamyl alcohol, which is produced in yeast [59]. Moreover, alcohols such as cinnamyl alcohol [28], vanillyl alcohol [41], and isoamyl alcohol [59] are among the others that are also used in pharmaceutical, textile, and chemical industries. This apart, diminishing availability of fossil fuel and increasing concerns of environmental problems have also prompted scientists to find an alternative source for hydrocarbon-based biofuel production [31].

Synthesis of hydrocarbons using microorganisms   247

8.4.4 Esters

Neutral lipid compounds like wax esters have their broad range of commercial applications, including personal care products, antiinflammatory drugs, candles, detergents, lubricants, printing inks, plastics, resins, and varnishes. Research had shown that hydrocarbon ester was synthesized in engineered *P. pastoris* by the incorporation of alcohol O-acyltransferase to generate a variety of volatile esters, including isobutyl acetate ester and isopentyl acetate ester [60]. Other researchers have shown to produce wax esters of C28–C36 chain length from microbes such as *Escherichia coli* and *Saccharomyces cerevisiae* [24,52]. A different study on *Spirogyra* sp. had shown that this particular species of algae were able to produce different kinds of esters successfully [69]. Some of the examples of ester synthesized in *Spirogyra* are eicosanoic acid methyl ester, hexadecanoic acid ethyl ester, octadecanoic acid methyl ester, pentadecanoic acid methyl ester, and tetradecanoic acid hexadecyl ester. Other esters of commercial values were reported to be derived from *Euglena gracilis*. *Euglena* species is being considered for large-scale biomass production in an open pond cultivation system. *Euglena gracilis* synthesizes esters such as wax esters from paramylon, when it is brought from aerobic to anaerobic conditions, and its lipid yield has shown to reach up to the 50% of cellular dry weight. Wax esters are the esters composed of saturated fatty acids and fatty alcohols. Some of the examples including hexadecyl palmitate, myristic acid, linoleic acid, palmitic acid, and stearic acid are also extracted from *Euglena* species [70,71].

8.4.5 Aromatic hydrocarbons/polycyclic aromatic hydrocarbon/ nitrogen-containing heterocyclic hydrocarbons

Aromatic compounds are an important class of chemicals used as dyes, organic solvents, and precursors in the processing of foods, pharmaceuticals, polymers, etc. These are currently generated from nonrenewable sources. Current research has been focused on the production of aromatic compounds from renewable sources [72]. Engineered *Escherichia coli* and *Pseudomonas putida* strains were used to develop sustainable processes to synthesize anthranilate, cyclohexadiene-trans-diols, 2-phenylethanol, *p*-hydroxycinnamic acid, *p*-hydroxystyrene, and *p*-hydroxybenzoate along with other useful chemicals [73]. With such engineered bacterial species even aromatic amino acids were successfully produced. Few amino acids that are produced are phenylalanine (Phe), tyrosine (Tyr), tryptophan, and folic acid [72]. The shikimate pathway is one of the metabolic pathways used by microbes to produce these aromatic amino acids and folates from phosphoenolpyruvate and erythrose 4-phosphate. The study reported that engineered *E. coli* cells produce different aromatic hydrocarbons such as phenyl-lactic acid, 4-hydroxyphenyllactic acid (4HPLA), 2-phenylethanol, 2-(4-hydroxyphenyl) ethanol (4HPE) (also known as tyrosol), phenylacetic acid, and 4-hydroxyphenylacetic acid (4HPAA) using the shikimate pathway. In another study, *Chlorella* species were used for biomass conversion into aromatic hydrocarbons. The study involved the catalytic pyrolysis of the algal species for the production of aromatic hydrocarbons, such as benzene, toluene, ethylbenzene, styrene, indane, naphthalene, etc. [74,75]. Other algal species such as *Schizochytrium limacinum*, *Arthrospira platensis*, and *Nannochloropsis oculata* were used for the synthesis of aromatic hydrocarbons for the production of biofuels through pyrolysis [76]. The yeast strain *S. cerevisiae* and fungal strain *Rhodotorula rubra* biosynthesize aromatic hydrocarbons of commercial value such as cinnamoyl-CoA and *p*-coumaroyl-CoA from amino acids (L–Phe and L–Tyr) and flavonoids (pinocembrin and naringenin) [77].

8.4.6 Lipids/phospholipid/phenolic lipids/neutral lipids

Phospholipids have structures similar to the triglycerides that compose the largest fraction of most algal oils. In different scientific reports, it has been mentioned that photosynthetic microalgae (*Nannochloropsis oculata* and *Chlorella*), brown algae (*Fucus serratus*, *Fucus vesiculosus*, and *Pelvetia canaliculata*), and *Chlamydomonas* species are considered a resource of the major phospholipid compounds, such as phosphatidylcholine, phosphatidylethanolamine, lysophosphatidylcholine, lyso-ethanolamine, etc. [78,79]. *Chlorella vulgaris* and *Dunaliella salina* are other such species reported to produce carotenoids.

Moreover, microorganisms are also a good source of lipids. Single-cell oils (SCO) produced by microorganisms from agro-industrial surpluses and residues are the current interest of research studies. Lipids, which are rarely found in the plant or animal kingdom, are also shown to be produced from microorganisms. Among such microorganisms, oleaginous Zygomycetes are of special interest, which are capable of producing lipids containing g-linolenic acid. In these cells, the accumulation of lipids is triggered by the exhaustion of nitrogen from the culture medium, which allows the conversion of sugar to storage lipids. In recent investigations, newly isolated oleaginous strains *Mortierella isabellina* and *Cunninghamella echinulata* were used to produce cellular lipid in notable quantities when glucose was used as a sole carbon source in nitrogen-limited media [17]. In growing microorganisms for efficient accumulation of the maximum amount of oil, it is necessary to ensure that the growth medium is limited by the availability of nitrogen. This ensures the restriction in the proliferation of cells at an early point of the growth cycle so that new cells can no longer be synthesized due to the absence of nitrogen. Generally, it is considered that in glucose media the lipid content of a microorganism is sufficient to produce a saleable quantity. The species, which are regarded as oil-bearing, are a small number of yeasts and molds (Table 8.3) [18]. Phenolic lipids derived from microalgal species include polyphenols. Phenolic hydrocarbons, such as bromophenols, terpenoids, coumarins, colpol, phlorotannins, and vanillic acid, were reported to be derived from algal species of *Rhodophyceae*, *Phaeophyceae*, *Chlorophyceae*, and *Spirulina maxima* [80,81].

8.5 Conclusions and future perspectives

The occurrence of hydrocarbons in microorganisms and the synthesis of such hydrocarbons by the various microorganisms have been known for decades. However, the driving force for developing hydrocarbon-based renewable fuels and edible feedstocks has led to many types of research in the recent past. To protect the environment and to meet the increasing energy consumption, renewable sources of energy need to be explored and adopted. During the past few decades, commendable progress has been made in developing such technologies, which can generate hydrocarbons. Researchers are simultaneously carrying out lab-scale studies to improve the technologies for the sustainable production of hydrocarbons using microbes. Further improvements in these processes are also needed. We foresee that the developments and advances in the production of hydrocarbons using microbes and their efficiencies by which biomasses are converted into hydrocarbon-based biofuels and other energy-rich molecules are crucial for the economy. Both the yield of production and the quality of microbes can be improved by using the tools of molecular biology and genetic engineering for a desired class of hydrocarbon.

Acknowledgment

Financial support received from the Science and Engineering Research Board (SERB), Department of Science and Technology, Government of India (Grant No. CRG/ 2018/002479), for this work is highly acknowledged.

References

[1] T.R. New, T. Eisner, M. Eisner, M. Siegler, Secret weapons. Defenses of insects, spiders, scorpions, and other many-legged creatures, J. Insect Conserv. 11 (2) (2007) 217.

[2] R.K. Thauer, Biochemistry of methanogenesis: a tribute to Marjory Stephenson:1998 Marjory Stephenson prize lecture, Microbiology 144 (9) (1998) 2377–2406.

[3] P.J. Considine, N. Flynn, J.W. Patching, Ethylene production by soil microorganisms, Appl. Environ. Microbiol. 33 (4) (1977) 977–979.

[4] B. Khara, et al., Production of propane and other short-chain alkanes by structure-based engineering of ligand specificity in aldehyde-deformylating oxygenase, Chembiochem 14 (10) (2013) 1204–1208.

[5] Y.-N. Zheng, et al., Optimization of fatty alcohol biosynthesis pathway for selectively enhanced production of C12/14 and C16/18 fatty alcohols in engineered Escherichia coli, Microb. Cell Factories 11 (1) (2012) 65.

[6] N. Ladygina, E.G. Dedyukhina, M.B. Vainshtein, A review on microbial synthesis of hydrocarbons, Process Biochem. 41 (5) (2006) 1001–1014.

[7] C.C. Lee, Y. Hu, M.W. Ribbe, Vanadium nitrogenase reduces CO, Science (New York, N.Y.) 329 (5992) (2010) 642.

[8] Y. Hu, C.C. Lee, M.W. Ribbe, Extending the carbon chain: hydrocarbon formation catalyzed by vanadium/molybdenum nitrogenases, Science (New York, N.Y.) 333 (6043) (2011) 753–755.

[9] Z.-Y. Yang, D.R. Dean, L.C. Seefeldt, Molybdenum nitrogenase catalyzes the reduction and coupling of CO to form hydrocarbons, J. Biol. Chem. 286 (22) (2011) 19417–19421.

[10] L.P. Wackett, et al., Genomic and biochemical studies demonstrating the absence of an alkane-producing phenotype in Vibrio furnissii M1, Appl. Environ. Microbiol. 73 (22) (2007) 7192–7198.

[11] D.J. Lea-Smith, et al., Contribution of cyanobacterial alkane production to the ocean hydrocarbon cycle, Proc. Natl. Acad. Sci. (2015) 201507274.

[12] Y.J. Zhou, E.J. Kerkhoven, J. Nielsen, Barriers and opportunities in bio-based production of hydrocarbons, Nat. Energy 3 (11) (2018) 925–935.

[13] J. Wang, et al., Biosynthesis of aromatic polyketides in microorganisms using type II polyketide synthases, Microb. Cell Factories 19 (1) (2020) 110.

[14] M.K. Julsing, et al., Functional analysis of genes involved in the biosynthesis of isoprene in *Bacillus subtilis*, Appl. Microbiol. Biotechnol. 75 (6) (2007) 1377–1384.

[15] C. Liu, et al., Engineering and manipulation of a mevalonate pathway in *Escherichia coli* for isoprene production, Appl. Microbiol. Biotechnol. 103 (2019) 239–250.

[16] H. Kaneko, et al., Lipid composition of 30 species of yeast, Lipids 11 (12) (1976) 837–844.

[17] S. Papanikolaou, et al., Lipid production by oleaginous Mucorales cultivated on renewable carbon sources, Eur. J. Lipid Sci. Technol. 109 (11) (2007) 1060–1070.

[18] C. Ratledge, Microorganisms for lipids, Acta Biotechnol. 11 (5) (1991) 429–438.

[19] G.V. Subhash, S.V. Mohan, Sustainable biodiesel production through bioconversion of lignocellulosic wastewater by oleaginous fungi, Biomass Convers. Biorefin. 5 (2) (2015) 215–226.

[20] R. Radakovits, et al., Genetic engineering of algae for enhanced biofuel production, Eukaryot. Cell 9 (4) (2010) 486–501.

[21] S. Moulin, et al., Continuous photoproduction of hydrocarbon drop-in fuel by microbial cell factories, Sci. Rep. 9 (1) (2019) 13713.

[22] S.R. Bashandy, M.H. Abd-Alla, M.F.A. Dawood, Alleviation of the toxicity of oily wastewater to canola plants by the N2-fixing, aromatic hydrocarbon biodegrading bacterium *Stenotrophomonas maltophilia*-SR1, Appl. Soil Ecol. 154 (2020) 103654.

[23] J.C. Ge, S.K. Yoon, N.J. Choi, Using canola oil biodiesel as an alternative fuel in diesel engines: a review, Appl. Sci. 7 (9) (2017) 881.

[24] K. Srirangan, et al., Biochemical, genetic, and metabolic engineering strategies to enhance coproduction of 1-propanol and ethanol in engineered *Escherichia coli*, Appl. Microbiol. Biotechnol. 98 (2014) 9499.

[25] Y.-X. Cao, et al., Heterologous biosynthesis and manipulation of alkanes in *Escherichia coli*, Metab. Eng. 38 (2016) 19–28.

[26] F. Liew, et al., Metabolic engineering of *Clostridium autoethanogenum* for selective alcohol production, Metab. Eng. 40 (2017) 104–114.

[27] M. Saini, et al., Metabolic engineering of Escherichia coli for production of n-butanol from crude glycerol, Biotechnol. Biofuels 10 (1) (2017) 173.

[28] C. Zhang, et al., Efficient biosynthesis of cinnamyl alcohol by engineered *Escherichia coli* overexpressing carboxylic acid reductase in a biphasic system, Microb. Cell Factories 19 (1) (2020) 163.

[29] L. Liu, et al., Assessment of enduracidin production from sweet sorghum juice by *Streptomyces fungicidicus* M30, Ind. Crop. Prod. 137 (2019) 536–540.

[30] J. Zhang, et al., Exploiting endogenous CRISPR–Cas system for multiplex genome editing in *Clostridium tyrobutyricum* and engineer the strain for high-level butanol production, Metab. Eng. 47 (2018) 49–59.

[31] C.-Y. Bu, et al., One-pot biosynthesis of furfuryl alcohol and lactic acid via a glucose coupled biphasic system using single *Bacillus coagulans* NL01, Bioresour. Technol. 313 (2020) 123705.

[32] J. Clomburg, et al., The isoprenoid alcohol pathway, a synthetic route for isoprenoid biosynthesis, Proc. Natl. Acad. Sci. 116 (2019) 12810–12815.

[33] R.C. Coates, et al., Characterization of cyanobacterial hydrocarbon composition and distribution of biosynthetic pathways, PLoS One 9 (1) (2014) e85140.

[34] M.G. Kalyuzhnaya, A.W. Puri, M.E. Lidstrom, Metabolic engineering in methanotrophic bacteria, Metabol. Eng. 29 (2015) 142–152.

[35] M. Wang, et al., Biosynthesis of medium chain length alkanes for bio-aviation fuel by metabolic engineered *Escherichia coli*, Bioresour. Technol. 239 (2017) 542–545.

[36] J.-W. Lee, N.P. Niraula, C.T. Trinh, Harnessing a P450 fatty acid decarboxylase from *Macrococcus caseolyticus* for microbial biosynthesis of odd chain terminal alkenes, Metabol. Eng. Commun. 7 (2018), e00076.

[37] M. Surger, A. Angelov, W. Liebl, Distribution and diversity of olefins and olefin-biosynthesis genes in gram-positive bacteria, Biotechnol. Biofuels 13 (1) (2020) 70.

[38] N. Belay, L. Daniels, Production of ethane, ethylene, and acetylene from halogenated hydrocarbons by methanogenic bacteria, Appl. Environ. Microbiol. 53 (7) (1987) 1604.

[39] U. Constantine, L. Shi-weng, N. Maurice, Characterization of short-chains hydrocarbons produced by Arthrobacter Nitroguajacolius strain IHBB9963 isolated from Shule River soil, Int. J. Sci. Res. 6 (2017), https://doi.org/10.21275/ART2017614.

[40] M.R.Z. Passarini, et al., Undecane production by cold-adapted bacteria from Antarctica, Extremophiles 24 (6) (2020) 863–873.

[41] Z. Chen, et al., Establishing an artificial pathway for de novo biosynthesis of vanillyl alcohol in *Escherichia coli*, ACS Synth. Biol. 6 (2020) 1784–1792.

[42] J.T. Boock, et al., Engineered microbial biofuel production and recovery under supercritical carbon dioxide, Nat. Commun. 10 (1) (2019) 587.

[43] Z. Rahman, et al., Enhanced production of n-alkanes in *Escherichia coli* by spatial organization of biosynthetic pathway enzymes, J. Biotechnol. 192 (2014) 187–191.

[44] Z. Fatma, et al., Model-assisted metabolic engineering of *Escherichia coli* for long chain alkane and alcohol production, Metab. Eng. 46 (2018) 1–12.

[45] M. Adamczyk, M. Sajdak, Pyrolysis behaviours of microalgae *Nannochloropsis gaditana*, Waste Biomass Valoriz. 9 (11) (2018) 2221–2235.

[46] E. Santillan-Jimenez, et al., Co-processing of hydrothermal liquefaction algal bio-oil and petroleum feedstock to fuel-like hydrocarbons via fluid catalytic cracking, Fuel Process. Technol. 188 (2019) 164–171.

[47] R.A. Ahmed, et al., Bioenergy application of *Dunaliella salina* SA 134 grown at various salinity levels for lipid production, Sci. Rep. 7 (1) (2017) 8118.

[48] K. Liu, S. Li, Biosynthesis of fatty acid-derived hydrocarbons: perspectives on enzymology and enzyme engineering, Curr. Opin. Biotechnol. 62 (2019) 7–14.

[49] N. Duongbia, et al., Acidic hydrolysis performance and hydrolyzed lipid characterizations of wet Spirulina platensis, Biomass Convers. Biorefin. 9 (2) (2019) 305–319.

[50] H.D. Kawale, N. Kishore, Production of hydrocarbons from a green algae (Oscillatoria) with exploration of its fuel characteristics over different reaction atmospheres, Energy 178 (2019) 344–355.

[51] H.S.H. Nguyen, et al., Direct hydrodeoxygenation of algal lipids extracted from Chlorella alga, J. Chem. Technol. Biotechnol. 92 (4) (2016) 741–748.

[52] L. Wenning, et al., Establishing very long-chain fatty alcohol and wax ester biosynthesis in *Saccharomyces cerevisiae*, Biotechnol. Bioeng. 114 (5) (2020) 1025–1035.

[53] E. Peralta-García, et al., Lipid characterization of red alga *Rhodymenia pseudopalmata* (Rhodymeniales, Rhodophyta), Phycol. Res. 65 (1) (2016) 58–68.

[54] G. Massé, S.T. Belt, S.J. Rowland, Biosynthesis of unusual monocyclic alkenes by the diatom *Rhizosolenia setigera* (Brightwell), Phytochemistry 65 (8) (2004) 1101–1106.

[55] S. Sugumar, et al., Biodiesel production from the biomass of *Dunaliella salina* green microalgae using organic solvent, Mater. Today Proc. 33 (2020), https://doi.org/10.1016/j.matpr.2020.04.652.

[56] K. Cho, et al., Use of phenol-induced oxidative stress acclimation to stimulate cell growth and biodiesel production by the oceanic microalga *Dunaliella salina*, Algal Res. 17 (2016) 61–66.

[57] D. Sorigué, et al., Microalgae synthesize hydrocarbons from long-chain fatty acids via a light-dependent pathway, Plant Physiol. 171 (4) (2016) 2393.

[58] L. d'Espaux, et al., Engineering high-level production of fatty alcohols by *Saccharomyces cerevisiae* from lignocellulosic feedstocks, Metab. Eng. 42 (2017) 115–125.

[59] E. Espinosa Vidal, et al., Biosynthesis of higher alcohol flavour compounds by the yeast *Saccharomyces cerevisiae*: impact of oxygen availability and responses to glucose pulse in minimal growth medium with leucine as sole nitrogen source, Yeast 32 (1) (2020) 47–56.

[60] W. Siripong, et al., Metabolic engineering of *Pichia pastoris* for production of isobutanol and isobutyl acetate, Biotechnol. Biofuels 11 (1) (2018) 1.

[61] P. Xu, et al., Engineering *Yarrowia lipolytica* as a platform for synthesis of drop-in transportation fuels and oleochemicals, Proc. Natl. Acad. Sci. U. S. A. 113 (2016).

[62] T. Fujii, T. Ogawa, H. Fukuda, Preparation of a cell-free, isobutene-forming system from *Rhodotorula minuta*, Appl. Environ. Microbiol. 54 (2) (1988) 583–584.

[63] B.J. Rawlings, Biosynthesis of fatty acids and related metabolites, Nat. Prod. Rep. 15 (3) (1998) 275–308.

[64] M. Berlanga, Brock biology of microorganisms, in: M.T. Madigan, J.M. Martinko (Eds.), International Microbiology, eleventh ed., Vol. 8, 2005, pp. 149–150.

[65] R. Fall, S.D. Copley, Bacterial sources and sinks of isoprene, a reactive atmospheric hydrocarbon, Environ. Microbiol. 2 (2) (2000) 123–130.

[66] C.T. Walsh, Revealing coupling patterns in isoprenoid alkylation biocatalysis, ACS Chem. Biol. 2 (5) (2007) 296–298.

[67] R.W. Evans, M. Kates, Lipid composition of halophilic species of Dunaliella from the dead sea, Arch. Microbiol. 140 (1) (1984) 50–56.

[68] J.A. Frias, J.E. Richman, L.P. Wackett, C29 olefinic hydrocarbons biosynthesized by Arthrobacter species, Appl. Environ. Microbiol. 75 (6) (2009) 1774–1777.

[69] E.I. Abdel-Aal, A.M. Haroon, J. Mofeed, Successive solvent extraction and GC–MS analysis for the evaluation of the phytochemical constituents of the filamentous green alga *Spirogyra longata*, Egypt. J. Aquat. Res. 41 (3) (2015) 233–246.

[70] I. Shimada, et al., Catalytic cracking of wax esters extracted from *Euglena gracilis* for hydrocarbon fuel production, Biomass Bioenergy 112 (2018) 138–143.

[71] Y. Liu, M. Inaba, K. Matsuoka, Catalytic deoxygenation of hexadecyl palmitate as a model compound of euglena oil in H_2 and N_2 atmospheres, Catalysts 7 (11) (2017) 333.

[72] D. Koma, et al., Production of aromatic compounds by metabolically engineered *Escherichia coli* with an expanded shikimate pathway, Appl. Environ. Microbiol. 78 (17) (2012) 6203.

[73] G. Gosset, Production of aromatic compounds in bacteria, Curr. Opin. Biotechnol. 20 (6) (2009) 651–658.

[74] S. Thangalazhy-Gopakumar, et al., Catalytic pyrolysis of green algae for hydrocarbon production using H + ZSM-5 catalyst, Bioresour. Technol. 118 (2012) 150–157.

[75] K. Wang, et al., Fast pyrolysis of microalgae remnants in a fluidized bed reactor for bio-oil and biochar production, Bioresour. Technol. 127 (2013) 494–499.

[76] Y. Zhou, C. Hu, Catalytic thermochemical conversion of algae and upgrading of algal oil for the production of high-grade liquid fuel: a review, Catalysts 10 (2) (2020) 145.

[77] D. Huccetogullari, Z.W. Luo, S.Y. Lee, Metabolic engineering of microorganisms for production of aromatic compounds, Microb. Cell Factories 18 (1) (2019) 41.

[78] A.Y. Manisali, A.K. Sunol, G.P. Philippidis, Effect of macronutrients on phospholipid production by the microalga *Nannochloropsis oculata* in a photobioreactor, Algal Res. 41 (2019) 101514.

[79] J. Harwood, Membrane Lipids in Algae, 2006, pp. 53–64.

[80] Y. Freile-Pelegrin, D. Robledo, Bioactive phenolic compounds from algae, in: Bioactive Compounds from Marine Foods: Plant and Animal Sources, 2014, pp. 113–129.

[81] H.A. el Baky, F. El-Baz, G. El Baroty, Production of phenolic compounds from *Spirulina maxima* microalgae and its protective effects in vitro toward hepatotoxicity model, Afr. J. Pharm. Pharmacol 3 (2009) 133–139.

CHAPTER 9

Metabolic engineering approaches for high-yield hydrocarbon biofuels

Kalyan Gayen
Department of Chemical Engineering, National Institute of Technology Agartala, Agartala, Tripura, India

Contents

9.1 Introduction	254
9.2 Microbial metabolic pathways involved in hydrocarbon biosynthesis	255
9.2.1 Isoprenoid biosynthesis pathway	255
9.2.2 Fatty acid biosynthesis pathway	256
9.2.3 Polyketide biosynthesis pathway	257
9.3 Metabolic engineering to improve yield of the hydrocarbon biofuels	258
9.3.1 Isoprenoid-derived biofuels	258
9.3.2 Fatty acid-derived biofuels	258
9.3.3 Polyketide-derived biofuels	261
9.4 Toxicity stress of hydrocarbons to microbial cells	261
9.5 Use of lignocellulosic materials as feedstock	262
9.6 Bioconversion of CO_2 to hydrocarbons	263
9.7 Challenges and future directions	265
9.8 Conclusions	265
References	266

Abbreviations

AAR	acyl–ACP reductase
ACC	acetyl–CoA carboxylase
ACP	acyl carrier protein
ACT	acetoacetyl–CoA transferase
ADO	aldehyde deformylating oxygenase
ALE	adaptive laboratory evolution
CDP-ME	4-(cytidine 50-diphospho)-2-C-methylerythritol
CDP-MEK	4-(cytidine 50-diphospho)-2-C-methylerythritol kinase
CDP-MEP	2-phospho-4-(cytidine 50-diphospho)-2-C-methylerythritol
DGA1	diacylglyceride acyl-transferase
DMAPP	dimethylallyl pyrophosphate
DXP	deoxyxylulose 5-phosphate pathway
DXR	deoxyxylulose-5-phosphate reductoisomerase
DXS	deoxyxylulose-5-phosphate synthase
EDP	Entner–Doudoroff pathway
FAP	fatty acid photodecarboxylase

Hydrocarbon Biorefinery
https://doi.org/10.1016/B978-0-12-823306-1.00005-4

Copyright © 2022 Elsevier Inc.
All rights reserved.

FAS	fatty acid synthase
FFA	free fatty acid
FPP	farnesyl diphosphate
G3P	glyceraldehyde-3-phosphate
GPP	geranyl pyrophosphate
GPPS	geranyl pyrophosphate synthase
HMG-CoA	3-hydroxy-3-methylglutaryl-CoA
HMGR	3-hydroxy-3-methylglutaryl-CoA reductase
HMGS	3-hydroxy-3-methylglutaryl-CoA synthase
IDI	isoprenyl diphosphate isomerase
IPP	isopentenyl pyrophosphate
MECP	2-C-methylerythritol-2,4-cyclodiphosphate
MECS	2-C-methylerythritol-2,4-cyclodiphosphate synthase
MEP	2-C-methyl-D-erythritol-4-phosphate
MEPC	2-C-methylerythritol-4-phosphate cytidyltransferase
MK	mevalonate kinase
MVA	mevalonate pathway
PKS	polyketide synthases
PMD	pyrophosphomevalonate decarboxylase
PMK	phosphomevalonate kinase
PPC	phosphoenolpyruvate carboxylase
PPP	pentose pyrophosphate

9.1 Introduction

Microbial production of short- and long-chain hydrocarbons has been attracting growing attention as a potential biofuel for reducing the dependency on fossil fuels and contributing toward solving the problem of climate change [1]. Although several potential wild-type oleaginous strains have shown impactful performance in accumulating a considerable amount of lipids, these microbial factories remained economically infeasible. For example, isoprenoid-derived hydrocarbon biofuels can naturally be produced from plants [2]. However, it is not feasible to obtain industrial-scale production of this hydrocarbon biofuels because of the low growth rate of plants, seasonal production, and harvesting conditions. Microorganisms are the excellent potential alternative to tackle these obstacles. Their fast growth rate and cheap media requirements make them best fit for industrial-scale production of hydrocarbon metabolites. Microorganisms are also advantageous to target the production of specific metabolites by genetic manipulations. To this end, metabolic engineering demonstrated significant examples of altering the flux of targeted biofuel-related metabolic pathways (e.g., isoprenoids, fatty acids, and polyketide biosynthesis pathways) in the metabolism of various strains [3].

Even though metabolic engineering can help to build high-yield hydrocarbon biofuels producing microbial cell factories, the cost and availability of feedstocks for fermentation processes is still a major challenge. At this end, various current research efforts demonstrated successful production of biofuels such as algae-based biofuels, bioethanol, and biobutanol from cheaper and naturally abundant feedstocks such as lignocellulosic materials and CO_2 at laboratory scale [4–6]. Such efforts could be applied to fermentation processes to synthesize hydrocarbon biofuels. In this chapter, we review the recent developments to enhance the microbial production of hydrocarbon biofuels using metabolic engineering strategies by manipulating suitable microorganisms' metabolism. We also highlighted the challenges and opportunities for using cheaper and abundant feedstocks in the fermentation processes of hydrocarbon biofuels production.

9.2 Microbial metabolic pathways involved in hydrocarbon biosynthesis

There are three naturally evolved metabolic pathways, namely pathways of isoprenoids, fatty acids, and polyketides biosynthesis, known for microbial biosynthesis of hydrocarbons (Fig. 9.1). Interestingly, these pathways depend on a common precursor, namely acetyl-CoA, during the biosynthesis of hydrocarbons from sugars [3].

9.2.1 Isoprenoid biosynthesis pathway

A broad range of isoprenoids such as cyclic and branched-chain alkenes, alkanes, and alcohols is synthesized via isoprenoid biosynthesis pathways. The isoprenoid biosynthesis pathway is initiated via two metabolic routes, first from isopentenyl pyrophosphate (IPP) and dimethylallyl pyrophosphate (DMAPP), which are generated through either mevalonate (MVA) pathway from acetyl-CoA or deoxyxylulose 5-phosphate (DXP) pathway from glyceraldehyde-3-phosphate (G3P) and pyruvate [3, 7, 8]. Glycolysis is a primary pathway that produces precursors for the isoprenoids pathway. Moreover, other pathways such as Entner-Doudoroff pathway (EDP), pentose pyrophosphate (PPP), amino acid metabolism, and fatty acid metabolism also produce precursors for the isoprenoids synthesis [9].

The MVA pathway includes six enzymes, namely (i) acetoacetyl-CoA transferase (ACT), (ii) 3-hydroxy-3-methylglutaryl-CoA synthase (HMGS), (iii) 3-hydroxy-3-methylglutaryl-CoA reductase (HMGR), (iv) mevalonate kinase (MK), (v) phosphomevalonate kinase (PMK), and (vi) pyrophosphomevalonate decarboxylase (PMD) that generate IPP from acetyl-CoA (Fig. 9.2) [10]. IPP transforms into its isomer DMAPP via an isomerization step by an enzyme, namely IPP isomerase (IDI).

IPP and DMAPP can also be synthesized via another pathway, namely DXP pathway (Fig. 9.2) [11]. This pathway starts with one molecule of both, pyruvate and G3P, to synthesize IPP and DMAPP through seven biochemical conversions catalyzed by the enzymes, namely (i) deoxyxylulose-5-phosphate synthase (DXS), (ii) deoxyxylulose-5-phosphate reductoisomerase (DXR), (iii) 2-C-methylerythritol-4-phosphate

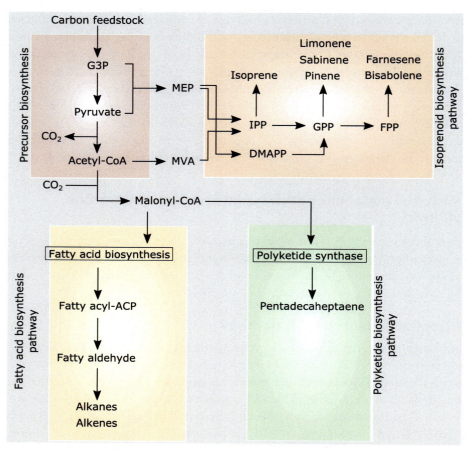

Fig. 9.1 Metabolic pathways for microbial biosynthesis of hydrocarbons. Microorganisms can consume biomass, carbon dioxide, methane, and methanol as carbon feedstock for producing precursors for biosynthesis pathways of isoprenoids, fatty acids, and polyketides. Please refer to the Abbreviations section for full forms of abbreviations used in this figure.

cytidyltransferase (MEPC), (iv) 4-(cytidine 50-diphospho)-2-*C*-methylerythritol kinase (CDP-MEK), (v) 2-*C*-methylerythritol-2,4-cyclodiphosphate synthase (MECS), (vi) isoprenyl diphosphate isomerase (IDI), and (vii) geranyl pyrophosphate synthase (GPPS).

9.2.2 Fatty acid biosynthesis pathway

Hydrocarbons such as alkanes and alkenes can be synthesized from fatty acids [12]. This pathway for hydrocarbon biosynthesis is initiated by synthesizing fatty acids from acetyl-CoA and malonyl-CoA (Fig. 9.2). This bioconversion is catalyzed by fatty acid synthase (FAS). FAS produces free fatty acids (FFAs) or fatty acyl-ACP (acyl-acyl carrier protein), which is transformed into hydrocarbons such as alkane and alkene via an intermediate, fatty aldehyde.

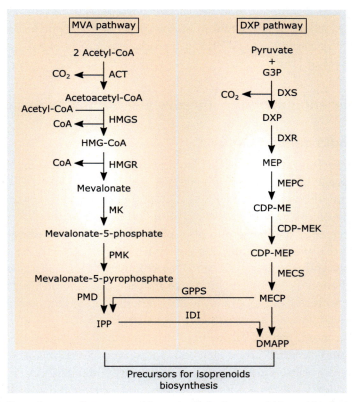

Fig. 9.2 Metabolic pathways of precursors biosynthesis for isoprenoid-based biofuels. MVA and DXP pathways are two known pathways to produce precursors for isoprenoids biosynthesis. Please refer to the Abbreviations section for full forms of abbreviations used in this figure.

Several enzymes such as OleT$_{JE}$ (cytochrome P450 peroxygenase), UndA and UndB [non-heme iron(II)-dependent oxidases], FAP (fatty acid photodecarboxylase), Gmlox1 (lipoxygenase I), AAR (acyl-ACP reductase), and ADO (aldehyde-deformylating oxygenase) are responsible for converting FFAs/fatty aldehyde/fatty acyl-ACP to alkane and alkene. For detailed information of catalytic activities of these enzymes, a recently published review can be referred [13]. The discovery of these enzymes has provided an excellent opportunity to perform metabolic engineering to many advantageous fast-growing hosts.

9.2.3 Polyketide biosynthesis pathway

Polyketides are a group of natural products that are synthesized by bacteria, fungi, and plants. They illustrate significant importance in producing medicines and hydrocarbon biofuels. Polyketides are synthesized from acetyl-CoA by polyketide synthases (PKSs). Three kinds of PKSs have been identified: (i) type I PKSs are involved in producing polyketides such as macrolides, polyethers, and polyene (e.g., erythromycin A) [14], (ii) type II PKSs catalyze the biosynthesis of polycyclic aromatic polyketides (e.g., tetracenomycin C) [15], and (iii) type III PKSs can produce mono- or bicyclic aromatic polyketides (e.g., flavolin) [16].

9.3 Metabolic engineering to improve yield of the hydrocarbon biofuels

9.3.1 Isoprenoid-derived biofuels

Microbial hosts such as *Escherichia coli* [17], yeast [18], and cyanobacteria [19] were tested to produce isoprene by engineering their MVA and DXP pathways. *E. coli* was found much superior to others with a higher titer of isoprene. Although cyanobacteria showed low productivity of isoprene, this microorganism has capability to fix carbon dioxide to biofuel. Monoterpenes, another group of isoprenoids, were produced using a metabolically modified *E. coli* strain from geranyl pyrophosphate (GPP) by monoterpene synthases. GPP is synthesized from IPP and DMAPP (Fig. 9.1). *E. coli* is mainly examined for producing monoterpenes such as limonene and sabiene. *Yarrowia lipolytica* and *Saccharomyces cerevisiae* were also explored for the biosynthesis of monoterpenes. However, the titers of monoterpenes were found much lower than *E. coli*.

A plant isoprenoid, namely farnesene, was synthesized by engineered *S. cerevisiae*. Further, in this study, central carbon metabolic pathways in *S. cerevisiae* have been manipulated to enable the acetyl–CoA (isoprenoid precursor) with a lesser ATP requirement, reduced carbon loss via CO_2 emission reactions, and ameliorated redox balance. The metabolically modified strain was able to produce 25% more farnesene with 75% less oxygen [20]. It was recently demonstrated that enhancing the flux of glucose 6-phosphate (G6P) toward the oxidative pentose phosphate pathway from glycolysis in *S. cerevisiae* increased NADP supply for isoprenoid production [21]. In another effort, metabolisms of *E. coli* and *S. cerevisiae* were metabolically modified to produce bisabolene that was finally converted to bisabolene (a substitute to D2 diesel) by chemical dehydrogenation [22].

Using *Rhodobacter sphaeroides* as host, Orsi et al. replaced the native MEP or DXP pathway by a heterologous MVA pathway (Table 9.1 for a description). Upregulation of the MVA pathway led to higher titers of isoprenoids compared to control strain. Authors also noticed that expressing only the MVA pathway was more effective for enhanced isoprenoid titers than keeping both pathways (MEP and MVA) [23].

9.3.2 Fatty acid-derived biofuels

Several studies have been conducted to enhance the intracellular pool of acetyl–CoA that is the key element of fatty acid biosynthesis and fatty acid–derived hydrocarbons (e.g., alkenes and alkanes) [24]. These metabolic engineering-based strategies were mainly

Metabolic engineering for hydrocarbon biofuels   259

Table 9.1 Brief descriptions.

Terms	Description
Metabolic engineering	Metabolic engineering can be stated as targeted manipulation in cellular networks that comprise metabolic, gene regulatory, and signaling networks to realize desired objectives like the improved synthesis of metabolites including biofuels.
Engineered microorganisms	Microorganisms, those native cellular networks have been modified to achieve specific objectives.
Heterologous pathways	Heterologous pathways include both natural and artificial biosynthetic routes that may come from other organisms. The expression of a heterologous pathway can help to target the production/consumption of specific metabolites in the host organism.
Host microorganisms	Microorganisms can adapt the heterologous pathways and still can grow.

implemented to enhance the acetyl-CoA in hosts such as *E. coli*, *S. cerevisiae*, and *Y. lipolytica*. Generally, the biosynthesis of acetyl-CoA is highly regulated and competes with the synthesis of several fermentation metabolites such as acetate, ethanol, and lactate. However, it was noticed that the removal of these competing metabolites enforced a slow growth and low consumption of substrate in *E coli*. It was also observed that the elimination of ethanol production led to slow growth and low consumption of glucose in *S. cerevisiae* [25]. These hurdles were resolved by introducing overexpression of a chimeric ATP:citrate lyase-malic enzyme-malate dehydrogenase-citrate transporter (Ctp1) that resulted in a 20% increase in fatty acid production [26]. In another example, *Y. lipolytica* naturally synthesizes a large pool of acetyl-CoA under nitrogen depleting conditions. To decouple the acetyl-CoA production from nitrogen depletion, overexpression of a carnitine acetyltransferase (Cat2) and a mitochondrial carnitine acetyltransferase (Yat1) aided to transport mitochondrial acetyl-CoA to cytoplasm with the increased titer of hydrocarbons [27]. The above examples illustrate that increase in acetyl-CoA pool can help to increase the titer of fatty acid-derived hydrocarbons.

Bioconversion of acetyl-CoA to malonyl-CoA by acetyl-CoA carboxylase (ACC) is considered as a rate-limiting reaction in several organisms for fatty acid biosynthesis. Several efforts have been conducted to improve the fatty acid biosynthesis by increasing the intracellular pool of malonyl-CoA [28]. Various metabolic engineering strategies have been implemented to improve the malonyl-CoA pool in *E. coli* as well as in yeasts (*S. cerevisiae* and *Y. lipolytica*): (i) by overexpression of ACC subunits that led to overproduction of fatty acids and alkenes [29], (ii) by overexpression of ACC1 and diacylglyceride acyl-transferase (DGA1) [30], (iii) by bypassing ACC pathway by expressing methylmalonyl-CoA carboxyltransferase and phosphoenolpyruvate carboxylase (PPC) [31], and (iv) by overcoming a feedback inhibition by saturated acyl-CoAs via overexpressing Δ9-stearoyl-CoA desaturase, or via overexpressing a Δ9-desaturase (OLE1) to enhance the fatty acyl-CoA pool [32].

Cyanobacteria is one of the microorganisms that can naturally produce alkanes. The metabolites involved in fatty acid pathway are transformed to alkanes and alkenes by an acyl-acyl carrier protein reductase and an aldehyde decarbonylase. The heterologous expression of alkane operon in *E. coli* was also investigated. Engineered *E. coli* could produce and secrete a mixture of C13–C17 alkenes and alkanes [33]. Kallio et al. also used *E. coli* as host to express a synthetic heterologous propane pathway to synthesize propane from glucose via butyrate. Propane is a major constituent in liquid petroleum gas [34]. In another study, the metabolism of *E. coli* was reconstructed to produce more alkanes present in gasoline (propane, butane, pentane, heptane, and nonane). Heptane and nonane were produced through the fatty acid biosynthesis route, while propane, butane, and pentane were synthesized through reverse-β-oxidation [35] (Table 9.2 for more examples).

Table 9.2 Hydrocarbon biofuels production from engineered microorganisms.

	Hydrocarbons	Chain length	Engineered microorganisms	References
Isoprenoid-derived biofuels	Isopentenols	C5	*E. coli*	[36]
	Isoprene	C5	*E. coli, S. cerevisiae, S. elongatus*	[17–19]
	Limonene	C10	*E. coli, Y. lipolytica, Synechococcus* sp.	[37–39]
	Pinene	C10	*E. coli*	[40]
	Sabinene	C10	*E. coli*	[41]
	Farnesol	C15	*E. coli, S. cerevisiae*	[42, 43]
	Farnesene	C15	*S. cerevisiae*	[20]
	Bisabolene	C15	*E. coli, S. cerevisiae, Streptomyces venezuelae, Synechococcus* sp.	[22, 37, 39, 44]
Fatty acid-derived biofuels	Short-chain alkanes (e.g., propane, butane, pentane, heptane, and nonane)	C3–C13	*E. coli*	[34, 35]
	Long-chain alkanes (e.g., Tetradecane, Pentadecane, Hexadecane, and Octadecane)	C13–C17	*E. coli, S. cerevisiae, Y. lipolytica*	[33, 45–47]
	Long-chain alkenes	C11–C19	*E. coli, S. cerevisiae*	[48, 49]
Polyketide-derived biofuels	Pentadecaheptaene	C15	*E. coli*	[50]
	α-Olefin (e.g., 1-hexene or butadiene)	C4–C6	*E. coli*	[51]

9.3.3 Polyketide-derived biofuels

The pathway of type I PKS (SgcE) and its cognate thioesterase (SgcE10) were engineered in *E. coli* to overproduce pentadecaheptaene. This study highlighted the dependency of pentadecaheptaene production on the proportion of SgcE10 and SgcE [50]. Moreover, a synthetic PKS comprising many naturally occurring PKSs was engineered in *E. coli* to produce α-olefin, such as 1-hexene or butadiene [51]. These efforts highlight the potential of PKSs to produce a variety of hydrocarbons.

9.4 Toxicity stress of hydrocarbons to microbial cells

As discussed above, various studies suggested that precise metabolic engineering methods can help to enhance the theoretical yield of hydrocarbons production in the cell. However, several terpenes and other hydrocarbons cause toxicity to growth of microbial cells that could hinder achieving maximum theoretical yield of selected microbial strains [52]. One solution could be application of transporter and tolerance engineering. Despite positive outcomes, this approach involves a significant amount of cellular energy to tolerate this stress [53]. This solution can be advantageous for the hydrocarbons containing high importance even though produced in small amounts (e.g., pharmaceutically important metabolites). However, the production of hydrocarbon biofuels should be comparable with the current petroleum fuels in terms of production cost and physiochemical properties. Therefore, approaches, such as finding low-toxic hydrocarbon biofuels with similar physiochemical properties to petroleum fuels, can benefit large-scale production. For example, careful screening of metabolic engineered strain coupled with growth could lead to finding low-toxic hydrocarbon without product inhibition. More research efforts, such as introducing the parallel pathways to microbial cells for multiple hydrocarbons, can improve the physicochemical properties of hydrocarbon biofuels. This approach can produce and blend several hydrocarbons to achieve properties similar to petroleum fuels [54]. Collectively, precise metabolic engineering efforts with controlled toxicity and blending of hydrocarbons can make the production of hydrocarbon biofuels feasible at the industrial scale. Moreover, along with metabolic engineering efforts, the extraction of hydrocarbons from broth can help to achieve even higher titers at the process engineering level.

Apart from product toxicity, an ideal microbial cell factory is required to withstand other stresses such as osmolarity and hypoxia, while scaling-up from laboratory-scale to industrial-scale production. The scale-up of hydrocarbon biofuel production can be enhanced using adaptive laboratory evolution (ALE). ALE is the method to generate adaptive microbial cells under different stresses in a gradual manner [55]. Growing microbial cells in stressed growth conditions for a prolonged time causes random mutations in the cells that work in favor of cells' survival in harsh conditions. This method may not be essential for growing metabolically engineered cells in the laboratory but can be crucial for pilot and large-scale microbial cultures.

9.5 Use of lignocellulosic materials as feedstock

Microbial production of several biofuels has been tested over the utilization of lignocellulosic materials from biomass. Scientists and industrials have tried to improve the processes such as the breakdown of complex sugars in lignocellulose to simple sugars, removal of toxic compounds from lignocellulose, identification of biomass with low impact on human and animal food materials, and productivities of biofuels [6, 56, 57]. Linking these efforts with hydrocarbon-producing strains can provide a definite advantage to reduce the production cost of hydrocarbon biofuels. Low cost and a vast abundance of lignocellulosic materials can benefit the microbial production of hydrocarbons. Despite numerous advantages of lignocellulosic materials, utilization of these as feedstock for hydrocarbon biofuels demonstrates several challenges related to converting complex sugars to simple fermentable sugars. For industrial production of hydrocarbon biofuels, the combinations of physical, chemical, and biology pretreatment processing of lignocellulosic materials connected to the hydrocarbon fermentation process can be improved to reach economic feasibility [58]. Moreover, connecting the pretreatment of feedstocks and fermentation in series can increase the processing time and decrease productivity. This issue can be resolved by performing enzymatic degradation of lignocellulose and fermentation simultaneously, which can reduce process time and significantly decrease process cost. Several biofuel-based industries have been using such an approach for ethanol production on the consumption of lignocellulosic materials.

Alkali pretreatment method was found superior over other methods in terms of the final yield of released fermentable sugars. However, this method has various challenges, such as fermentation inhibitors, high-energy requirement, high cost, long process time, and damage of equipment due to corrosion [59]. However, biological pretreatment does not require costly chemicals, works in mild environmental conditions, and reduces process cost of pretreatment. In general, microorganisms capable of synthesizing ligninolytic and cellulolytic enzymes can be used for the biological pretreatment process. Several factors such as the type of microorganisms and substrate and growth conditions affect the efficiency of the biological delignification process. Currently, two methods, namely separated hydrolysis and simultaneous hydrolysis and fermentation, have been used for bioethanol production that could be implemented for the production of hydrocarbon biofuels as well [60].

In biological pretreatment, several filamentous fungi have been investigated to degrade the lignocellulosic materials. Some common examples are *Trametes versicolor* [61], *Phanerochaete chrysosporium* [62], and *Pleurotus ostreatus* [63] that were used to remove lignin from lignocellulosic biomass such as beech wood and cotton stalks. Even though fungi demonstrated the potential to pretreat the biomass, their slow growth increases the processing time, hence the process cost. This drawback makes fungi less desirable for industrial-scale biofuel production. However, identification of suitable species and metabolic perturbations can make fungi a more promising option for pretreating lignocelluloses. The use of lignin-degrading bacteria can be an alternative option. Bacteria, from different phylum such as Actinobacteria, Firmicutes, and Proteobacteria, have been tested to degrade the plant biomass into fermentable sugars. *Streptomyces viridosporus* T7A could degrade synthetic kraft lignin, lignin, polyethylene plastic, and aromatic dyes. Other *Streptomyces* species, such as *Streptomyces flavovirens*, *Streptomyces badius* ATCC 39117, *Streptomyces cyaneus* CECT 3335, *Streptomyces psammoticus*, and *Amycolatopsis* sp. ATCC 39116/75iv2, have shown the potential for depolymerization and mineralization activities on lignocellulosic materials [64, 65]. Apart from fungi and bacteria, direct delignification enzymes can be employed to pretreat the complex plant biomasses. Some examples of such enzymes are cellulases, laccase, manganese peroxidase, and lignin peroxidase [66–68]. In general, many factors such as moisture content and substrate concentration, type and nature of lignocellulosic biomass, culturing time, culturing temperature, and types of microbial strains, and pH can affect the biological pretreatments. Other cheaper and naturally abundant feedstocks for microbial production of hydrocarbons can be CO_2, methane, and methanol. However, more research efforts to engineer the strains metabolism toward consuming these advantageous feedstocks are required.

9.6 Bioconversion of CO_2 to hydrocarbons

Oxygenic photosynthesizers, like cyanobacteria, can consume naturally abundant CO_2 and convert it to several organic compounds via Calvin cycle [69]. Cyanobacteria is a suitable candidate for converting CO_2 to hydrocarbons using metabolic engineering approaches. Moreover, the photosynthetic biosynthesis pathway of cyanobacteria can be introduced to heterotrophic bacteria like *E. coli* for the production of hydrocarbon biofuels. The biosynthesis of hydrocarbons from CO_2 is advantageous compared to lignocellulosic materials because of no requirement of feedstock pretreatment, which is a cost- and time-expensive step for lignocellulosic materials.

Production of two hydrocarbons, namely alkanes and isoprenoids, has been examined significantly in cyanobacteria cells [70, 71]. Cyanobacteria take a common route, like other alkane-producing organisms, through decarbonylation of fatty aldehydes [45]. The involvement of enzymes such as fatty aldehyde decarbonylase and fatty acyl-ACP reductase and related gene sequences was found in the decarbonylation pathway [72]. Cyanobacteria can also be employed to synthesize fatty acids of preferred chain lengths such as C_8, C_{10}, and C_{12} to compensate the jet fuel and gasoline. To increase such specificity of hydrocarbons, different heterologous genes encoding different acyl-ACP thioesterases such as *FatB*, *FatB2*, and *FatB1* from *Arabidopsis*, *Cuphea hookeriana*, and *Umbellularia californica*, respectively, can be introduced to cyanobacteria metabolism. In addition to the insertion of an adequate acyl-ACP thioesterase, a high synthesis of free fatty acid in the cells will be needed to optimize a high yield of hydrocarbons from cell factories. At this end, increasing the activity of the gene like ACC can play a crucial role in enhancing the production of free fatty acids because the product of ACC gene catalyzes a critical rate-limiting step in fatty acid ACP biosynthesis. ACC gene contains many subunits such as AccA, AccB, AccC, and AccD, and not all cyanobacteria genomes include all these genes. Another strategy can be to insert heterologous genes from other organisms to enhance the synthesis of free fatty acids in cells. Along with the insertion of ACC, inactivation of AAS can reduce the re-thioesterification of free fatty acids, hence will enhance the net synthesis of free fatty acids [73].

Several studies illustrated the capabilities of cyanobacteria to produce a wide range of linear, branched, and cyclic alkanes. Biosynthesis of some of alkanes such as methyl and ethylalkanes is exclusively present in cyanobacterial cells. For instance, cyanobacterial strain, *Anabaena cylindrica* can produce $C_9–C_{16}$ n-alkanes under high salt (NaCl) concentration in the media [74]. Moreover, *Microcoleus vaginatus* was demonstrated to produce several n-alkanes and branched alkanes. More exploration efforts to identify and investigate the different cyanobacterial strains will lead us to an increasing variety of alkanes in terms of chain lengths and structures.

Isoprenoids have a higher-octane rating compared to n-alkanes, hence, they can be employed for high-performance gasoline engines. Cyanobacteria can naturally produce carotenoids such as geranyl pyrophosphate, geranylgeranyl pyrophosphate, and farnesyl pyrophosphate. These carotenoids can be employed directly as biofuel after some processing; however, carotenoids can be transformed to isoprenoids for more efficient use. Above carotenoids can be used as precursors for monoterpenes, sesqui- and triterpenes, and di- and tetraterpenes [75]. Although the production of isoprene was not reported naturally in cyanobacteria, however, insertion of isoprene synthase especially from plants demonstrated the biosynthesis of isoprene hydrocarbons in cyanobacteria cells [76]. The engineered carotenoid pathway in cyanobacteria, for producing isoprene, monoterpene, or sesquiterpene, can be established by moderate efforts through the insertion of a single isoprene synthase gene or different terpene synthase genes. However, manipulating carotenoid pathway for synthesizing pinene and farnesene will be more beneficial as these isoprenoids are considered as precursors of jet fuels and diesel fuels.

The capability of cyanobacteria to cultivate in low to high concentrations of CO_2 makes them a suitable candidate for producing biofuels by consuming the CO_2 produced from coal-based industries. Despite several advantages, considering cyanobacteria as biofuel producers brings many challenges. For example, (i) the dependency on the light will affect the biofuel production in the dark; (ii) harvesting light for growth is a slow process in the photosynthetic organism that will affect the productivity of biofuel production; and (iii) harvesting and extracting hydrocarbons can be a costly process as the current microalgae-based biofuel industry is facing this challenge. The above challenges should be addressed for establishing the economically feasible hydrocarbon-based biofuels from CO_2.

9.7 Challenges and future directions

Continuous reliance on petroleum-based fuels as a primary energy source has been increasing the level of greenhouse gases in our environment. These harmful gases started showing destructed effects on climate as well as on our health. Utilization of hydrocarbon biofuels produced from lignocellulosic materials and CO_2 as vehicular fuel can contribute to lower the greenhouse gases in the environment. In addition to utilizing cheaper feedstocks, naturally efficient or metabolically engineered microbial cell factories would enhance the yield and productivity of the fermentation process to produce hydrocarbon biofuels that directly will reduce the production cost. These efforts will provide better opportunities to scale-up microbial production of hydrocarbons at the industrial level.

Recently, numerous studies have been conducted to improve the biosynthesis of hydrocarbons using fast-growing microbial hosts. To date, the most successful hosts for performing metabolic engineering toward producing hydrocarbon biofuels are *E. coli*, *S. cerevisiae*, and *Y. lipolytica*. Although these hosts are capable of adapting synthetic heterologous hydrocarbon-producing pathways from different microorganisms and plants, none of the efforts could achieve the economically feasible titer of hydrocarbons. Many challenges still remain: (i) achieving a theoretical maximum yield of specific hydrocarbons, (ii) producing hydrocarbons with selective chain length, (iii) finding the solutions of hydrocarbon toxicity to producers, and (iv) engineering hosts to consume low-cost feedstocks.

Identification of hydrocarbons that are nontoxic to microbial culture and with similar properties as constituents of current petroleum fuels can be helpful to design a high titer producing hydrocarbon microbial cell factories. Moreover, cheaper and naturally abundant feedstocks such as CO_2 [77], methane [78], and methanol [79] can be considered for microbial production of hydrocarbons. In addition, the consumption of CO_2 and methane for the biosynthesis of hydrocarbon will reduce the greenhouse gases in the environment. Both gases and methanol are produced from natural gas or coal in abundance. Another potential feedstock can be lignocellulosic materials [57, 80]. Plenty of studies have been conducted to utilize cheap and abundant lignocellulosic materials for producing biofuels such as ethanol and butanol [56]. These feedstocks can reduce the production cost of hydrocarbons to make them economically feasible. However, feeding these feedstocks to metabolically engineered cell factories is still a major challenge. Therefore, there a clear need to develop novel strain manipulation strategies to make hosts capable of utilizing cost-effective feedstocks.

9.8 Conclusions

Recent developments in metabolic engineering can be deployed to design high hydrocarbon-producing microbial strains for achieving economically feasible biosynthesis of hydrocarbon biofuels. Combining these efforts with the production of other fermentative biofuels such as ethanol and butanol can make it possible to provide cheaper fuels to all the sectors of transportation, from road vehicles to airplanes. Moreover, the use of cheaper and abundant feedstocks, such as lignocellulosic materials and CO_2, for biofuel production will help reduce the production costs and emission of greenhouse gases in the environment. Instead of choosing one feedstock, building multifeedstock-based production plants will help tackle the shortages of sessional biomass-based feedstocks.

References

[1] J. Jaroensuk, P. Intasian, W. Wattanasuepsin, N. Akeratchatapan, C. Kesornpun, N. Kittipanukul, et al., Enzymatic reactions and pathway engineering for the production of renewable hydrocarbons, J. Biotechnol. 309 (2020) 1–19. https://linkinghub.elsevier.com/retrieve/pii/S0168165619309460. (Accessed February 2020).

[2] H.K. Lichtenthaler, The 1-deoxy-D-xylulose-5-phosphate pathway of isoprenoid biosynthesis in plants, Annu. Rev. Plant. Physiol. Plant. Mol. Biol. 50 (1999) 47–65.

[3] J.C. Liao, L. Mi, S. Pontrelli, S. Luo, Fuelling the future: microbial engineering for the production of sustainable biofuels, Nat. Rev. Microbiol. 14 (2016) 288–304.

[4] Y. Chen, C. Xu, S. Vaidyanathan, Influence of gas management on biochemical conversion of CO2 by microalgae for biofuel production, Appl. Energy 261 (2020) 114420. https://linkinghub.elsevier.com/retrieve/pii/S0306261919321075. (Accessed March 2020).

[5] M. Kumar, T.K. Bhowmick, S. Saini, K. Gayen, Current status and challenges in biobutanol production, in: O. Konur (Ed.), Bioenergy and Biofuels, first ed., CRC Press, Boca Raton, 2018, pp. 237–262. https://www.taylorfrancis.com/books/9781138032828. (Accessed 2 January 2018).

[6] M. Kumar, K. Gayen, Developments in biobutanol production: new insights, Appl. Energy 88 (2011) 1999–2012. Elsevier Ltd https://doi.org/10.1016/j.apenergy.2010.12.055. (Accessed January 2011).

[7] B.M. Lange, T. Rujan, W. Martin, R. Croteau, Isoprenoid biosynthesis: the evolution of two ancient and distinct pathways across genomes, Proc. Natl. Acad. Sci. U. S. A. 97 (2000) 13172–13177.

[8] Y.J. Zhou, E.J. Kerkhoven, J. Nielsen, Barriers and opportunities in bio-based production of hydrocarbons, Nat. Energy 3 (2018) 925–935. https://doi.org/10.1038/s41560-018-0197-x. (Accessed January 2018).

[9] M. Kanehisa, S. Goto, Y. Sato, M. Furumichi, M. Tanabe, KEGG for integration and interpretation of large-scale molecular data sets, Nucleic Acids Res. 40 (2012) 109–114.

[10] Miziorko HM. Enzymes of the mevalonate pathway of isoprenoid biosynthesis. Arch. Biochem. Biophys. Elsevier Inc.; 2011;505:131–143. https://doi.org/10.1016/j.abb.2010.09.028. Accessed 2011.

[11] W.N. Hunter, The non-mevalonate pathway of isoprenoid precursor biosynthesis, J. Biol. Chem. 282 (2007) 21573–21577.

[12] B.F. Pfleger, M. Gossing, J. Nielsen, Metabolic engineering strategies for microbial synthesis of oleochemicals, Metab. Eng. 29 (2015) 1–11.

[13] Liu K, Li S. Biosynthesis of fatty acid-derived hydrocarbons: perspectives on enzymology and enzyme engineering. Curr. Opin. Biotechnol.; Elsevier Ltd, 2020;62:7–14. https://doi.org/10.1016/j.copbio.2019.07.005. Accessed 2020.

[14] J. Staunton, K.J. Weissman, Polyketide biosynthesis: a millennium review, Nat. Prod. Rep. 18 (2001) 380–416.

[15] A. Das, C. Khosla, Biosynthesis of aromatic polyketides in bacteria, Acc. Chem. Res. 42 (2009) 631–639.

[16] B.S. Moore, Discovery of a new bacterial polyketide biosynthetic pathway, ChemBioChem 2 (2001) 35–38.

[17] G.M. Whited, F.J. Feher, D.A. Benko, M.A. Cervin, G.K. Chotani, J.C. McAuliffe, et al., Development of a gas-phase bioprocess for isoprene-monomer production using metabolic pathway engineering, Ind. Biotechnol. 6 (2010) 152–163. http://www.liebertpub.com/doi/10.1089/ind.2010.6.152. (Accessed June 2010).

[18] X. Lv, F. Wang, P. Zhou, L. Ye, W. Xie, H. Xu, et al., Dual regulation of cytoplasmic and mitochondrial acetyl–CoA utilization for improved isoprene production in Saccharomyces cerevisiae, Nat. Commun. 7 (2016). Nature Publishing Group.

[19] X. Gao, F. Gao, D. Liu, H. Zhang, X. Nie, C. Yang, Engineering the methylerythritol phosphate pathway in cyanobacteria for photosynthetic isoprene production from CO2, Energ. Environ. Sci. 9 (2016) 1400–1411. Royal Society of Chemistry.

[20] A.L. Meadows, K.M. Hawkins, Y. Tsegaye, E. Antipov, Y. Kim, L. Raetz, et al., Rewriting yeast central carbon metabolism for industrial isoprenoid production, Nature 537 (2016) 694–697. Nature Publishing Group https://doi.org/10.1038/nature19769. (Accessed January 2016).

[21] S. Kwak, E.J. Yun, S. Lane, E.J. Oh, K.H. Kim, Y.S. Jin, Redirection of the glycolytic flux enhances Isoprenoid production in Saccharomyces cerevisiae, Biotechnol. J. 15 (2020) 1–10.

[22] P.P. Peralta-Yahya, M. Ouellet, R. Chan, A. Mukhopadhyay, J.D. Keasling, T.S. Lee, Identification and microbial production of a terpene-based advanced biofuel, Nat. Commun. 2 (2011). Nature Publishing Group.

[23] E. Orsi, J. Beekwilder, D. van Gelder, A. van Houwelingen, G. Eggink, S.W.M. Kengen, et al., Functional replacement of isoprenoid pathways in Rhodobacter sphaeroides, J. Microbial. Biotechnol. 13 (2020) 1082–1093.

[24] J. Nielsen, Synthetic biology for engineering acetyl coenzyme a metabolism in yeast, MBio 5 (2014) 5–7.

[25] Z. Dai, M. Huang, Y. Chen, V. Siewers, J. Nielsen, Global rewiring of cellular metabolism renders Saccharomyces cerevisiae Crabtree negative, Nat. Commun. 9 (2018) 1–8. Springer US https://doi.org/10.1038/s41467-018-05409-9. (Accessed January 2018).

[26] Y.J. Zhou, N.A. Buijs, Z. Zhu, J. Qin, V. Siewers, J. Nielsen, Production of fatty acid-derived oleochemicals and biofuels by synthetic yeast cell factories, Nat. Commun. 7 (2016) 11709. Nature Publishing Group.

[27] P. Xua, K. Qiao, W.S. Ahn, G. Stephanopoulos, Engineering Yarrowia lipolytica as a platform for synthesis of drop-in transportation fuels and oleochemicals, Proc. Natl. Acad. Sci. U. S. A. 113 (2016) 10848–10853.

[28] Johnson AO, Gonzalez-Villanueva M, Wong L, Steinbüchel A, Tee KL, Xu P, et al. Design and application of genetically-encoded malonyl-CoA biosensors for metabolic engineering of microbial cell factories. Metab. Eng. Elsevier Inc.; 2017;44:253–64. https://doi.org/10.1016/j.ymben.2017.10.011. Accessed 2017.

[29] R.M. Lennen, D.J. Braden, R.M. West, J.A. Dumesic, B.F. Pfleger, A process for microbial hydrocarbon synthesis: overproduction of fatty acids in Escherichia coli and catalytic conversion to alkanes, Biotechnol. Bioeng. 106 (2010) 193–202.

[30] Qiao K, Imam Abidi SH, Liu H, Zhang H, Chakraborty S, Watson N, et al. Engineering lipid overproduction in the oleaginous yeast Yarrowia lipolytica. Metab. Eng. Elsevier; 2015;29:56–65. https://doi.org/10.1016/j.ymben.2015.02.005. Accessed 2015.

[31] Shin KS, Lee SK. Introduction of an acetyl-CoA carboxylation bypass into Escherichia coli for enhanced free fatty acid production. Bioresour. Technol. Elsevier Ltd; 2017;245:1627–33. https://doi.org/10.1016/j.biortech.2017.05.169. Accessed 2017.

[32] d'Espaux L, Ghosh A, Runguphan W, Wehrs M, Xu F, Konzock O, et al. Engineering high-level production of fatty alcohols by Saccharomyces cerevisiae from lignocellulosic feedstocks. Metab. Eng. Elsevier Inc.; 2017;42:115–125. https://doi.org/10.1016/j.ymben.2017.06.004. Accessed 2017.

[33] P. Xu, K. Qiao, W.S. Ahn, G. Stephanopoulos, Engineering Yarrowia lipolytica as a platform for synthesis of drop-in transportation fuels and oleochemicals, Proc. Natl. Acad. Sci. 113 (2016) 10848–10853. http://www.pnas.org/lookup/doi/10.1073/pnas.1607295113. (Accessed 27 September 2016).

[34] P. Kallio, A. Pásztor, K. Thiel, M.K. Akhtar, P.R. Jones, An engineered pathway for the biosynthesis of renewable propane, Nat. Commun. 5 (2014) 4–11.

[35] Sheppard MJ, Kunjapur AM, Prather KLJ. Modular and selective biosynthesis of gasoline-range alkanes. Metab. Eng. Elsevier; 2016;33:28–40. https://doi.org/10.1016/j.ymben.2015.10.010. Accessed 2016.

[36] Y. Zheng, Q. Liu, L. Li, W. Qin, J. Yang, H. Zhang, et al., Metabolic engineering of Escherichia coli for high-specificity production of isoprenol and prenol as next generation of biofuels, Biotechnol. Biofuels 6 (2013) 1–13.

[37] J. Alonso-Gutierrez, E.-M. Kim, T.S. Batth, N. Cho, Q. Hu, L.J.G. Chan, et al., Principal component analysis of proteomics (PCAP) as a tool to direct metabolic engineering, Metab. Eng. 28 (2015) 123–133. https://linkinghub.elsevier.com/retrieve/pii/S109671761400161X. (Accessed March 2015).

[38] X. Cao, Y.B. Lv, J. Chen, T. Imanaka, L.J. Wei, Q. Hua, Metabolic engineering of oleaginous yeast Yarrowia lipolytica for limonene overproduction, Biotechnol. Biofuels 9 (2016) 1–11. BioMed Central.

[39] F.K. Davies, V.H. Work, A.S. Beliaev, M.C. Posewitz, Engineering limonene and bisabolene production in wild type and a glycogen-deficient mutant of Synechococcus sp. PCC 7002, Front. Bioeng. Biotechnol. 2 (2014) 1–11.

[40] S. Sarria, B. Wong, H.G. Martín, J.D. Keasling, P. Peralta-Yahya, Microbial synthesis of pinene, ACS Synth. Biol. 3 (2014) 466–475.

[41] H. Zhang, Q. Liu, Y. Cao, X. Feng, Y. Zheng, H. Zou, et al., Microbial production of sabinene—a new terpene-based precursor of advanced biofuel, Microb. Cell Fact. 13 (2014) 1–10.

[42] M. Muramatsu, C. Ohto, S. Obata, E. Sakuradani, S. Shimizu, Alkaline pH enhances farnesol production by Saccharomyces cerevisiae, J. Biosci. Bioeng. 108 (2009) 52–55. https://linkinghub.elsevier.com/retrieve/pii/S1389172309001108. (Accessed July 2009).

[43] C. Wang, S.-H. Yoon, A.A. Shah, Y.-R. Chung, J.-Y. Kim, E.-S. Choi, et al., Farnesol production from Escherichia coli by harnessing the exogenous mevalonate pathway, Biotechnol. Bioeng. 107 (2010) 421–429. http://doi.wiley.com/10.1002/bit.22831. (Accessed 15 October 2010).

[44] R.M. Phelan, O.N. Sekurova, J.D. Keasling, S.B. Zotchev, Engineering terpene biosynthesis in Streptomyces for production of the advanced biofuel precursor bisabolene, ACS Synth. Biol. 4 (2015) 393–399. https://pubs.acs.org/doi/10.1021/sb5002517. (Accessed 17 April 2015).

[45] A. Schirmer, M.A. Rude, X. Li, E. Popova, S.B. del Cardayre, Microbial biosynthesis of alkanes, Science 329 (2010) 559–562. https://www.sciencemag.org/lookup/doi/10.1126/science.1187936. (Accessed 30 July 2010).

[46] Cao Y-X, Xiao W-H, Zhang J-L, Xie Z-X, Ding M-Z, Yuan Y-J. Heterologous biosynthesis and manipulation of alkanes in Escherichia coli. Metab. Eng. 2016;38:19–28. https://linkinghub.elsevier.com/retrieve/pii/S109671761630043X. Accessed 2016 Nov.

[47] Y.J. Zhou, N.A. Buijs, Z. Zhu, D.O. Gómez, A. Boonsombuti, V. Siewers, et al., Harnessing yeast peroxisomes for biosynthesis of fatty-acid-derived biofuels and chemicals with relieved side-pathway competition, J. Am. Chem. Soc. 138 (2016) 15368–15377. https://pubs.acs.org/doi/10.1021/jacs.6b07394. (Accessed 30 November 2016).

[48] Y. Liu, C. Wang, J. Yan, W. Zhang, W. Guan, X. Lu, et al., Hydrogen peroxide-independent production of α-alkenes by OleTJE P450 fatty acid decarboxylase, Biotechnol. Biofuels 7 (2014) 28. http://biotechnologyforbiofuels.biomedcentral.com/articles/10.1186/1754-6834-7-28. (Accessed January 2014).

[49] Y.J. Zhou, Y. Hu, Z. Zhu, V. Siewers, J. Nielsen, Engineering 1-alkene biosynthesis and secretion by dynamic regulation in yeast, ACS Synth. Biol. 7 (2018) 584–590. https://pubs.acs.org/doi/10.1021/acssynbio.7b00338. (Accessed 16 February 2018).

[50] Q. Liu, K. Wu, Y. Cheng, L. Lu, E. Xiao, Y. Zhang, et al., Engineering an iterative polyketide pathway in Escherichia coli results in single-form alkene and alkane overproduction, Metab. Eng. 28 (2015) 82–90. Elsevier https://doi.org/10.1016/j.ymben.2014.12.004. (Accessed January 2015).

[51] Francisco S, Katz L, Steen EJ, Keasling JD. Producing Alpha-Olefins Using Polyketide Synthases. 2013. https://patents.google.com/patent/US20130267696. Accessed 2013.

[52] J. Sikkema, J.A.M. De Bont, B. Poolman, Mechanisms of membrane toxicity of hydrocarbons, Microbiol. Rev. 59 (1995) 201–222.

[53] Z. Gong, J. Nielsen, Y.J. Zhou, Engineering robustness of microbial cell factories, Biotechnol. J. 12 (2017) 1700014. http://doi.wiley.com/10.1002/biot.201700014. (Accessed October 2017).

[54] T.C.R. Brennan, C.D. Turner, J.O. Krömer, L.K. Nielsen, Alleviating monoterpene toxicity using a two-phase extractive fermentation for the bioproduction of jet fuel mixtures in Saccharomyces

cerevisiae, Biotechnol. Bioeng. 109 (2012) 2513–2522. http://doi.wiley.com/10.1002/bit.24536. (Accessed October 2012).

[55] M. Dragosits, D. Mattanovich, Adaptive laboratory evolution—principles and applications for biotechnology, Microb. Cell Fact. 12 (2013) 64. http://microbialcellfactories.biomedcentral.com/articles/10.1186/1475-2859-12-64. (Accessed January 2013).

[56] M. Kumar, K. Gayen, Biobutanol: The Future Biofuel, in: C. Baskar, S. Baskar, R.S. Dhillon (Eds.), Biomass Conversion: The Interface of Biotechnology, Chemistry, and Materials Science, Springer Berlin Heidelberg, Berlin, Heidelberg, 2012, pp. 221–236. https://doi.org/10.1007/978-3-642-28418-2_7. (Accessed January 2012).

[57] Kumar M, Goyal Y, Sarkar A, Gayen K. Comparative economic assessment of ABE fermentation based on cellulosic and non-cellulosic feedstocks. Appl. Energy. Elsevier Ltd; 2012;93:193–204. http://linkinghub.elsevier.com/retrieve/pii/S0306261911008853. Accessed 2012 May.

[58] V.B. Agbor, N. Cicek, R. Sparling, A. Berlin, D.B. Levin, Biomass pretreatment: fundamentals toward application, Biotechnol. Adv. 29 (2011) 675–685. https://linkinghub.elsevier.com/retrieve/pii/S0734975011000607. (Accessed November 2011).

[59] Á.T. Martínez, F.J. Ruiz-Dueñas, M.J. Martínez, J.C. del Río, A. Gutiérrez, Enzymatic delignification of plant cell wall: from nature to mill, Curr. Opin. Biotechnol. 20 (2009) 348–357. https://linkinghub.elsevier.com/retrieve/pii/S0958166909000603. (Accessed June 2009).

[60] M.M. Ishola, A. Jahandideh, B. Haidarian, T. Brandberg, M.J. Taherzadeh, Simultaneous saccharification, filtration and fermentation (SSFF): a novel method for bioethanol production from lignocellulosic biomass, Bioresour. Technol. 133 (2013) 68–73. https://linkinghub.elsevier.com/retrieve/pii/S0960852413001624. (Accessed April 2013).

[61] Bari E, Taghiyari HR, Naji HR, Schmidt O, Ohno KM, Clausen CA, et al. Assessing the destructive behaviors of two white-rot fungi on beech wood. Int. Biodeter. Biodegr.. 2016;114:129–40. https://linkinghub.elsevier.com/retrieve/pii/S0964830516302050. Accessed 2016 Oct.

[62] J. Shi, M. Chinn, R. Sharmashivappa, Microbial pretreatment of cotton stalks by solid state cultivation of Phanerochaete chrysosporium, Bioresour. Technol. 99 (2008) 6556–6564. https://linkinghub.elsevier.com/retrieve/pii/S0960852407009911. (Accessed September 2008).

[63] E.A. Adebayo, D. Martinez-Carrera, Oyster mushrooms (Pleurotus) are useful for utilizing lignocellulosic biomass, Afr. J. Biotechnol. 14 (2015) 52–67. http://academicjournals.org/journal/AJB/article-abstract/AED32D349437.

[64] K.N. Niladevi, P. Prema, Mangrove Actinomycetes as the source of ligninolytic enzymes, Actinomycetologica 19 (2005) 40–47. http://joi.jlc.jst.go.jp/JST.JSTAGE/saj/19.40?from=CrossRef. (Accessed January 2005).

[65] R. Moya, M. Hernández, A.B. García-Martín, A.S. Ball, M.E. Arias, Contributions to a better comprehension of redox-mediated decolouration and detoxification of azo dyes by a laccase produced by Streptomyces cyaneus CECT 3335, Bioresour. Technol. 101 (2010) 2224–2229. https://linkinghub.elsevier.com/retrieve/pii/S0960852409015661. (Accessed April 2010).

[66] D.W.S. Wong, Structure and action mechanism of Ligninolytic enzymes, Appl. Biochem. Biotechnol. 157 (2009) 174–209. http://link.springer.com/10.1007/s12010-008-8279-z. (Accessed 26 May 2009).

[67] M. Sundaramoorthy, M.H. Gold, T.L. Poulos, Ultrahigh (0.93Å) resolution structure of manganese peroxidase from Phanerochaete chrysosporium: Implications for the catalytic mechanism, J. Inorg. Biochem. 104 (2010) 683–690. https://linkinghub.elsevier.com/retrieve/pii/S0162013410000498. (Accessed June 2010).

[68] L.I. Trubitsina, S.V. Tishchenko, A.G. Gabdulkhakov, A.V. Lisov, M.V. Zakharova, A.A. Leontievsky, Structural and functional characterization of two-domain laccase from Streptomyces viridochromogenes, Biochimie 112 (2015) 151–159. https://linkinghub.elsevier.com/retrieve/pii/S0300908415000644. (Accessed May 2015).

[69] C. Jansson, T. Northen, Calcifying cyanobacteria—the potential of biomineralization for carbon capture and storage, Curr. Opin. Biotechnol. 21 (2010) 365–371. https://linkinghub.elsevier.com/retrieve/pii/S0958166910000595. (Accessed June 2010).

[70] J. Kirby, J.D. Keasling, Metabolic engineering of microorganisms for isoprenoid production, Nat. Prod. Rep. 25 (2008) 656. http://xlink.rsc.org/?DOI=b802939c. (Accessed January 2008).

[71] J.D. Keasling, H. Chou, Metabolic engineering delivers next-generation biofuels, Nat. Biotechnol. 26 (2008) 298–299. http://www.nature.com/articles/nbt0308-298. (Accessed March 2008).

[72] D.M. Warui, N. Li, H. Nørgaard, C. Krebs, J.M. Bollinger, S.J. Booker, Detection of formate, rather than carbon monoxide, as the stoichiometric coproduct in conversion of fatty aldehydes to alkanes by a cyanobacterial aldehyde decarbonylase, J. Am. Chem. Soc. 133 (2011) 3316–3319. https://pubs.acs.org/doi/10.1021/ja111607x. (Accessed 16 March 2011).

[73] D. Kaczmarzyk, M. Fulda, Fatty acid activation in cyanobacteria mediated by acyl–acyl carrier protein synthetase enables fatty acid recycling, Plant Physiol. 152 (2010) 1598–1610. http://www.plantphysiol.org/lookup/doi/10.1104/pp.109.148007. (Accessed March 2010).

[74] P. Bhadauriya, R. Gupta, S. Singh, P.S. Bisen, N-alkanes variability in the diazotrophic cyanobacterium Anabaena cylindrica in response to NaCl stress, World J. Microbiol. Biotechnol. 24 (2008) 139–141. http://link.springer.com/10.1007/s11274-007-9439-y. (Accessed 10 January 2008).

[75] S.A. Agger, F. Lopez-Gallego, T.R. Hoye, C. Schmidt-Dannert, Identification of sesquiterpene synthases from *Nostoc punctiforme* PCC 73102 and Nostoc sp. strain PCC 7120, J. Bacteriol. 190 (2008) 6084–6096. https://jb.asm.org/content/190/18/6084. (Accessed 15 September 2008).

[76] P. Lindberg, S. Park, A. Melis, Engineering a platform for photosynthetic isoprene production in cyanobacteria, using Synechocystis as the model organism, Metab. Eng. 12 (2010) 70–79. https://linkinghub.elsevier.com/retrieve/pii/S1096717609000871. (Accessed January 2010).

[77] W. Pang, D. Hou, J. Ke, J. Chen, M.T. Holtzapple, J.K. Tomberlin, et al., Production of biodiesel from CO2 and organic wastes by fermentation and black soldier fly, Renew. Energy 149 (2020) 1174–1181. https://www.sciencedirect.com/science/article/abs/pii/S0960148119315903. (Accessed April 2020).

[78] J.M. Clomburg, A.M. Crumbley, R. Gonzalez, Industrial biomanufacturing: the future of chemical production, Science 355 (6320) (2017) aag0804, https://doi.org/10.1126/science.aag0804.

[79] J. Schrader, M. Schilling, D. Holtmann, D. Sell, M.V. Filho, A. Marx, et al., Methanol-based industrial biotechnology: current status and future perspectives of methylotrophic bacteria, Trends Biotechnol. 27 (2009) 107–115.

[80] M. Kumar, K. Gayen, Developments in biobutanol production: new insights, Appl. Energy 88 (2011) 1999–2012. http://linkinghub.elsevier.com/retrieve/pii/S0306261910005751. (Accessed June 2011).

CHAPTER 10

Oligomerization of bio-olefins for bio-jet fuel

Joshua Gorimbo[a], Mahluli Moyo[b], and Xinying Liu[b]
[a]Zhijiang College, Zhejiang University of Technology, Shaoxing, Zhejiang, China
[b]Institute for the Development of Energy for African Sustainability (IDEAS), University of South Africa (UNISA), Florida Campus, Johannesburg, South Africa

Contents

10.1 Introduction	271
10.1.1 Background	271
10.1.2 Jet-fuel specification	273
10.1.3 Sources of bio-olefin	275
10.1.4 Olefin oligomerization reactions	278
10.2 Bio-jet-fuel production pathways	278
10.2.1 Bio-jet fuel from hydro-processing of lipid feedstock	279
10.2.2 Bio-jet fuel by Fischer-Tropsch synthesis	280
10.2.3 Bio-jet fuel from catalytic reforming of sugars	281
10.2.4 Alcohol-to-bio-jet fuel	283
10.3 Oligomerization of olefins	283
10.3.1 Types of catalysts used in the oligomerization reaction	284
10.3.2 Reaction mechanism	284
10.3.3 Reaction conditions with the selectivity to products	286
10.4 Aromatization of hydrocarbons from oligomerization	287
10.5 Economics of bio-jet-fuel production	289
10.6 Conclusions	290
Acknowledgments	290
References	291

Abbreviations

ASTM American Society for Testing and Materials
ATF aviation turbine fuel
FTS Fischer–Tropsch synthesis

10.1 Introduction

10.1.1 Background

The oligomerization of bio-olefins to jet fuel has become a crucial process in the aviation industry's strategy to address energy concerns and global warming issues. The world's energy-related CO_2 emissions are forecast to grow at an average of 0.6% per year

Hydrocarbon Biorefinery
https://doi.org/10.1016/B978-0-12-823306-1.00010-8

Copyright © 2022 Elsevier Inc.
All rights reserved.

between 2018 and 2050 [1]. Currently, aviation fuel is mainly produced from petroleum-based sources and contributes 4%–5% of greenhouse gases released, but these emissions are increasing [2]. However, the International Air Transport Association reported that the global civil aviation industry released 815 metric tons of CO_2 in 2016, which accounted for 2.5% of global energy-related carbon dioxide emissions [3]. The aviation industry recorded an 8% numerical increase in flights between 2014 and 2017 and has projected that the number of flights will increase by 42% by 2040 [4]. The European Union has set goals to attain a low carbon economy by the year 2050. The plan indicates that it aims to achieve a 60% reduction in greenhouse gas emissions by 2040 (from 1990 levels) and to use sustainable fuel sources for aviation [4].

The Carbon Offset and Reduction Scheme for International Aviation continues to be the International Air Transport Association's primary focus encouraging countries to apply the Carbon Offset and Reduction Scheme for International Aviation regulations. As per the International Air Transport Association 2018 review, the association aims to (i) halve CO_2 emissions in the aviation industry by 2050, relative to 2005 levels and (ii) introduce a cap on net CO_2 from 2020 [3]. Jet fuels derived from renewable feedstocks can help cushion dependency on petroleum-based energy sources, effectively lessening greenhouse gas emissions. The widespread deployment of sustainable aviation fuels is now a key pursuit in the aviation industry. Therefore research laboratories, governments, corporations involved in oil-refining, firms producing biofuel, and agricultural institutions should be focusing on developing commercially sustainable and cost-effective processes for the production of renewable jet fuels that have low greenhouse gas emissions.

Different methods can be employed in producing bio-jet fuel, with the correct specifications defined by the American Society for Testing and Materials (ASTM). Achieving a net decrease in atmospheric carbon dioxide, using biomass for jet-fuel production, may be a major contributor to achieving these targets. In addition, renewable fuels for the aviation industry must have the following attributes [5,6].

- It should be "drop-in," meaning it must be used without altering existing engines.
- It should be available in sufficient quantities.
- It should be comparatively cheaper than or equal to conventional petroleum fuels.
- It should originate from nonfood crop raw materials.
- It should produce comparatively less greenhouse gas emissions.

Alternative jet fuel initiatives implemented by the Department of Defense Air Force USA once aimed to use blended fuel and alternative fuel with a 1:1 ratio [7]. The USA Air Force is also working toward ensuring 50% of all local aviation fuels from alternative fuel blends by 2025 [8]. Projections in the Annual Energy Outlook indicated that the percentage of biofuels blended into US fuel would increase from 7.3% in 2019 to a plateau of 9.0% in 2040 [9]. It is also suggested that several biofuel technologies will attain economic viability and market competitiveness in the period 2025–2030 [10].

Working groups, such as the Commercial Aviation Alternative Fuels Initiative, focus on augmenting environmental sustainability and energy security in the aviation industry by utilizing alternative jet fuels [11]. Commercial Aviation Alternative Fuels Initiative established a structure that includes participants from governments and industries that invest in sustainable aviation fuel, and it has concluded the certification of production pathways for bio-jet fuel and provided commercialization models [12]. According to the Paris agreement goals and initiatives, the development of the biofuel producing industry, construction of pilot plants and refineries for biofuel production aims at increasing the use of renewable energy. A coalition of airline companies, manufacturers of engines and aircraft, and producers of fuel is developing sustainable aviation fuels for commercial aircraft [11].

Sustainable utilization of biomass results in no net increase in atmospheric CO_2. Therefore the increased substitution of fossil-based jet fuels with biomass-based fuel would help reduce global warming. The choice of feedstock is a crucial factor in the synthesis of biofuel. Based on the various types of feedstocks, nonedible oils, algal oils, waste animal fats, and waste cooking oils are promising raw materials for biofuel production. The production of biofuel from nonedible sources is attractive because they are a renewable and sustainable source that guarantees food security.

10.1.2 Jet-fuel specification

Fuel certification and specification compliance is a crucial stage before fuel can be used in commercial flights. Among all the scientifically proven biofuel synthesis routes, only six have been certified by the ASTM to date [4]. Fuels utilized in the aviation industry are highly specialized with precise properties and qualities [13]. Almost all aviation fuels used to date are derived from crude oil in refineries.

Jet-fuel hydrocarbons range from C_8 to C_{16} and are spiked with additives to improve or introduce certain vital properties, including freezing point, smoke point, cetane number, etc. A specific range of hydrocarbons is governed by the product specification. The kerosene-type jet fuel is Jet fuel A, while Jet A-1 is a subset that has a distribution of carbon number ranging from C_8 to C_{16}. The naphtha-type jet fuel (for instance, Jet B) has a carbon number in the range of C_5 to C_{15}. The precise specifications for Jet fuel A and Jet A-1 include a higher flash point of more than 38°C with 210°C autoignition temperature. The detailed specifications of these jet fuels are shown in Table 10.1.

Aviation turbine fuel (ATF) powers most modern-generation jet engine aircraft. The engines have to function within predetermined parameters and specified altitude and flight speed ranges. Therefore jet fuel has to be of consistent quality and airworthy. Major civil JET-fuel grades include Jet A-1, Jet A, and Jet B. Jet fuel TS-1 is manufactured based on the Russian standard GOST 10227 and is suitable for use in cold weather. Product

Table 10.1 The most important specifications of jet fuel [14,15].

Fuel	Jet A	Jet A-1	TS-1	Jet B
Specification	ASTM D 1655	DEF STAN 91-91	GOST 10227	CGSB-3.22
Acidity, mg KOH/g	0.10	0.015	0.7 (mg KOH/100 mL)	0.10
Aromatics, % vol. max	25	25.0	22 (% mass)	25.0
Sulfur, mass %	0.30	0.30	0.25	0.40
Sulfur, mercaptan, mass%	0.003	0.003	0.005	0.003
Distillation, °C				
Initial boiling point	–	Report	150	Report
10% recovered, max	205	205	165	Report
50% recovered, max	Report	Report	195	min 125, max 190
90% recovered, max	Report	Report	230	Report
Endpoint	300	300	250	270
Vapor pressure, kPa, max	–	–	–	21
Flash point, °C, min	38	38	28	–
Density, 15°C, kg/m^3	775–840	775–840	min 774@20°C	750–801
Freezing point, °C, max	−40	−47.0	−50 (chilling point)	−51
Viscosity, −20°C, mm^2/sec, max	8	8.0	8.0@ −40°C	–
Net heat of combustion, MJ/kg. min	42.8	42.8	42.9	42.8
Smoke point, mm	18	19.0	25	20
Naphthalenes, vol% max	3.0	3.00	–	3.0
Copper corrosion, 2h@100°C, max rating	No.1	No.1	Pass (3h@100°C)	No.1
Thermal stability				
Filter pressure drop, mmHg max	25	25	–	25
Visual tube rating, max	<3	<3	–	3.0
Static test 4h@150°C, mg/100 max	–	–	18	–
Existent gum, mg/100 mL max	7	7	5	–

It provides the minimum property requirements for Jet A, Jet A-1, TS-1, and Jet B. It also lists acceptable additives for use in civil operated engines and aircraft. Report: results not comparable or no standardized values, report what you find.

specification is a mechanism by which fuel companies control crucial properties for reliable and satisfactory performance. The main jet-fuel specifications are as follows.

- ASTM D 1655: Specifications for the USA's three commercial jet fuels (Jet A, Jet A-1, and Jet B).
- GB 6537-2006 Chinese standard: Jet-fuel manufactured in China is now all designated RP-3 (renamed No. 3 Jet fuel).
- Defense standard 91-91: (DEN STAN) The UK Defense Ministry maintains this specification.
- GOST 10227, Russian specification: TS-1 major jet-fuel type used in Russian aircraft.
- CGSB-3.22-93: Standards Council of Canada.

There are many individual national specifications, with major similarities between the ASTM and DEF STAN specifications, which are typically based on the US and UK specifications. The group of fuel manufacturing companies has brought together the most restrictive requirements from DEF, STAN 91-91, ASTM, and D 1655 to form a joint checklist. Jet fuels are a specialized type of petroleum-based fuel, and the International Air Transport.

Association has documented specifications for ATF. Efforts are being made to introduce jet fuel from synthesized isoparaffins [16] and hydro-processed fatty acid esters and fatty acids [17,18]. In 2009, ASTM International developed the ASTM standard specification D7566, which caters to ATF made from these renewable synthesized hydrocarbons [19]. These specifications guide semisynthetic jet fuel produced by mixing conventional jet fuel with synthetic blending fractions. The semisynthetic jet fuels with specifications under D7566 contain vital properties and have a chemical makeup and performance similar to that of conventional jet fuels. The specification defined the properties of fuel synthesized from biomass, coal, and natural gas via processes like Fischer-Tropsch synthesis (FTS).

10.1.3 Sources of bio-olefin

The main factor in the successful implementation of bio-jet fuel is feedstock availability at a sustainable scale and reasonable price. The first step in biofuel production is the production of bio-olefins, which can be reaped from any potential carbon-containing renewable feedstock. Four types of bio-olefin feedstocks are discussed in this section: bio-oils, sugars, alcohol, and gas. The cost of aviation fuel derived from biomass relative to fossil-based fuel is one of the main obstacles to its greater market penetration. The cost incurred in procuring feedstock gives an idea of the cost of the final bio-based aviation fuel. The potential feedstocks used for producing bio-jet fuel are classified as solids, liquids such as oils, and gas-based feedstock, depending on the production pathway used.

10.1.3.1 Olefins from bio-oils feedstock

In bio-jet-fuel production, bio-oils are first processed to yield light olefins. These olefins are then processed into biofuel using a series of processes. Raw bio-oil has some undesirable characteristics (unstable, high viscosity, high acidity, and low heating value) due to its high oxygen content, which hinders its direct use as an engine fuel [20,21]. Therefore the upgrading of bio-oil is a necessary step to produce bio-jet fuel. Various processes are used for bio-oil upgrading, as follows:
- catalytic cracking [22,23]
- ketonization [24]
- hydro-deoxygenation [25]
- steam-reforming followed by FTS [26]

The feedstocks considered for the production of bio-oil include waste cooking oil, algal oil (algae possesses exponential growth potential and high oil content), pyrolysis oil (heating biomass in the absence of oxygen that produces pyrolysis gas, biochar, and pyrolysis oil), and plant oil (including rapeseed, canola, soybean, palm oils, and corn oil) [18,27]. These feedstocks can be converted to bio-oil via hydrothermal liquefaction, thermal treatment or pyrolysis, and hydro-processed esters and fatty acids. The oil-derived jet-fuel competes with biodiesel and hydro-processed renewable diesel for the availability of feedstock. The generalized equation for bio-oil to olefins is shown below [28] (La-modified zeolite, atmospheric pressure, 500–650°C).

$$\text{Bio} - \text{oils}\left(C_xH_yO_x\right) \xrightarrow[\text{and deoxygenation}]{\text{catalytic cracking}} C_2H_4 + C_3H_6 + C_4H_8 \qquad (10.1)$$

10.1.3.2 Olefins from sugar

Bioengineering sugarcane could be the most productive way of producing bio-jet fuel [29]. Feedstock for sugar includes sucrose derived from sugarcane or sugar beet, sugar from corn starch, and lignocellulosic sugar derived from the hydrolysis of hemicellulose and cellulose [5,7]. Olefins can be produced by microbial conversion of biomass–derived fermentable sugars, such as in the case of Lee, Niraula, and Trinh (2018) [30]. In their study, they demonstrated direct olefin biosynthesis from glucose in a recombinant *Escherichia coli* (see Eq. 10.2). The recombinant *E. coli* yielded odd-chain terminal alkenes, comprising C_{11}, C_{13}, C_{15}, and C_{17} olefins from glucose [30].

$$\text{Glucose} \xrightarrow[\text{decarboxylase}]{\text{P450 fatty acid}} \text{olefins} \qquad (10.2)$$

Another well-documented pathway is direct fermentation of sugar to short-chain alcohols with yeast followed by the conversion of alcohols to olefins by dehydration reaction [20,29,31] (see Eq. 10.3).

$$\text{Sugars} \xrightarrow{\text{fermentation}} \text{Bio} - \text{alcohol} \xrightarrow{\text{dehydration}} C_2 - C_4 \qquad (10.3)$$

10.1.3.3 Olefins from alcohols

Biomass as a starting material can be processed via many conversion pathways to yield alcohol (e.g., isopropanol, butanol, and ethanol). Biomass feedstock includes fermentable sugars [27,32,33]:

- sugarcane and sugar beet
- hydrolyzed grain starch derived from corn or wheat
- hydrolyzed polysaccharides derived from lignocellulosic biomass
- wood processed via thermochemical conversion.

The produced alcohol undergoes additional processing into olefin. The olefin produced will be oligomerized to yield hydrocarbons with a jet-fuel-range chain length (C_8–C_{16}). This process is termed alcohol-to-jet process.

$$\text{Bio} - \text{alcohol} \xrightarrow{\substack{\text{catalytic} \\ \text{dehydration}}} \text{olefins} \qquad (10.4)$$

Eq. (10.4) depicts the catalytic dehydration reaction of alcohols to yield olefins of corresponding carbon numbers. Such is the case in linalool dehydration to limonene, terpinolene, and 2,6-dimethyloctatrienes [34]. Generally the dehydration reaction is achieved using heating (25–140°C) in the presence of strong acids such as concentrated phosphoric or sulfuric acid temperatures.

10.1.3.4 Olefins from gaseous feedstock

Gas-based feedstock includes biogas and syngas. Biogas is reaped from animal manure, landfills, and wastewater via anaerobic digestion of organic matter. However, synthesis gas is produced from biomass via gasification. The biogas predominantly consists of methane and carbon dioxide. It can be converted to synthesis gas via steam reforming, partial oxidation, dry reforming, and the reverse water gas shift reaction [35,36]:

$$\text{Steam reforming} \quad CH_4 + H_2O \rightarrow CO + H_2 \qquad (10.5)$$

$$\text{Partial oxidation} \quad CH_4 + \frac{1}{2}O \rightarrow CO + H_2 \qquad (10.6)$$

$$\text{Dry reforming} \quad CH_4 + CO_2 \rightarrow CO + H_2 \qquad (10.7)$$

$$\text{Water} - \text{gas shift} \quad CO_2 + H_2 \leftrightarrow CO + H_2O \qquad (10.8)$$

The produced synthesis is then converted to olefins using the Fischer-Tropsch reaction [37]. This process yields a wide spectrum of products, such as olefins, paraffins, and alcohols. Of interest are the olefin and reaction conditions (220–350°C, Fe or Co-based catalyst, 1–20 bar), which can be tailored to the desired product [37–39].

$$\text{Paraffins } (2n + 1)H_2 + nCO \rightarrow C_nH_{2n+2} + nH_2O \tag{10.9}$$

$$\text{Olefins } 2nH_2 + nCO \rightarrow C_nH_{2n} + nH_2O \tag{10.10}$$

$$\text{Alcohols } 2nH_2 + nCO \rightarrow C_nH_{2n+2}O + (n-1)H_2O \tag{10.11}$$

10.1.4 Olefin oligomerization reactions

The previous section provides a cursory outline of the technologies used to convert different biomass into bio-based light olefins. The oligomerization process is then used to convert the bio-olefins to the desired liquid hydrocarbon biofuels. The oligomerization process entails oligomer complexes where oligomers are formed from their monomers with a finite degree of polymerization, typically 3–5 [40]. Olefins comprise all aliphatic hydrocarbons, including both acyclic and cyclic types that possess single or multiple carbon-to-carbon double bonds [41]. The oligomerization reaction can be achieved thermally or with the aid of a catalyst.

Several acid catalysts are active in oligomerization, such as organometallic catalysts, solid phosphoric acid, alumina-silica, sulfonic acid resins, zeolites, and metal-modified zeolite. Lower olefin is the main feedstock used in the oligomerization reaction. Due to the high temperature needed to activate ethylene, propene, and other heavy olefins, C_4 and C_5 olefins are preferred [40,42]. Linear light α-olefins are of significance in most chemical industries, as they are versatile intermediates and building blocks for a variety of downstream processes. These light olefins (mainly C_2–C_6) can be oligomerized to higher molecular weight olefins through a series of catalyzed oligomerization and isomerization reactions. Many different catalytic systems are used for the oligomerization of olefins, and these are discussed in the subsequent section, together with the product spectrum. The product spectrum is made up of mixtures of olefins, paraffins, cycloalkanes, and aromatics. The selectivity in this spectrum may be governed by the reaction conditions and the nature of the catalyst employed.

10.2 Bio-jet-fuel production pathways

This section presents four main routes for producing sustainable aviation fuels. These routes include hydro-processed fatty acid esters and free fatty acid, FTS, hydro-processing of fermented sugars, and alcohol-to-bio-jet fuel. The use of sustainable aviation fuels is presently at a minimum, and the technology used for these needs to be incentivized. Sustainable aviation fuels can also make a significant contribution to alleviating the impact on climate change.

10.2.1 Bio-jet fuel from hydro-processing of lipid feedstock

Hydro-processed esters and fatty acids are a subset of hydro-treated vegetable oil. These are renewable fuels derived from the oil-based feedstock. Hydro-processed fatty acid esters and fatty acids and hydro-treated vegetable oil fuels are not contaminated with aromatics and sulfur, and they have a high cetane number. Oil-based feedstocks are processed using hydrogen to yield green diesel. The product undergoes additional separation to yield bio-jet-fuel. As per the ASTM, the highest blending ratio should be 50% [4].

The hydro-processed fatty acid esters and fatty acids process are used to hydro-treat triglycerides, saturated fatty acids, and unsaturated fatty acids into a plethora of hydrocarbons. In this process, triglycerides react with hydrogen under high pressure (see Fig. 10.1A and B). Hydrocarbon chains are chemically equivalent to those of petroleum-based fuels. The process involves catalytic hydrogenation of unsaturated triglycerides and fatty acids into saturated fatty acids [43].

Since feedstock availability is the principal challenge for hydro-processed fatty acid esters and fatty acids/hydro-treated vegetable oil production, there is ongoing research on new resources for triglycerides. For example, algae with an oil content of 20%–80% [44], water hyacinth [45], and jatropha oil [46] are potential feedstocks. South Africa has a problem with water hyacinth which grows wild in water bodies, just like algae [47]. Therefore biofuel production using water hyacinth has raised interest. Studies done in Egypt using *Eichhornia crassipes* (a type of water hyacinth found in Egypt) indicated that variable lipid compositions (6.79%–10.45%) could be obtained, which upon further processing using transesterification reaction can produce biodiesel (3.22%–6.36%) [48].

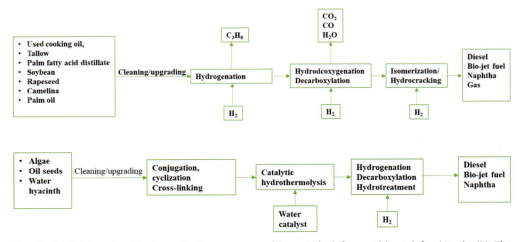

Fig. 10.1 (A) The simplified production process of bio-jet fuel from oil-based feedstock. (B) The catalytic hydrothermolysis process.

The commonly used routes of converting oil-based feedstock to fuel are hydrogenated esters and fatty acids (Fig. 10.1A), as well as the catalytic hydrothermolysis pathway (see Fig. 10.1B). The raw materials for hydro-processed fatty acid esters and fatty acids range from oils derived from vegetables, discarded cooking oil, and fats derived from animals, whereas in the catalytic hydrothermolysis process, oils are derived from algae, water hyacinth, or oil plant.

10.2.2 Bio-jet fuel by Fischer-Tropsch synthesis

Biomass is first processed to produce synthesis gas via gasification. The synthesized gas is then converted into bio-crude via FTS. The bio-crude is further processed and separated to obtain biomass-derived aviation fuel. The maximum allowable blending ratio of bio-jet fuel with conventional jet fuel is 50% by volume. FTS is an industrially proven route of converting CO and H_2 into hydrocarbons and other compounds, as shown in Fig. 10.2. It is one of the six pathways approved by ASTM International to produce sustainable aviation fuels.

Biomass-to-liquid process refers to the process of transforming biomass into fuels via a thermochemical pathway, usually FTS. The biomass-to-liquid process produces hydrocarbons with carbon numbers ranging from C_1 to C_{19+}. These hydrocarbons include n-alkanes, isoalkanes, cycloalkanes, olefins, and aromatics, which are further fractioned, processed, and separated to obtain the desired products, such as jet fuel. Research show that selectivity toward light olefins (C_2–C_5) could be as high as 22.24% [39]. Therefore experimental conditions can be tailored to obtain the desired product distribution. Of interest are the light olefins, which can further undergo oligomerization to produce bio-jet fuel.

Biomass is processed to synthetic gas via gasification (Eq. 10.12) and then into bio-derived fuel.

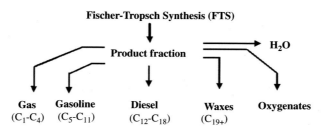

Fig. 10.2 Typical products from FTS. The selectivity toward these products varies with process conditions and the catalyst used [49].

$$CH_xO_{y(biomass)} + O_{2(air)} + H_2O_{(steam)} \rightarrow CH_{4(g)} + CO_{(g)} + CO_{2(g)} + H_{2(g)}$$
$$+ H_2O_{(unreacted\ steam)} + C_{(char,\ tar)} \qquad (10.12)$$

The gasification reaction happens in an oxidizing atmosphere, as given by Eq. (10.12). Products from the gasifier are then introduced to the FT reactor. Eqs. (10.13)–(10.15) indicate the main products during FTS. The flow diagram is given in Fig. 10.3.

$$\text{Paraffins } (2n+1)H_2 + nCO \rightarrow C_nH_{2n+2} + nH_2O \qquad (10.13)$$

$$\text{Olefins } 2nH_2 + nCO \rightarrow C_nH_{2n} + nH_2O \qquad (10.14)$$

$$\text{Alcohols } 2nH_2 + nCO \rightarrow C_nH_{2n+2}O + (n-1)H_2O \qquad (10.15)$$

The iron-based catalysts (FeCuKSiO$_2$) showed a decrease in olefin selectivity from 22.24% to 14.83% for an increase in pressure from 1.85 bar (abs) to 20.85 bar (abs) [39]. Gorimbo et al. [39] reported a drop in the alkene/alkane ratio with an increasing carbon number. Optimum conditions for olefin production are therefore crucial.

10.2.3 Bio-jet fuel from catalytic reforming of sugars

Bio-jet fuels can be derived from sugars via two pathways. The first pathway is the direct sugar to hydrocarbons or the direct fermentation of sugar to jet fuel, which is also referred to as synthetic isoparaffin derived by the fermentation of sugars. The second pathway is referred to as the aqueous-phase reforming process. It produces hydrogen from sugars [51].

The process flow diagram for the direct sugar to hydrocarbons process is shown in Fig. 10.5, which depicts the biological processing of sugars to hydrocarbons, and these are further processed to bio-jet fuels. The direct sugar to hydrocarbons process does not produce an alcohol intermediate (as in the alcohol to jet process) but yields paraffin-type fuels directly from the fermentation of sugars [43]. The technology uses genetic manipulation and screening technologies that allow modified microbes to metabolize sugar directly into the desired products [52]. A detailed conversion pathway of direct sugar to

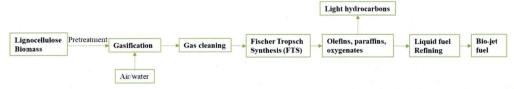

Fig. 10.3 Biomass to bio-jet-fuel production route via FTS. Various technologies are integrated for the commercial production of bio-jet fuel. For instance, the high-temperature Fischer-Tropsch synthesis yielded 74% hydrocarbon biofuel, with the bio-jet-fuel fraction accounting for 16.1 mass% [50]. Around 21.8% of transportation fuel is jet fuel.

hydrocarbons is provided by Davis et al. [53] with the six main steps, as shown below. The direct sugar to hydrocarbons is generally less energy-intensive, due to the low fermentation temperature, with the maximum blending ratio limit of 10% [43].

(i) pretreatment and conditioning
(ii) hydrolysis by enzymes
(iii) hydrolysate clarification
(iv) biological conversion
(v) product purification
(vi) hydro-processing

The aqueous-phase reforming route is shown in Fig. 10.4. This process transforms soluble sugars into chemical intermediates, specifically aldehydes, ketones, acids, furans, alcohols, and some oxygenated hydrocarbons. Additional processing of these intermediates yields jet-fuel range hydrocarbons [55]. The aqueous-phase reforming reaction uses a mild temperature if a catalyst is employed. Platinum and nickel-based catalysts are often used that are supported on metal oxides and carbonaceous materials [51]. The production efficiency and catalyst stability are, however, the main challenges.

The process commences with the feed preconditioning and enzymatic hydrolysis of biomass to get C_5 and C_6 sugars mainly [43]. Acid condensation entails the conversion of ketones, alcohols, aldehydes, and acids to paraffins and isoolefins where a catalyst is used. The process includes the dehydration of oxygenates to olefins, oligomerization of the lower olefins to heavier ones, cracking, cyclization, and aromatization which occur via dehydrogenation of larger olefins. The condensation by base functions transforms ketones, alcohols, and aldehydes to olefins via catalyzed direct condensation. In both the direct sugar to hydrocarbons and aqueous-phase reforming process, the biomass feedstock is initially transformed into solubilized sugars. This conversion process is typically done via biomass pretreatment and enzymatic hydrolysis (see Fig. 10.4).

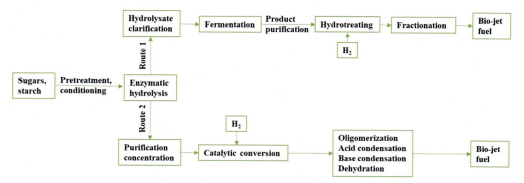

Fig. 10.4 Route 1 shows the direct sugar to hydrocarbon conversion process; route 2 shows the aqueous-phase reforming process [54].

10.2.4 Alcohol-to-bio-jet fuel

Alcohols undergo a series of processes to convert into hydrocarbon biofuels, as depicted in Fig. 10.5. The alcohol-to-jet-fuel process transforms both lower carbon chain alcohols including methanol (CH_3OH), ethanol (C_2H_5OH), butanol (C_4H_9OH) or higher carbon chain alcohols to jet-fuel range paraffins with a chain length of C_{12}/C_{16} [52]. Commercial production often uses ethanol, butanol, or isobutanol as the intermediate to convert alcohol to jet fuel via a sequence of reactions: dehydration followed by oligomerization, hydro-processing, and distillation [56]. The alcohol is produced conventionally from sugars or starch either by fermentation or through other routes, such as industrial microbiology and from algae.

Corporations that are exploring the alcohol-to-jet-fuel route (such as Gevo) have invented a proprietary integrated fermentation technology made up of a yeast biocatalyst that effectively produces isobutanol from sugars. The produced isobutanol is transformed into isoparaffinic kerosene (IPK), which is a bio-jet-fuel blend-stock [52]. Additionally, some processes have been developed to draw out sugars from biomass and further process them into bio-n-butanol, a precursor molecule in the synthesis of a broad array of biofuels and chemicals [52].

The bio-jet fuel from the alcohol production process has four upgrading steps. The first step is dehydration of alcohol to produce alkenes. The second step involves alkene oligomerization, which happens with the aid of a catalyst to yield the middle distillate. The oligomerization process could be catalyzed by either homogeneous or heterogeneous catalysts. In the third step, the middle distillates undergo hydrogenation. The final step involves distillation to separate the jet-fuel range hydrocarbons [57,58].

10.3 Oligomerization of olefins

Oligomerization of olefins involves the catalytic conversion of olefinic monomers into oligomers. An oligomer is made up of repeating units and the number of these units or monomers is finite. Dimers (two monomers), trimers (three monomers), tetramers (three monomers), pentamers, hexamers, and heptamers are examples of oligomers. The process of oligomerization also brings about branching, which gives excellent cold-flow properties in the case of hydrogenated kerosene [59].

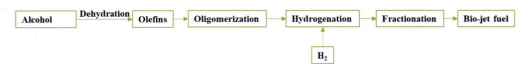

Fig. 10.5 Alcohol-to-jet-fuel production route.

10.3.1 Types of catalysts used in the oligomerization reaction

The catalysts in olefin oligomerization, particularly the light olefins to higher oligomers in the jet-fuel range, continue to be a field of interest for both the industry and academia. Transition metals (both homogeneous and heterogeneous) have shown their ability to be an effective catalyst in the oligomerization of light olefins (see Table 10.2). Oligomerization reactions function to enhance lower olefin streams, which emanate from various pathways mentioned in the above section, into more usable heavier olefins.

The oligomerization reaction is catalyzed by heterogeneous and homogeneous acid-catalysts to yield high-value chemicals [70]. Table 10.2 indicates that C_3 and C_4 olefins are the main feedstocks in oligomerization reactions. Oligomerization of ethylene has been frequently performed by employing homogeneous catalytic systems, namely Ni, Mo, Ti, or Zr complexes, which are identical to polymerization catalysts [60]. The main products are valuable building blocks used in various industrial applications.

10.3.2 Reaction mechanism

The proposed reaction mechanism that describes how olefin reacts to produce oligomers is shown in Fig. 10.6. Condensation of two monomeric olefins $(C_x^=, C_y^=)$ gives rise to a single long chain olefin $C_{x+y}^=$ where subscripts x and y are carbon numbers of a given olefin. For instance, propylene $C_3^=$ feed will give rise to oligomers with multiples of three $C_6^=$, $C_9^=$, $C_{12}^=$, $C_{15}^=$, etc. The resulting olefins may undergo skeletal isomerization.

Fig. 10.6 shows how various products are obtained from the oligomerization of ethylene on a Ni-based catalyst. The oligomerization process is governed by the coordination chemistry on the nickel catalyst sites. These active sites function as both oligomerization initiator sites (as in the case of ethylene) and additional oligomerization sites that yield butenes [71]. The C_4 and C_6 olefins could be utilized via co-oligomerization reactions over acid sites (the mechanism that involves carbenium ions) to yield octenes or higher branched olefins. The oligomers in the carbon range of C_4–C_{10} can participate in further acid catalyzed reactions, which lead to the formation of heavy hydrocarbons. The formation of oligomers follows a geometrical sequence and is controlled by the starting olefin. The most accepted mechanism in the literature is the Ni-hybrid mechanism and, possibly, the metallacycle mechanism [72]. The metallacycle mechanism involves the cycloaddition reaction between a metal alkylidene (usually a transition metal catalyst) complex and the olefin to yield an intermediate metallacycle [73]. The metallacycle then breaks up oppositely to give a new carbene and releases a metathesis product (new olefin) (Fig. 10.7). This kind of olefin metathesis makes use of metal carbene as the initiator. Olefin metathesis transforms the carbon-carbon double bond with the aid of metathesis catalysts into new double-double bonds with different functional groups. Olefin metathesis finds application in industries: (i) Shell's higher

Table 10.2 Catalysts used in the oligomerization reaction.

Catalyst	Olefin feed	Reactor type	Condition	Product selectivity	Reference
SiO_2-Al_2O_3 supported Ni_2 P (xNi_2 P/ASA)	C_2	Parr high-pressure batch reactor	35 bar, 230°C	Heavy oligomers $C_4^=$, $C_6^=$, $C_8^=$, $C_{10+}^=$	[60]
ZSM-5 zeolite	C_3– C_6	High-pressure, fixed tubular bed	30–100-bar, 200–300°C	Distillate-range olefins with a petrochemical-type structure	[61]
C84–3 solid phosphoric acid	C_4	Two packed-bed reactor systems	150°C, 3.8 MPa and a liquid hourly space velocity (LHSV) of $0.7\,h^{-1}$	Gasoline-range product (78%–82%)	[62]
Nickel-substituted synthetic mica-montmorillonite (Ni-SMM)	C_3, C_4	Stainless steel tube reactor	218°F, 600 psig, and 1.0 LHSV	C_{6+}, C_{12}–C_{18} and C_{21+}	[63]
Micro/mesoporous zeotypes (HZSM-5(51)-PZSi and HZSM-5(31)-CoT)	C_4	Fixed-bed reactor	200–250°C, 30–40 bar	C_6–C_{24} (C_6–C_{10} naphtha type) (C_{10}–C_{24} diesel-type products)	[64]
Cp($C_5H_4SiMe_2$tol) ZrMe$_2$] (Cp = C_5H_5; tol = p-C_6H_5Me)	C_3	Glass batch reactor	−20°C to +70°C	C_4 to C_{17}	[65]
$TiCl_4$-$Et_3Al_2Cl_3$, $TiCl_4$-Et_2AlCl, $TiCl_4$-Et_3Al and $Ti(OBu)_4$-$Et_3Al_2Cl_3$	C_2	Autoclave (PAAR)	80 ± 5°C, 200 psig	$C_4^=$, $C_6^=$, $C_8^=$, $C_{10}^=$, $C_{12}^=$, $C_{14}^=$, $C_{16}^=$, $C_{18}^=$ $C_{20+}^=$,	[66]
L1–4/Ni2 + -Mica procatalysts	C_2	Autoclave	50–70°C, 0.4–0.7 bar	C_6–C_{22}	[67]
ZSM-5 zeolite (ammonia form)	C_2– C_4	Fixed-bed continuous flow reactor	atmospheric pressure, 200–450°C	C_1–C_8	[68]
Mesoporous aluminosilicates	C_4	Tubular, fixed-bed reactor	200°C, 30 bar, WHSV of $2.2\,g\,g_{cat}^{-1}\,h^{-1}$	C_6–C_{24}	[69]

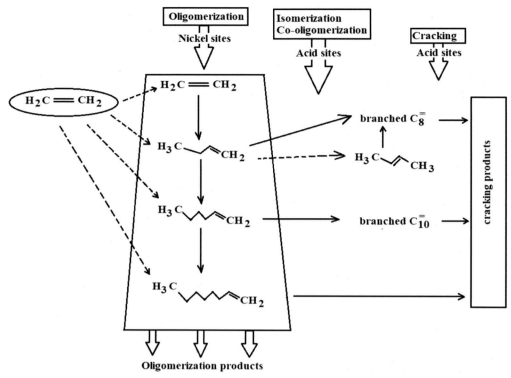

Fig. 10.6 The ethylene oligomerization route using a nickel-based catalyst, with isomerization as the side reaction [71].

olefin process, which converts ethylene to linear alpha olefins [72] and (ii) olefin conversion technology with propylene synthesized from 2-butene and ethylene [75].

The proposed catalytic cycle (Fig. 10.7) shows ethylene being oligomerized to an even number of carbon-containing olefins with the aid of a transition metal catalyst. The active form of the catalyst is first formed from the precatalyst upon its first reaction with the ethylene molecule. The oligomerization of lower olefins (C_2, C_3, and C_4) happens in the presence of metal catalysts. The obtained oligomers are intermediates used in the production of jet fuels.

10.3.3 Reaction conditions with the selectivity to products

In the oligomerization process, the ethylene is selectively oligomerized to an even number of carbon-containing higher olefins with the aid of a catalyst (mostly heterogeneous). Some of the well-known and commercially applied catalysts for olefin oligomerization are shown in Table 10.2, together with their unique optimum conditions. At fairly low pressure and elevated temperature, olefin equilibration is achieved due to the low molecular weight of the products. At elevated pressure, the equilibrium product selectivity

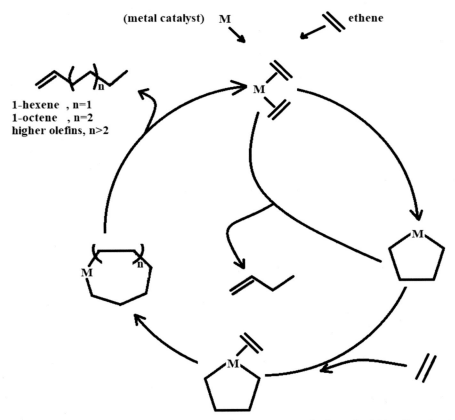

Fig. 10.7 The proposed metallacycle mechanism oligomerization of ethene [74]. Metallacycles appear frequently as reactive intermediates. *(Modified from L. Liu, Z. Liu, C. Ruihua, X. He, B. Liu B, Unraveling the effects of H$_2$, N substituents and secondary ligands on Cr/PNP-catalyzed ethylene selective oligomerization, Organometallics 21 (October) (2018) 3893–900.)*

favors olefins with a higher molecular weight [61]. The elevated partial pressure of ethylene ensures satisfactory reaction rates and increased linearity of the alpha-olefin products [73]. The reaction conditions are optimized to produce olefins in the desired carbon range, which are later converted to jet-fuel grade. Generally alpha olefins possessing an even number of carbon atoms are comparatively much cheaper than those with an odd number of carbons. This is attributed to the ease with which the former is easily produced from ethylene using Shell's higher olefin process [72].

10.4 Aromatization of hydrocarbons from oligomerization

Aromatics are high-energy dense compounds with high octane numbers, making this class of compound essential in jet fuel production. The addition of these aromatics to jet fuels also enhances their elastomer swelling properties [79]. However, aromatics are absent in synthetic aviation fuels except those obtained from the biomass-to-liquid process. So an additional aromatization route must be added. The aromatic components in jet fuel are governed by the specifications of the ATF. Fig. 10.8 shows a pathway for lower octane number olefins undergoing aromatization to yield lighter, higher octane number compounds.

Aromatization refers to the process of converting nonaromatic hydrocarbons to an aromatic ring. This reaction can be catalyzed by the ZSM-5 catalyst loaded with Pt, Ga, or Cr_2O_3 on alumina [77,78,80,81]. Light aromatic hydrocarbons, such as benzene, toluene, and dimethylbenzene, are important feedstocks in most industrial processes to produce organic chemicals. They are also used in fuels to augment the octane number of gasoline [77, 78]. Therefore more effective utilization of the huge amounts of light hydrocarbons contained in biomass is crucial.

Aromatization of light paraffins with the aid of an acidic zeolite catalyst follows a sequence of reactions, such as dehydrogenation reaction, followed by oligomerization, then cyclization and aromatization. Undesirable side reactions, such as cracking and coking, also occur [77]. Jet fuels contain fractions of n-paraffins, isoparaffins, olefins, naphthenes, and aromatics in different proportions. AJF 1 fuel has a significant amount of aromatics with very low paraffin content [82]. Jet fuels contain all groups of compounds. Although a high aromatic content will increase the formation of soot, aromatics are necessary (until a certain level is reached) to avoid leaks in the seals in the fuel system. The aromatic content in jet fuels for engine certification is typically between 15 and 23 vol % [83].

It is important to emphasize that the presence of olefins in jet fuels is undesirable, as these are the most reactive class of hydrocarbons. Several of the alternative jet fuels studied do not have a balanced composition, as additional processes such as aromatization are crucial. The chemical composition of known alternative jet fuels is shown in Table 10.3. The nine jet fuels are labeled as AJF 1-9, and three jet fuels that are available commercially are labeled as CJF 1, 2, and 3.

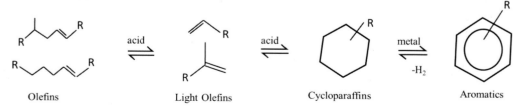

Fig. 10.8 Heavier olefins undergo cracking to yield lighter ones, which then undergo oligomerization to give a significant composition of cyclic products. The reaction is catalyzed by acid and metal catalysts [76–78].

Table 10.3 General composition of jet fuels (wt% of total quantified fuel) [82].

Fuel	n-paraffin	Iso-paraffin	Olefin	Naphthene	Aromatic	Total
AJF1	–	0.2	4.3	34.4	59.4	98.3
AJF2	–	96.4	0.2	1.3	–	97.9
AJF3	44.0	6.9	5.1	32.9	8.8	97.7
AJF4	–	99.8	–	0.2	–	100
AJF5	11.7	87.3	0.1	0.9	–	100
AJF6	4.0	82.9	12.4	0.6	0.1	100
AJF7	19.6	79.9	0.1	0.1	–	99.7
AJF8	9.1	89.4	0.1	0.7	–	99.3
AJF9	12.8	86.9	0.1	0.3	–	100.1
CJF1	28.1	38.8	1.2	15.1	14.4	97.6
CJF2	37.5	42.2	6.6	11.5	2.6	100.4
CJF3	–	81.2	0.3	4.9	13.0	99.4

The naphthalene and aromatic hydrocarbon compositions of ATF determine their burning behavior and the tendency to produce smoke [13]. These compositions and properties are incorporated in ATF specification D 1655. The US Environmental Protection Agency governs aromatic composition, and the California Air Resources Board controls the total polynuclear aromatics and aromatic hydrocarbon content in the diesel fuel used in on-road vehicles [13].

10.5 Economics of bio-jet-fuel production

One challenge with the production of biofuel is finding suitable and sustainable feedstock. The price of bio-based aviation fuel is a function of the nature of the feedstock available, its price, and the process involved. These factors determine market penetration. In Europe, where most biofuel production is done, bio-derived jet-fuel is synthesized from discarded cooking oil. It is reported to be priced between €950 and €1015/ton, whereas the average cost of aviation fuel derived from fossil sources is around €600/ton [4]. This comparison indicates the need to improve the process in order to reduce production costs. Furthermore, the bio-based raw materials that comply with sustainability objectives, for instance, discarded cooking oil utilized in the hydro-processed fatty acid esters and fatty acids procedure, also find use in biofuel production in the road fuel sector.

To determine the cost of production of bio-jet fuel obtained from biomass, the cost of processing feedstock to active ingredients and the conversion of those ingredients to the desired product has to be determined. For instance, in synthesizing bio-jet fuel from alcohols, the cost associated with synthesizing different alcohols needs to be evaluated. The cost of the fuel upgrading processes then has to be considered in order to evaluate the overall alcohol to jet-fuel conversion process and approximate its commercial viability. The economics of different biomass to jet biofuel processes has been examined in the literature [7], but more studies need to be done to make processes economically viable. The general conclusion is that feedstocks of raw materials are the most significant variables in terms of the commercial viability of bio-jet-fuel production.

10.6 Conclusions

Sustainable bio-jet fuel provides an opportunity to address climate change issues. Several conversion technologies explored to convert biomass into fuels have been tried and documented, and a few of them are being applied commercially. In this chapter, a review of the oil, sugar, alcohol, and gas feedstock upgrading processes was discussed. The commercial viability of bio-jet-fuel production is governed by the availability of feedstock. Feedstocks from agricultural waste and forests, as well as algal biomass, are the main raw materials for the production of alcohol fuels and can furnish large amounts of material for converting into jet fuel. Waste cooking oils, plant oils, animal fats, algal oil, and pyrolysis oils are some of the raw materials for oil-related conversion pathways. Forest and agricultural waste can serve as the main feedstock source for syngas or biogas generation. There are plenty of sources of sugars, such as lignocellulosic materials or grain sugar, which can be fermented to synthesize fuels. Both the conversion of technology and feedstock cost contribute a significant percentage of the production cost of the process.

To produce drop-in sustainable jet fuel derived from biomass, the variation in the physicochemical characteristics between biofuel produced and conventional jet fuel must be reduced. Typical biomass to bio-jet-fuel processes has been discussed, with special attention being given to olefin production routes and oligomerization. One advantage of this route is that it has been tested at a commercial level, and the risk associated with scaling-up is therefore reduced. Commercial oligomerization processes have utilized both homogeneous and heterogeneous catalysts. Sustainable bio-jet fuel addresses environmental issues and assists in creating a new industry and employment. However, a significant amount of work is needed to achieve the Paris Agreement goals, and hence collaborations among experts from the aviation and biofuel industry and academics are required. They should continue to work together to determine the best processes for the effective use of biomass feedstock to produce renewable aviation fuels.

Acknowledgments

The authors are grateful for the financial support provided by the University of South Africa (UNISA), Zhijiang College of Zhejiang University of Technology, National Research Foundation (NRF) of South Africa, and the Institute for the Development of Energy for African Sustainability (IDEAS) research unit at UNISA.

References

[1] International Energy Outlook, International Energy Outlook 2019, 2019.

[2] J. Larsson, A. Kamb, J. Nässén, J. Åkerman, Measuring greenhouse gas emissions from international air travel of a country's residents methodological development and application for Sweden, Environ. Impact Assess. Rev. 72 (May 2018) (2018) 137–144, https://doi.org/10.1016/j.eiar.2018.05.013.

[3] International Air Transport Association, Annual Review, 2018.

[4] EASA, European Aviation Environmental Report, 2019.

[5] J. Han, L. Tao, M. Wang, Biotechnology for Biofuels Well-to-wake analysis of ethanol-to-jet and sugar-to-jet pathways, Biotechnol. Biofuels (2017) 1–15.

[6] M. Colket, T. Edwards, S. Williams, N.P. Cernansky, D.L. Miller, F.L. Dryer, et al., Identification of Target Validation Data for Development of Surrogate Jet Fuels, 2008, pp. 1–12. January.

[7] W. Wang, L. Tao, J. Markham, Y. Zhang, E. Tan, L. Batan, et al., Review of Biojet Fuel Conversion Technologies Review of Biojet Fuel Conversion Technologies, 2016.

[8] K. Blakeley, DOD Alternative Fuels: Policy, Initiatives and Legislative Activity, CRS Rep Congr, 2012.

[9] AEO, Anual energy outlook 2020 [Internet]. U.S Energy Information Administraction, 2019, Available from: https://www.eia.gov/outlooks/aeo/.

[10] J. Xu, Z. Yuan, S. Chang, Long-term cost trajectories for biofuels in China projected to 2050, Energy 160 (2018) 452–465, https://doi.org/10.1016/j.energy.2018.06.126.

[11] CAAFI, Commercial Aviation Alternative Fuels Initiative, 2006.

[12] International Air Transport Association, IATA Guidance Material for Sustainable Aviation Fuel Management [Internet], 2015, pp. 1–37. Available from: https://www.iata.org/whatwedo/environment/Documents/IATAGuidanceMaterialforSAF.pdf.

[13] R.A.K. Nadkarni, Guide to ASTM Test Methods for the Analysis of Petroleum, second ed., ASTM International, West Conshohocken, 2007, pp. 28–37.

[14] DSTAN, Ministry of Defence: Defence Standard 91-91 Turbine, Vol. 2015, 2015.

[15] C.Y. Wei, P.Y. Chiu, P.N. Hou, H. Matsuda, G.U. Hung, Aviation fuels: Technical Review, Vol. 42, Chevron Products Company, 2017.

[16] C.J. Chuck, in: G. ACA (Ed.), Biofuels for Aviation: Feedstocks, Technology and Implementation, Elsevier, London, 2016, pp. 295–311.

[17] L. Mendes, D. Souza, P.A.S. Mendes, D.A.G. Aranda, Oleaginous feedstocks for hydro-processed esters and fatty acids (HEFA) biojet production in southeastern Brazil : a multi-criteria decision analysis, Renew. Energy 149 (2020) 1339–1351, https://doi.org/10.1016/j.renene.2019.10.125.

[18] L. Starck, L. Pidol, N. Jeuland, T. Chapus, P. Bogers, J. Bauldreay, Production of hydroprocessed esters and fatty acids (HEFA) – optimisation of process yield, Oil Gas Sci. Technol. 10 (71) (2016) 1–13.

[19] M. Rumizen, Special Airworthiness Information Bulletin: FAA Alternative Fuels Program Staff, Federal Aviation Administration, 2016.

[20] D.M. Fatih, Biorefineries for biofuel upgrading: a critical review, Appl. Energy 86 (Suppl. 1) (2009) S151–S161, https://doi.org/10.1016/j.apenergy.2009.04.043.

[21] P.M. Mortensen, J.D. Grunwaldt, P.A. Jensen, K.G. Knudsen, A.D. Jensen, A review of catalytic upgrading of bio-oil to engine fuels, Appl. Catal. A: Gen. 407 (1–2) (2011) 1–19, https://doi.org/10.1016/j.apcata.2011.08.046.

[22] Z. Zhang, P. Bi, P. Jiang, M. Fan, S. Deng, Q. Zhai, et al., Production of gasoline fraction from bio-oil under atmospheric conditions by an integrated catalytic transformation process, Energy 90 (2015) 1922–1930, https://doi.org/10.1016/j.energy.2015.07.009.

[23] M.F. De Miguel, M.J. Groeneveld, S.R.A. Kersten, C. Geantet, G. Toussaint, N.W.J. Way, et al., Hydrodeoxygenation of pyrolysis oil fractions: process understanding and quality assessment through co-processing in refinery units, Energy Environ. Sci. 4 (3) (2011) 985–997.

[24] C.A. Görtner, J.C. Serrano-Ruiz, D.J. Braden, J.A. Dumesic, Catalytic upgrading of bio-oils by ketonization, ChemSusChem 2 (12) (2009) 1121–1124.

[25] M.V. Bykova, D.Y. Ermakov, V.V. Kaichev, O.A. Bulavchenko, A.A. Saraev, M.Y. Lebedev, et al., Ni-based sol-gel catalysts as promising systems for crude bio-oil upgrading: Guaiacol

hydrodeoxygenation study, Appl. Catal. B Environ. 113–114 (2012) 296–307, https://doi.org/10.1016/j.apcatb.2011.11.051.

[26] Z.X. Wang, T. Dong, L.X. Yuan, T. Kan, X.F. Zhu, Y. Torimoto, et al., Characteristics of bio-oil-syngas and its utilization in Fischer-Tropsch synthesis, Energy Fuels 21 (4) (2007) 2421–2432.

[27] W. Wang, L. Tao, Bio-jet fuel conversion technologies, Renew. Sustain. Energy Rev 53 (2016) 801–822, https://doi.org/10.1016/j.rser.2015.09.016.

[28] F. Gong, Z. Yang, C. Hong, W. Huang, S. Ning, Z. Zhang, et al., Selective conversion of bio-oil to light olefins: controlling catalytic cracking for maximum olefins, Bioresour. Technol. 102 (19) (2011) 9247–9254, https://doi.org/10.1016/j.biortech.2011.07.009.

[29] A. Kang, T.S. Lee, Converting sugars to biofuels: ethanol and beyond, Bioengineering 2 (4) (2015) 184–203.

[30] J.W. Lee, N.P. Niraula, C.T. Trinh, Harnessing a P450 fatty acid decarboxylase from *Macrococcus caseolyticus* for microbial biosynthesis of odd chain terminal alkenes, Metab. Eng. Commun. 7 (July) (2018) 1–9, https://doi.org/10.1016/j.mec.2018.e00076.

[31] V. Zacharopoulou, A.A. Lemonidou, Olefins from biomass intermediates: a review, Catalysts 8 (1) (2018) 2.

[32] C.A. Cardona, Fuel ethanol production : process design trends and integration opportunities, Bioresour. Technol. 98 (2007) 2415–2457.

[33] A.D. Rosenschein, C.H. Road, Energy analysis of ethanol production from sugarcane in Zimbabwe, Biomass Bioenergy I (4) (1992) 241–246.

[34] T.J. Korstanje, J.T.B.H. Jastrzebski, R.J.M. Kleingebbink, Mechanistic insights into the rhenium-catalyzed alcohol-to-olefin dehydration reaction, Chem A: Eur. J. 19 (39) (2013) 13224–13234.

[35] J.M. Lavoie, Review on dry reforming of methane, a potentially more environmentally-friendly approach to the increasing natural gas exploitation, Front Chem. 2 (Nov) (2014) 1–17.

[36] O. Onel, A.M. Niziolek, C.A. Floudas, Optimal production of light olefins from natural gas via the methanol intermediate, Ind. Eng. Chem. Res. 55 (11) (2016) 3043–3063.

[37] J. Gorimbo, X. Lu, X. Liu, D. Hildebrandt, D. Glasser, A long term study of the gas phase of low pressure Fischer-Tropsch products when reducing an iron catalyst with three different reducing gases, Appl. Catal. A: Gen. 534 (2017) 1–11, https://doi.org/10.1016/j.apcata.2017.01.013.

[38] J. Gorimbo, X. Lu, X. Liu, D. Hildebrandt, D. Glasser, A long term study of the gas phase of low pressure Fischer-Tropsch products when reducing an iron catalyst with three different reducing gases, Appl. Catal. A Gen. 534 (2017) 1–11.

[39] J. Gorimbo, A. Muleja, X. Liu, D. Hildebrandt, Fischer–Tropsch synthesis: product distribution, operating conditions, iron catalyst deactivation and catalyst speciation, Int. J. Ind. Chem. 9 (4) (2018) 317–333, https://doi.org/10.1007/s40090-018-0161-4.

[40] O. Muraza, Maximizing Diesel Production through Oligomerization: A Landmark Opportunity for Zeolite Research, 2015.

[41] F.A. Kekul, U.G. Friedrich, Union I, Chemistry A. Review of Basic Organic Chemistry, 1896.

[42] T.J. GricusKofke, R.J. Gorte, A temperature-programmed oligomerization desorption study of ole fin in H-ZSM-5, J Catal. 115 (1989) 233–243.

[43] H. Wei, W. Liu, H. Chen, X. Chen, Renewable bio-jet fuel production for aviation: a review, Fuel 254 (June) (2019).

[44] S. Khan, R. Siddique, W. Sajjad, G. Nabi, K.M. Hayat, P. Duan, et al., Biodiesel production from algae to overcome the energy crisis. Elsevier enhanced reader, Hayati J. Biosci. 24 (24) (2017) 163–167.

[45] M.J. Rani, M. Murugan, P. Subramaniam, E. Subramanian, A study on water hyacinth *Eichhornia crassipes* as oil sorbent, J. Appl. Nat. Sci. (2014), https://doi.org/10.31018/jans.v6i1.389.

[46] I.V. Kanna, A. Devaraj, K. Subramani, Bio diesel production by using Jatropha – the fuel for future, Int. J. Ambient Energy 41 (2018) 289–295, https://doi.org/10.1080/01430750.2018.1456962.

[47] R. Arp, G. Fraser, M. Hill, Quantifying the economic water savings benefit of water hyacinth (*Eichhornia crassipes*) control in the Vaalharts irrigation scheme, Water SA 43 (1) (2017) 58–66.

[48] S.M.M. Shanab, Water hyacinth as non-edible source for biofuel production water hyacinth as non-edible source for biofuel production, Waste Biomass Valoriz. 9 (2) (2017) 255–264.

[49] M. Moyo, Cobalt and Iron Supported on Carbon Spheres Catalysts for Fischer-Tropsch Synthesis, University of the Witwatersrand, 2012.

[50] A. De Klerk, Fischer–Tropsch fuels refinery design, Energy Environ. Sci. 4 (2011) 1177–1205.

[51] I. Coronado, M. Stekrova, M. Reinikainen, P. Simell, L. Lefferts, A review of catalytic aqueous–phase reforming of oxygenated hydrocarbons derived from biorefinery water fractions, Int. J. Hydrog. Energy 41 (2016) 11003–11032, https://doi.org/10.1016/j.ijhydene.2016.05.032.

[52] A. Milbrandt, C. Kinchin, R. Mccormick, A. Milbrandt, C. Kinchin, R. Mccormick, The Feasibility of Producing and Using Biomass-Based Diesel and Jet Fuel in the United States, 2013. December.

[53] R. Davis, M. Biddy, E. Tan, S. Jones, Biological Conversion of Sugars to Hydrocarbons Technology Pathway, 2013.

[54] F.G. Naghdi, L.M. Gonzalez, W. Chan, P.M. Schenk, Progress on lipid extraction from wet algal biomass for biodiesel production, Microb. Biotechnol. 9 (6) (2016) 718–7726.

[55] R. Davis, L. Tao, C. Scarlata, E.C.D. Tan, J. Ross, J. Lukas, et al., Process Design and Economics for the Conversion of Lignocellulosic Biomass to Hydrocarbons: Dilute-Acid and Enzymatic Deconstruction of Biomass to Sugars and Catalytic Conversion of Sugars to Hydrocarbons Process Design and Economics for the Conversion, 2015. March.

[56] D. Chiaramonti, M. Prussi, M. Buffi, D. Tacconi, Sustainable bio kerosene: process routes and industrial demonstration activities in aviation biofuels, Appl. Energy 136 (2014) 1–8, https://doi.org/10.1016/j.apenergy.2014.08.065.

[57] J.C. Serrano-ruiz, E.V. Ramos-fern, S.-E. Antonio, From biodiesel and bioethanol to liquid hydrocarbon fuels : new hydrotreating and advanced microbial technologies, Energy Environ. Sci. 5 (2012) 5638–5652.

[58] V.F. Tret'yakov, Y.I. Makarfi, K.V. Tretyakov, N.A. Frantsuzova, R.M. Talyshinskii, The catalytic conversion of bioethanol to hydrocarbon fuel: biocatalysis, Catal. Ind. 2 (4) (2010) 402–420.

[59] A. De Klerk, Hydroprocessing peculiarities of Fischer-Tropsch syncrude, Catal. Today 130 (2008) 439–445.

[60] M. Shin, Y. Suh, Ethylene oligomerization over SiO_2-Al_2O_3 supported Ni2P catalyst, Chemcatchem Commun. (2020) 135–140.

[61] R.J. Quam, L.A. Green, S.A. Tabak, F.J. Krambeck, Chemistry of olefin oligomerization over ZSM-5 catalyst, Ind. Eng. Chem. Res. 27 (1988) 565–570.

[62] A. De Klerk, D.J. Engelbrecht, H. Boikanyo, Oligomerization of Fischer-Tropsch olefins: effect of feed and operating conditions on hydrogenated motor-gasoline quality, Ind. Eng. Chem. Res. 43 (2004) 7449–7455.

[63] P.G. Berclk, K.J. Metzger, H.E. Swift, Oligomerization of C3-C4 olefins using nickel-aluminosilicate catalyst, Ind. Eng. Chem. Res. 17 (3) (1978) 214–219.

[64] A.F. Silva, A. Fernandes, M.M. Antunes, M.F. Ribeiro, C.M. Silva, A.A. Valente, Ole fi n oligomerisation over nanocrystalline MFI-based micro/mesoporous zeotypes synthesised via bottom–up approaches, Renew. Energy 138 (2019) 820–832, https://doi.org/10.1016/j.renene.2019.02.019.

[65] H. ZCC, H. Me, J. Saßmannshausen, K. Karu, Propene oligomerisation at ambient temperature with [Cp(C5H4 SiMe 2 tol)], Inorg. Chim. Acta 487 (October 2018) (2019) 177–183.

[66] P. Dutta, S.C. Roy, L.N. Nandi, P. Samuel, S.M. Pillai, B.D. Bhat, et al., Synthesis of lower olefins from methanol and subsequent conversion of ethylene to higher olefins via oligomerisation, J. Mol. Catal. A: Chem. 223 (2004) 231–235.

[67] H. Kurokawa, K. Miura, K. Yamamoto, T. Sakuragi, Oligomerization of ethylene to produce linear α-olefins using heterogeneous catalyst prepared by immobilization of α-diiminenickel(II) complex into fluorotetrasilicic mica interlayer, Catalysts 3 (1) (2013) 125–136.

[68] P. Borges, R.R. Pinto, M.A.N.D.A. Lemos, F. Lemos, Light olefin transformation over ZSM-5 zeolites a kinetic model for olefin consumption, Appl. Catal. A Gen. 324 (2007) 20–29.

[69] A.F. Silva, A. Fernandes, M.M. Antunes, P. Neves, S.M. Rocha, M.F. Ribeiro, et al., TUD-1 type aluminosilicate acid catalysts for 1-butene oligomerisation, Fuel 209 (April) (2017) 371–382, https://doi.org/10.1016/j.fuel.2017.08.017.

[70] P. Breuil, L. Magna, H. Olivier-bourbigou, P. Breuil, L. Magna, H.O. Role, Role of homogeneous catalysis in oligomerization of olefins: focus on selected examples based on group 4 to group 10 transition metal complexes to cite this version: HAL id: HAL-01119632, Catal Lett. 145 (2015) 173–192.

[71] A. Finiels, F. Fajula, V. Hulea, Nickel-based solid catalysts for ethylene oligomerization – a review, Catal. Sci. Technol. 4 (2014) 2412–2426.

[72] W. Keim, Oligomerization of ethylene to alpha-olefins: discovery and development of the shell higher olefin process (SHOP), Angew Chem. Int. Ed. 52 (2013) 12492–12496.

[73] B.D. Gupta, A.J. Elias, Basic Organometallic Chemistry, Universities Press, 2010, pp. 315–342.

[74] L. Liu, Z. Liu, C. Ruihua, X. He, B. Liu, Unraveling the effects of H2, N substituents and secondary ligands on Cr/PNP-catalyzed ethylene selective oligomerization, Organometallics 21 (October) (2018) 3893–3900.

[75] J.C. Mol, Industrial applications of olefin metathesis, J. Mol. Catal. A Chem. 213 (2004) 39–45.

[76] F. Zaera, Selectivity in hydrocarbon catalytic reforming: a surface chemistry perspective, Appl. Catal. A Gen. 229 (1–2) (2002) 75–91.

[77] W.J.H. Dehertog, G.F. Fromen, A catalytic route for aromatics production from LPG, Appl. Catal. A: Gen. 189 (February) (1999) 63–75.

[78] Y. Ono, Transformation of lower alkanes into aromatic hydrocarbons over ZSM-5 zeolites, Catal. Rev. Sci. Eng. 34 (3) (1992) 179–226.

[79] M. Selam, W. Ahmed, M. Bohra, H. Mohamed, R. Hussain, N. Elbashir, Role of aromatics in synthetic fuels, Energy Environ. 5339 (2014), https://doi.org/10.5339/qfarc.2014.EESP0487.

[80] T. Inui, F. Okazumi, Propane conversion to aromatic hydrocarbons on Pt/H-ZSM-5 catalysts, J. Catal. 367 (1984) (2000) 366–367.

[81] L.I.U. Ru-ling, Z.H.U. Hua-qing, W.U. Zhi-wei, Q.I.N. Zhang-feng, F. Wei-bin, W. Jian-guo, Aromatization of propane over Ga-modified ZSM-5 catalysts, J. Fuel Chem. Technol. 43 (8) (2015) 961–969, https://doi.org/10.1016/S1872-5813(15)30027-X.

[82] P.A. Pires, Y. Han, J. Kramlich, M. Garcia-Perez, Chemical composition and fuel properties of alternative jet fuels, Bioresources 13 (2) (2018) 2632–2657.

[83] B. Brem, L. Durdina, F. Siegerist, P. Beyerle, K. Bruderer, T. Rindlisbacher, et al., Effects of fuel aromatic content on non-volatile particulate emissions of an in-production aircraft gas turbine, Environ. Sci. Technol. (2015) 2–24.

SECTION 3

Conversion of biomass-derived compounds to hydrocarbon biofuels

CHAPTER 11

Carbon-carbon (C—C) bond forming reactions for the production of hydrocarbon biofuels from biomass-derived compounds

Olusola O. James[a] and Sudip Maity[b]

[a]Chemistry Unit, Faculty of Pure & Applied Sciences, Kwara State University, Malete, Kwara State, Nigeria
[b]Gasification, Catalysis and CTL Research Group, Central Institute of Mining and Fuels Research (Digwadih Campus), Dhanbad, Jharkhand, India

Contents

11.1 Introduction	298
11.2 Lignocellulose to hydrocarbon biofuels	299
11.2.1 Hydrocarbon fuel (drop-in fuel) targets	301
11.3 Upgrading of fermentation fuels to drop-in fuels	301
11.3.1 Dehydration-oligomerization approach	302
11.3.2 Dehydrogenation-aldol condensation approach	304
11.3.3 C—C coupling reactions for upgrading acetone to hydrocarbon biofuels	305
11.4 Upgrading of sugar dehydration intermediates to drop-in fuels	306
11.4.1 Levulinic acid as a platform for hydrocarbon biofuels	308
11.4.2 Furfural as a platform for hydrocarbon biofuels	312
11.4.3 5-hydroxylmethylfurfural (HMF) as a platform for hydrocarbon biofuels	317
11.5 Conclusions	321
References	323

Abbreviations

2–MTHF	2-methyltetrahydrofuran
ABE	acetone–butanol–ethanol
DMF	dimethylformamide
DMTHF	dimethyltetrahydrofuran
FAME	fatty acid methyl ester
GVL	γ–valerolactane
HAP	hydroxyapatite
HMF	hydroxylmethyl furfural
MTBK	methylisobutylketone

Hydrocarbon Biorefinery
https://doi.org/10.1016/B978-0-12-823306-1.00002-9

Copyright © 2022 Elsevier Inc.
All rights reserved.

11.1 Introduction

Although human wants are many, food, shelter, and clothing are generally regarded as the tripartite necessities of life. Activities toward meeting these needs involve expending energy in various forms. In recent times, energy resources for meeting our needs are oftentimes in conflict with the needs we intend to meet [1]. Early human civilizations exploited animal muscles and biomass. Fossil fuels displaced biomass over time being a more portable and abundant energy resources. Extensive exploitation of fossil fuels also midwifed industrialization and mechanization of virtually all human activities [2]. Petroleum-derived liquid fuels are ubiquitous and daily consumption is increasingly driven by population growth, the quest toward higher living standards, infrastructure developments, etc. [3].

However, owing to its being a finite resource, there is increasing fear of a gap in the demand-supply of petroleum in the future [4]. Moreover, global warming had been intricately linked to increasing CO_2 content in the atmosphere [5]. Unchecked CO_2 emission may lead to global warming crossing a threshold beyond which a drastic climate change will be inevitable [6]. The urgency of slowing down global warming is being advocated at several national, regional, and international summits and conferences. Carbon-free energy resource options include solar, wind, geothermal, hydroelectricity, and nuclear. Although constituting a small fraction of the global energy mix, biomass will remain an important component of the global energy resources. Biomass is considered a carbon-neutral energy resource because it is in a close loop with the natural carbon cycle through photosynthesis. Therefore replacing fossil fuels with biomass is expected to slow down CO_2 emission and help stem the rise of global temperature and avoid drastic climate change [7]. However, biomass in its native forms does not fit into established applications of fossil fuels. Thus different biomass processing into fossil fuel mimics had been developed.

Table 11.1 shows the biomass-derived fuels (biofuels) that are mimics of fossil fuels. Processing of biomass to petroleum-derived liquid fuel mimics attracted greater devotion than the solid and gaseous counterparts. Presently, the transportation sector depends on petroleum-derived liquid fuels, and it is desirable to create biomass-derived substitutes [8]. The conventional liquid biofuels are produced from edible or food biomass. This had generated concern about competition with food supply. To prevent competition with food, the use of vegetable oil from nonedible oilseeds is being promoted. Also, bioethanol and hydrocarbon biofuel production from non-food biomass (cellulose and crop residues) is being encouraged. Lignocellulose is the most abundant biomass resource for hydrocarbon biofuel production [9]. Thus this chapter briefly explores the processes involved in obtaining hydrocarbon biofuels from lignocellulose.

Carbon-carbon (C—C) bond forming reactions   299

Table 11.1 Biomass-derived fuels mimics of fossil fuels.

Mimics	Processes or transformations carried out on biomass	Biomass-derived fuels (biofuels)
Coal (solid fuels)	Pyrolysis	Charcoal
	Compaction/densification	Briquettes
Petroleum (liquid fuels)	Sugar extraction, fermentation, and distillation	Bioethanol and biobutanol
	Sugar extraction, dehydration, condensation, and hydrodeoxygenation	Green gasoline, jet fuel, and diesel
	Fast pyrolysis and hydrodeoxygenation	Green gasoline
	Gasification and Fischer-Trospch synthesis	Green gasoline, jet fuel, and diesel
	Oil extraction and transesterification	Biodiesel
	Oil extraction, hydrocracking, and hydrodeoxygenation	Green gasoline and diesel
Natural gas (gaseous fuels)	Bio-digestion	Bio-methane
	Gasification and methanation	Biosyngas and methane

11.2 Lignocellulose to hydrocarbon biofuels

Lignocellulosic biomass is a composite material made up of three polymers (cellulose, hemicellulose, and lignin) and other extractable components. Early predispositions at valorization of lignocellulose can be traced back to the extraction of cellulose from wood for pulp and paper production. For cellulosic ethanol biofuel, the target is not cellulose but fermentable sugars, whereas lignocellulose biomass is only subjected to pretreatment to make the cellulose accessible and susceptible to hydrolysis under mild conditions. The resulting sugars (majorly C_6 and C_5) can then be subjected to fermentation to ethanol or dehydration to platform chemicals or biofuel intermediates [10]. Pretreatment methods were initially aimed at the destruction of lignin, but recent trends in biorefinery aimed at full exploitation of lignocellulose by selective deconstruction and extraction of lignin. This allows for the fractionation of lignocellulose into cellulose, hemicellulose, and lignin and subsequent conversions of the component polymers into niche intermediates and fuels/chemicals (Fig. 11.1) [11].

Extensive work had been carried out on biochemical (hydrolysis and fermentation) and thermochemical (hydrolysis and dehydration) transformations of cellulose and hemicellulose into biofuels or intermediates. Ethanol is the most popular liquid biofuel, yet its current fuel application hinged on being an octane booster. Table 11.3 shows the structure of common intermediates from bio/thermochemical (hydrolysis and dehydration/fermentation) transformation of cellulose and hemicellulose. H/C and C/O ratios are important indices of the energy density of a compound. Although the

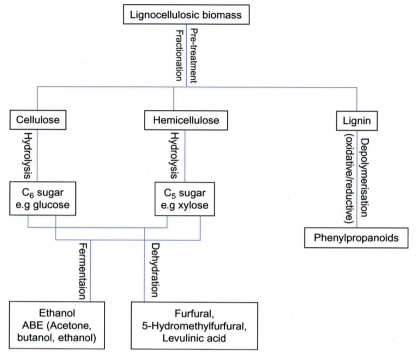

Fig. 11.1 Processes/unit operations involved in lignocellulose-based hydrocarbon biofuels.

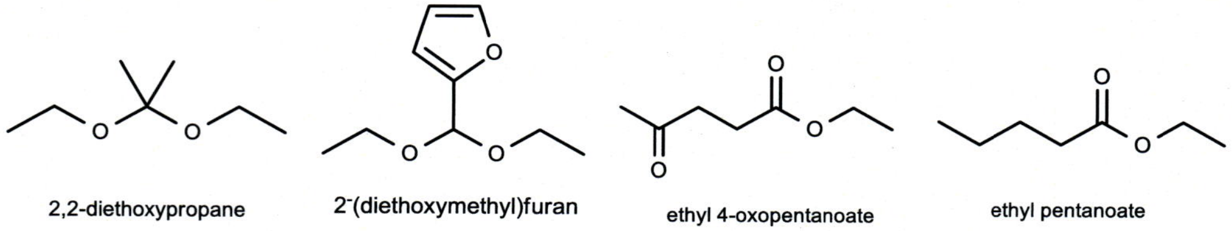

Fig. 11.2 Biomass-derived acetal, ketal, and ester fuel additives.

C/O ratio → ∞ is usually a target, the compound containing one or two oxygen atoms may be desirable to obtain a high octane number, provided oxygen functional groups that create acidic hydrogen are avoided. The preferred oxygen functional groups are ester, ether, acetal, and ketal. Acetone ketals, furfural acetals, and levulinate and valerate esters of ethanol (Fig. 11.2) are octane and cetane number enhancer candidates. Like ethanol, the potential application of the biomass-derived oxygenates seems to be limited to fuel additives. In order to promote increase in the contribution of biofuel to global fuel consumption, it is imperative to convert biomass-derived intermediate into molecules that are similar or identical to molecules present in gasoline, jet fuel, and diesel. This will entail increasing the carbon chain length and reducing the oxygen content. In the next section,

Table 11.2 Common products of bio/thermochemical transformation of cellulose and hemicellulose.

Fermentation biofuels			Sugar dehydration intermediates		
Ethanol	Butanol	Acetone	Levulinic acid	Furfural	5-Hydroxylmethylfurfural

we will discuss hydrodeoxygenation and C—C bond forming reactions for obtaining drop-in fuels from the molecules in Table 11.2.

11.2.1 Hydrocarbon fuel (drop-in fuel) targets

Global biofuel demand is projected to increase, and the proportion of nonconventional biofuels is expected to rise rapidly over the next two decades. Bio-jet fuel is of particular interest to the aviation industry as a commitment to lower its CO_2 emission [12,13]. It is therefore desirable to produce drop-in fuels from biomass-derived biofuels and intermediates, as shown in Table 11.2. Some important properties of drop-in fuels are shown in Table 11.3. Conversions of ethanol or any of the sugar-derived intermediates will require C—C couplings and hydrogenation/deoxygenation. The processes will be discussed in this section, starting with fermentation biofuels and later the dehydration intermediates.

11.3 Upgrading of fermentation fuels to drop-in fuels

Technologies for the production of fermentation biofuels serve as platforms for the initial defunctionalization of the highly functionalized biomass feedstocks prior to their subsequent conversions to hydrocarbon drop-in fuels. Ethanol has the most mature production technology among the fermentation biofuels and especially where there are existing infrastructures, which makes it, in particular, a desirable precursor to other fuels. Ethanol can be upgraded to drop-in fuels via C—C bond forming reactions in two ways: dehydration to ethylene followed by oligomerization and dehydrogenation followed by Aldol condensation. Butanol and acetone (ABE fermentation products),

Table 11.3 Some properties of drop-in fuels.

Fuel	Density ($g\,cm^{-3}$)	Heating value ($kJ\,cm^{-3}$)	Carbon chain length	Boiling point (°C)	Flash point (°C)	Max aromatic content (v%)	Cetane number
Gasoline	0.735	31.8	4–12	40–215	−45	35	−
Jet (A-1)	0.808	34.1	8–16	140–300	>38	25	−
Diesel	0.850	36.2	10–22	170–370	>52	35	>40

although having less mature technology and production volume compared to ethanol, can be upgraded to drop-in fuels via the same C—C bond forming reactions as ethanol. Butanol and ethanol are homologues, and therefore they share upgrading chemistries, while only Aldol condensation is the application for upgrading acetone. Ethanol/butanol dehydration is generally catalyzed by acid, but ethylene/butylene oligomerization to higher olefins can be catalyzed by acids or transition metal complexes, while Aldol condensation of aldehydes and ketones can be acids or bases that are catalyzed.

11.3.1 Dehydration-oligomerization approach

Dehydration of alcohol is a common example of the elimination reaction in many organic chemistry textbooks. The generally agreed mechanisms of ethanol dehydration are shown in Fig. 11.3A. In the unimolecular E1 elimination mechanism, initiation protonation of OH weakens the C—O bond and facilitates its cleavage to generate a carbenium ion intermediate, whereas subsequent elimination of a β-H$^+$ gives ethylene. In the bimolecular mechanism, E2, β-H$^+$ abstraction, and C—O cleavage occur simultaneously over acid-base pairs in a concerted manner. The unimolecular conjugate base mechanism, E1$_{CB}$, followed the reversed order of the steps in the E1 mechanism. Strong and weak Bronsted acid catalysts favor the E1 and E1$_{CB}$ mechanisms, respectively, while moderate acid strength favors the E2 mechanism. Strongly acidic catalyst lowers the activation energy for carbenium ion formation and affords lower dehydration temperature. The mechanism of butanol dehydration is similar to that of ethanol but with a little twist (Fig. 11.3B). The E1 mechanism of butanol dehydration leads to two isomers. The initial primary carbenium ion intermediate readily isomerizes to more stable secondary and tertiary carbenium ions. Subsequently, β-H$^+$ elimination gives the respective isomers and the product distribution is dictated by the order of stability of the carbenium ion intermediate.

Strong liquid acid catalysts are highly corrosive; hence, they are less suitable for industrial alcohol dehydration. Strongly acidic solids are prone to degradation by hydrolysis; moreover, due to the leveling effect, the strongest acidic species in the presence of water is hydroxonium ion. Thus water tolerance is a key criterion in the selection of solid catalysts for alcohol dehydration. Catalyst deactivation due to coke formation is another important factor during alcohol dehydration. The olefin product of the dehydration is prone to acid site side reactions leading to coke formation on the catalyst surface [14].

The resulting alkenes (ethylene or butylene) can be oligomerized to increase the carbon chain lengths. Oligomerization involves one or more consecutive addition reactions. Oligomerizations of olefins are established processes in the refinery and petrochemical

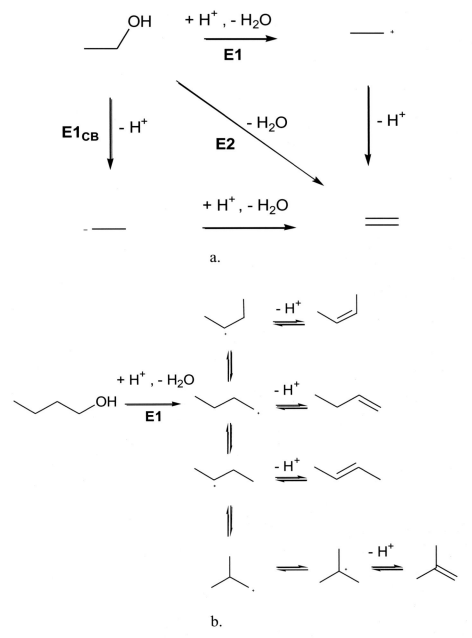

Fig. 11.3 Mechanisms of ethanol and butanol dehydration. (A) Ethanol. (B) Butanol.

industry. Industrial olefins oligomerization processes are usually carried out using acid or transition metal complex catalysts. Since oligomerization is discussed in another chapter in this book, elaborating on it here will amount to repetition, but it is emphasized as a viable C—C bonding forming reaction for the conversion of biomass-derived olefins into drop-in fuels.

11.3.2 Dehydrogenation-aldol condensation approach

Dehydrogenation-Aldol condensation is another approach for achieving C—C coupling of fermentation fuels (alcohols) toward obtaining drop-in hydrocarbon biofuels. The approach is based on Guerbet condensation. A chart of Guerbet condensation of ethanol to higher alcohols is presented in Fig. 11.4.

The reaction proceeds through the dehydrogenation of two ethanol molecules to acetaldehydes, followed by Aldol condensation to a β-hydroxyaldehyde. Dehydration to the

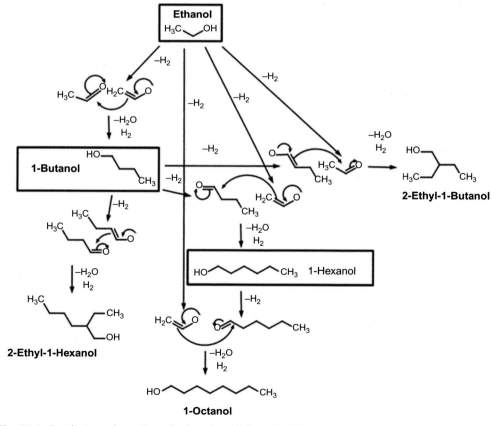

Fig. 11.4 Guerbet condensation of ethanol to higher alcohols.

β-hydroxyaldehyde is followed by full hydrogenation to butanol. Further C—C coupling of butanol-ethanol and butanol-butanol following the dehydrogenation and Aldol condensation steps gives linear and branched hexanol and octanol, respectively [15]. Linear hexanol and octanol are produced only when the nucleophile originates from ethanol. Nucleophiles from butanol or higher alcohols lead to branching at α carbon of the counterpart electrophile. Thus Guerbet condensation is a plausible route for upgrading ethanol to C_{4-8} alcohols in which further C—C coupling via dehydration and oligomerization can lead to diesel or jet drop-in fuels. Although it has been established that after the first dehydrogenation step, the remaining steps are energetically downhill, the cascade of reaction steps with different activation barriers makes the overall process nontrivial.

Guerbet condensations usually take place at temperature $> 300°C$ over basic solid catalysts, such as MgO, calcined hydrotalcites (e.g., MgxAlOy), hydroxyapatite (HAP, $Ca_5(PO_4)_3(OH)$), basic zeolites (e.g., ion-exchanged and impregnated zeolite X). However, at such high temperatures, ethanol is prone to side reactions, such as dehydration, etherification, Lebedev condensation, esterification, ketone–aldehyde isomerization, and decarbonylation. Also, high C_{6-8} alcohol selectivity requires high ethanol conversion (~80%), which is attainable at high temperature in competition with side reactions [15]. Hence recent efforts on Guerbet condensation are focused on selective ethanol to butanol conversion at lower temperatures. The promotion of the basic solid catalysts with dehydrogenation/hydrogenation metals enhances Guerbet condensation [16,17]. Guerbet condensation of ethanol at $<200°C$ had been reported using homogeneous Ru- and Ir-based organometallic catalysts [18]. Overall, the dehydrogenation-Aldol condensation route for the C—C coupling of fermentation fuels is less mature compared to the dehydration-oligomerization route. Because of the low yield of butanol production by ABE fermentation, ethanol-to-butanol via Guerbet coupling seems a more economically viable route to biobutanol.

11.3.3 C—C coupling reactions for upgrading acetone to hydrocarbon biofuels

Acetone is one of the products of ABE fermentation. Acetone possesses a carbonyl group with α-hydrogens that makes it a ready substrate for C—C coupling via self- or cross-Aldol condensation. The self-condensation of acetone will be discussed here, whereas examples of its cross-condensation with other substrates will be examined in the next sections. Self-condensations of acetone are well documented in the literature. Aldol dimerization of acetone forms diacetone alcohol, which readily dehydrates to mesityl oxide (Fig. 11.5). Selective hydrogenation of mesityl oxide gives methyl isobutyl ketone (MIBK) a very important industrial solvent [19]. MIBK may be hydrodeoxygenated to 2-methylpentane, a gasoline-range hydrocarbon biofuel, although it has a less impressive octane rating of 73. The hydrodeoxygenation followed initial hydrogenation of the carbonyl group to alcohol, followed by dehydration creating unsaturation, and subsequent

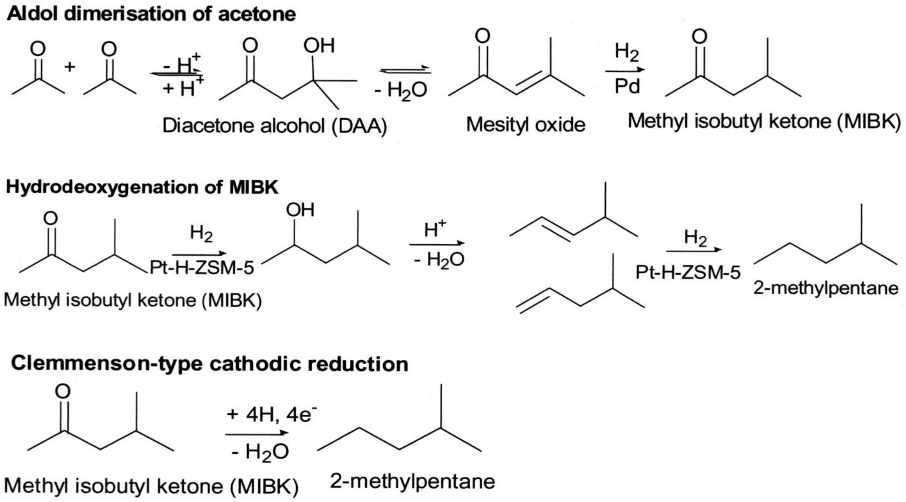

306 Hydrocarbon biorefinery

Fig. 11.5 Self-Aldol condensation of acetone and hydrocarbon biofuel products.

hydrogenation of the unsaturation. It takes place over metal/acid sites of bifunctional catalysts. The deoxygenation can also be accomplished via a Clemmensen–type cathodic reduction. Depending on the catalysts used, the Aldol trimer of acetone can give three C_9 products (Fig. 11.6). Base-catalyzed condensation gives linear and cyclic aliphatic unsaturated ketones while acid-catalyzed condensation gives aromatic hydrocarbon mesitylene [20], which is useful as an octane number enhancer or as an aromatic blending stock. Hydrogenation and hydrodeoxygenation of the unsaturated ketones will yield iso- and cyclononanes [21], which are within the gasoline and jet fuel range.

Because acetone and its self-Aldol dimer and trimers are capable of acting as multiple nucleophiles, the process is prone to many side reactions, such as cyclization, aromatization, and coke formation. Aldol condensation of acetone is usually carried out with partial hydrogenation to reduce the level of unsaturation to minimize the extent of side reactions. However, the ability of acetone to act as multiple nucleophiles is an advantage for cross–Aldol condensation with carbonyl compounds lacking α-hydrogen. Such cross–Aldol condensation will be examined in the next section.

11.4 Upgrading of sugar dehydration intermediates to drop-in fuels

We had discussed C—C couplings of fermentation fuels toward obtaining drop–in fuels in the previous section. Here we will examine C—C coupling reactions for upgrading intermediates derived from dehydration of sugars to drop-in fuels. The focus is on levulinic acid, furfural, and 5-hydroxymethylfurfural. These molecules have a lower C/O ratio compared to the fermentation-derived intermediates. Hence more intensive

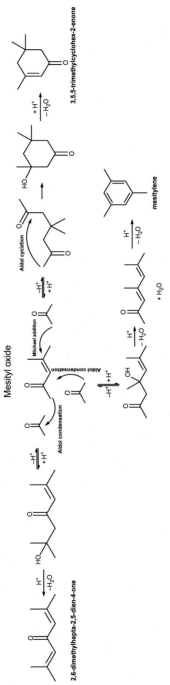

Fig. 11.6 Self-Aldol condensation trimer of acetone.

deoxygenation is required to upgrade to saturated hydrocarbons. Because the intermediates are C5–C6 molecules, the corresponding C5–C6 hydrocarbons are at best limited to gasoline fuel uses. However, the molecules possess functionality that allows interesting C—C coupling reactions for increasing their carbon chain length toward jet fuel and diesel ranges. We will examine the C—C coupling reactions starting with levulinic acid.

11.4.1 Levulinic acid as a platform for hydrocarbon biofuels

Levulinic acid has been the subject of several studies as a platform chemical for a plethora of chemicals and biofuels. With a C/O ratio of 1.67 and an acidic functional group, levulinic acid is unsuitable as a drop-in fuel. Levulinic acid has ketone and carboxylic acid groups, both of which should be fully reduced to obtain hydrocarbon (pentane). Generally carboxylic acids and esters are tougher to reduce compared to ketones or aldehydes. Laboratory reduction of carboxylic acids and esters is usually achieved using aggressive reagents (e.g., $LiAlH_4$) under moisture-free conditions, while ketones/aldehydes can be reduced using a milder reducing agent such as $NaBH_4$. Similarly, catalytic reduction of carboxylic acids and esters are more demanding than that of ketones or aldehydes. This point will be illustrated in the hydrodeoxygenation of levulinic acid in the next section.

11.4.1.1 Hydrodeoxygenation of levulinic acid toward hydrocarbon biofuels

A chart of some reported hydrodeoxygenation of levulinic acid is shown in Fig. 11.7. Hydrodeoxygenation of levulinic acid readily gives valerolactone (GVL). Despite the improved C/O ratio, GVL is not suitable as a drop-in fuel, but it possesses excellent solvent properties that facilitate dissolution and conversion of cellulose and lignocellulose into sugars. Hydrodeoxygenation of levulinic acid to GVL over supported Cu-Ni/SiO_2 takes place at 250°C at ambient hydrogen pressure. The reaction had also been accomplished at low hydrogen pressure (<5 bar) over different supported metal catalysts (metals: Ru, Cu, Pd, Pt, Ni, etc. supports: SiO_2, Al_2O_3, ZrO_2, etc.) and even using a liquid hydrogen carrier, such as formic acid. However, further hydrodeoxygenation of GVL to 2-methyltetrahydrofuran (2-MTHF) does not take place readily at low pressure; it requires high hydrogen pressure ≥15 bar. Direct Hydrodeoxygenation of levulinic acid to 2-MTHF was achieved at very high hydrogen pressure (70 bar).

High-pressure hydrogenolysis of fatty acid methyl esters (FAME) is a common practice in the oleo-chemical industry for the production of fatty alcohol. Production of fatty alcohol had evolved from the use of $CuO/CuCr_2O_4$ at 250–400°C in the early 1930s through Cu-Fe-Al oxide and the currently popular supported Ru-Sn catalysts, which permits FAME hydrogenolysis to fatty alcohol at a much lower temperature. Other bimetallic catalysts made up a platinum-group metal (Pt, Pd, Ir, Rh, and Ru) and any one of Sn, Mo, and Re was found to display high FAME hydrogenolysis performance [22]. Because of their high cost of platinum-group metals, hydrogenation metals (Ni, Co, Fe, and Cu) can be used, albeit at the expense of lower selectivity or activity.

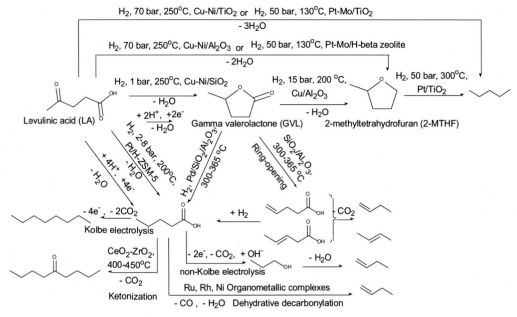

Fig. 11.7 Hydrodeoxygenation routes for levulinic acid.

The active sites for activation of the ester or carboxylic acid group had been attributed to the bimetallic interfaces. It is no surprise that the direct conversion of levulinic acid to 2-MTHF was reportedly achieved over Pt-Mo/H-β-zeolite [23] at milder reaction conditions (130°C and 50 bar) compared to that over Cu-Ni/Al$_2$O$_3$-ZrO$_2$ (250°C and 70 bar) [24]. The GVL-2-MTHF conversion takes place via the 1,4-pentanediol intermediate in which its subsequent dehydration to 2-MTHF was catalyzed by the acid sites of the zeolite. 2-MTHF with an octane rating of 83 qualifies at least as a gasoline blending stock. The noble metal on acid support (Pt/H-ZSM-5) converts levulinic acid to valeric acid. The Pt/H-ZSM-5 could not reduce the carboxylic acid group. This highlighted the importance of the Pt-Mo bimetallic interface for the reduction of the carboxylic group. But TiO$_2$ support is critical for selective hydrogenolysis of furan. Perhaps the use of TiO$_2$ as a support for the bimetallic catalysts (Cu-Ni and Pt-Mo) may afford direct hydrodeoxygenation of levulinic acid to pentane. The Ketone group can also be converted to the methylene group via the same mechanism as the reduction of MIBK to 2-methylpentane, as shown in Fig. 11.6. Similarly, as explained for MIBK, levulinic acid to valeric acid conversion can be accomplished via the cathodic process over a Pb cathode in acid media, but valerolactone is obtained in alkaline media. Anodic oxidative decarboxylation (Kolbe electrolysis) of valeric acid using a Pt anode gives n-butyl radicals, two of which readily couple together to octane. The cathodic reduction of levulinic acid to valeric acid can be coupled with the anodic decarboxylation of valeric acid to octane in a single electro-chemical cell. This provides a one-pot system for achieving deoxygenation and C—C coupling [25]. Although octane has a low octane number, it can be isomerized over suitable catalysts, such as $Pt/Cl^-/Al_2O_3$. Another C—C coupling reaction for valeric acid is ketonization. This is a high-temperature process when two carboxylic acid groups are condensed to form a ketone with the discharge of a carbon dioxide molecule. The high temperature makes the process prone to many side reactions. The reaction is usually carried out over metal oxide catalysts and the highest valeric acid ketonization selectivity reported is over CeO_2-ZrO_2 mixed oxide [26]. GVL had also been reported to undergo ketonization reaction over SiO_2/Al_2O_3 [27]. GVL ketonization did not lead to C—C bond formation; it occurs via the opening of the lactone ring with the introduction of unsaturation on the alkyl chain forming unsaturated acids (pent-3-enoic acid and pent-2-enoic acid). In the presence of hydrogen, the unsaturated acids are hydrogenated to valeric acid. But in the absence of hydrogen, they were found to be more susceptible to decarboxylation than valeric acid. The unsaturated acids were decarboxylated to butene isomers. Valeric acid can also be converted to butane via anodic oxidative decarboxylation in alkaline media or dehydrative decarbonylation using organometallic catalysts [28]. Irrespective of the routes, butene can be used as feedstock to jet or diesel range fuel via oligomerization. Other C—C coupling reactions that can enable jet and diesel fuel ranges will be examined in the next section.

11.4.1.2 C—C coupling of levulinic acid derivatives and hydrodeoxygenation to hydrocarbon biofuels

In addition to Kolbe electrolysis coupling and ketonization shown in Fig. 11.7, there are other C—C coupling options for obtaining carbon chain length in the range of jet and diesel fuels. Levulinic acid and some of its derivatives in Fig. 11.7 have functional groups that can be exploited for C—C coupling reactions. These reactions are classified into two groups: those involving self-condensations (Fig. 11.8) and those involving cross-condensations with methyl isobutyl ketone (MIBK). The cross-condensation options are not limited to MIBK; other ketones or aldehydes can be exploited. MIBK is chosen here for illustration because it is currently a commercial solvent and as discussed in Section 11.3.3, it can be produced from renewable acetone. The ketone group in levulinic acid can be exploited for C—C coupling via Aldol condensation. The acid group of levulinic acid makes base-catalyzed aldol dimerization problematic, but acid-catalyzed dimerization had been reported [29]. Hydrodeoxygenation of the condensed products afforded C_{10} hydrocarbons. Similarly, the self-condensation of GVL had been reported to take place via Claisen condensation [30]. The reaction is similar to Aldol condensation, except that a strong base catalyst is required to generate a nucleophile from GVL (ester in general) than from a ketone. Because of the presence of lactone and furan rings in the condensation product, $Pt-MoO_x/TiO_2$ may be recommended for its selective

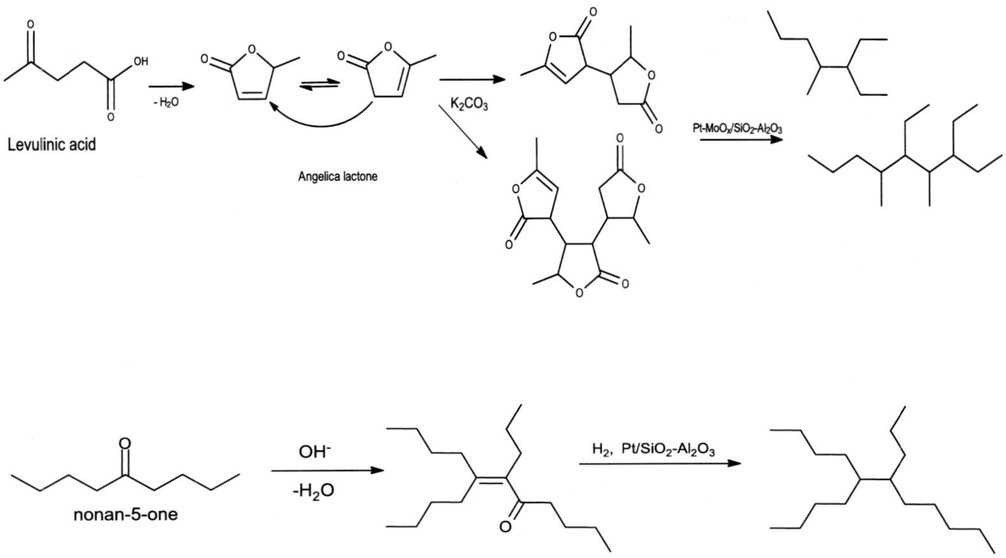

Carbon-carbon (C—C) bond forming reactions   311

Fig. 11.8 Self-condensation of levulinic acid derivatives and their hydrodeoxygenations.

hydrodeoxygenation to C_{10} hydrocarbon. Angelica lactone, a GVL replica but with a double bond, was somewhat unnotican ed. during early investigations on levulinic acid but came to the limelight during elucidation of the mechanism of hydrogenation of levulinic acid to GVL, where it was identified as an intermediate for the alternate path for levulinic acid–GVL [31]. It has attracted attention since, especially for C—C couplings to obtain jet and diesel fuel range. Angelica lactone exists as two isomers, one in an α,β-unsaturated carbonyl form and the other in a nonconjugated form. The nonconjugated form readily transforms into a nucleophile in the presence of a weak base while the α,β-unsaturated carbonyl form acts as an electrophile for Michael addition of the two isomers [32,33]. Thus unlike GVL, which requires a strong base and moisture-free conditions because of self-condensation, angelica lactone can undergo multiple self-condensations at mild conditions. Self-condensation of angelica lactone can give gasoline, jet, and diesel range carbon chains [33]. To prevent excessive aromatization, microporous and strongly acidic supports should be avoided to achieve selective hydrodeoxygenation to jet and diesel range hydrocarbon biofuels. Although 5-nonanone (ketonization product of valeric acid) can be readily hydrodeoxygenated to nonane [34] and then can be used as gasoline fuel or blending stock, self-condensation of 5-nonanone gives the C_{18} product, which can be hydrodeoxygenated to hydrocarbon useful as jet and diesel fuels.

Cross-condensations between levulinic acid and its derivatives are possible. Moreover, several cross-condensations between levulinic acid or its derivatives and other biomass-derived ketones and aldehydes are equally possible. A few examples are shown

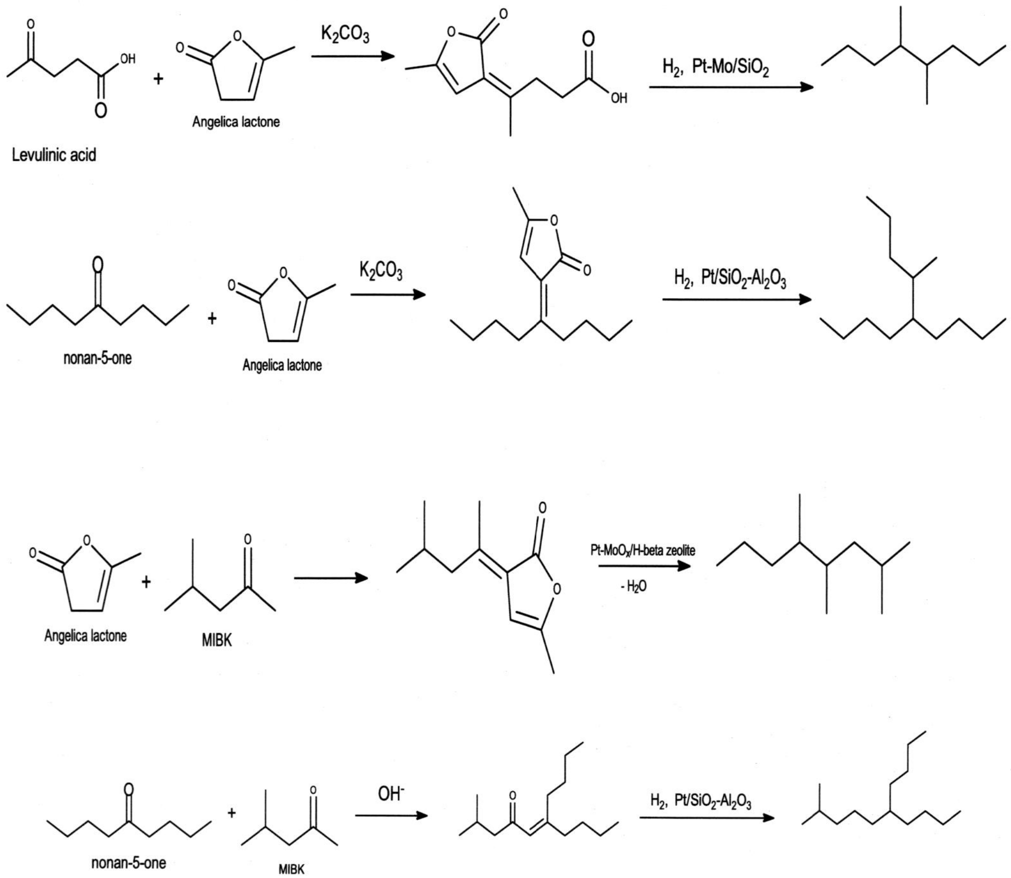

Fig. 11.9 Cross-condensation of levulinic acid and its derivatives with methyl isobutyl ketone (MIBK).

in Fig. 11.9. Depending on the fuel properties of target, there may be a number of self- or cross-condensation options to select from. Selective hydrodeoxygenation of the condensed product(s) is the ultimate measure of the success of the quest to target biofuels.

11.4.2 Furfural as a platform for hydrocarbon biofuels

Furfural and levulinic acid are five-carbon compounds. Pentane is the terminal saturated hydrocarbon obtainable from both. Furfural has higher C/O but lower H/C ratios compared to levulinic acid. This does not make hydrogen consumption for transforming furfural to pentane any lesser than that of levulinic acid. Compared to levulinic acid, absence of carboxylic acid group in furfural facilitates its hydrodeoxygenation under milder conditions. Moreover, hydrodeoxygenation of furfural leads to a greater number of

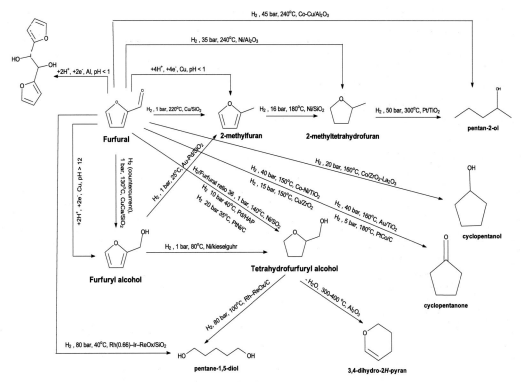

Fig. 11.10 Hydrodeoxygenation routes for furfural.

intermediate products than that of levulinic acid. A chart of hydrodeoxygenation routes for furfural is presented in Fig. 11.10.

11.4.2.1 Hydrodeoxygenation of furfural toward hydrocarbon biofuels

Fig. 11.10 is the extract from various literature on catalytic and cathodic reduction of furfural. The aldehyde group of furfural is readily reduced over supported copper catalysts. At ambient hydrogen pressure, the aldehyde group is converted to alcohol over copper supported on basic support materials at 130°C [35]. Hydrogenolysis of the resulting furfuryl alcohol to 2-methylfuran at ambient temperature and hydrogen pressure was recently reported over the Au-Pd/SiO$_2$ catalyst [36]. The synergy between weak acid sites on SiO$_2$ and the Cu nanoparticles had been shown to favor selective furfural to 2-methylfuran over the Cu/SiO$_2$ catalysts at ambient hydrogen pressure at a higher temperature (>200°C).

Cathodic reduction of furfural to 2-methylfuran and furfuryl alcohol is also possible over a copper cathode at acidic and alkaline pH, respectively. The use of an aluminum instead of a copper cathode at acidic pH will afford a furfural dimer pinacol product [37].

Activation of the aromatic of furfural takes place readily over supported nickel catalysts. Conversion of furfuryl alcohol to tetrahydrofurfuryl alcohol at ambient hydrogen pressure and 80°C was reported to take place over Ni/kieselguhr [38]. However, at a slightly higher temperature (140°C), the silica-supported nickel catalyst had been found to be selective for direct furfural to tetrahydrofurfuryl alcohol [39]. This conversion had also been reported at ambient temperatures on the Pd/HAP [40] and PtNi catalyst but at higher hydrogen pressures [41]. 1,5-pentanediol, a valuable chemical for the polymer industry, can be obtained from tetrahydrofurfuryl alcohol or directly from furfural at very high hydrogen pressure [42]. However, in the absence of hydrogen, tetrahydrofurfuryl can be dehydrated to 3,4-dihydropyran [43], a potential gasoline blending feedstock.

Furfural to 2-methylfuran conversion is possible at ambient hydrogen pressure over a Cu/SiO_2 catalyst. Subsequently, hydrogenation of 2-methylfuran to 2-methyltetrahydrofuran can be carried out at 180°C and 16 bar hydrogen pressure over Ni/SiO_2. Direct hydrodeoxygenation of furfural at 210°C over $Cu-Ni/Al_2O_3$ had been reported to give a mixture of 2-methylfuran and 2-methyltetrahydrofuran [44]. The 2-methylfuran selectivity is higher than that of 2-methyltetrahydrofuran. Perhaps tandem conversions using separate Cu/SiO_2 and Ni/SiO_2 beds may afford higher 2-methyltetrahydrofuran yield at ambient hydrogen pressure. However, it takes drastically higher hydrogen pressure (35 bar) to achieve direct furfural to 2-methyltetrahydrofuran conversion at about the same temperature. Even higher hydrogen pressure (45 bar) was needed to achieve direct furfural to pentan-2-ol conversion [45]. This alcohol may be used directly as fuel or as feedstocks to other fuels via processes discussed earlier for butanol. Another important hydrodeoxygenation product of furfural is cyclopentanone. A fascination about this reaction is that it must be carried out in an aqueous medium; hence, the solution from furfural production can be used as feed for cyclopentanone production without the energy-demanding distillation step. Furfural to cyclopentanone conversion takes place at a high hydrogen pressure (20–40 bar) over supported base metal (Co-Ni, Cu, Au) catalysts at 150–160°C [46–49]. The aqueous medium requirement makes leaching-induced catalyst deactivation a problem. Cyclopentanone is a valuable chemical currently used in the synthesis of pharmaceuticals, fungicides, rubber chemicals, flavors, and fragrances. It is also used in the synthesis of polyamides via δ-valerolactam. Cyclopentanone, being a cyclic ketone, can undergo C—C coupling via Aldol condensation. This and other C—C coupling reactions of furfural and its derivatives will be examined in the next section.

11.4.2.2 C—C coupling of furfural derivatives and their hydrodeoxygenation to hydrocarbon biofuels

While the cathodic pinacol coupling of furfural is already depicted in Fig. 11.10, Pinacol coupling and other condensation of furfural and some of its derivatives in Fig. 11.10 will be discussed here. First, we look at self-condensation reactions in Fig. 11.11; then we will

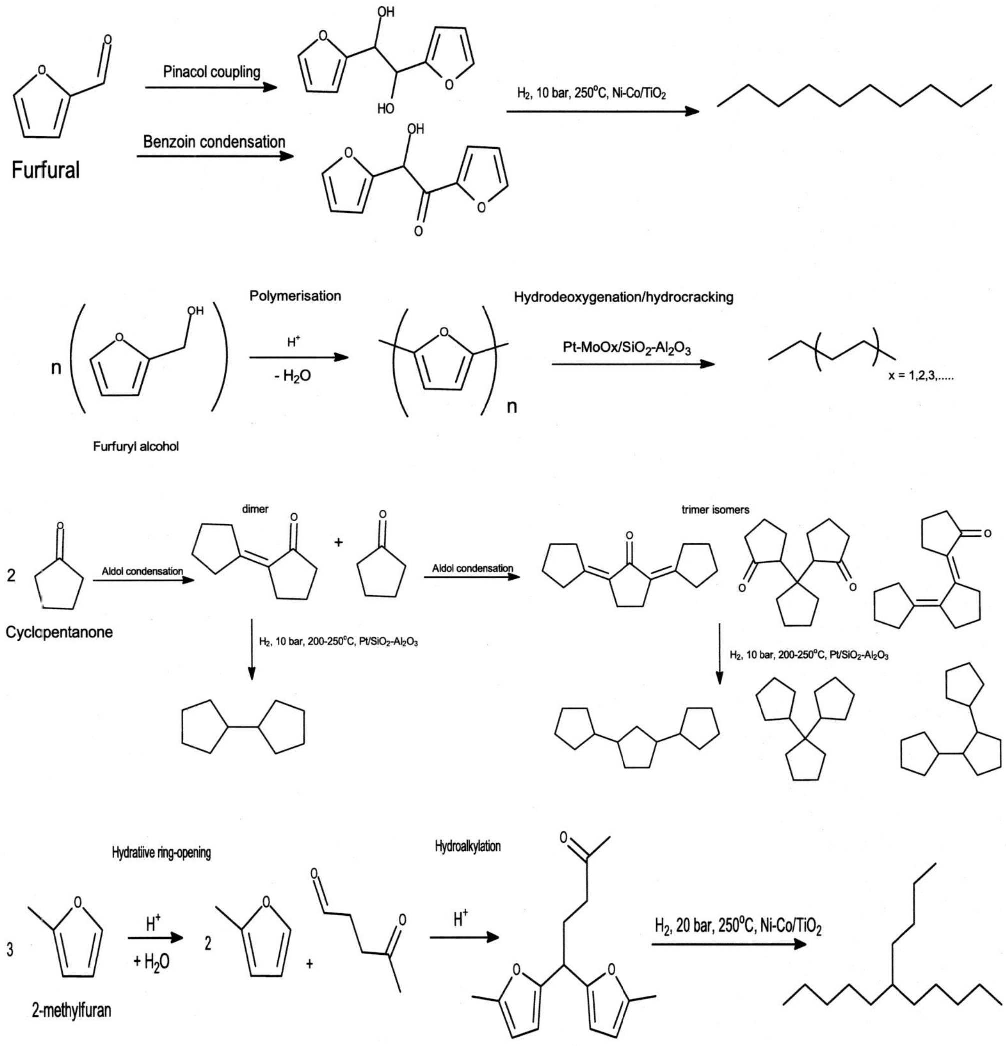

Carbon-carbon (C—C) bond forming reactions 315

Fig. 11.11 Self-condensation of furfural derivatives and their hydrodeoxygenation.

examine some cross–condensation C—C bond forming reactions. Pinacol coupling is a reductive dimerization of aldehydes and ketones via C—C bond formation. The reaction requires stoichiometric metallic reagents, low valent metal ion (Ti^{3+}, V^{3+} etc.) as a catalyst, and a metallic coreductant (e.g., Mg, Mn, Al, etc.). The reaction is usually difficult to achieve in aqueous media, because the low valent cation catalysts are often unstable in an aqueous environment. But VCl_3 catalyzed Pinacol coupling with Al as coreductant had been reported by Xu and Hirao [50]. This is in tandem with reported selective Pinacol coupling of furfural over an Al cathode. Similar dimerization of furfural had been achieved via benzoin condensation [51]. Here the dimerization takes place via the umpolung mechanism. It can be catalyzed by cyanide or thiazolium ions. Selective hydrodeoxygenation of the resulting furfural dimers (Pinacol product or furoin) gives decane, which is a gasoline or jet range fuel. As explained earlier, hydrogenation of the furan ring readily takes place over the supported nickel catalyst at ambient hydrogen pressure. But the ring-opening of the furan ring requires high hydrogen pressure and temperature and is also facilitated by TiO_2 or ZrO_2 support. This informs the prescribed catalyst and reaction conditions for the hydrodeoxygenation. The polymerization of furfuryl alcohol is a long-standing problem for its storage. The reaction is catalyzed by acid. This problem can be exploited for fuel production. Passing aqueous furfuryl alcohol solution containing the catalytic amount of H_2SO_4 over a bed of glass beads is reported to give oligomers [52]. Saturation of the furan rings in the oligomer can be achieved over the supported nickel catalyst at ambient hydrogen pressure and 100°C. The oligomer may be converted into saturated hydrocarbon via cascade hydrogenation, hydrogenolysis, and hydrocracking. Depending on the catalyst used, some cyclization and aromatization may be beneficial for meeting the targeted fuel specification. Cyclopentanone is a cyclic analogue of acetone; it can undergo similar dimerization and trimerization like acetone via Aldol condensation [53]. Cyclopentanone dimer and trimer will give C_{10} and C_{15} intermediates which are precursors to cycloalkane jet and diesel range fuels after full hydrodeoxygenation. In the absence of the furan ring, Pt/SiO_2-Al_2O_3 may enable hydrodeoxygenation with minimal chain cracking. With an octane number of 74, 2-methylfuran may qualify as gasoline fuel or blending stock. Three 2-methylfuran molecules can self-condense into a C_{15} intermediate. The condensation is acid-catalyzed via an initial hydrative ring-opening of 2-methylfuran. Then the condensation takes place via hydroalkylation [54]. The aldehyde group of 4-oxopentanal is linked to 2-methylfuran via C—C bond formation. Ni-Co/TiO_2 could enable selective hydrodeoxygenation of the hydroalkylation product into branched hydrocarbon biofuels that will be suitable for use in jet and diesel engines.

Fig. 11.12 shows some C—C couplings between furfural and its derivatives. Hydroalkylation of 2-methylfuran and furfural offers another route for C—C coupling to a C_{15} intermediate. The prospective alkane product is identical to that discussed earlier for self-condensation of 2-methylfuran. Cross-Aldol condensation between furfural and acetone had been reported [55]. The cyclopentanone analogue is shown in Fig. 11.12. Like acetone, a cyclopentanone molecule can form C—C bonds with one or two furfural molecules that give C_{10} and C_{15} intermediates, respectively [53,56]. The long carbon molecule may be imagined in hydroalkylation of 2-methylfuran with 5-nanonone. Friedel-Craft alkylation of mesitylene with furfuryl alcohol has been reported as another strategy to obtain diesel range biofuel [57]. The resulting mesitylmethylfurfural has to be hydrodeoxygenated before it can be useful as diesel fuel. But the use of tetrahydrofurfuryl alcohol instead of furfural will afford a one-step reaction to mesitylenetetrahydrofuran, an oxygenate that meets the C/O ratio

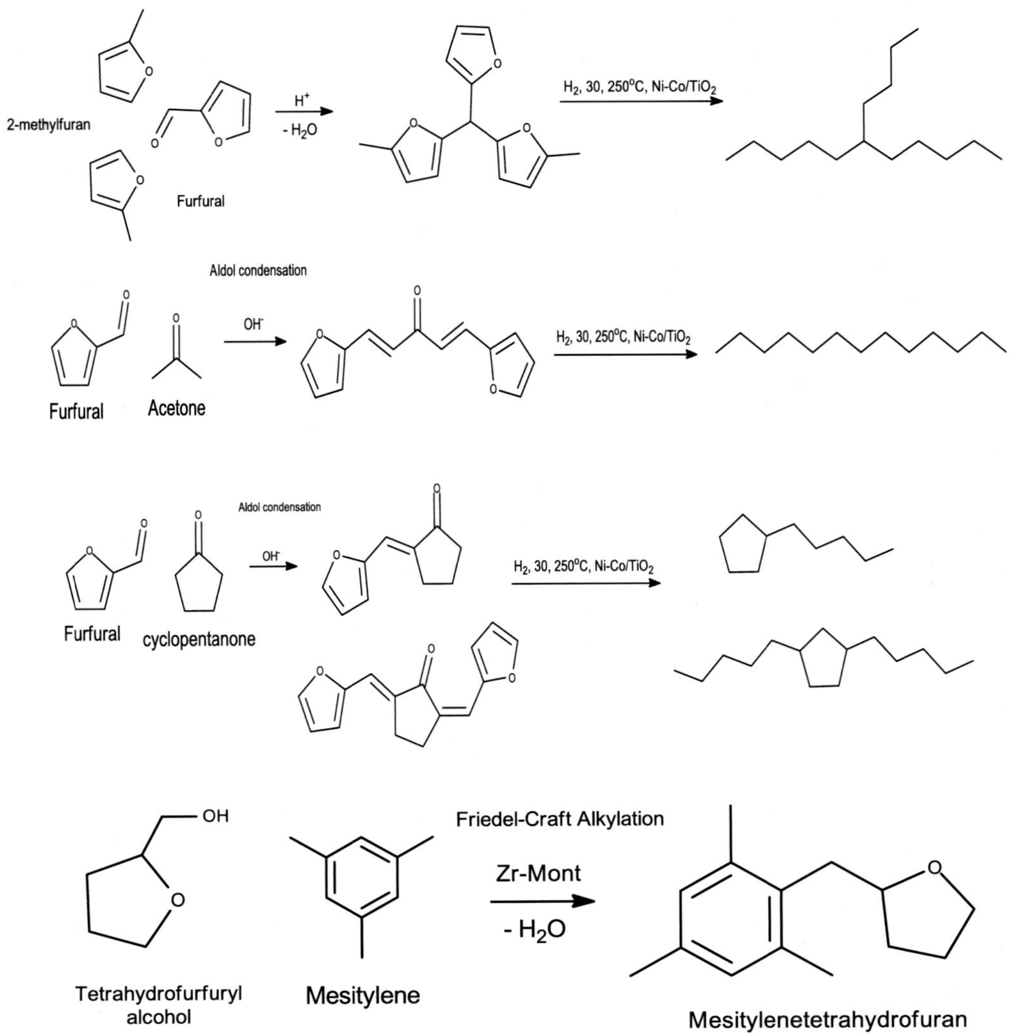

Fig. 11.12 Cross-condensation of furfural derivatives and their hydrodeoxygenation.

requirement for jet fuel application. Several pairs and C—C coupling reactions can be selected to obtain carbon chain length in the range of jet and diesel fuel.

11.4.3 5-Hydroxylmethylfurfural (HMF) as a platform for hydrocarbon biofuels

Because of its potential transformation into many valuable chemicals and fuels, 5-hydromethylfurfural (HMF) has been described as a sleeping giant. The veracity of this

318 Hydrocarbon biorefinery

description notwithstanding, the fundamental chemistry of HMF is not unique. It shares a lot of similarities with furfural. HMF has a higher H/C but a lower C/O ratio compared to furfural. Hydrogen consumption per molecule for hydrodeoxygenation to saturated hydrocarbon will be higher for HMF compared to furfural. Thus higher hydrogen pressure may be required for HMF for similar transformations as for furfural.

11.4.3.1 Hydrodeoxygenation of 5-hydroxymethylfurfural toward hydrocarbon biofuels

Literature extracts on HMF catalytic and cathodic reduction are presented in Fig. 11.13. As discussed in Section 11.4.2.1, furanic aldehyde is readily reduced to furfuryl alcohol, and the hydrogenolysis of hydroxyl groups of the furfuryl alcohol to methyl group takes place readily over the supported copper catalyst at ambient hydrogen pressure. Here for HMF, these transformations over copper catalysts take place at higher hydrogen pressure [58–60]. Precious metal catalysts (e.g., Ru and Pd) allow the same transformations at lower temperatures. They usually use promoters to enhance the performance of copper catalysts. Similarly, as for furfural, a cathodic reductive of HMF is selective to DHMF at neutral pH and selective to DMF at acidic pH [37]. Selectivity toward the dimerization of HMF via Pinacol coupling is very low. Over a Zn cathode, after DMF formation, hydrative ring-opening takes place forming hexane-2,5-dione [61]. However, hexane-2,5-dione is very reactive in aqueous media and prone to side reactions. Methyl isobutyl ketone (MIBK) is efficient in the extraction of hexane-2,5-dione from the aqueous phase. So a biphasic system can make the cathodic process viable for the production of hexane-2,5-dione. HMF to 3-methylcyclopentanone conversion is analogous to furfural to cyclopentanone conversion. This transformation takes place over supported Cu, Co, and Au as similarly reported for furfural [62,63]. Using methanol as a hydrogen source, DMF, DMTHF, and hexan-2-ol were reportedly obtained over a Cu/hydrotalcite catalyst under the supercritical condition at 300°C [64].

As discussed for furfural, hydrogenation of furanic rings can take place on nickel nanoparticles. We expect to see nickel catalyst catalyze DHMF to DHMTHF (2,5-dihydroxymethylfurfural) and direct HMF to DHMTHF. But there appears to be a gap in the literature in this respect. However, the use of the supported palladium catalysts had been reported for direct HMF to DHMTHF conversion at low temperatures and high hydrogen pressure. Direct HMF to DMTHF (2,5-dimethyltetrahydrofuran) conversion takes place over similar catalysts and reaction conditions as furfural to 2-methyltetrahydrafuran conversion [65]. Ring-opening and hydrogenolysis of DHMTHF to 1,6-hexanediol has been reported under the same reaction conditions and using a similar catalyst as its furfural analogue. $Pt-WO_x/TiO_2$ has been demonstrated as an alternative catalyst for DHMTHF to 1,6-hexanediol conversion [66]. In contrast to the situation in furfural, where direct furfural to 1,5-pentanediol conversion was achieved over one catalyst ($Rh(0.66)–Ir–ReOx/SiO_2$), direct HMF to 1,6-hexanediol

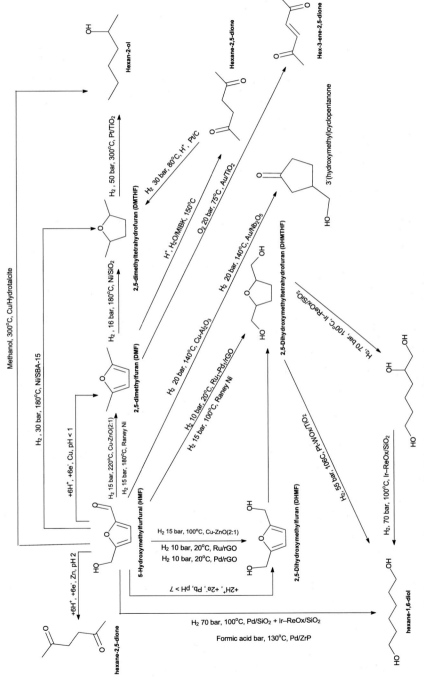

Fig. 11.13 Hydrodeoxygenation routes for 5-hydroxymethylfurfural.

conversion was only achieved using two catalyst beds (Pd/SiO$_2$ + Ir–ReOx/SiO$_2$). The first catalyst bed (Pd/SiO$_2$) converts HMF to DHMTHF while subsequent DHMTHF to 1,6-hexanediol conversion takes place on the second catalyst bed (Ir–ReOx/SiO$_2$) [67]. Obviously, the hydrodeoxygenation of furfural and HMF has many similarities and a few differences. This is no surprise because furfural and HMF are structurally similar. Note sharing on conversion processes on the two molecules will be beneficial for accelerated progress on their valorization efforts. Next, we examine C—C coupling reactions of HMF and its derivatives.

11.4.3.2 C—C coupling of 5-hydroxymethylfurfural derivatives and hydrodeoxygenation to hydrocarbon biofuels

Although awareness of DMF gained attention since 2007 as a potential gasoline fuel and having better fuel properties than ethanol [68], little progress has been made in its commercial production and adoption. Despite its well-published higher energy density and C/O ratio compared to ethanol, the improvement on the H/C ratio of DMF relative to HMF is only marginal. Improvement in the H/C ratio is more significant in DMTHF than in DMF. Hence DMTHF is tipped as a biofuel with energy density comparable to that of petroleum-derived gasoline. Other HMF derivatives that can be processed into gasoline fuels are 3-(hydroxymethyl)cyclopentanone and hexane-2,5-dione (Fig. 11.14). With methylcyclopentane having an octane rating of 80, it can give motivation to explore hydrogenolysis of 3-(hydroxymethyl)cyclopentanone to methylcyclopentane. Hexane-2,5-dione is very reactive, as its cyclization in the presence of a weak base, such as K$_2$CO$_3$, and hydrogenolysis can give a C$_6$ mixture that is useful as gasoline fuel.

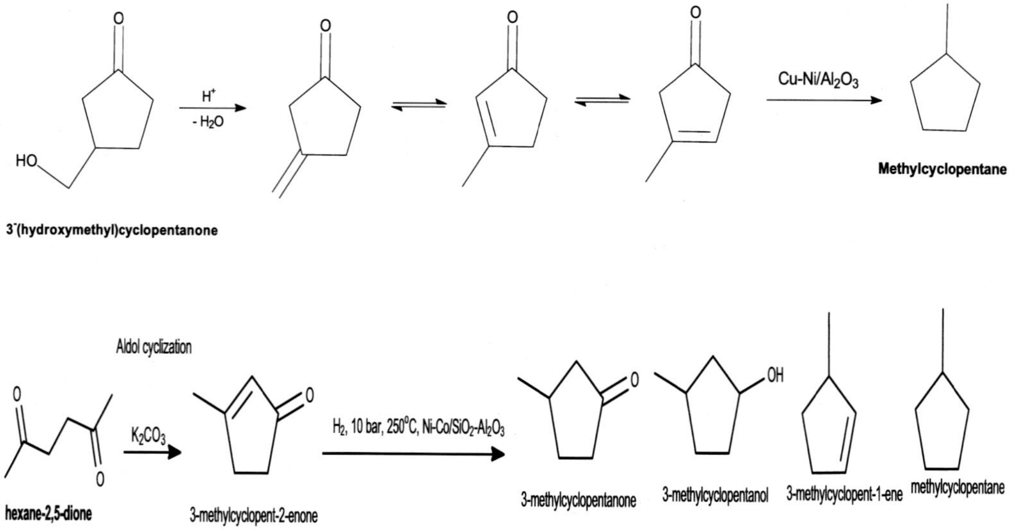

Fig. 11.14 Methylcyclopentane from HMF derivatives.

With DMF, DMTHF, and methylcyclopentane as gasoline fuel candidates, C—C coupling of HMF, its derivatives, and suitable molecules are required to obtain jet and diesel range fuel candidates. Cathodic pinacol coupling of HMF is less efficient compared to that of furfural. However, HMF dimerization via benzoin condensation had been demonstrated alongside furfural. Hydrogenolysis of the furoin product gives dodecane [51]. Methylcyclopentane offers the opportunity to C_{12} and C_{18} intermediates via Aldol condensation. Selective hydrodeoxygenation of the intermediates can give jet and diesel fuel cycloalkanes. Hydrogenation metal(s) on weakly acidic support materials is recommended as prospective hydrodeoxygenation catalysts. Hexane-2,5-dione is an acetone dimer. In addition to Aldolcyclization, Aldol oligomerization of hexane-2,5-dione and its cyclized derivative is readily feasible. The proposed path to dimer isomers is shown in Fig. 11.15. It should be noted that like acetone, hexane-2,5-dione is prone to runaway Aldol condensation leading to an extensive carbon chain network. Hence tandem Aldol condensation of hexane-2,5-dione and partial hydrogenation should be carried out to minimize uncontrollable chain growth. Due to the absence of alpha–hydrogen atom bonded to the carbonyl carbon, HMF cannot undergo self-Aldol condensation. But it can undergo cross-condensation with acetone, hexane-2,5-dione, cyclopentanone, levulinic acid, and 3-(hydrooxymethyl)cyclopentanone to give C_{11}–C_{18} intermediates. Selective hydrodeoxygenation of the intermediates will give hydrocarbons that are useful as jet and diesel fuels. While HMF cannot undergo self-hydroalkylation, it can undergo cross-hydroalkylation with 2-methylfuran to give a C_{16} product. Also, it can be linked with mesitylene via Friedel-Craft alkylation. Hydrodeoxygenation of the resulting intermediate will give a C_{15} alkylbenzene, which can be useful as an aromatic blending feedstock.

11.5 Conclusions

At present, our transportation sector depends largely on petroleum-derived liquid fuels. There is an increasing commitment to increase the proportion of biofuel use in the transportation sector. The focus is on hydrocarbon biofuels that are suitable for use in existing combustion engines. Lignocellulose is the most abundant biomass resource. The hydrocarbon biofuel production from lignocellulose involves preliminary operations/processes, which include: fractionation, hydrolysis, and fermentation/dehydration. The fuels/intermediates at this phase include ethanol, butanol, acetone, levulinic acid, furfural, and 5-hydroxymethylfurfural. The fuel properties (H/C and C/O ratios) of the fuels/intermediates are unsuitable for use in engines but can be enhanced by hydrodeoxygenation. The hydrodeoxygenated intermediates are only suitable as gasoline fuel. To obtain jet and diesel fuels, the C—C coupling of two or more of these biomass-derived molecules is required. Different types of C—C bond forming reactions for achieving the desired carbon chain lengths were examined. The functional groups present in the molecules dictate the types of the C—C bond forming reaction that is suitable in each case. However, Aldol condensation proved very versatile and applicable in all the six

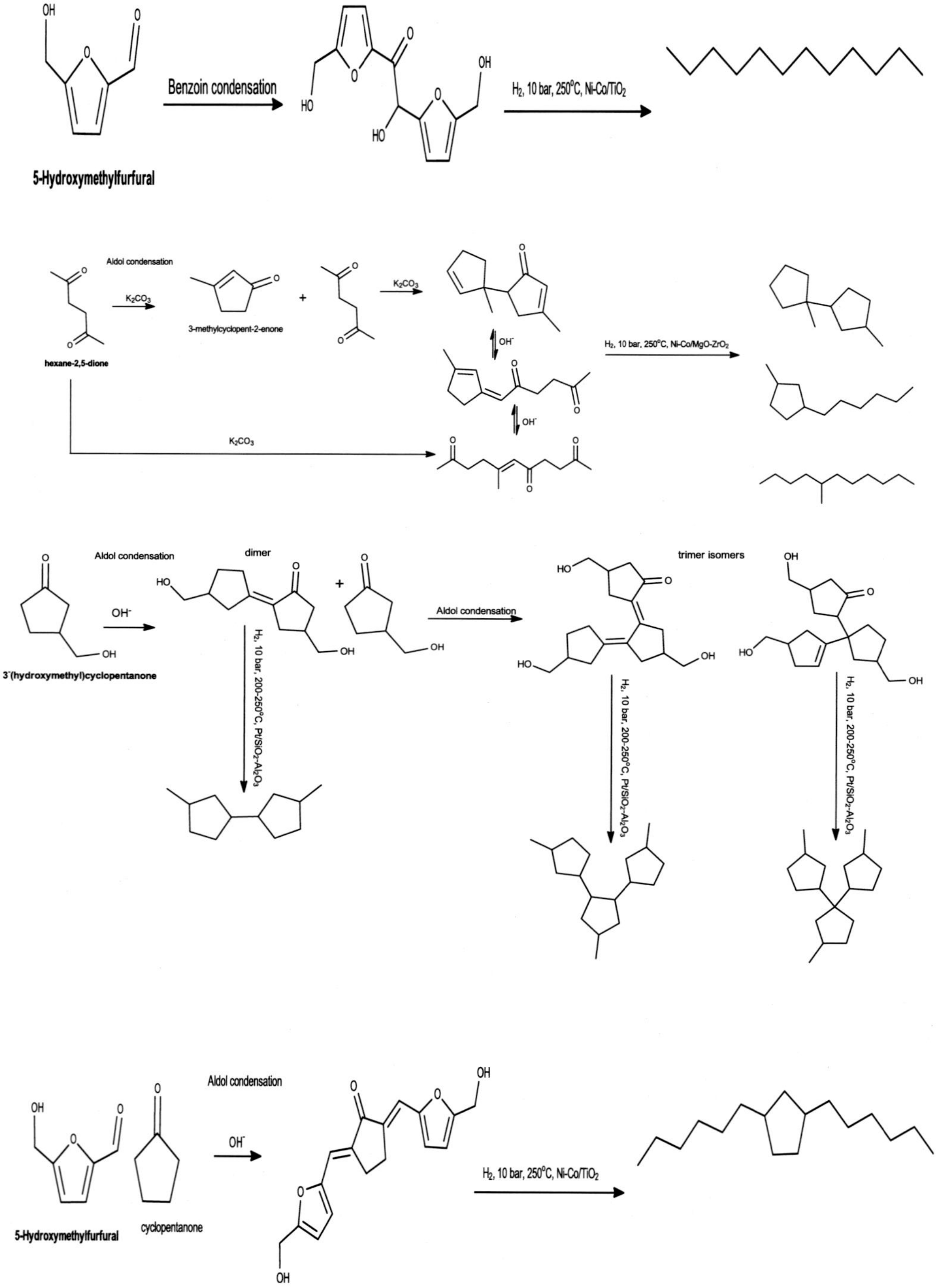

322 Hydrocarbon biorefinery

Fig. 11.15 C—C couplings of 5-hydroxymethylfurfural derivatives and their hydrodeoxygenation.
(Continued)

Fig. 11.15, cont'd

molecules considered. Other useful C—C bonding forming reactions include oligomerization, pinacol coupling, benzoin condensation, ketonization, hydroalkylation, and Friedel–Craft alkylation. The condensed products require varied levels of hydrodeoxygenation before they can be useful as drop-in fuels. Catalyst selection and reaction conditions for hydrodeoxygenation are dictated by the functional groups present in the condensed molecules.

References

[1] H. Krugman, J. Goldemberg, Technol. Forecast. Soc. Chang. 24 (1983) 45–60.
[2] K.-H. Erb, S. Gingrich, F. Krausmann, H. Haberl, J. Ind. Ecol. 12 (5–6) (2008) 686–703.
[3] P.G. Levi, J.M. Cullen, Environ. Sci. Technol. 52 (2018) 1725–1734.
[4] J.J. MacKenzie, Nonrenew. Resour. 7 (2) (1998) 97–100.
[5] T.R. Anderson, E. Hawkins, P.D. Jones, Endeavour 40 (3) (2016) 178–187.
[6] S. Solomon, G.-K. Plattner, R. Knutti, P. Friedlingstein, PNAS 106 (6) (2009) 1704–1709.
[7] E. Holden, G. Gilpin, Sustainability 5 (2013) 3129–3149.
[8] J.A. Melero, J. Iglesia, A. Garcia, Energy Environ. Sci. 5 (2012) 7393–7420.
[9] Y. Liu, Y. Nie, X. Lu, X. Zhang, H. He, F. Pan, L. Zhou, X. Liu, X. Ji, S. Zhang, Green Chem. 21 (2019) 3499–3535.
[10] R.C. Kuhad, Crit. Rev. Biotechnol. 13 (2) (1993) 151–172.

[11] T. Renders, S. Van den Bosch, S.-F. Koelewijn, W. Schutyser, B.F. Sels, Energy Environ. Sci. 10 (2017) 1551–1557.

[12] H. Wei, W. Liu, X. Chen, Q. Yang, J. Li, H. Chen, Fuel 254 (2019) 115599.

[13] W.-C. Wang, L. Tao, Renew. Sust. Energ. Rev. 53 (2016) 801–822.

[14] N.M. Eagan, M.D. Kumbhalkar, J.S. Buchanan, J.A. Dumesic, G.W. Huber, Nat. Rev. (Chemistry) 3 (2019) 223–249.

[15] S. Veibel, J.I. Nielsen, Tetrahedron 23 (1967) 1723–1733.

[16] J. Apuzzo, S. Cimino, L. Lisi, RSC Adv. 8 (2018) 25846–25855.

[17] X. Wu, G. Fang, Z. Liang, W. Leng, K. Xu, D. Jiang, J. Ni, X. Li, Catal. Commun. 100 (2017) 15–18.

[18] H. Aitchison, R.L. Wingad, D.F. Wass, ACS Catal. 6 (10) (2016) 7125–7132.

[19] M.A. Alotaibi, E.F. Kozhevnikova, I.V. Kozhevnikov, J. Catal. 293 (2012) 141–144.

[20] G.S. Salvapati, K.V. Ramanamurthy, M. Janardanarao, J. Mol. Catal. 54 (1989) 9–30.

[21] E. Ligner, F. Meunier, A. Travert, S. Maury, N. Cadran, Catal. Sci. Technol. 4 (2014) 2480–2483.

[22] J. Pritchard, G.A. Filonenko, R. van Putten, E.J.M. Hensen, E.A. Pidko, Chem. Soc. Rev. 44 (2015) 3808–3833.

[23] T. Mizugaki, K. Togo, Z. Maeno, T. Mitsudome, K. Jitsukawa, K. Kaneda, ACS Sustain. Chem. Eng. 4 (3) (2016) 682–685.

[24] I. Obregob, I. Ganadarias, N. Miletic, A. Ocio, P.L. Arias, ChemSusChem 8 (20) (2015) 3483–3488.

[25] P. Nilges, T.R. dos Santos, F. Harnisch, U. Schroder, Energy Environ. Sci. 5 (1) (2012) 5231–5235.

[26] A.A. Shutilov, M.N. Simonov, Y.A. Zaytseva, G.A. Zenkovets, I.L. Kinet, Catalogue 54 (2) (2013) 184–192.

[27] J.Q. Bond, D.M. Alonso, R.M. West, J.A. Dumesic, Langmuir 26 (21) (2010) 16291–16298.

[28] A. Chatterjee, S.H.H. Eliasson, V.R. Jense, Catal. Sci. Technol. 8 (2018) 1487–1499.

[29] Z. Li, J. Zhang, M.M. Nielsen, H. Wang, C. Chen, J. Xu, Y. Wang, T. Deng, X. Hou, ACS Sustain. Chem. Eng. 6 (5) (2018) 5708–5711.

[30] C.S. Santos, C.C.S.P. Soares, A.S. Vieira, A.C.B. Burtoloso, Green Chem. 21 (2019) 6441–6450.

[31] O. Mamum, M. Saleheen, J.Q. Bond, A. Heyden, J. Phys. Chem. C 121 (3) (2017) 18746–18761.

[32] G.A. Kraus, B. Roth, Tetrahedron Lett. 18 (36) (1977) 3129–3132.

[33] M. Mascal, S. Dutta, I. Gandarias, Angew. Chem. Int. Ed. 53 (2014) 1854–1857.

[34] J.C. Serrano-Ruiz, D. Wang, J.A. Dumesic, Green Chem. 12 (2010) 574–577.

[35] J. Wu, Y. Shen, C. Liu, H. Wang, C. Geng, Z. Zhang, Catal. Commun. 6 (9) (2005) 633–637.

[36] O.F. Aldosari, J. Saudi Chem. Soc. 23 (7) (2019) 938–946.

[37] P. Nilges, U. Schroder, Energy Environ. Sci. 6 (2013) 2925–2931.

[38] H.E. Hoydonckx, W.M. Van Rhijn, W. Van Rhijn, D.E. De Vos, P.A. Jacobs, Furfural and Derivatives, Wiley-VCH Verlag GmbH & Co. KGaA, Weinheim, 2012.

[39] N. Merat, C. Godama, A. Gaset, J. Chem, Technol. Biotechnol. 48 (1990) 145–159.

[40] C. Li, G. Xu, X. Liu, Y. Zhang, Y. Fu, Ind. Eng. Chem. Res. 56 (31) (2017) 8843–8849.

[41] J. Wu, X. Zhang, Q. Chen, L. Chen, Q. Liu, C. Wang, L. Ma, Energy Fuel 34 (2) (2020) 2178–2184.

[42] S. Koso, I. Furikado, A. Shimao, T. Miyazawa, K. Kunimori, K. Tomishige, Chem. Commun. (2009) 2035–2037.

[43] L. Li, K.J. Barnett, D.J. McClelland, D. Zhao, G. Liu, G.W. Huber, Appl. Catal. B: Environ. 245 (2019) 62–70.

[44] Z. Zhang, Z. Pei, H. Chen, K. Chen, Z. Hou, X. Lu, P. Quyang, J. Fu, Ind. Eng. Chem. Res. 57 (12) (2018) 4225–4230.

[45] B. Seemala, R. Kumar, C.M. Cai, C.E. Wyman, P. Christopher, React. Chem. Eng. 4 (2019) 261–267.

[46] Y. Li, X. Guo, D. Liu, X. Mu, X. Chen, Y. Shi, Catalysts 8 (2018) 193, https://doi.org/10.3390/cata18050193.

[47] Y. Zhang, G. Fan, L. Yang, F. Li, Appl. Catal. A: Gen. 561 (2018) 117–126.

[48] M. Dohade, P.L. Dhepe, Catal. Sci. Technol. 8 (2018) 5259–5269.

[49] G.-S. Zhang, M.-M. Zhu, Q. Zhang, Y.-M. Liu, H.-Y. He, Y. Cao, Green Chem. 18 (2016) 2155–2164.

[50] X. Xu, T. Hirao, J. Org. Chem. 70 (2005) 8594–8596.

[51] H. Zang, E.Y.X. Chen, Int. J. Mol. Sci. 16 (4) (2015) 7143–7158.

[52] M. Choura, N.M. Belgacem, A. Gandini, Macromolecules 29 (11) (1996) 3839–3850.

[53] D. Liang, G. Li, Y. Liu, J. Wu, X. Zhang, Catal. Commun. 81 (2016) 33–36.

[54] A. Corma, O. de la Torre, M. Renz, Energy Environ. Sci. 5 (2012) 6328–6344.

[55] L. Faba, E. Diaz, S. Ordonez, Appl. Catal. B: Environ. 113-114 (2012) 201–211.

[56] M. Hronec, K. Fulajtarova, T. Liptaj, M. Stolcova, N. Pronayova, T. Sotak, Biomass Bioenergy 63 (2014) 291–299.

[57] S.H. Shinde, C.V. Rode, ACS Omega 3 (5) (2018) 5491–5501.

[58] Y. Zhu, X. Kong, H. Zheng, G. Ding, Y. Zhu, Y.-W. Li, Catal. Sci. Technol. 5 (2015) 4208–4217.

[59] J. Tan, J. Cui, Y. Zhu, X. Cui, Y. Shi, W. Yan, Y. Zhao, ACS Sustain. Chem. Eng. 7 (12) (2019) 10670–10678.

[60] X. Kong, Y. Zhu, H. Zheng, F. Dong, Y. Zhu, Y.-W. Li, RSC Adv. 4 (2014) 60467–60472.

[61] J.J. Roylance, K.-S. Choi, Green Chem. 18 (2016) 2956–2960.

[62] R. Ramos, A. Crigorropoulos, N. Perret, M. Zanella, A.P. Katsoulidis, T.D. Manning, J.B. Claridge, M.J. Rosseinsky, Green Chem. 19 (2017) 1701–1713.

[63] J. Ohyama, R. Kanao, A. Esaki, A. Satsuma, Chem. Commun. 50 (2014) 5633–5636.

[64] T.S. Hansen, K. Barta, P.T. Anastas, P.C. Ford, A. Riisager, Green Chem. 14 (2012) 2457–2461.

[65] S. Chen, C. Ciotonea, K.D.O. Vigier, F. Jerome, R. Wojcieszak, F. Dumeignil, E. Marceau, S. Royer, ChemCatChem 12 (2020) 2050–2059.

[66] J. He, S.P. Burt, M. Ball, D. Zhao, I. Hermans, J.A. Dumesic, G.W. Huber, ACS Catal. 8 (2) (2018) 1427–1439.

[67] B. Xiao, M. Zheng, X. Li, J. Pang, R. Sun, H. Wang, X. Pang, A. Wang, X. Wang, T. Zhang, Green Chem. 18 (2016) 2175–2184.

[68] Y. Roman-Leshkov, C.J. Barrett, Z.Y. Liu, J.A. Dumesic, Nature 447 (2007) 982–986.

CHAPTER 12

Production of long-chain hydrocarbon biofuels from biomass-derived platform chemicals: Catalytic approaches and challenges

Sudipta De
KAUST Catalysis Center (KCC), King Abdullah University of Science and Technology, Thuwal, Saudi Arabia

Contents

12.1 Introduction	328
12.2 Strategies for C–C bond formation	329
12.2.1 Aldol condensation	329
12.2.2 Hydroxyalkylation-alkylation (HAA)	331
12.2.3 Ketonization	332
12.2.4 Pinacol coupling	334
12.2.5 Other routes of C–C coupling	335
12.3 Strategies for oxygen removal from oxygenated fuel precursors	336
12.4 Examples of long-chain hydrocarbon biofuels derived from different feedstocks	341
12.4.1 Hydrocarbon biofuels from furan-based compounds	341
12.4.2 Hydrocarbon biofuels from levulinic acid and its derivatives	344
12.4.3 Hydrocarbon biofuels from lignin	347
12.5 Perspectives and challenges	348
12.6 Conclusions	350
References	350

Abbreviations

2–MF	2-methylfuran
AL	angelica lactone
GVL	γ-valerolactone
HAA	hydroxyalkylation–alkylation
HDO	hydrodeoxygenation
HMF	5-hydroxymethylfurfural
LA	levulinic acid
RO	ring–opening
RS	ring–saturation

Hydrocarbon Biorefinery
https://doi.org/10.1016/B978-0-12-823306-1.00001-7

Copyright © 2022 Elsevier Inc.
All rights reserved.

12.1 Introduction

Bioethanol and biodiesel are the most widely used liquid biofuels for a long time [1]. Bioethanol is usually produced from edible sugars via fermentation processes, while biodiesel is produced from triglycerides in the presence of alcohol (usually methanol or ethanol) via transesterification reaction. However, to reduce the consumption of edible biomass and proper utilization of nonedible biomass, modern biorefinery research has been dedicated to the exploitation of lignocellulosic biomass to obtain various organic chemicals and transportation fuels. Although the consumption of bioethanol and biodiesel has increased exponentially in the last few years due to their simple and well-known production technologies [2], these biofuels have several drawbacks such as low energy density, low oxidation stability, poor cold flow properties, and corrosive nature of biodiesel, which restrict their use as transportation fuels [3, 4]. Although a small amount of oxygen is acceptable in road fuels, aviation fuel must have a minimum gravimetric energy density of $42.8\,MJ\,kg^{-1}$ (according to the properties of Jet A-1 fuel) and therefore, oxygen content has to be minimized as much as possible [5]. None of the currently commercialized biofuels have this minimum energy density due to their high oxygen content. Besides, the high water solubility of bioethanol is a big problem for its blending with gasoline as it increases the risk of engine damage. To mitigate these problems, it is important to find out new pathways for the synthesis of hydrocarbon biofuels with high energy content (e.g., gasoline, jet fuel, and diesel). These biofuels have properties similar to current transportation fuels and compatible with the present infrastructure of the transportation sector.

Conversion of biomass-derived platform molecules, such as furfural, 5-hydroxymethylfurfural (HMF), and levulinic acid (LA), into liquid hydrocarbon biofuels is an interesting approach. However, these platform molecules have multiple oxygen functionalities that need to be selectively removed to improve their fuel properties. Hydrodeoxygenation (HDO) is a well-known strategy to convert oxygenated precursors into hydrocarbon biofuel components with a higher H/C ratio. For example, partial or complete HDO of HMF gives 2,5-dimethylfuran or hexane, which have higher energy density. Since biomass-based platform chemicals usually contain five or six carbon units (as they are derived from pentose or hexose sugars), their final HDO products are C_5 or C_6 alkanes. These low-molecular-weight products cannot be used as road transportation fuels or fuel additives because of their low octane number and high volatility. Therefore, a new approach to increase the carbon numbers via the C–C coupling reaction is important to obtain hydrocarbon biofuels with high octane numbers.

This chapter summarizes the strategies for upgrading biomass-derived platform molecules into high carbon-number hydrocarbon biofuels developed in the past few years. The first section is focused on different C–C bond forming reactions for the synthesis of oxygenated fuel precursors. Suitable catalysts and reaction conditions for these reactions

Production of long-chain hydrocarbon biofuels **329**

are briefly discussed. The subsequent section is focused on general strategies for oxygen removal from the oxygenated fuel precursors via HDO techniques, with a special emphasis on catalyst features and reaction mechanisms. For the details about HDO techniques, readers may refer to some excellent reviews published in the past on this topic [6–10]. This chapter briefly summarizes the C–C coupling and HDO reactions for the synthesis of long-chain hydrocarbon biofuels from different biomass feedstocks. General perspectives, associated challenges, and probable solutions are also discussed at the end of the chapter.

12.2 Strategies for C–C bond formation

The C–C coupling reaction is a prerequisite step to upgrade the platform molecules into liquid hydrocarbon biofuels with the desired carbon number. There are different strategies for C–C coupling reactions which include aldol condensation, hydroxyalkylation-alkylation (HAA), ketonization, pinacol coupling, Michael addition, Robinson annulation, and so on. In this section, catalytic strategies and reaction conditions of these reactions are discussed.

12.2.1 Aldol condensation

This is one of the most often used techniques to increase the carbon chain length of small molecules. This type of reaction occurs between two carbonyl compounds having a reactive α-H on at least one of the carbonyl groups. The reaction usually requires acid or base catalysts to activate α-H. There are several reports on the aldol condensation of furfurals with carbonyl compounds, usually acetone, to synthesize the fuel precursors. Reactions can happen both in the presence and in the absence of solvents at a temperature ranging from room temperature to 100°C. The solvent is usually not required when acetone is used because it can act as both reactant and solvent. Depending on the chain structure of targeted alkanes (e.g., branched or cyclic), oxygenated fuel precursors are synthesized by selecting appropriate substrates in aldol condensation. Scheme 12.1 shows the formation of targeted oxygenates of desired structures by selecting different substrates via both cross- and self-aldol condensation.

Both homogeneous and heterogeneous catalysts are reported for this reaction. Among homogeneous catalysts, an aqueous NaOH solution is very effective to form aldol products from the reaction between furfural (or its derivatives) and ketones [11]. To achieve high yield, an organic phase is usually used with the NaOH solution, which efficiently extracts the aldol adducts from the aqueous solution due to their low solubility in water. Sometimes, inorganic salts (e.g., NaCl) are also added to the aqueous solution to improve the partition coefficient of the aldol product. Product distribution can be tuned by changing the molar ratio of the furfural precursor to ketone and the amount of base.

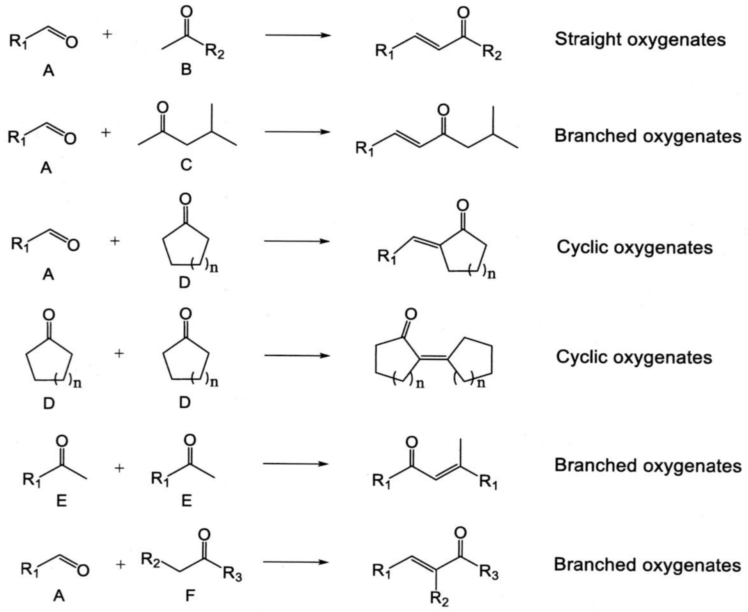

A = aldehydes (furfural, HMF, butanal); B = straight ketones (acetone, 2-pentanone, 2-heptanone);
C = branched ketone (methyl isobutyl ketone); D = cyclic ketones (cyclopentanone, cyclohexanone, isophorone);
E = straight ketones (acetone, methyl isobutyl ketone); F = chain ketones (3-pentanone, angelica lactone, alkyl levulinate).

Scheme 12.1 Aldol condensation reactions of different carbonyl compounds.

For example, condensation between furfural and acetone gives both single- and double-condensation products in high yields. On the other hand, only single condensation products are observed when HMF is used. It is because the degradation of HMF generates acidic by-products that reduce the basicity of the solution and restrict the formation of the second condensation product.

Due to the problems of environmental pollution and equipment corrosion, the use of heterogeneous catalysts has gained more attention these days. Basic metal oxides [12], basic zeolites [13], phosphates [14], and layered double hydroxides [15] are generally used as heterogeneous catalysts. One big advantage of these catalysts is that the strength and number of basic sites can be easily adjusted according to the nature of the reaction. Apart from that, these solid catalysts can be used as supports to make bifunctional catalysts by introducing active metals (such as Pd, Pt, Ni, Ru, etc.), which perform both aldol condensation and HDO reaction simultaneously. However, one drawback of the solid base catalysts is their poor hydrothermal stability that causes the leaching of active sites in aqueous media, and therefore, they need organic solvents. Thus, the development of hydrothermally stable catalysts is critical to run the process in aqueous media. In this respect, mixed oxides such as $MgO-Al_2O_3$, $MgO-ZrO_2$, and $CaO-ZrO_2$ are highly promising as the extent of leaching is very low [16]. The basic strength of these mixed oxides generally depends on the combination and molar ratio of the two oxides. It is observed that the mixed metal oxides with medium strength and an optimal ratio of acid/base site (1:1.1) are suitable to obtain C_{15} adduct from HMF and acetone. On the other hand, stronger basic sites assist the retro–aldol condensation. The activity of mixed oxides can be further improved by grafting them on mesoporous carbon supports. For example, in the reaction between furfural and acetone, Mg—Zr mixed oxide supported on high-surface-area graphite achieved 96.5% furfural conversion with 87.8% selectivity for C_8 and C_{13} adducts [17].

12.2.2 Hydroxyalkylation-alkylation (HAA)

This process is usually used to upgrade furan-based molecules to produce the precursors of branched alkanes. Furan and 2-methylfuran (2-MF) were mostly used by different groups. Huber's group first reported that furan can act as a nucleophile to couple with electrophiles, such as HMF and other furfurals, via the HAA process in the presence of H_2SO_4 to form different oxygenated intermediates with higher carbon number [18]. The reaction undergoes by the protonation of the alcohol or the carbonyl oxygen (in HMF or furfural) and the subsequent reaction with furan to form difurylmethane (C_9) that further reacts with another furan to yield trifurylmethane (C_{13}). Similarly, reaction with HMF produces C_{10}, C_{14}, and C_{18} precursors (Scheme 12.2).

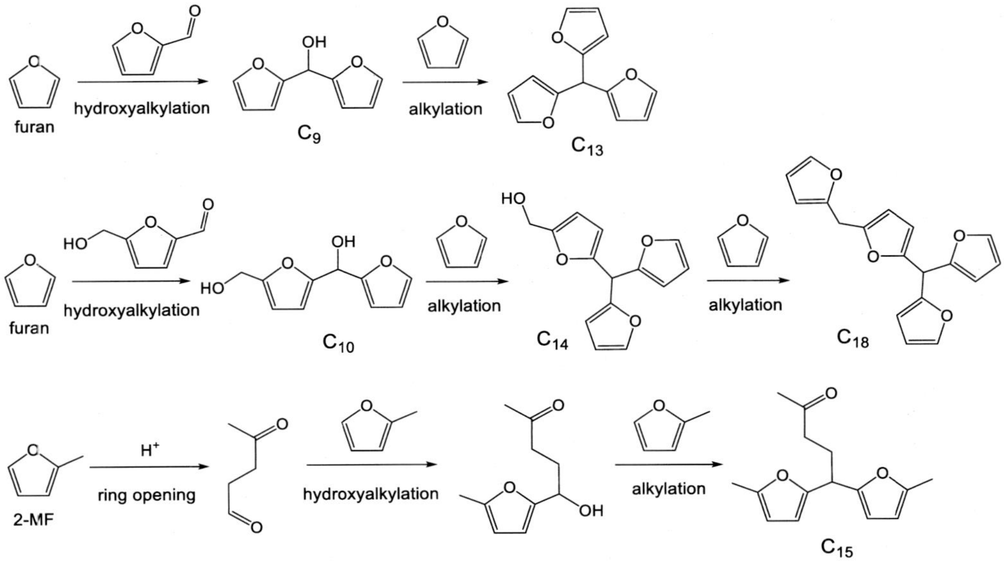

Scheme 12.2 Hydroxyalkylation-alkylation reaction of furan and 2-methylfuran (Sylvan).

332   Hydrocarbon biorefinery

2-MF also undergoes a similar reaction with a carbonyl compound to form different precursor molecules for linear and branched-chain hydrocarbons. Compared to furan, the use of 2-MF is more promising in HAA reaction because of several reasons. First of all, it is more reactive and selective in the HAA reaction. Secondly, the methyl group protects one of the reactive α-positions and therefore, the reaction of 2-MF with carbonyl compounds restricts the formation of undesired polymers. Finally, the extra methyl group makes the molecule more hydrophobic which facilitates easy separation of products from the aqueous phase. Corma and coworkers developed a new technique named "Sylvan process" where three molecules of 2-MF (Sylvan) are involved in the HAA reaction to produce a C_{15} precursor (Scheme 12.2) [19–21]. Under the acidic conditions, 2-MF undergoes the ring-opening reaction via hydrolysis to form 4-oxopentanal. It further reacts with two molecules of 2-MF through the HAA reaction to produce the C_{15} precursor, which subsequently undergoes the HDO reaction to produce a C_{15} branched alkane. One major problem with 2-MF is that currently there is no known process to produce 2-MF industrially. Therefore, the first step is to develop a cost-effective and efficient method for 2-MF production to make the Sylvan process industrially successful.

HAA reactions can undergo both in the presence and absence of solvents at a temperature ranging from room temperature to $120\,°C$. Brønsted acidic catalysts are usually used for these reactions. Among homogeneous catalysts, both inorganic acids (e.g., HCl and H_2SO_4) and organic acids (e.g., p-toluenesulfonic acid) are used [19, 21]. Among solid acid catalysts, resins (e.g., Amberlyst-15, Amberlyst-36, Nafion-115, and Nafion-212,) [19, 21–23], zeolites (e.g., H-Y, H-USY, and H-ZMS-5) [21, 23], phosphates (e.g., zirconium phosphate) [23], and acid functionalized carbonaceous materials (e.g., AC-SO$_3$H) [23] are reported by different groups. Collective results indicate that HAA reaction is dependent on the acid strength of the catalysts and strong acid sites are more effective for this process. Nafion-212 was found to be the best among all catalysts in terms of catalytic performance and stability.

12.2.3 Ketonization

Ketonization is usually applied to synthesize a ketone from two carboxylic acids via decarboxylation. The main goal of ketonization is to remove the oxygen from biofuels. The reaction has recently received increasing attention in biorefinery since it can increase the carbon number of biomass-derived small molecules and can remove the reactive acid groups. Self-condensation of LA is a very well-known example in biorefinery, where two molecules of LA give C_9 oxygenated precursor (nonane-2,5,8-trione, Scheme 12.3) for the production of 5-nonanone and finally nonane after HDO reaction [24, 25]. This reaction provides a new approach to directly transform LA to C_9 hydrocarbons by losing only one carbon. Ketonization is also an effective pathway to upgrade fatty acids to produce a heavy fatty ketone in the presence of a base catalyst. For example, ketonization of lauric acid ($C_{12}H_{24}O_2$) over MgO catalyst at 400 °C can give the corresponding ketone (a precursor of diesel-range fuel) with an excellent selectivity of 97% at 95% conversion (Scheme 12.3) [26].

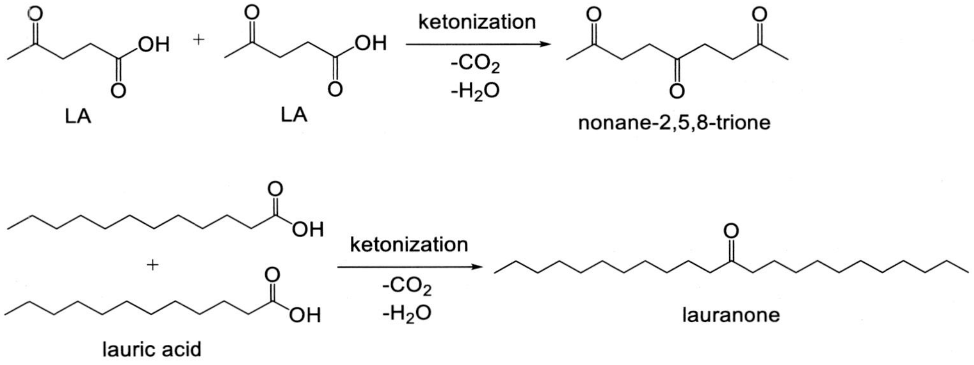

Scheme 12.3 Ketonization reaction of LA and lauric acid to their corresponding ketones.

Among different metal oxide-based catalysts (acidic, basic, and amphoteric), the weakly basic and amphoteric CeO_2 were found to be very efficient for the effective removal of acid functionalities of bio-oils via the ketonization reaction [27]. The catalyst also works well for acetic acid to produce acetone. Other amphoteric metal oxides including ZrO_2 and TiO_2 also show good performance. It may be noted that these metal oxides (CeO_2, TiO_2, and ZrO_2) are reducible in nature and have oxygen vacancy in their lattice, which is highly important for decarboxylation. Highly acidic zeolites are also known to catalyze the ketonization reaction efficiently. Most of the acid-catalyzed ketonization reactions are carried out in the gas phase at relatively higher temperatures (200–400 °C). High temperatures are specially required when alcohol groups containing molecules are also present in the reaction. Under the reaction condition, two parallel reactions are observed: (i) ketonization and (ii) reversible esterification. The esterification is more favorable at lower temperatures, while ketonization dominates at higher temperatures. However, high temperatures can be problematic in some cases. For instance, the ketonization product of LA (nonane-2,5,8-trione) is highly reactive at high temperatures, and therefore, it is difficult to achieve high selectivity to nonene [28]. Therefore, the design of new catalysts is essential, which can work in water at a lower temperature. This is particularly important for LA because of its low vapor pressure and high water solubility.

In spite of several studies on ketonization reaction to understand its mechanism, there are still some debates. An excellent review by Resasco et al. has summarized all the possible mechanisms of ketonization over different catalysts [28]. Briefly, ketonization activity can be related to several potential characteristics of a catalyst which include (i) oxygen

334 Hydrocarbon biorefinery

vacancies, (ii) acid sites, (iii) basic sites, (iv) redox properties, and (v) coordinatively unsaturated cations.

12.2.4 Pinacol coupling

In this route, furan derivatives with an aldehyde group on the α-position and aromatic aldehydes undergo the self- or cross-coupling reaction to form condensation products (Scheme 12.4). The reaction proceeds through a radical pathway in the presence of reductants under mild reaction conditions. Huang et al. reported the pinacol coupling of furanic aldehydes to produce C_{10}–C_{14} precursors using metallic powders (e.g., Al, Mg, and Zn) as reductants in different mediums including H_2O and aqueous solutions of NaOH, KOH, or NH_4Cl [29]. Using Zn in 10% aqueous solution of NaOH, 99% furfural conversion, and 96% selectivity for the targeted C_{10} product were achieved at 50°C. However, when the optimized conditions for furfural were used to convert HMF, polymerized products with molecular weights of 200–800 were obtained, which was probably due to the low stability of HMF. Alternatively, 5-methylfurfural (5-MF) can be used instead of HMF as 5-MF is more stable. Using 5-MF as starting material, 85% yield of the corresponding C_{12} dimer was obtained.

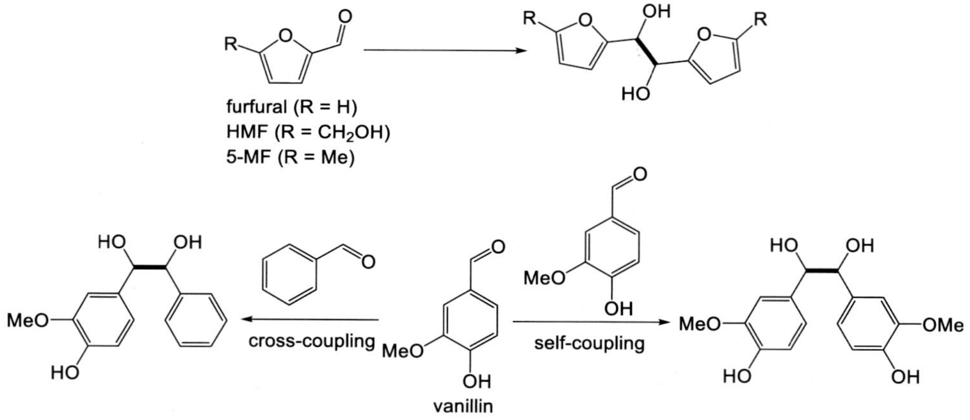

Scheme 12.4 Self- and cross-condensation of furanic and other aromatic aldehydes via pinacol coupling.

Pinacol coupling can be successfully employed for upgrading varieties of aromatic aldehydes derived from lignin. Vanillin is an important example, which can undergo pinacol coupling to form an intermediate for C_{14} alkane [30]. The reaction is carried out both chemically and electrochemically using Zn powder and Zn electrode as catalysts, respectively. There are also examples of cross pinacol coupling between vanillin and other aldehydes (e.g., benzaldehyde) under different reaction conditions.

12.2.5 Other routes of C–C coupling

Apart from the above-discussed pathways, other reactions, such as Michael addition, Robinson annulation, olefin addition, and oligomerization, are also known as viable techniques to upgrade different platform molecules to hydrocarbon biofuels. Michael addition is an excellent technique to form highly branched ketones from a β-dicarbonyl compound and an α,β-unsaturated carbonyl compound [31]. The main problem of this process is that the starting compounds cannot be directly obtained from biomass; rather they need to be synthesized by an aldol condensation reaction. Transition metal chlorides are known to work well as catalysts for Michael's addition reaction because of their ability to form chelate complexes with β-dicarbonyl compounds. For example, using $CoCl_2 \cdot 6H_2O$ as a catalyst, 82.9% conversion of 4-(2-furanyl)-3-butene-2-one (FA, the aldol condensation product of furfural and acetone) and 75.1% yield of furfural-acetone-diketone (FAD, Michael product of FA and 2,4-pentanedione) was achieved (Scheme 12.5) [31]. When the Michael addition product contains an active α-H atom, an intramolecular aldol condensation can be performed to synthesize cyclic oxygenates. The consecutive process of Michael addition followed by intramolecular aldol condensation is named as Robinson annulation, which eventually gives a six-membered ring with a carbon chain (Scheme 12.5) [32]. The obtained high-quality alkylcyclohexanes (after HDO) can be easily mixed with traditional fuels to improve their combustion properties.

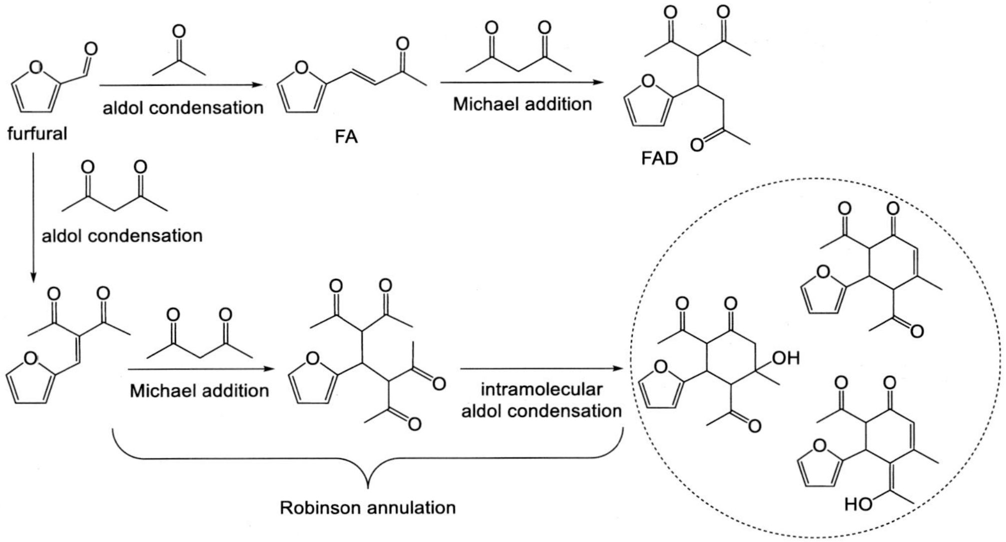

Scheme 12.5 Michael addition and Robinson annulation of furfural with carbonyl compounds.

Oligomerization is another approach to produce long-chain alkene precursors from low-molecular-weight alkenes. For upgrading LA to liquid hydrocarbon biofuels, oligomerization is a good pathway (Scheme 12.6) [33]. However, first LA needs to be converted into butenes via γ-valerolactone (GVL) and pentenoic acid. Production of GVL from LA is a well-established process, where hydrogenation catalysts (such as Pd, Ru, Ni, Cu, etc.) are successfully used [34]. GVL can undergo ring-opening reaction to give pentenoic acid, which is subsequently decarboxylated to give butenes. Oligomerization of butenes takes place usually in the presence of acidic catalysts, such as H-ZSM-5, Amberlyst resins, etc., at a temperature ranging from 170 °C to 220 °C.

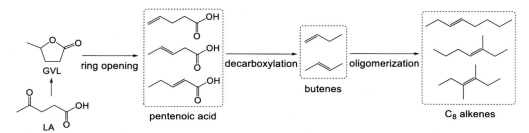

Scheme 12.6 Production of higher alkenes from GVL via oligomerization reaction.

12.3 Strategies for oxygen removal from oxygenated fuel precursors

The most important step in the synthesis of liquid hydrocarbon biofuels is the removal of oxygen from the oxygenated precursors via the HDO process. This process typically occurs via hydrogenolysis of C–O bonds and the subsequent elimination of oxygen in the form of water. The fuel precursors obtained from C–C coupling reactions usually contain different oxygen functionalities (such as C=O, C–OH, and C–O–C bonds) and unsaturation (C=C bond). Therefore, a series of reactions including dehydration, hydrogenation, and hydrogenolysis is necessary to improve the H/C ratio. HDO process is the most challenging part of the process of hydrocarbon biofuels production because the main goal of HDO is to achieve the effective removal of oxygen with minimum H_2 consumption. The catalyst design thus plays a crucial role in this process. Until now, several catalysts containing noble metals (e.g., Pd, Pt, Ru, Rh, Re, etc.) and nonnoble metals (e.g., Fe, Ni, Cu, Mo, etc.) have been reported for the HDO process [6, 9]. Initially, HDO catalysts were designed by the formulation of hydrodesulfurization (HDS) catalysts that are mainly used in the petroleum industries for removing sulfur impurities. Traditional industrial catalysts based on sulfided Co–Mo–Ni supported on Al_2O_3 show good HDO performance but they suffer from the following serious problems. (i) The maintenance of stable catalytic activity needs the addition of sulfur, which results in sulfur contamination and H_2S emissions. (ii) An unavoidable by-product in the HDO process is

Production of long-chain hydrocarbon biofuels   337

water that converts Al_2O_3 into boehmite (AlO(OH)). (iii) The high Lewis acidity of Al_2O_3 induces coke formation which results in the deactivation of the catalyst [8]. In this context, alternative supports, such as SiO_2, CeO_2, TiO_2, and mesoporous carbons, are of good choice due to their low acidity and coke-resistance ability [35]. Among sulfur-free HDO catalysts, metal carbides, metal phosphides, and metal nitrides are also widely used for their comparable activities [35]. The latter studies in this direction have focused more on investigating the reaction network and catalyst requirement in each step of the HDO process, with a special emphasis on understanding the substrate-catalyst interaction. It is important to mention here that most of the fundamental studies on HDO catalyst design use model compounds containing the functional group of interest because it is much easier to study the mechanism with model compounds rather than the real feedstocks.

Since various types of reactions are involved in an HDO process, the catalyst requirement is also different for each reaction. Therefore, the choice of a suitable catalyst should consider the optimization of several factors to achieve higher selectivity of the targeted products. From Table 12.1, it is clear that a bifunctional catalyst containing an acidic site and a metallic counterpart is essential for all the necessary steps to

Table 12.1 Reactions involved in the HDO process and their corresponding catalysts.

Reaction	Example	Catalyst requirement
Dehydration of alcohols		Brønsted acidic catalysts including inorganic acids. Activity increases with the acidity
Hydrogenation		Bifunctional catalysts consisting of metals (Pd, Ru, Pt, Ni, Cu, etc.) and acid sites
Conversion of ethers		Combination of metal catalyst and acid under harsh conditions
C–O hydrogenolysis		Bimetallic catalysts containing noble metal and group 6 or 7 metals (e.g., Pt—Re, Pt—W, Pd—W, Rh—Mo, Rh—Re, Ir—Re, etc.)
C–C dissociation and isomerization		Combination of metal (e.g., Ru, Ir, and Rh) and acid
Decarboxylation and decarbonylation	R_1—COOH $\longrightarrow$ R_1—H + CO_2 R_1—CHO $\longrightarrow$ R_1—H + CO	Transition metals are typically used, but acid or base can also promote this reaction

perform together. Important reactions including hydrogenation of different functional groups (such as C=C, C=O, and −COOH), C−C dissociation, decarboxylation, and decarbonylation occur on metal sites. On the other hand, acid sites promote different reactions, such as dehydration, isomerization, hydration, and hydrolysis. It is important to note here that the HDO process often requires high H_2 pressure that can also cause overhydrogenation or carbon loss by C−C bond cleavage. Bifunctional catalysts can perform the C−O bond cleavage at a lower H_2 pressure than those required for standalone metal nanoparticles, which is a significant advantage [36].

Bifunctional catalysts can be formulated in two different ways. The most common strategy is dispersing the metal components on solid acid support (e.g., Al_2O_3, zeolite, Nafion, etc.). In this case, the major focus is given on the textural, electronic, and acid–base properties of the support, which can influence the overall catalyst design. Importantly, the type of acidic sites determines the reaction pathway since different acid sites have different roles in the HDO process. For example, the Lewis acid site binds electron-rich species to the catalyst surface whereas the Brønsted acid site provides protons to the intermediates [8]. It is worth mentioning that although the acidic supports are of immense importance for some of the reactions, they also trigger the coke formation and catalyst deactivation. In this regard, oxophilic supports, such as ZrO_2, TiO_2, and Nb_2O_5, exhibited less susceptibility for carbon deposition and higher selectivity to aromatic compounds in the phenolics HDO process [37]. Another type of bifunctional catalyst is the bimetallic catalyst, which is composed of a reducing metal (such as Pd, Pt, Rh, etc.) and an oxophilic metal (such as Re, W, etc.) [7]. The acidic −OH groups strongly attached to the oxophilic metals provide protons to the reactants to form carbenium ions [38]. Close proximity of these two types of metal sites can give high activity for hydrogenolysis. Experimental studies proved that ReO_x species cover reducible metals (e.g., Pt), resulting in the direct interaction of both sites with each other [39].

Among several catalysts explored in HDO processes, niobium-supported solid acid catalysts have attracted immense attention because of their better activity under mild conditions as compared to the other catalysts supported on Al_2O_3, SiO_2, or H-ZSM-5. For example, $Pd/NbOPO_4$ shows almost full conversion of furfural-acetone aldol adduct into octane with 94% yield at 170°C and 20 bar H_2 [40]. Combined experimental and DFT calculation results indicate that the Nb−O−Nb chains in both NbO_x and $NbOPO_4$ are very efficient to break the C−O bond on the tetrahydrofuran ring. It is also observed that different states of NbO_x have a different activity, where NbO_x with a lower coordination number of Nb is suitable for the HDO of furanics [41]. Briefly, Nb-based catalysts have the following features: (i) high mesoporosity and surface areas for homogeneous metal dispersion, (ii) high oxygen affinity to activate C−O bonds, (iii) synergistic effect between NbO_x and metal for effective oxygen removal, (iv) desired acidity of NbO_x to assist the necessary acid-catalyzed reactions, and (v) high stability and water resistance ability.

It is often seen that the furan ring is a part of the structure of many fuel precursors since furanic platform chemicals are used as the main carbon sources. To achieve open-chain alkanes, the furan ring needs to be opened and it is very important to know the ring-opening (RO) mechanism in the reaction network and its impact on the total HDO process. Another event associated with the ring-opening process is ring-saturation (RS) and it is also important to know the sequence of these two events in order to determine the selectivity of final products. There are several reports on the catalytic hydrogenolysis of lignocellulose-derived aromatics to linear alcohols/polyols. However, the sequence of RO and RS steps remained a controversial topic [21, 42]. Sutton et al. experimentally proved that it is hard to open the ring once it is fully saturated [43]. As can be seen in Scheme 12.7, the formation of **3** by the full hydrogenation of the furan ring is a dead-end for the production of alkanes under mild conditions. However, RO of **2** to a polyketone (**4**) prior to RS can produce nonane (**5**) in high yield via a lower energy route. In support of this result, Chen and Vlachos performed combined studies based on DFT and high-resolution electron energy loss spectroscopy (HREELS) analysis to study the hydrogenolysis of saturated (e.g., furan, furfural, and furfuryl alcohol) and unsaturated (e.g., tetrahydrofurfuryl alcohol) heterocyclic compounds on iridium, which also revealed that unsaturation promotes RO, while full saturation causes slower kinetics

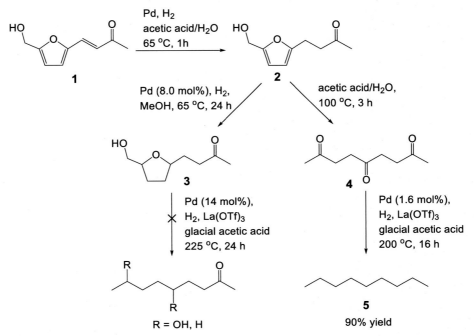

Scheme 12.7 The divergent pathway leads to different products during the HDO of HMF-acetone adduct.

and hinders the hydrogenolysis [44, 45]. However, in contrast to the earlier report by Sutton et al. [43] this study showed that selective partial RS can kinetically enhance the hydrogenolysis process.

Contradictory results were reported by Song et al. where it was shown that RO is possible even after full saturation of the furan ring in presence of a metal triflate promoter ($M(OTf)_x$, where $M = La$, Ce, Sm, Yb, Sc, Fe, Al, and Hf) [46]. However, the RO step is observed to be successful only in the case of metal triflate with strong Lewis acidity. The entire process proceeds through two main steps: (i) Pd/C-catalyzed saturation of the furan ring and exocyclic double bond at low temperatures, and (ii) triflate-catalyzed HDO of saturated furan intermediate at high temperatures. Up to 92% of nonane was obtained from HMF-acetone adduct using Pd/C–Hf(OTf)$_4$ combination. The high activity of Hf(OTf)$_4$ is related to the high charge density at Hf^{4+} center that facilitates RO by activating the C–O bonds. Followed by this work, Dutta and Saha also used Hf(OTf)$_4$ promoter for the HDO of furylmethanes to produce alkanes [47]. For example, in the case of C_{15} trifuran (**1**), the furan rings can be first hydrogenated by Pd/C to fully saturated cyclic ethers, which can then undergo facile RO and deoxygenation to form the alkanes (Fig. 12.1). The RO step is believed to be assisted by the formation of an intermediate between nucleophilic ethereal O and Hf^{4+} center. Additionally, O atoms of the $-SO_3$ group can form a cyclic intermediate (**4**) to cleave the β-C–H bonds, which facilitates the RO of Hf-bonded ether to form Hf-bonded alcohol (**5**). It is observed that the furan ring without $-CH_3$ group first undergoes RO, which could be due to the formation of a stable adduct between Hf^{4+} and furan O in absence of steric

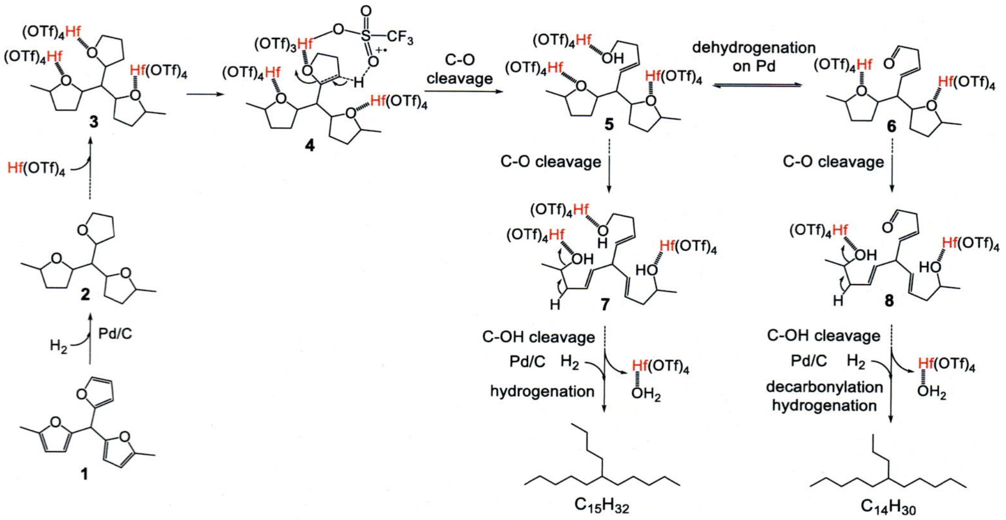

Fig. 12.1 Reaction pathways for the synthesis of alkanes from C_{15} trifuran using Pd/C–Hf(OTf)$_4$ as catalyst.

Production of long-chain hydrocarbon biofuels   **341**

hindrance. Marks et al. also established a similar mechanism for the hydrogenolysis of 2-methyltetrahydropyran to *n*-hexane using metal triflates [48]. From all these results, it is clear that the whole HDO process of furan compounds involving ring hydrogenation, RO, and deoxygenation depends on various factors including ring substitution, catalyst nature (e.g., charge density on the metal sites, acidity, etc.), and reaction conditions.

12.4 Examples of long-chain hydrocarbon biofuels derived from different feedstocks

12.4.1 Hydrocarbon biofuels from furan-based compounds

Among furan-based feedstocks, furfural and HMF are widely used monomers to produce liquid hydrocarbon biofuels. Aldol condensation is one of the most effective routes to upgrade these monomers into long-chain hydrocarbon biofuels, where acetone is mainly used as a reactive α-H containing unit. Several studies have reported the synthetic strategy of hydrocarbon biofuels via aldol condensation of furfural and HMF with acetone and other ketones [49–55]. In 2005, Dumesic et al. first reported the production of C_7–C_{15} alkanes from biomass-derived carbohydrates using a multistep process (Fig. 12.2) [49–51]. According to the protocol, the first step is the conversion of sugars to furfural and HMF, which undergo aldol condensation in the next step with acetone to form long-chain oxygenated fuel precursors. Improved yield of aldol adducts can be achieved by using a biphasic reactor, where the organic solvent continuously extracts the products from the aqueous phase. To perform the HDO of these precursors, they designed a four-phase reactor system made of (i) an aqueous inlet stream containing the water-soluble large organic reactant, (ii) a hexadecane alkane inlet stream, (iii) a hydrogen inlet gas stream, and (iv) a solid HDO catalyst (Pt/SiO_2-Al_2O_3). As the oxygenated precursors undergo dehydration/hydrogenation reactions, they become more hydrophobic, and the hexadecane alkane stream removes them from the catalyst before they further react to form coke. Both single and double condensation products were obtained when acetone was used. Product distribution can be tuned by changing the ratio of furfural to acetone. Different bifunctional catalysts, such as Pd/Al_2O_3 (at 100–140°C and 25–52 bar H_2) and $Pt/NbOPO_4$ (at 255–295°C and 60 bar H_2), were used to obtain a mixture of linear alkanes. For HMF-acetone systems, 58%–69% yield of C_7–C_{15} alkanes (based on carbon) was achieved from fructose, while 79%–94% carbon yield was achieved from various furfurals.

Furfural can be condensed with LA (or its ester) to produce precursors for diesel-range hydrocarbon biofuels. For example, Olson and Heide patented a process to convert cellulose into fuel components, where furfural and ethyl levulinate are converted into C_9–C_{15} alkanes [56]. Following this work, Li and coworkers reported an energy-efficient tandem three-step process to directly convert raw lignocellulosics (e.g., corncob) into long-chain branched alkanes (Fig. 12.3) [57]. In the first step, hemicellulose and cellulose

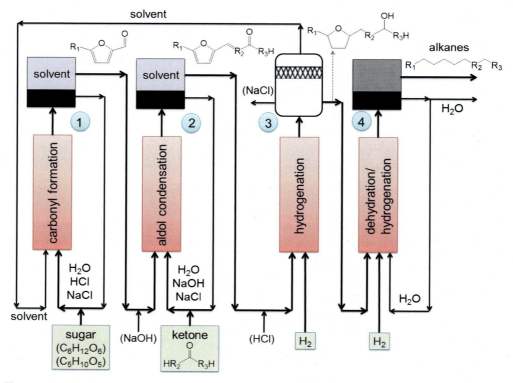

Fig. 12.2 Step-wise synthesis of liquid hydrocarbon biofuels from biomass-derived carbohydrates.

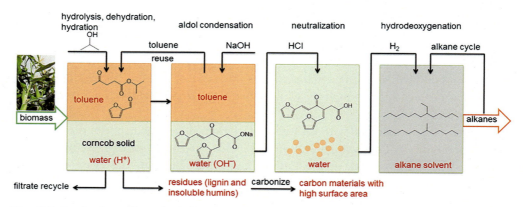

Fig. 12.3 Production of branched alkanes from corncob biomass.

fractions are converted into furfural and isopropyl levulinate through an acid-catalyzed hydrolysis–dehydration–hydration reaction in a water/toluene biphasic system with added isopropanol. The addition of isopropanol enables the spontaneous formation of LA ester and its transfer into the organic phase. As a result, LA ester coexists with furfural in the same phase, which makes their double aldol condensation (the second step) effective in the toluene phase. In the third step, HDO of double condensation adduct was performed in an organic solvent (e.g., cyclohexane) using $Pd/NbOPO_4$ catalyst to produce 91.4% C_{15} alkane along with 5.6% C_{14} alkane at 220 °C and 60 bar H_2.

Other than aldol condensation, furfural can undergo reductive self-condensation via pinacol coupling reaction in presence of a base and a metal, such as Al, Zn, or Mg (in powder form) to produce C_{10} fuel precursor (Scheme 12.8) [29]. Subsequent HDO of this dimer over Pt/AC and Pd/C catalysts gives C_8–C_{10} alkanes and the yield depends on the nature of the acidic counterpart ($NbOPO_4$, $TaOPO_4$, and WO_3) of the catalysts. Using $Pt/AC–TaOPO_4$ as the catalyst, a maximum molar yield of 83.6% was achieved from an aqueous solution of the dimer at 300 °C. This strategy can also be applied to other substrates (such as 5-methylfurfural, anisaldehyde, and vanillin) to obtain C_{12} and C_{14} alkanes. Similar to pinacol coupling, HMF can be upgraded to C_{12} alkane through a self-coupling reaction via organocatalysis and subsequent HDO via metal–acid tandem

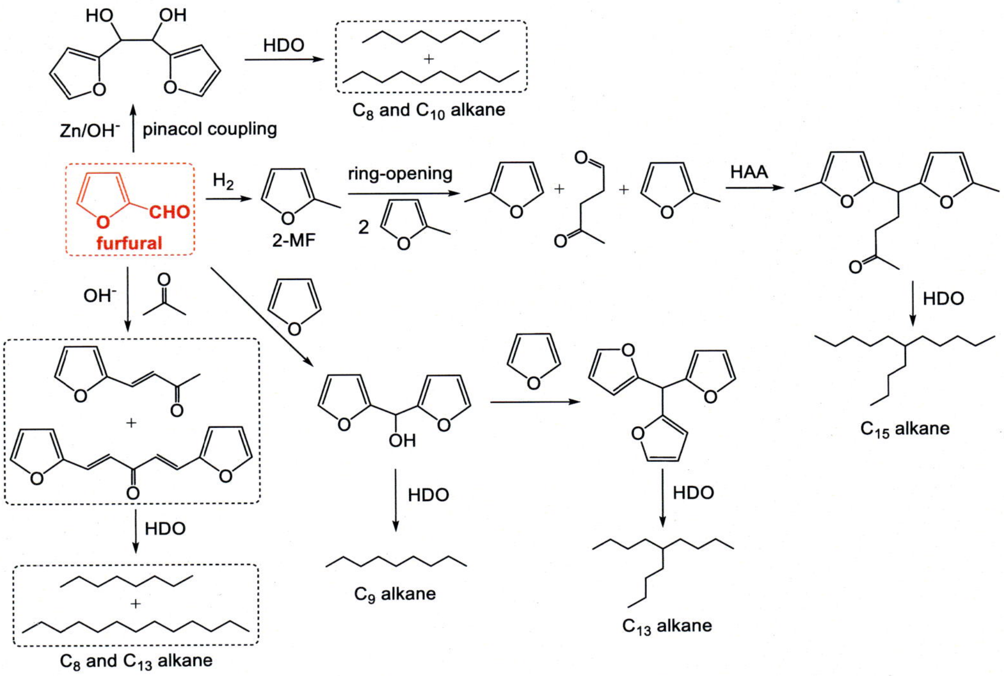

Scheme 12.8 Different routes for the conversion of furfural into liquid hydrocarbon biofuels.

catalysis [58]. In the first step, HMF is converted to 5,5′-dihydroxymethyl-furoin (DHMF) in 91% yield using N-heterocyclic carbenes at 60°C. In the second step, DHMF is converted to linear alkanes with 78% carbon yield and 64% selectivity to n-$C_{12}H_{26}$ using a bifunctional catalyst containing Pd/C, acetic acid, and La(OTf)$_3$ at 250°C.

Huber et al. developed a method to form C–C bond between furfural and furan to obtain linear and branched C_9–C_{13} alkanes (Scheme 12.8) [18]. Corma et al. established the Sylvan process for the conversion of 2-MF (Sylvan) into diesel-range branched alkanes via a two-step process, i.e., HAA followed by HDO [19–21]. In their initial study, 2-MF was trimerized in the presence of an acid (e.g., H_2SO_4) to form 5,5–bis (sylvyl)-2-pentanone at 60°C (Scheme 12.8) [19]. Subsequent HDO with Pt/C–Pt/TiO$_2$ catalyst at 350°C and 50 bar H_2 resulted in 87% overall yield of diesel from Sylvan. This strategy can be further extended to the reactions between 2-MF and other carbonyl compounds (such as butanal, furfural, HMF, 5-methylfurfural, acetone, hydroxyacetone, and cyclopentanone) to obtain branched alkanes with different carbon numbers [20, 23, 59–62].

12.4.2 Hydrocarbon biofuels from levulinic acid and its derivatives

The United States Department of Energy identified LA as 1 of the 12 important biomass-derived building blocks with a wide range of potentials in chemical production [63]. LA has two main functional groups, i.e., carboxylic acid and ketone, which make LA highly reactive for multiple transformations. LA can be converted into hydrocarbon biofuels by a number of catalytic routes involving ketonization, condensation, lactonization, and other reactions.

12.4.2.1 Direct route

Since LA contains five carbons, the coupling of two or more LA can give rise to the precursors of long-chain hydrocarbon biofuels. For example, acid-catalyzed ketonization of LA via decarboxylation is a promising route to generate nonane-2,5,8-trione, which can be further converted into nonane by HDO reactions [25]. Karimi et al. reported aqueous phase conversion of LA to nonane using Red Mud as a catalyst [25]. Since Red Mud contains an alkaline mixture of Fe_2O_3, TiO_2, and various complex sodium alumino-silicates, it can be used as a multifunctional catalyst for different reactions including reductions, oxidations, and acid/base mediated transformations. The active phase of the catalyst is made of reduced FeO_x, aluminate, silicate, and carbide phases that are generated by either prereduction of the Red Mud in the presence of formic/acetic acid or in situ reduction with H_2. At 365°C, the total conversion of LA could be achieved with a 76% yield of alkanes with high selectivity for C_9 alkanes.

Due to the presence of active α-H, LA can undergo self-aldol condensation catalyzed by either base or acid to form C_{10} fuel precursors. For example, mixed MgZr oxide can form two condensation products of LA. The first one is from the self-condensation of LA, while the other one is from the condensation between LA and its lactone obtained via cyclization [64]. However, a maximum of 30% LA conversion could be achieved with a 21% yield of C_{10} adduct. This is because the $-COOH$ group competes with the $-C=O$ group for the basic sites and deactivates the catalyst by neutralizing the basicity. As an alternative, Brønsted acid (e.g., trichloroacetic acid) and Lewis acid (e.g., $ZnCl_2$) can promote the reaction with C_{10} yield of 50.9% (Scheme 12.9) [65].

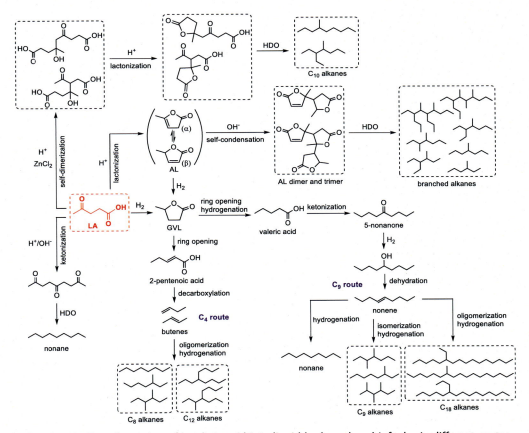

Scheme 12.9 Transformation of levulinic acid into liquid hydrocarbon biofuels via different routes.

12.4.2.2 Conversion via angelica lactone

Angelica lactone (AL) is obtained via intramolecular dehydration of LA in the presence of acids. It is more reactive than LA due to the ring unsaturation. Under relatively mild conditions and in the presence of alkali catalysts, AL can be quantitatively converted to C_{10} or C_{15} intermediates, which are the precursors of branched alkanes (Scheme 12.9) [66–68].

Several catalysts including Pt-ReO$_x$/C, Ir-ReO$_x$/SiO$_2$, Pd/Al$_2$O$_3$, and Pd/C can be used for the HDO reaction. The advantages of this process are as follows. (i) It requires relatively low H$_2$ pressure to achieve an almost quantitative yield of hydrocarbons. (ii) The process does not require any organic solvent. (iii) The number of steps involved in the production of liquid alkanes is less. (iv) The atom efficiency of this process is about 90% as the catalyst can be easily recovered from the reaction medium and regenerated. Besides self-condensation, AL can also undergo a cross-coupling reaction with other carbonyl-containing compounds (e.g., furfural) to produce C$_9$ and C$_{10}$ alkanes [69]. Various metal oxides (both acidic and basic) including Mn$_2$O$_3$, La$_2$O$_3$, MgO, ZnO, Fe$_2$O$_3$, Cr$_2$O$_3$, Nb$_2$O$_5$, and TiO$_2$ are used for the solvent-free condensation reaction, among which Mn$_2$O$_3$ showed the highest activity and stability. The obtained C$_{10}$ oxygenate was hydrogenated and hydrodeoxygenated over Pd-based catalysts (Pd/C and Pd-FeO$_x$/SiO$_2$) with up to 96% carbon yield of C$_9$ and C$_{10}$ alkanes. The addition of iron in the Pd/SiO$_2$ catalyst avoids undesired decarboxylation and decarbonylation reaction and increases the HDO activity.

12.4.2.3 Conversion via γ-valerolactone

LA can be converted to γ-valerolactone (GVL) by hydrogenation via either AL intermediate or 4-hydroxypentanoic acid intermediate [70]. GVL is considered a fuel additive because of its high energy density, calorific value similar to ethanol, and low vapor pressure. But the use of GVL as a fuel additive is restricted by some limitations, such as high water solubility and relatively lower energy density as compared to traditional fossil fuels. Two possible routes are reported for upgrading GVL into liquid hydrocarbon biofuels. The first route proceeds through the formation of butenes, which is designated as the C$_4$ route (Scheme 12.9). Bond et al. developed a combined catalytic route for the conversion of GVL to liquid hydrocarbon biofuels using a dual reactor system [33]. In the first reactor, GVL is converted to pentenoic acid via acid-catalyzed RO and then decarboxylated in the presence of solid acid catalysts (e.g., SiO$_2$–Al$_2$O$_3$) to form butenes. Over 99% yield of butenes was obtained from an aqueous solution of GVL at 375 °C and 35 bar H$_2$. The gas mixture produced in the first reactor was then transferred into the second rector to form C$_8$–C$_{16}$ hydrocarbons via oligomerization over an acid catalyst (e.g., H-ZSM-5 or Amberlyst-70). The temperature of the second reactor was kept relatively low (150–170 °C) to avoid the undesired cracking reactions and improve the yield of higher alkenes.

The second route follows the reductive RO of GVL to pentanoic acid, which undergoes ketonization to produce 5-nonanone. Hydrogenation of 5-nonanone followed by dehydration produces C$_9$ alkene (nonene). This route is designated as C$_9$ route (Scheme 12.9). Nonene can be further converted to n-nonane or other branched C$_9$ alkanes via isomerization. Dumesic group developed a method to produce 5-nonanone in high yields (90%) from a concentrated aqueous solution of GVL by using a dual catalyst bed in a single reactor [24, 71]. In the first catalyst bed, pentanoic acid is produced from GVL via RO and hydrogenation over the bifunctional Pd/Nb_2O_5 catalyst at 325 °C. In the second catalyst bed, ketonization of pentanoic acid is carried out over $Ce_{0.5}Zr_{0.5}O_2$ catalyst at 425 °C. The hydrophobic nature of 5-nonanone makes its separation from water spontaneous. 5-Nonanone can be transformed into *n*-nonane via hydrogenation/dehydration using a bifunctional catalyst such as Pt/Nb_2O_5. Alternatively, the carbon number of alkanes can be increased via oligomerization of nonene using an acid catalyst, which produces a mixture of C_{18}–C_{27} alkenes that can be eventually hydrogenated to diesel-range fuels [72]. It was estimated that 100 kg of GVL can produce about 50 kg of liquid hydrocarbon biofuels retaining more than 90% of its initial energy content.

12.4.3 Hydrocarbon biofuels from lignin

The lignin has a higher C/O ratio and energy density compared to cellulose and hemicellulose. This factor makes it a perfect candidate for the production of hydrocarbon biofuels. The depolymerized monomers and dimers of lignin generally contain 7–18 carbon, which makes them appropriate precursors for the production of aviation fuel [73]. However, the stable polymeric structure and irregular arrangement of lignin prevent its chemical depolymerization and selective conversion challenge.

Lignin structure has a high content of the methoxy group. The removal of the methoxy group is thus essential for its full conversion. But the presence of a methoxy group is sometimes advantageous if the methyl group is transferred to the aromatic ring. This can be easily done through transalkylation or disproportionation reaction, which produces alkylphenols or alkylbenzenes [74]. In this way, both carbon loss and H_2 consumption can be reduced. Additionally, the HDO products of these alkylphenols and alkylbenzenes have high octane numbers, which are suitable for fuel blend.

Apart from transferring a methyl group to the aromatic ring, two lignin-degraded monomers can be coupled to obtain hydrocarbon biofuels with increased carbon number [73, 75–78]. For example, vanillin can be converted to vanillin alcohol and *p*-cresol, which are condensed to form high carbon number cycloalkanes using Ru@Al catalyst [77]. Compared to lignin model compounds, the structure of real lignin is more complicated and its conversion is thus challenging. The dissociation energy of the C–O bond ($\sim$218–314 kJ mol^{-1}) is lower than that of the C–C single bond ($\sim$384 kJ mol^{-1}). Therefore, it is theoretically possible to selectively cleave C–O–C bonds in the lignin structure by keeping C–C bonds intact [73]. However, the selection of proper catalysts and reaction conditions is necessary. Wang et al. used a highly concentrated aqueous solution of $ZnCl_2$ (63%) to depolymerize softwood technical lignin at 120–200 °C (Scheme 12.10) [75]. Zn^{2+} ions in such a concentrated solution selectively coordinate with oxygen atoms of C–O–C bonds to cause easier cleavage of key linkages of lignin under mild conditions.

Alkylphenols (both monomeric and dimeric) are obtained as major products that can be converted into cycloalkanes using Ru/C catalyst. Some linear alkanes are also found, indicating that Ru/C can assist RO through hydrogenolysis. Other dimers different from lignin subunits are also obtained, confirming the occurrence of coupling reactions. Under certain conditions, the cleavage of ether bonds releases benzylic carbonium ions that subsequently react with aromatic species to give coupling products.

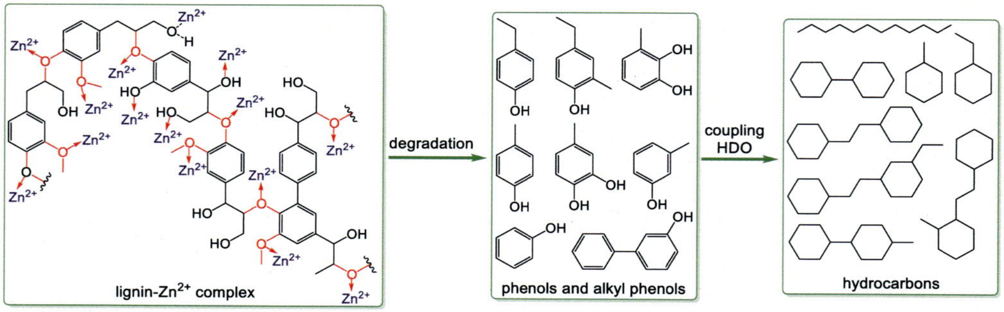

Scheme 12.10 Conversion of lignin to hydrocarbon biofuels using ZnCl₂ solution and Ru/C catalyst.

12.5 Perspectives and challenges

Traditional aviation fuel typically consists of n-alkanes (20%), isoalkanes (40%), cycloalkanes (20%), aromatic hydrocarbons (20%), small amounts of olefins (<5%), and sulfur (<3000 ppm) [79]. Each component plays an important role in aviation fuel. Therefore, to replace the existing aviation fuel with 100% biomass-derived aviation fuel, we need to produce a mixture of n-, iso-, and cycloalkanes along with aromatics from biomass. This can be achieved by converting all three components of lignocellulose (hemicellulose, cellulose, and lignin) to aviation fuel. At the same time, achieving a high atom economy is equally important. Several strategies have been developed to utilize cellulose and hemicellulose to produce plenty of n- and isoalkanes. On the other hand, lignin is converted into cycloalkanes and aromatics. But the catalytic technologies for the clean conversion of raw lignin still lack fundamental understanding.

Another goal is to make biofuels economically competitive with conventional aviation fuel. The development of a highly integrated process is thus required, that can avoid expensive biomass pretreatment steps and allow "one-pot" conversion. The design of robust, recyclable, and multifunctional catalysts is required to combine many reactions in one step. A well-known multifunctional catalyst, Pd/NbOPO₄, can perform HAA and HDO steps efficiently to produce C₁₅ alkanes from 2-MF and furfural [62]. However, the catalyst undergoes phase transformation, and sintering of Pd particles is observed when the reactions are operated in batch mode (80 °C for HAA and 200 °C, 40 bar H₂ for HDO). The catalyst deactivates by the carbon deposition resulted from the severe

Production of long-chain hydrocarbon biofuels 349

polymerization of large oxygenated HAA products due to their low solubility in alkane solvent (cyclohexane). Therefore, a suitable catalyst that can operate at mild conditions without compromising catalytic activity is required. In this regard, transition metal carbides based on molybdenum and tungsten exhibit high selectivity toward C–O bond scission in HDO reactions under relatively mild conditions [80, 81]. In addition, they are highly resistant to metal sintering, which keeps the active metals well-dispersed under the reductive environment. One problem of transition metal carbides is their high sensitivity toward water and oxygen that causes catalyst deactivation through the formation of metal oxides or oxycarbides. The deactivation by oxygen-poisoning can be prevented by using metal modifiers, which are deposited on transition metal carbide surfaces to lower the binding energy for oxygen. Alternatively, modification of the carbide surface to be hydrophobic and spatial confined carbides are also proposed to stabilize the carbide catalysts [81]. The solvent is another crucial component of the process. Therefore, it will be highly economical if the reactions are performed without a solvent or in the presence of a reusable solvent.

A big obstacle in implementing the HDO process at a commercial level is the high demand for external H_2. Consumption of H_2 can be minimized in many ways. For example, combined metal hydrolysis and HDO, combined in situ reforming and HDO, water-assisted in situ HDO, catalytic transfer hydrogenation/hydrogenolysis, and nonthermal plasma technology are gaining increased attention, which can successfully avoid the use of external H_2 [82]. Considering the low price and high availability of organic by-products (e.g., alcohols, polyols, and acids) produced from thermal conversion of biomass, they can serve as promising H_2 sources for in situ HDO reactions. Recently, an excellent strategy of avoiding external H_2 has been reported, where water provides the required H_2 during the in situ HDO of guaiacol [83]. Ni supported on CeO_2/carbon was used as a catalyst for multiple advantages. First, CeO_2 provides oxygen adsorption sites and facilitates H_2 release from the water. Second, activated carbon has an extremely high surface area to support and stabilize cluster-range Ni species. Finally, the hydrophobic nature of activated carbon can avoid the possibility of catalyst deactivation by water poisoning. Activation of water is believed to occur on CeO_2 oxygen vacancies, while the C–O bond cleavage is facilitated by Ni clusters, and the subsequent steps take place on the Ni–CeO_2 interface. The dispersion of CeO_2 on carbon reduces the particle size of CeO_2 and maximizes the Ni–CeO_2 interface, which eventually increases the activity. Another effective way of minimizing H_2 consumption is magnetic heating by using ferromagnetic nanoparticles, which release heat through hysteresis losses by exposing them to high-frequency alternative magnetic fields [84]. The close contact between the catalyst and the heating agent significantly increases the catalytic activity. Selective HDO can be performed using FeC@Ru nanoparticles at a low catalyst loading and a very low pressure of H_2 (3 bar). The process is benefited from the high heating power of the $Fe_{2.2}C$ nanoparticles and the high activities of Ru. In most cases, all these new techniques are applied to model oxygenated compounds, which is pivotal to design working catalysts rationally. Now, a major question arises—are these new strategies applicable to raw biomass feedstocks? To find the solution, we need more case studies of these techniques using the open literature already available on complex feedstocks.

12.6 Conclusions

This chapter discussed different routes for the production of hydrocarbon biofuels from important platform chemicals derived from three major components of lignocellulose biomass. The catalyst development is one of the main tasks for future biorefineries to effectively utilize the abundant biomass resources. Remarkable efforts have been made in the last two decades to develop new catalytic routes for the conversion of highly oxygenated biomass feedstocks, where HDO is a key technique. Several fundamental studies were done using model compounds to understand the role of catalysts and the HDO reaction mechanism. Successful efforts were also made for real biomass precursors. However, the complex nature of biomass remains a major hurdle for the selective transformation of raw feedstocks into desired products. Another major concern is the cost of hydrocarbon biofuels that directly depends on the cost of hydrogen production. Therefore, the search for cheap technologies and resources for hydrogen production is an important task to commercialize the biomass conversion processes.

References

[1] R. Luque, L. Herrero-Davila, J.M. Campelo, J.H. Clark, J.M. Hidalgo, D. Luna, J.M. Marinas, A.A. Romero, Biofuels: a technological perspective, Energ. Environ. Sci. 1 (2008) 542–564.

[2] C.R. Soccol, L.P. de Souza Vandenberghe, A.B.P. Medeiros, S.G. Karp, M. Buckeridge, L.P. Ramos, A.P. Pitarelo, V. Ferreira-Leitão, L.M.F. Gottschalk, M.A. Ferrara, E.P. da Silva Bon, L.M.P. de Moraes, J. de Amorim Araújo, F.A.G. Torres, Bioethanol from lignocelluloses: status and perspectives in Brazil, Bioresour. Technol. 101 (2010) 4820–4825.

[3] J.R. Regalbuto, Cellulosic biofuels—got gasoline? Science 325 (2009) 822–824.

[4] M.J. Climent, A. Corma, S. Iborra, Conversion of biomass platform molecules into fuel additives and liquid hydrocarbon fuels, Green Chem. 16 (2014) 516–547.

[5] C.J. Chuck, J. Donnelly, The compatibility of potential bioderived fuels with jet A-1 aviation kerosene, Appl. Energy 118 (2014) 83–91.

[6] S. De, B. Saha, R. Luque, Hydrodeoxygenation processes: advances on catalytic transformations of biomass-derived platform chemicals into hydrocarbon fuels, Bioresour. Technol. 178 (2015) 108–118.

[7] Y. Nakagawa, S. Liu, M. Tamura, K. Tomishige, Catalytic total hydrodeoxygenation of biomass-derived polyfunctionalized substrates to alkanes, ChemSusChem 8 (2015) 1114–1132.

[8] W. Jin, L. Pastor-Pérez, D. Shen, A. Sepúlveda-Escribano, S. Gu, T. Ramirez Reina, Catalytic upgrading of biomass model compounds: novel approaches and lessons learnt from traditional hydrodeoxygenation—a review, ChemCatChem 11 (2019) 924–960.

[9] S. Kim, E.E. Kwon, Y.T. Kim, S. Jung, H.J. Kim, G.W. Huber, J. Lee, Recent advances in hydrodeoxygenation of biomass-derived oxygenates over heterogeneous catalysts, Green Chem. 21 (2019) 3715–3743.

[10] S. Dutta, Hydro(deoxygenation) reaction network of lignocellulosic oxygenates, ChemSusChem 13 (2020) 2894–2915.

[11] R. Xing, A.V. Subrahmanyam, H. Olcay, W. Qi, G.P. van Walsum, H. Pendse, G.W. Huber, Production of jet and diesel fuel range alkanes from waste hemicellulose-derived aqueous solutions, Green Chem. 12 (2010) 1933–1946.

[12] J. Cueto, L. Faba, E. Díaz, S. Ordóñez, Performance of basic mixed oxides for aqueous-phase 5-hydroxymethylfurfural-acetone aldol condensation, Appl. Catal. Environ. 201 (2017) 221–231.

[13] J.-F. Zhang, Z.-M. Wang, Y.-J. Lyu, H. Xie, T. Qi, Z.-B. Si, L.-J. Liu, H.-Q. Yang, C.-W. Hu, Synergistic catalytic mechanism of acidic silanol and basic alkylamine bifunctional groups over SBA-15 zeolite toward aldol condensation, J. Phys. Chem. C 123 (2019) 4903–4913.

[14] W. Li, M. Su, T. Yang, T. Zhang, Q. Ma, S. Li, Q. Huang, Preparation of two different crystal structures of cerous phosphate as solid acid catalysts: their different catalytic performance in the aldol condensation reaction between furfural and acetone, RSC Adv. 9 (2019) 16919–16928.

[15] O. Kikhtyanin, D. Kadlec, R. Velvarská, D. Kubička, Using mg-Al mixed oxide and reconstructed hydrotalcite as basic catalysts for aldol condensation of furfural and cyclohexanone, ChemCatChem 10 (2018) 1464–1475.

[16] L. Faba, E. Díaz, S. Ordóñez, Aqueous-phase furfural-acetone aldol condensation over basic mixed oxides, Appl. Catal. Environ. 113-114 (2012) 201–211.

[17] L. Faba, E. Díaz, S. Ordóñez, Improvement on the catalytic performance of mg–Zr mixed oxides for furfural–acetone aldol condensation by supporting on mesoporous carbons, ChemSusChem 6 (2013) 463–473.

[18] A.V. Subrahmanyam, S. Thayumanavan, G.W. Huber, C–C Bond formation reactions for biomass-derived molecules, ChemSusChem 3 (2010) 1158–1161.

[19] A. Corma, O. de la Torre, M. Renz, N. Villandier, Production of high-quality diesel from biomass waste products, Angew. Chem. Int. Ed. 50 (2011) 2375–2378.

[20] A. Corma, O. de la Torre, M. Renz, High-quality diesel from hexose- and pentose-derived biomass platform molecules, ChemSusChem 4 (2011) 1574–1577.

[21] A. Corma, O. de la Torre, M. Renz, Production of high quality diesel from cellulose and hemicellulose by the sylvan process: catalysts and process variables, Energ. Environ. Sci. 5 (2012) 6328–6344.

[22] S. Li, N. Li, G. Li, L. Li, A. Wang, Y. Cong, X. Wang, T. Zhang, Lignosulfonate-based acidic resin for the synthesis of renewable diesel and jet fuel range alkanes with 2-methylfuran and furfural, Green Chem. 17 (2015) 3644–3652.

[23] G. Li, N. Li, Z. Wang, C. Li, A. Wang, X. Wang, Y. Cong, T. Zhang, Synthesis of high-quality diesel with furfural and 2-methylfuran from hemicellulose, ChemSusChem 5 (2012) 1958–1966.

[24] J.C. Serrano-Ruiz, D.J. Braden, R.M. West, J.A. Dumesic, Conversion of cellulose to hydrocarbon fuels by progressive removal of oxygen, Appl. Catal. Environ. 100 (2010) 184–189.

[25] E. Karimi, I.F. Teixeira, L.P. Ribeiro, A. Gomez, R.M. Lago, G. Penner, S.W. Kycia, M. Schlaf, Ketonization and deoxygenation of alkanoic acids and conversion of levulinic acid to hydrocarbons using a red mud bauxite mining waste as the catalyst, Catal. Today 190 (2012) 73–88.

[26] A. Corma, M. Renz, C. Schaverien, Coupling fatty acids by ketonic decarboxylation using solid catalysts for the direct production of diesel, lubricants, and chemicals, ChemSusChem 1 (2008) 739–741.

[27] L. Deng, Y. Fu, Q.-X. Guo, Upgraded acidic components of bio-oil through catalytic ketonic condensation, Energy Fuel 23 (2009) 564–568.

[28] T.N. Pham, T. Sooknoi, S.P. Crossley, D.E. Resasco, Ketonization of carboxylic acids: mechanisms, catalysts, and implications for biomass conversion, ACS Catal. 3 (2013) 2456–2473.

[29] Y.-B. Huang, Z. Yang, J.-J. Dai, Q.-X. Guo, Y. Fu, Production of high quality fuels from lignocellulose-derived chemicals: a convenient C–C bond formation of furfural, 5-methylfurfural and aromatic aldehyde, RSC Adv. 2 (2012) 11211–11214.

[30] H. Zang, K. Wang, M. Zhang, R. Xie, L. Wang, E.Y.X. Chen, Catalytic coupling of biomass-derived aldehydes into intermediates for biofuels and materials, Cat. Sci. Technol. 8 (2018) 1777–1798.

[31] Y. Jing, Q. Xia, X. Liu, Y. Wang, Production of low-freezing-point highly branched alkanes through Michael addition, ChemSusChem 10 (2017) 4817–4823.

[32] Y. Jing, Q. Xia, J. Xie, X. Liu, Y. Guo, J.-J. Zou, Y. Wang, Robinson annulation-directed synthesis of jet-fuel-ranged alkylcyclohexanes from biomass-derived chemicals, ACS Catal. 8 (2018) 3280–3285.

[33] J.Q. Bond, D.M. Alonso, D. Wang, R.M. West, J.A. Dumesic, Integrated catalytic conversion of γ-valerolactone to liquid alkenes for transportation fuels, Science 327 (2010) 1110–1114.

[34] K. Yan, Y. Yang, J. Chai, Y. Lu, Catalytic reactions of gamma-valerolactone: a platform to fuels and value-added chemicals, Appl. Catal. Environ. 179 (2015) 292–304.

[35] X. Li, X. Luo, Y. Jin, J. Li, H. Zhang, A. Zhang, J. Xie, Heterogeneous sulfur-free hydrodeoxygenation catalysts for selectively upgrading the renewable bio-oils to second generation biofuels, Renew. Sustain. Energy Rev. 82 (2018) 3762–3797.

[36] J.A. Hunns, M. Arroyo, A.F. Lee, J.M. Escola, D. Serrano, K. Wilson, Hierarchical mesoporous Pd/ZSM-5 for the selective catalytic hydrodeoxygenation of m-cresol to methylcyclohexane, Cat. Sci. Technol. 6 (2016) 2560–2564.

[37] C.A. Teles, P.M. de Souza, A.H. Braga, R.C. Rabelo-Neto, A. Teran, G. Jacobs, D.E. Resasco, F.B. Noronha, The role of defect sites and oxophilicity of the support on the phenol hydrodeoxygenation reaction, Appl. Catal. Environ. 249 (2019) 292–305.

[38] M. Chia, Y.J. Pagán-Torres, D. Hibbitts, Q. Tan, H.N. Pham, A.K. Datye, M. Neurock, R.J. Davis, J.A. Dumesic, Selective hydrogenolysis of polyols and cyclic ethers over bifunctional surface sites on rhodium–rhenium catalysts, J. Am. Chem. Soc. 133 (2011) 12675–12689.

[39] Y.T. Kim, J.A. Dumesic, G.W. Huber, Aqueous-phase hydrodeoxygenation of sorbitol: a comparative study of Pt/Zr phosphate and PtReOx/C, J. Catal. 304 (2013) 72–85.

[40] Q.-N. Xia, Q. Cuan, X.-H. Liu, X.-Q. Gong, G.-Z. Lu, Y.-Q. Wang, Pd/NbOPO4 multifunctional catalyst for the direct production of liquid alkanes from aldol adducts of furans, Angew. Chem. Int. Ed. 53 (2014) 9755–9760.

[41] F. Xue, D. Ma, T. Tong, X. Liu, Y. Hu, Y. Guo, Y. Wang, Contribution of different NbOx species in the hydrodeoxygenation of 2,5-dimethyltetrahydrofuran to hexane, ACS Sustain. Chem. Eng. 6 (2018) 13107–13113.

[42] Y. Nakagawa, M. Tamura, K. Tomishige, Catalytic reduction of biomass-derived furanic compounds with hydrogen, ACS Catal. 3 (2013) 2655–2668.

[43] A.D. Sutton, F.D. Waldie, R. Wu, M. Schlaf, L.A. 'Pete' Silks, J.C. Gordon, The hydrodeoxygenation of bioderived furans into alkanes, Nat. Chem. 5 (2013) 428–432.

[44] G.R. Jenness, W. Wan, J.G. Chen, D.G. Vlachos, Reaction pathways and intermediates in selective ring opening of biomass-derived heterocyclic compounds by iridium, ACS Catal. 6 (2016) 7002–7009.

[45] W. Wan, G.R. Jenness, K. Xiong, D.G. Vlachos, J.G. Chen, Ring-opening reaction of furfural and tetrahydrofurfuryl alcohol on hydrogen-predosed iridium(111) and cobalt/iridium(1 1 1) surfaces, ChemCatChem 9 (2017) 1701–1707.

[46] H.-J. Song, J. Deng, M.-S. Cui, X.-L. Li, X.-X. Liu, R. Zhu, W.-P. Wu, Y. Fu, Alkanes from bioderived furans by using metal triflates and palladium-catalyzed hydrodeoxygenation of cyclic ethers, ChemSusChem 8 (2015) 4250–4255.

[47] S. Dutta, B. Saha, Hydrodeoxygenation of furylmethane oxygenates to jet and diesel range fuels: probing the reaction network with supported palladium catalyst and hafnium triflate promoter, ACS Catal. 7 (2017) 5491–5499.

[48] Z. Li, R.S. Assary, A.C. Atesin, L.A. Curtiss, T.J. Marks, Rapid ether and alcohol C–O bond hydrogenolysis catalyzed by tandem high-valent metal triflate + supported Pd catalysts, J. Am. Chem. Soc. 136 (2014) 104–107.

[49] G.W. Huber, J.N. Chheda, C.J. Barrett, J.A. Dumesic, Production of liquid alkanes by aqueous-phase processing of biomass-derived carbohydrates, Science 308 (2005) 1446–1450.

[50] J. Dumesic, G. Huber, J. Chheda, C. Barrett, Stable, Aqueous-Phase, Basic Catalytsts and Reactions Catalyzed Thereby, US Pat., 2008. 20080058563.

[51] R.M. West, Z.Y. Liu, M. Peter, J.A. Dumesic, Liquid alkanes with targeted molecular weights from biomass-derived carbohydrates, ChemSusChem 1 (2008) 417–424.

[52] J.C. Gordon, L.A. Silks, A.D. Sutton, R. Wu, M. Schlaf, F. Waldie, R. West, D.I. Collias, Compounds and Methods for the Production of Long Chain Hydrocarbons from Biological Sources, WO Pat., 2013. 2013040311.

Production of long-chain hydrocarbon biofuels **353**

[53] M. Chatterjee, K. Matsushima, Y. Ikushima, M. Sato, T. Yokoyama, H. Kawanami, T. Suzuki, Production of linear alkane via hydrogenative ring opening of a furfural-derived compound in supercritical carbon dioxide, Green Chem. 12 (2010) 779–782.

[54] J. Yang, N. Li, S. Li, W. Wang, L. Li, A. Wang, X. Wang, Y. Cong, T. Zhang, Synthesis of diesel and jet fuel range alkanes with furfural and ketones from lignocellulose under solvent free conditions, Green Chem. 16 (2014) 4879–4884.

[55] L. Faba, E. Díaz, S. Ordóñez, One-pot aldol condensation and hydrodeoxygenation of biomass-derived carbonyl compounds for biodiesel synthesis, ChemSusChem 7 (2014) 2816–2820.

[56] E.S. Olson, C. Heide, Multiproduct Biorefinery for Synthesis of Fuel Components and Chemicals from Lignocellulosics Via Levulinate Condensations, WO Pat., 2010. WO2010141950A2.

[57] C. Li, D. Ding, Q. Xia, X. Liu, Y. Wang, Conversion of raw lignocellulosic biomass into branched long-chain alkanes through three tandem steps, ChemSusChem 9 (2016) 1712–1718.

[58] D. Liu, E.Y.X. Chen, Integrated catalytic process for biomass conversion and upgrading to C12 furoin and alkane fuel, ACS Catal. 4 (2014) 1302–1310.

[59] G. Li, N. Li, S. Li, A. Wang, Y. Cong, X. Wang, T. Zhang, Synthesis of renewable diesel with hydroxyacetone and 2-methyl-furan, Chem. Commun. 49 (2013) 5727–5729.

[60] G. Li, N. Li, X. Wang, X. Sheng, S. Li, A. Wang, Y. Cong, X. Wang, T. Zhang, Synthesis of diesel or jet fuel range cycloalkanes with 2-methylfuran and cyclopentanone from lignocellulose, Energy Fuel 28 (2014) 5112–5118.

[61] G. Li, N. Li, J. Yang, L. Li, A. Wang, X. Wang, Y. Cong, T. Zhang, Synthesis of renewable diesel range alkanes by hydrodeoxygenation of furans over Ni/Hβ under mild conditions, Green Chem. 16 (2014) 594–599.

[62] Q. Xia, Y. Xia, J. Xi, X. Liu, Y. Zhang, Y. Guo, Y. Wang, Selective one-pot production of high-grade diesel-range alkanes from furfural and 2-methylfuran over Pd/NbOPO4, ChemSusChem 10 (2017) 747–753.

[63] A. Morone, M. Apte, R.A. Pandey, Levulinic acid production from renewable waste resources: bottlenecks, potential remedies, advancements and applications, Renew. Sustain. Energy Rev. 51 (2015) 548–565.

[64] L. Faba, E. Díaz, S. Ordóñez, Base-catalyzed condensation of Levulinic acid: a new biorefinery upgrading approach, ChemCatChem 8 (2016) 1490–1494.

[65] Z. Li, J. Zhang, M.M. Nielsen, H. Wang, C. Chen, J. Xu, Y. Wang, T. Deng, X. Hou, Efficient C–C bond formation between two levulinic acid molecules to produce C10 compounds with the cooperation effect of Lewis and Brønsted acids, ACS Sustain. Chem. Eng. 6 (2018) 5708–5711.

[66] M. Mascal, S. Dutta, I. Gandarias, Hydrodeoxygenation of the Angelica lactone dimer, a cellulose-based feedstock: simple, high-yield synthesis of branched C7–C10 gasoline-like hydrocarbons, Angew. Chem. Int. Ed. 53 (2014) 1854–1857.

[67] F. Chang, S. Dutta, M. Mascal, Hydrogen-economic synthesis of gasoline-like hydrocarbons by catalytic hydrodecarboxylation of the biomass-derived Angelica lactone dimer, ChemCatChem 9 (2017) 2622–2626.

[68] O.O. Ayodele, F.A. Dawodu, D. Yan, H. Dong, J. Xin, S. Zhang, Production of bio-based gasoline by noble-metal-catalyzed hydrodeoxygenation of α-Angelica lactone derived di/trimers, ChemistrySelect 2 (2017) 4219–4225.

[69] J. Xu, N. Li, X. Yang, G. Li, A. Wang, Y. Cong, X. Wang, T. Zhang, Synthesis of diesel and jet fuel range alkanes with furfural and Angelica lactone, ACS Catal. 7 (2017) 5880–5886.

[70] S. Dutta, I.K.M. Yu, D.C.W. Tsang, Y.H. Ng, Y.S. Ok, J. Sherwood, J.H. Clark, Green synthesis of gamma-valerolactone (GVL) through hydrogenation of biomass-derived levulinic acid using non-noble metal catalysts: a critical review, Chem. Eng. J. 372 (2019) 992–1006.

[71] J.C. Serrano-Ruiz, D. Wang, J.A. Dumesic, Catalytic upgrading of levulinic acid to 5-nonanone, Green Chem. 12 (2010) 574–577.

[72] D.M. Alonso, J.Q. Bond, J.C. Serrano-Ruiz, J.A. Dumesic, Production of liquid hydrocarbon transportation fuels by oligomerization of biomass-derived C9 alkenes, Green Chem. 12 (2010) 992–999.

[73] H. Wang, H. Ruan, H. Pei, H. Wang, X. Chen, M.P. Tucker, J.R. Cort, B. Yang, Biomass-derived lignin to jet fuel range hydrocarbons via aqueous phase hydrodeoxygenation, Green Chem. 17 (2015) 5131–5135.

[74] H. Wang, M. Feng, B. Yang, Catalytic hydrodeoxygenation of anisole: an insight into the role of metals in transalkylation reactions in bio-oil upgrading, Green Chem. 19 (2017) 1668–1673.

[75] H. Wang, L. Zhang, T. Deng, H. Ruan, X. Hou, J.R. Cort, B. Yang, ZnCl2 induced catalytic conversion of softwood lignin to aromatics and hydrocarbons, Green Chem. 18 (2016) 2802–2810.

[76] J.S. Yoon, Y. Lee, J. Ryu, Y.-A. Kim, E.D. Park, J.-W. Choi, J.-M. Ha, D.J. Suh, H. Lee, Production of high carbon number hydrocarbon fuels from a lignin-derived α-O-4 phenolic dimer, benzyl phenyl ether, via isomerization of ether to alcohols on high-surface-area silica-alumina aerogel catalysts, Appl. Catal. Environ. 142-143 (2013) 668–676.

[77] I. Yati, A.A. Dwiatmoko, J.S. Yoon, J.-W. Choi, D.J. Suh, J. Jae, J.-M. Ha, One-pot catalytic reaction to produce high-carbon-number dimeric deoxygenated hydrocarbons from lignin-derived monophenyl vanillin using Al2O3-cogelled Ru nanoparticles, Appl. Catal. A Gen. 524 (2016) 243–250.

[78] H. Wang, H. Ruan, M. Feng, Y. Qin, H. Job, L. Luo, C. Wang, M.H. Engelhard, E. Kuhn, X. Chen, M.P. Tucker, B. Yang, One-pot process for hydrodeoxygenation of lignin to alkanes using Ru-based bimetallic and bifunctional catalysts supported on zeolite Y, ChemSusChem 10 (2017) 1846–1856.

[79] F. Cheng, C.E. Brewer, Producing jet fuel from biomass lignin: potential pathways to alkyl-benzenes and cycloalkanes, Renew. Sustain. Energy Rev. 72 (2017) 673–722.

[80] Z. Lin, R. Chen, Z. Qu, J.G. Chen, Hydrodeoxygenation of biomass-derived oxygenates over metal carbides: from model surfaces to powder catalysts, Green Chem. 20 (2018) 2679–2696.

[81] J. Pang, J. Sun, M. Zheng, H. Li, Y. Wang, T. Zhang, Transition metal carbide catalysts for biomass conversion: a review, Appl. Catal. Environ. 254 (2019) 510–522.

[82] W. Jin, L. Pastor-Pérez, J. Yu, J.A. Odriozola, S. Gu, T.R. Reina, Cost-effective routes for catalytic biomass upgrading, Curr. Opin. Green Sustain. Chem. 23 (2020) 1–9.

[83] W. Jin, L. Pastor-Pérez, J.J. Villora-Picó, A. Sepúlveda-Escribano, S. Gu, T.R. Reina, Investigating new routes for biomass upgrading: "H2-free" hydrodeoxygenation using Ni-based catalysts, ACS Sustain. Chem. Eng. 7 (2019) 16041–16049.

[84] J.M. Asensio, A.B. Miguel, P.-F. Fazzini, P.W.N.M. van Leeuwen, B. Chaudret, Hydrodeoxygenation using magnetic induction: high-temperature heterogeneous catalysis in solution, Angew. Chem. Int. Ed. 58 (2019) 11306–11310.

CHAPTER 13

Bioeconomy of hydrocarbon biorefinery processes

Janakan S. Saral, R.S. Ajmal, and Panneerselvam Ranganathan
Department of Chemical Engineering, National Institute of Technology Calicut, Kozhikode, India

Contents

13.1	Introduction	355
13.2	Methodology of economic analysis and types of biorefinery	357
	13.2.1 Lignocellulosic biomass-based hydrocarbon biorefinery	358
	13.2.2 Microalgae-based hydrocarbon biorefinery	369
	13.2.3 Waste-based hydrocarbon biorefinery	374
13.3	Conclusions and future prospects	380
	Acknowledgment	380
	References	381

Abbreviations

GGE	gasoline gallon equivalent
GHG	greenhouse gas
HTL	hydrothermal liquefaction
IEA	International Energy Agency
IRR	internal rate of return
MFSP	minimum fuel selling price
MISR	metric for inspecting sales and reactant
MT	metric tons
NPV	net present value
ROI	return on investment
tpd	tons per day

13.1 Introduction

The world is still dependent on fossil fuels to power vehicles and to run factories. However, the world looks for alternative energy from renewable carbon sources since fossil fuels are running out of supply and associated with environmental issues. In this aspect, bio–based energy has shown an inevitable source to sustain the present and future requirements of mankind. Aiming for biofuel alone from biological feedstocks may not be an economically viable option for sustainable energy management. Thus, the integrated biorefinery concept was introduced in the literature. The utilization of whole biomass to obtain multiple products, such as biofuels, bioenergy, and biochemical, with minimization of the waste

Hydrocarbon Biorefinery
https://doi.org/10.1016/B978-0-12-823306-1.00011-X

Copyright © 2022 Elsevier Inc.
All rights reserved.

generated during the conversion process are focused on the integrated biorefinery concept. The International Energy Agency (IEA) Bioenergy Task 42 defines biorefining as "Biorefinery is the sustainable processing of biomass into a spectrum of marketable products and energy" [1]. A biorefinery can be defined as a framework or a structure in which biomass is utilized in an optimal manner to produce multiple products and tries to be self-sustaining and not harmful to the environment. This concept is used to obtain different products using various biomass feedstocks from lignocellulose materials, various macro and microalgal materials, and waste from food and municipal solids.

A biorefinery could be one of the crucial parts of the bio-based economy or bioeconomy, where different types of biomass are utilized and processed to diverse biogenic products [2]. Also, it mitigates greenhouse gas (GHG) emission as a renewable carbon resource at the same time providing opportunities for the society around. Further, the biorefinery is focused on creating multiple jobs and entrepreneurial openings which are more prominent in the rural areas. To advance the bioeconomy for sustainable development, a biorefinery is to be developed using the optimized and improved methods of biomass conversion. Also, the integration of biorefinery is imperative when the production of biofuels themselves is not profitable and sometimes not energetically favorable. In an integrated biorefinery with biofuel as the major product, by-products should be manufactured using the waste of the feedstock. The production of such value-added products can help in making the biorefinery economically feasible. Moreover, it also helps to decrease the selling price of biofuel to a more competitive price point. Furthermore, the by-products from hydrocarbon biorefinery should be utilized to produce steam and or electricity which can be used in the plant itself to make the whole process energetically feasible.

Various biorefinery concepts reported in the literature are based on different feedstocks, as shown in Table 13.1 [3]. These biorefineries are mostly concentrated on the production of biodiesel from oilseed crops and bioethanol from lignocellulose feedstock. Biofuels generated from the above biorefinery are not well-matched with the existing petroleum refinery setup. Thus, hydrocarbon biorefinery has been proposed in the literature which can produce renewable liquid hydrocarbon biofuels (diesel, jet, and gasoline) or drop-in fuels. The drop-in fuels, the upgraded bio-crude, are produced from biomass using thermochemical pathways like gasification, pyrolysis, and hydrothermal liquefaction. They are functionally equivalent to petroleum-related transportation fuels. The updrading processes used in a biorefinery, such as hydrotreating or hydrocracking are already existing in petroleum refinery. But these processes need to be improved further to match with biomass as feedstock and provide higher yields in a biorefinery. One of the effective options for improvement is coprocessing. Here, drop-in fuels are generated using the conventional petroleum refinery, which substantially lessens the capital costs. However, the economics of the whole process still hinders their capabilities and is a critical problem to improve the biorefinery. Thus, techno-economic analysis helps to understand the costs involved in the setup and running of a plant for a longer time. This chapter

Bioeconomy of hydrocarbon biorefinery processes   357

Table 13.1 Description of biorefinery types [3].

Type of biorefinery	Details
Conventional biorefinery	As per the prevailing industries, such as the sugar, vegetable oils, pulp and paper, and petrochemical industry.
Green biorefinery	Use wet biomass, for example, green grasses and crops.
Whole crop biorefinery	Uses dry or wet milling of biomass, such as cereals like corn and wheat.
Lignocellulosic biorefinery	As a result of the fractionation of lignocellulosic biomass composed by cellulose, hemicellulose, and lignin.
Marine-based biorefinery	Marine biomass as microalgae and macro-algae were used.
Two platform concept biorefinery	Considers platforms as sugar and syngas. The sugars are produced by fractionation of cellulose and hemicellulose and the syngas by thermochemical processes from lignin.
Thermochemical biorefinery	As per several methods, such as torrefaction, pyrolysis, gasification, etc.

thus aims to review the techno-economic analysis of various biorefinery processes for the synthesis of hydrocarbon biofuels reported in the literature. Based on the review, the perspectives and future aspects of techno-economic aspects of hydrocarbon biofuels are proposed.

13.2 Methodology of economic analysis and types of biorefinery

Fig. 13.1 shows the methodology involved in the techno-economic analysis of biorefinery. The main part of the techno-economic study is the development of the process design. First, a collection of literature data from various sources (reports, patents, and published experiment results) related to the biorefinery need to be reviewed and then obtain relevant technical and economic data related to various process units in the biorefinery. Then a preliminary process design is developed to calculate mass and energy flow in a biorefinery. After the simulation of various designs by defining various process units through any simulation software, material and energy flow including sensitivity analysis is obtained. Then economics and investment analysis of the plant are evaluated based on the sizing of the equipment. Later, the financial results of various process designs are calculated and the best design is selected which gives the best minimum fuel selling price of biofuel.

The techno-economic assessment accounts for operational and labor costs and maintenance costs along with the capital costs of all equipments in a plant. It basically determines an accurate net present value (NPV) or minimum fuel selling price (MFSP) for a certain rate of internal rate of return (IRR). These parameters are used to analyze whether it is profitable in investing in a biorefinery. Besides MFSP or NPV analysis using the discount cash flow method, a sensitivity analysis is also an important part of economic analyses of the biorefinery and crucial in deciding the viability of the project [4, 5]. In this

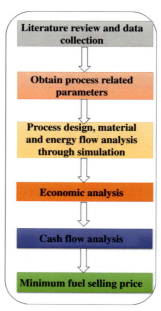

Fig. 13.1 Methodology of techno-economic analysis.

analysis, various parameters like tax rate, labor costs, capital costs, the yield of products, and the impact of the change on the MFSP or profit of the plant are investigated. This study helps in assessing the important factors in deciding the MFSP and hence makes the biorefinery more competitive. Softwares like Aspen Plus and SuperPro Designer are generally used for the detailed process design of a biorefinery. They can work on designing processes that can easily adapt to different scenarios. In this chapter, the techno-economic analysis of three types of hydrocarbon biorefinery is discussed based on the type of feedstocks. These biorefineries are lignocellulosic biomass-based biorefinery, microalgae-based biorefinery, and waste-based biorefinery, as depicted in Fig. 13.2.

13.2.1 Lignocellulosic biomass-based hydrocarbon biorefinery

Biorefinery based on lignocellulosic feedstock covers refining lignocellulosic biomass, such as wood, straw, etc., into a variety of products and fuels. Lignocellulosic feedstocks are widely available at a cheaper cost and become the utmost significant biomass resources of the future. In this section, very recent studies reported in the literature on the techno-economic analysis of the lignocellulosic material-based hydrocarbon biorefinery process model are reviewed. We will look into the biorefinery plant design which deals with different lignocellulosic biomass. The pathways involved in the production of hydrocarbon biofuels from lignocellulosic biomass are shown in Fig. 13.3. These pathways are

Bioeconomy of hydrocarbon biorefinery processes   359

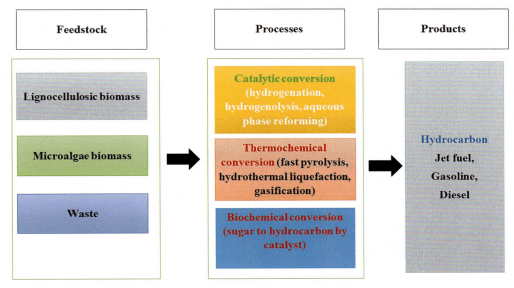

Fig. 13.2 Various hydrocarbon biorefinery concepts.

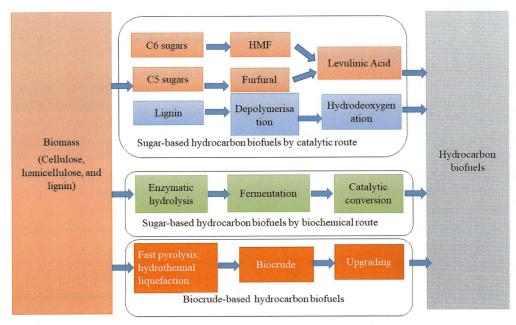

Fig. 13.3 Various possibilities for the production of hydrocarbons from lignocellulose biomass.

360 Hydrocarbon biorefinery

sugar-based hydrocarbon biofuels by catalytic conversion, sugar-based hydrocarbon biofuels by biochemical conversion, and bio–crude-based hydrocarbon biofuels by thermochemical conversion. The following section discusses the economic assessment of each pathway.

13.2.1.1 Sugar-based hydrocarbon biorefinery

In this biorefinery, biomass undergoes decomposition reactions to produce sugar-derived oligomer. This process usually performed using biphasic organic acids. These oligomers produce HMF and furfural by the dehydration reaction. Long-chain oxygen-containing intermediates are obtained by aldol condensation of furfural/HMF with acetone. Lastly, long-chain alkanes like C_8–C_{15} are generated by a series of hydrogenation, dehydration, and isomerization reactions [6]. These long-chain alkanes are the components of jet fuel. Another method to produce hydrocarbon biofuels from lignocellulosic biomass is a fermentation of sugars to alcohols, followed by catalytic conversion of alcohols to renewable hydrocarbon biofuels by hydrodeoxygenation (HDO). However, these processes are still in the research stage and no commercial stage is reported so far.

The feasibility of hydrocarbon biofuels generation from lignocellulosic biomass is shown in Table 13.2. Tao et al. studied the techno-economic analysis for the production of renewable liquid fuels from both corn mill and corn stover feedstocks [8]. In this study, ethanol is first produced from biomass via hydrolysis and fermentation. The ethanol is then upgraded to hydrocarbon biofuels via catalytic reactions, such as dehydration, olefin oligomerization, and hydrotreating. The main product was jet fuel with naphtha and diesel as coproducts in this work. The MFSP was $3.91/GGE and $5.37/GGE for corn mill and corn stover, respectively, by taking the nth-plant assumption (many such plants have been already established). Olcay et al. reported techno-economic analysis for the production of chemicals (furfural, hydroxymethylfurfural, and γ-valerolactone) and drop-in fuels (naphtha, jet fuel, and diesel) from cellulosic biomass through novel aqueous-phase processing [10]. The conceptual design and process simulation were used for economic assessment. The author's analyzed various methods of H_2 synthesis for the economic feasibly of commercial-scale production of hydrocarbon liquid biofuels. The MFSP of jet fuel for the different refinery system configurations was obtained as $0.26–1.67/L, which was 61% lower and 146% higher, than the average conventional jet fuel price. Also, it was reported that an MFSP mainly depends on the hydrogen source and the fuel cost.

Shen et al. reported a techno-economic study of high-energy-density jet fuel production from lignin along with ethanol production from corn stover [11]. In a scenario, a portion of lignin separated from the fermentation process is converted into jet fuel, and the rest is combusted to produce power and heat for the biorefinery process. The conversion of lignin to jet fuel is based on the hydrodeoxygenation (HDO) process. The minimum selling price of lignin jet fuel at a 10% discount rate was found to be in the range of $6.35–1.76/gal. The MFSP for the scenario of combusting complete lignin was $2.83/gal ethanol, with a total capital investment of $463.6 million, the variable

Table 13.2 Collected works on the techno-economic analysis of sugar-based hydrocarbon biofuels from lignocellulose feedstock.

Procedure description	Feedstock	Products	Minimum selling price of biofuel	Ref.
Method: biological conversion Capacity: 2000 MTPD	Lignocellulose	Renewable diesel and succinic acid	$9.55/gallon gasoline equivalent (GGE) for renewable diesel only $5.28/GGE with coproduction	Biddy et al. [7]
Method: biological conversion Capacity: 2000 MTPD	Corn mill, corn stover	Jet fuel, diesel, and naphtha	$3.91/GGE for corn mill; $5.37/GGE for corn stover	Tao et al. [8]
Method: catalytic method C_6 conversion into jet range fuels; C_5 conversion into tetrahydrofurfuryl alcohol and 1,2-pentanediol	Lignocellulose	Jet fuel range alkenes, tetrahydrofurfuryl alcohol, and 1,2-pentanediol	$2.37/kg 1,2-pentanediol	Byun and Han [9]
Method: catalytic method C_6 conversion into renewable liquid fuels; C_5 conversion into furfural, hydrofurfuryl alcohol	Lignocellulose	Chemicals (furfural, hydroxymethylfurfural, and γ-valerolactone) and fuels (naphtha, jet fuel, and diesel)	$1.00–6.31/gal	Olcay et al. [10]
Method: Biorefinery with lignin conversion Capacity: 2000 dry MTPD Tool: Aspen Plus	Corn stover	Ethanol and jet fuel	$6.35–$1.76/gal of lignin jet fuel	Shen et al. [11]
Method: Oil extraction and conversion Capacity: 19 million liter per year of jet fuel	Pennycress	Renewable jet fuel	Revenue M$7.80/year	Mousavi-Avval and Shah [12]
Method: biochemical conversion of biomass to isobutanol and upgraded to hydrocarbon biofuels	Douglas-fir forest residuals	Jet and gasoline fuels along with coproducts of activated carbon and lignosulphate	$2.07/L for jet fuel $2.46/L for gasoline-range fuel	Brandt et al. [13]

GGE, gasoline gallon equivalent.

operating cost of $88.2 million per year (including coproduct credit from selling electricity to the grid, $6.2 million per year), and fixed cost of $11.7 million per year. When 23.6% and 76.1% of lignin was considered for jet fuel production, the MESP was found to be $2.88 and $2.83/gal ethanol, respectively. Further, a sensitivity analysis was carried out to analyze the effect of jet fuel production along with ethanol in a biorefinery. It was reported that jet fuel yield and lignin flow rate can influence the MFSP of ethanol.

In a detailed work of Brandt et al. the techno–economic study of integrated biorefinery was performed where Douglas-fir forest residuals were used to produce hydrocarbon biofuels through the biochemical route [13]. In this route, isobutanol is produced from biomass, followed by the conversion of alcohol to jet fuel. In this route, the coproducts are generated from the lignin-abundant by-product. Three coproduct scenarios were studied to understand the selling price of jet fuel. These coproducts are activated carbon, lignosulfonate, and their combination. The minimum fuel selling price of alternative jet fuel (AJF) was found to be $2.07/L whereas the aviation fuel blendstock was found to have a minimum selling price of $2.46/L. It was found that the coproduct scenario does reduce the price of the product AJF than the baseline scenario. When both activated carbon and lignosulfonate are coproduced simultaneously, there is a substantial decrease in the minimum fuel selling price of around 28%, even though the total operating cost increases by around 30%. The sensitivity analysis was performed for all the scenarios. Total capital investment, fuel yield, and discount rate were the most important factors on the minimum selling price. The cost of feedstock was also found to be a key factor that has an impact on the minimum selling price of the fuel.

13.2.1.2 Bio-crude-based hydrocarbon biorefinery

The lignocellulosic feedstock is converted into biofuels, electricity, and chemicals by various thermochemical conversion processes. Fast pyrolysis and hydrothermal liquefaction are the most important thermochemical conversion methods and are widely used to produce hydrocarbon biofuels from the lignocellulosic biomass [14]. Pyrolysis is the thermal degradation of biomass without the presence of oxygen to produce a variety of products including a solid residue or biochar, liquid called bio–oil, and volatile and noncondensable gases. This process can be conducted either in situ or ex situ mode. In in situ fast pyrolysis, the catalyst is added in the same reactor along with biomass for upgrading, whereas ex situ case, the catalyst is placed in a separate reactor for pyrolysis vapors stabilization. Hydroprocessing is used for the upgradation of bio-oil in the presence of hydrogen. In hydrothermal liquefaction, wet biomass is decomposed into bio–oil, char, and noncondensable gases at elevated temperature and pressure. In literature, the techno–economic feasibility of various routes including in situ and ex situ fast pyrolysis and hydrothermal liquefaction are reported for the production of biofuel and different chemicals from various feedstocks. The schematic process flow diagram for thermochemical based hydrocarbon biorefinery from lignocellulosic biomass is shown in Fig. 13.4.

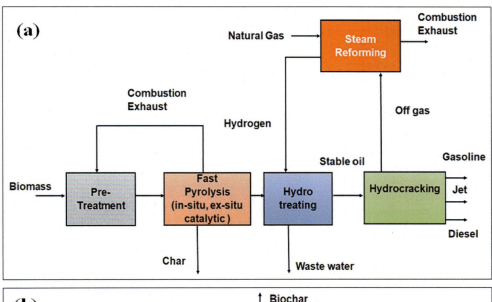
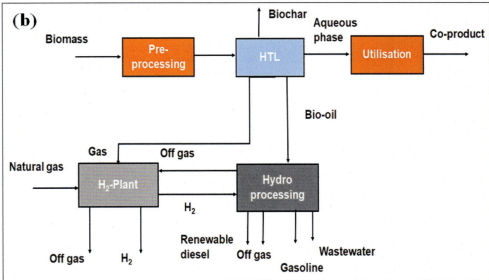

Fig. 13.4 Thermochemical routes in biorefinery from lignocellulosic biomass: (A) fast pyrolysis and (B) hydrothermal liquefaction.

In this section, an attempt is made to analyze these scenarios and calculation methods used for that purpose. Table 13.3 depicts the literature survey on the techno-economic feasibility of hydrocarbon biofuel generation from fast pyrolysis and hydroprocessing. From the data given in the table, it can be concluded that the biofuel produced by the fast pyrolysis pathway will have an MFSP in the range of $0.5–2/L. The credit to

364   Hydrocarbon biorefinery

Table 13.3 The techno-economic analysis of bio-crude-based hydrocarbon biorefinery using fast pyrolysis.

Process details	MFSP	References
The bio-crude production by fast pyrolysis of biomass using poplar wood chips Processing capacity: 2000 tpd	$2.04/GGE	Jones et al. [15]
The biofuels production from corn stover by fast pyrolysis, gasification and biochemical Processing capacity: 2000 tpd	$2.00–5.50/GGE	Anex et al. [16]
The bio-oil production by fast pyrolysis of poplar wood chips Processing capacity: 2000 tpd	$2.11/gal for hydrogen purchase and $3.09/gal for in situ hydrogen production	Wright et al. [17]
The bio-oil and electricity productions from fast pyrolysis of corn stover Processing capacity: 2000 dry tpd	$2.11/gal	Brown et al. [18]
The synthesis of biofuels from woody biomass by mild catalytic fast pyrolysis and partial deoxygenation Capacity: 2000 MTPD	$3.69/gal	Thilakaratne et al. [19]
The synthesis of hydrocarbons by catalytic fast pyrolysis from dried distillers grains with soluble along with ethanol production from corn though integrated process Capacity: 2000 MTPD	$2.27/gal for integrated process and $2.18/gal for ethanol production process only	Wang et al. [20]
The synthesis of biofuels by in situ and ex situ fast pyrolysis of woody biomass Capacity: 2000 MTPD	$1.11/L for in situ case and $1.13/L for ex situ case	Li et al. [21]
The synthesis of bio-oil and miniature electric power generation from fast pyrolysis of pine wood and bio-oil hydroprocessing Capacity: 72 MTPD	$6.25/gasoline gallon equivalent (GGE)	Shemfe et al. [22]
The synthesis of biofuels by integrated catalytic upgrading of pyrolysis oil from biomass Capacity: 2000 MTPD	$1.62/gal	Sharifzadeh et al. [23]
The synthesis of bio-oil from fast pyrolysis of pine wood Capacity: 1000 MTPD	$22.19/GJ with torrefaction and $16.89/GJ without torrefaction	Winjobi et al. [24]

Bioeconomy of hydrocarbon biorefinery processes   **365**

Table 13.3 The techno-economic analysis of bio-crude-based hydrocarbon biorefinery using fast pyrolysis—cont'd

Process details	MFSP	References
The synthesis of bio-oil and from fast pyrolysis of pine wood and bio-oil catalytic cracking via zeolite Capacity: 72 MTPD	£7.20/GGE	Shemfe et al. [25]
The renewable diesel and gasoline from Formate-assisted pyrolysis of pine sawdust Capacity: 2000 MTPD	$4.58/GGE	AlMohamadi et al. [26]
The synthesis of hydrocarbons by stabilization of bio-oil in petroleum refineries Capacity: 2000 MTPD	$2.85/gal for biofuels only and $2.77/gal for biofuels with mixed alcohol production	Li et al. [27]
The synthesis of biofuels from tail-gas reactive pyrolysis of guayule bagasse Capacity: 200 MTPD	1.88 $/L for gasoline, 1.84 $/L for jet fuel, and 1.91 $/L for diesel fuel	Sabaini et al. [28]
The synthesis of biofuels from municipal solid waste (MSW) by integrated intermediate pyrolysis and combined heat and power generation Capacity: 500 ton/h	The levelized cost of electricity is £0.03/kWh	Yang et al. [29]
The synthesis of biofuels from fast pyrolysis based process with two different bio-oil upgrading processes namely, two-stage hydroprocessing and electro-chemical conversion. Capacity: 2000 MTPD	$2.48/gal for bio-oil upgrading; $5.36/kWh electrochemical	Dang et al. [30]
The synthesis of renewable diesel and gasoline from fast pyrolysis of aspen wood chips Capacity: 2000 dry ton/day	$1.09/L for diesel and 1.04/L for gasoline	Patel et al. [31]
The synthesis of aviation biofuels from different routes namely, pyrolysis followed hydroprocessing, gasification followed by Fischer-Tropsch synthesis and pyrolysis followed zeolite cracking Capacity: 100 ton/h	$1.98/L for hydroprocessing, $2.32/L gasification followed by Fischer-Tropsch synthesis, and $2.21/L for zeolite cracking	Michailos and Bridgwater [32]
The synthesis of biofuels from corncob by fast pyrolysis, gasification and combustion Capacity: 96.81 ton/h	$305/t methanol by gasification, $80.1/MWh electricity by combustion and $1.47/GGE bio-oil	Brigagão et al. [33]

Continued

366 Hydrocarbon biorefinery

Table 13.3 The techno-economic analysis of bio-crude-based hydrocarbon biorefinery using fast pyrolysis—cont'd

Process details	MFSP	References
The synthesis of biofuels and coproducts by catalytic fast pyrolysis and hydrodeoxygenation Capacity: 2000 MTPD	$21/GJ for hydropyrolysis and $14/GJ for polygeneration product	Nguyen and Clausen [34]
The synthesis of renewable biofuels by integrated thermochemical process of pyrolysis, gasification, and combustion from Napier grass bagasse Capacity: 50 kg/h	$5.81/GGE ($1.45/L)	Mohammed et al. [35]
Method: Hydrodeoxygenation of fast pyrolysis of bio oil (water soluble and insoluble bio-oil) Capacity 10 MT/year	$0.406–1.465/kg	Bagnato and Sanna [36]
The synthesis of renewable jet fuels by fast pyrolysis of rice husk Capacity: 600 MTPD	$3.21/L	Liu and Wang [37]
Method: Pyrolysis of forest residues Pressure: 1 atm Temperature 500°C Bio-oil yield: (a) 22.86 wt% (noncatalytic) and (b) 21.64 wt% (catalytic)	**(a)** $0.9/L (noncatalytic 200 km) **(b)** $1.50/L (catalytic 200 km) **(c)** $4.36/GGE (noncatalytic 300 km) **(d)** $6.64 (catalytic 300 km)	van Schalkwyk et al. [38]

GGE, gallon gasoline equivalent; *MT*, metric tons; *tpd*, tons per day.

this discrepancy in the MFSP of the biofuel can be given for multiple reasons, like the difference in the biorefinery pathway, the assumptions considered for different parameters, and the difference in the techno–economic analysis and the uncertainty in the parameters of the process.

Brown et al. conducted the techno-economic study of biomass by both fast pyrolysis and by hydroprocessing [18]. In this study, gasoline, diesel fuel, and electricity produced from a 2000 MPTD capacity plant using corn stover as feedstock, the minimum fuel selling price obtained was $2.57/gal. From these analyses, it was reported that the economics of the system depends on the process yield, input cost, and output market value. This process route is not on par with petroleum-based fuels. By this method, 57.4 million gallons of transportation fuel and 223 million kWh of electricity were produced yearly. Even though the appreciable amount of revenue is created from the electric power that does not reflect in the overall economics due to high-end equipment cost involved in the process. Hence, it was reported that an in-depth examination is mandatory to lessen the equipment cost to make it economically viable.

For a biorefinery via pyrolysis using zeolite followed by hydrotreatment, Thilakaratne et al., Li et al., and Dutta et al. reported the MFSP of renewable hydrocarbons (diesel, gasoline, and jet fuel) $0.98/L, $1.11/L, and $0.91/L (all 2011$), respectively [21, 39, 40]. Upgraded bio-oil was compared to diesel range fuel which has a market value of $1.03/L. In summary, the MSFP of upgraded bio-oil was economically competitive with its fossil fuel counterpart compared to crude bio-oil.

Several previously reported literature considered the implementation of the solid biochar as a value-added by-product [41, 42]. These studies provided a positive result on improving the economic feasibility of fast pyrolysis-based biorefinery. More than 50% of the total revenue was the contribution from the activated carbon that was processed from solid biochar in a study by Kuppens et al. [41]. Winjobi et al. employed torrefaction as a pretreatment for fast pyrolysis to enhance bio-oil yield by decreasing biochar yield [24]. Do and Lim compared the economic viability of various conversion methods for biofuels production from empty fruit bunches using a hierarchical four-level economic potential approach [43]. The authors reported that fast pyrolysis-based hydrocarbon biofuels production is more promising compared to the other technologies like bioethanol-based jet fuel production and gasification routes because of the maximum economic potential, ROI, and IRR. Chen et al. studied process simulation followed by the techno-economic viability of mobile auto thermal pyrolysis system based biomass conversion at the local level [44]. This system was compared with other liquid biofuel production methods like fixed bed-based biomass pyrolysis plants and Fischer-Tropsch liquids production via biomass gasification. The author reported that the mobile pyrolysis system is the promising one when considered in a long-term economical aspect.

Recently, Michailos and Bridgwater compared the technical feasibility of three different aviation biofuel production routes from biomass. The biofuels routes are listed as follows: pyrolysis and bio-oil hydroprocessing method, gasification followed by Fischer-Tropsch synthesis, and pyrolysis followed by zeolite cracking. The authors reported that the lowermost minimum selling price was found for bio-fuel hydroprocessing way [32].

van Schalkwyk et al. inspected the techno-economic assessment of bio-oil production from forest residue using the pyrolysis method [38]. Both catalytic and noncatalytic pyrolysis methods were adopted in this study with biomass collecting radii of 100, 200, and 300 km from the biorefinery. The pyrolysis process was carried out at 500°C and 1 atm pressure. Aspen Plus economic analyzer was used for the economic calculation. The bio-oil yield obtained for a noncatalytic method was 22.86 wt%, while it was 21.64 wt% for the catalytic method. The minimum selling price of biofuel was $0.9/L (noncatalytic with 200 km), $1.50/L (catalytic with 200 km), $4.36/GGE (noncatalytic with 300 km), and $6.64/GGE (catalytic with 300 km). From the above details, noncatalytic with 300 km was the most economically feasible one. Even though several techno-economic analysis on biomass fast pyrolysis has been reported in the literature for

368 Hydrocarbon biorefinery

hydrocarbon biofuel generation, however, the techno-economic analysis with various coproducts is yet to be studied in an integrated approach to make the process viable.

Table 13.4 displays the literature outline on the techno-economic performance of hydrocarbon biofuels production from lignocellulosic biomass by hydrothermal liquefaction. The data shown in table that the minimum selling price of hydrocarbon biofuel in this process varies in the range of $0.77–1.2/L. A techno-economic study was conducted by PNNL for HTL of lignocellulosic biomass, followed by hydroprocessing of bio-oil. The minimum fuel selling price was reported to be around 0.53$/LGE [50]. This study considered processing 2000 MTPD (metric tons per day) of dry wood. Zhu et al. investigated the production of renewable fuels using a 'state-of-technology' of the HTL process using woody biomass (15 wt% dry biomass content) followed by hydroprocessing of bio-oil [45]. The MFSP was found to be 0.67 $/L of Gasoline Equivalent (LGE).

Table 13.4 Literature report on the techno-economic calculation for bio-crude-based hydrocarbon biorefinery using HTL.

Process details	Feedstock	Products	MFSP	References
Conversion process: HTL Processing capacity: 2000 dry tpd	Woody biomass	Diesel, gasoline, and jet fuel	$4.44/GGE	Zhu et al. [45]
Conversion process: HTL Processing capacity: 1000 tpd	Forestry residue	Jet fuel	€25.1/GJ bio-jet fuel	Tzanetis et al. [46]
Conversion process: HTL	Lignocellulosic biomass, glycerol, and water	Diesel, gasoline, and jet fuel	$4.32/GGE-coliquefaction; $3.11/GGE-woody biomass	Pedersen et al. [47]
Conversion process: HTL Processing capacity: 1500 tpd (dry)	Lignocellulosic residue	Diesel, gasoline, and jet fuel	€2.92/GGE	Magdeldin, Kohl and Järvinen [48]
Conversion process: HTL Processing capacity: 100 million liters per year	Forest residue	Diesel, gasoline, and jet fuel	$0.82/LGE–$0.90/LGE	Nie and Bi [49]

tpd, tons per day.

Pedersen et al. studied the production of hydrocarbon biofuels from hydrothermal liquefaction of lignocellulosic biomass followed by hydroprocessing of bio-oil [47]. The authors investigated three scenarios to find out the key parameters that affect MFSP: (i) wood-glycerol coliquefaction process followed by thermal cracking and hydroprocessing, (ii) HTL of wood only, and (iii) HTL of wood only with full hydrogenation of all compounds. The MFSP for these scenarios was found as 1.14 $/LGE, 0.82 $/LGE, and 0.94 $/LGE. The sensitivity analysis further showed that bio-crude yield, hydrogen cost, and feedstock costs are the crucial elements affecting the MFSP. Magdeldin et al. reported HTL of forest residue as woody biomass and hydroprocessing bio-oil for the production of hydrocarbon biofuels in Finland [48]. The economic analysis was performed for the production of the polygeneration of renewable liquid fuels, biochar, and hydrogen. The break-even price of the coproducts was found to be €1.03/kg of gasoline or €2.46/kg of hydrogen or €51.4/MWh of biochar. Nie and Bi reported the economic assessment of renewable hydrocarbon biofuels from forest residue through HTL and hydroprocessing of bio-oil in Canada [49]. The minimum selling price (MSP) of HTL biofuels was found to be $0.82/LGE–$0.90/LGE which is about 63%–80% higher than that of petroleum-derived transportation fuels. This study also focused on understanding numerous supply chain designs on the economics of the plant which planned to produce 10 million liters per year of hydrocarbon biofuel. The authors found that the scenario like the conversion of biomass to bio-oil and wood pellet before transporting to conversion facility shows lower operating costs but higher MFSP. Recently, two-stage sequential HTL, coliquefaction of HTL, and cosolvent HTL are the intensified process developed for HTL of lignocellulosic biomass. These improved processes can achieve a lower minimum selling price of fuel to compete with fossil fuels-based transportation fuels. However, there are many engineering challenges associated with the intensified process which may lead to lower technology readiness levels.

13.2.2 Microalgae-based hydrocarbon biorefinery

In this section, a techno-economic assessment for the generation of hydrocarbon biofuels from microalgae is discussed. Three vital constituents of microalgae are protein, lipids, and carbohydrates. These microalgae constituents can be transformed into various bio-products through thermochemical methods including biofuels, biochemicals, and value-added products. The thermochemical processes, such as fast pyrolysis and hydrothermal liquefaction, are used for the production of bio-crude. These are shown in Fig. 13.5. These are lipid extracted algae to bio-crude and defatted algae-bio-crude. Table 13.5 represents the literature studies on the techno-economic analysis of microalgae-based hydrocarbon biorefinery. As observed from the table, the MFSP lies in the range of $3.8–4.5/GGE for the hydrothermal liquefaction pathway.

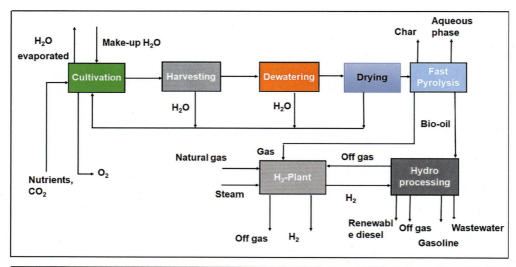

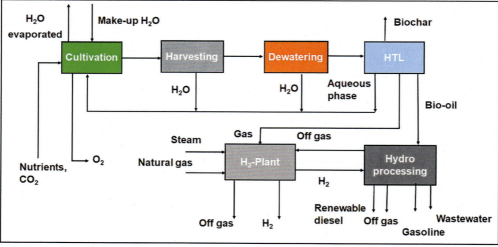

Fig. 13.5 Microalgae-based hydrocarbon biorefinery using (A) fast pyrolysis and (B) hydrothermal liquefaction.

Thilakaratne et al. carried out the techno-economic assessment study using microalgae remnants and catalytic pyrolysis [40]. The microalgae remnant is a low-cost material produced as a by-product of microalgae lipid extraction. The catalytic pyrolysis-based conversion method was employed for this study. Two pretreatment methods were tested before sending the feedstock to pyrolysis. The former method is thermal drying, and the latter is the mechanical dewatering of feedstock. By this method, aromatic hydrocarbons (benzene, toluene, and xylene) are produced, which in turn produces gasoline and diesel.

Table 13.5 Overview of literature on the techno-economics of microalgae-based hydrocarbon biorefinery.

Process details	Feedstock	Products	MFSP	References
Method: HTL Plant capacity: 608 tpd	Lipid extracted microalgae	Diesel, gasoline, and jet fuel	$2.07–7.11/GGE	Zhu et al. [51]
Method: HTL at 300°C, 15 min time, and 1:10 biomass/water ratio	Microalgae	Diesel, gasoline, and jet fuel	$4.49/GGE	Jones et al. [52]
Method: HTL Plant capacity: 2000 tpd	Defatted microalgae	Diesel, gasoline, and jet fuel	$2.57/gal	Ou et al. [53]
Method: microwave-assisted pyrolysis of algae Processing capacity of wastewater: 400 L/day cultivation volume: 1200 L	Wastewater cultivated microalgae	Diesel, gasoline, and jet fuel	$2.23/gal	Xin et al. [54]
Conversion method: HTL Processing capacity of wastewater: 227 million L/day Algal production rate: 25 g/m^2/day	*Chlorella vulgaris*	Diesel, gasoline, jet fuel, and syngas by catalytic hydrothermal gasification of HTL aqueous product	$6.62/gal	Juneja and Murthy [55]
Method: HTL Algal concentration: 20 wt% algae slurry HTL (PFR reactor): 350°C and 20 MPa	Microalgae with different species	Bio-crude	**(i)** Chlorella sp.—11.35$/GGE **(ii)** Scenedesmus sp.—12.03$/GGE **(iii)** Nannochloropsis sp.—11.03$/GGE	Jiang et al. [56]
Method: HTL with different aqueous phase treatment methods Capacity: 170 MTPD	Chlorella sp.	Diesel, gasoline, and jet fuel	$12.5/GGE for direct recycle; $13.8/GGE for CHG; $12.9/GGE for AD	Zhu et al. [57]
Method: two-stage sequential HTL Capacity: 170 MTPD	*Chlorella sorokiniana*	Diesel, gasoline, and jet fuel	$1.6/GGE for two-stage sequential HTL; $2.1/GGE for direct HTL	Gu et al. [58]

tpd, tons per day.

The capacity of the plant was 2000 MTPD. It was reported that mechanical dewatering is more efficient than thermal drying. Thermal drying-based catalytic fast pyrolysis has high operational costs since it requires natural gas for feedstock drying. The hike in the capital cost of mechanical drying can be reduced by developing low-cost equipment for mechanical drying.

Kumar et al. studied a comparative techno-economic analysis of algal thermochemical conversion for producing chemicals (diluent) for bitumen transport [59]. Here two thermochemical routes were analyzed for a plant with a capacity of 2000 tpd: hydrothermal liquefaction and fast pyrolysis. The product diluent consists of liquid naphtha or natural gas condensate. When these chemicals (diluents) are mixed with bitumen, it shrinks the density as well as the viscosity of bitumen suitable for pipeline flow. The HTL process was carried out at 350°C and 20.3 MPa, while fast pyrolysis was performed at 500°C and 0.102 MPa. The yield of bio-oil in HTL and fast pyrolysis were 40.3% and 55.9%, respectively. Aspen Icarus evaluator was used for economic analysis. The sensitivity analysis showed that the product yield was very much related to the product value and the cost of biomass utilized. Based on the pilot-scale data, algal biofuel production from wastewater was assessed on the basis of techno-economic feasibility. Xin et al. reported $2.33/gal as the estimated minimum fuel selling price of the bio-oil [54]. Also, they updated their calculation for the same process and reported the break-even price of about $1.85/gal [60]. Ranganathan and Savithri developed a theoretical design of biofuel processing and production using microalgae-wastewater feedstock through hydrothermal liquefaction, followed by bio-oil hydroprocessing [61]. The treatment of wastewater, harvesting and cultivation of microalgae culture, hydrothermal liquefaction of the biomass and hydroprocessing were considered. A steady state-based simulation was performed for obtaining energy and mass involvement. The techno-economic performance was further studied using the simulation results. It was found that the MFSP of the hydrocarbon biofuels was comparable with the previously published literature values and was found to be $4.3/GGE. Moreover, the most influential parameters of both economics and process were studied on the minimum fuel selling price.

A few studies are focusing on techno-economic assessment on a random basis with a proper evaluation of uncertainty of factors due to the deficiency of commercial knowledge and knowledge gap in the algae-based biorefinery. This uncertainty quantification could be one option for the analysis of the potential risks related to the techno-economic feasibility of a large-scale biofuel production. Jiang et al. reported techno-economic uncertainty quantification for algae-based bio-crude from the hydrothermal liquefaction process [56]. The authors performed uncertainty quantification for PNNL's continuous flow HTL reactor system using Monte Carlo sampling method. They studied uncertainty in bio-crude yield and key economic parameters. It was reported that the MFSP ranges from nearly $5/GGE to $16/GGE with an uncertainty up to ±12% when the price of algae ranges from $400/dry-ash free ton to $1800/dry-ash free ton. They found that the capital investment and composition of microalgae are the most influential parameters that affect the economic uncertainties of the process. Also, they noticed high feedstock cost and less conversion cost leading to high MFSP.

DeRose et al. investigated the techno-economic analysis of the conversion using low lipid content algae to produce renewable fuels [62]. This study determined MFSP requirements of $12.85/GGE for the biochemical pathway and $10.41/GGE for the thermochemical pathway. The sensitivity analysis showed that an MFSP of $3.85/GGE could be achieved by improving the process design. Recently, an intensified process of two-stage sequential HTL was developed by Washington State University to improve the conventional one-stage direct HTL process in terms of the quality of bio-crude. For evaluating the economic performance of this process, Gu et al. analyzed the techno-economic feasibility of two-stage sequential HTL of microalgae for the coproducts and biofuels and compared with the direct stage HTL process [58]. This process was also considered the subsequent upgrading of bio-crude to biofuel via hydroprocessing. The authors found the overall energy return on the investment was 6.73 for two-stage sequential HTL and 5.31 for direct HTL. The minimum fuel selling price of biofuels was found to be $1.61/GGE ($0.49/L) for two-stage sequential HTL and $2.10/GGE for direct HTL. A two-stage sequential HTL was able to achieve a lower MFSP than conventional HTL because of the high yield of biofuels obtained in a two-stage sequential HTL. Financial and process parameters were changed by certain factors to determine their impact on the overall MFSP of the biofuels. The results showed that the bio-oil production rate was the major factor in determining the MFSP. When more biofuels are produced lesser MFSP can be made. Algal feedstock price also plays an important role in determining the MFSP. So further studies need to be done in developing cheaper algae cultivation and pretreatment methods. The recycling of nutrients from the aqueous stream and production of coproducts like hydrocarbon biofuels are also factors in deciding the MFSP of the biofuel and hence the MFSP of sequential HTL is lesser than conventional HTL.

In the above studies, a single thermochemical conversion method of microalgae to produce biofuels was used. However, an integrated biorefinery needs to be developed to produce value-added products in addition to biofuels. This can be achieved by various methods which are a combination of hydrothermal liquefaction and hydrothermal gasification, lipid extraction combined with thermochemical residue conversion (Fig. 13.6A), protein extraction combined with thermochemical residue conversion (Fig. 13.6B), and protein and lipid extraction before thermochemical conversion (Fig. 13.6C). The hydrothermal liquefaction after protein removal from microalgae can lessen the nitrogen amount in bio-crude which can shrink the hydrogen requirement in the hydroprocessing of bio-oil.

Recently, Zhu et al. assessed the economic performance of different treatment methods for the liquid phase of microalgae hydrothermal liquefaction for proper utilization of the waste generated in the process [57]. These methods include the algal cultivation system, catalytic hydrothermal gasification, and anaerobic digestion. A minimum fuel selling price of biofuels was found to be \$12.5/GGE for direct recycling; \$13.8/GGE for CHG; \$12.9/GGE for anaerobic digestion. It was reported that the capital and operating investments are more for the case of catalytic hydrothermal gasification and the anaerobic digestion process has shown to be advantageous due to the low cost of aqueous phase treatment and lower energy requirement. The sensitivity analysis showed that bio-crude yield and plant-scale are important factors that influence the economics of microalgae-based biorefinery.

Cruce and Quinn performed the economic analysis of different biorefinery pathways (baseline HTL, protein removal, fragmentation-carbohydrate, and a small-scaled first-of-a-kind plant, i.e., first commercial plant coupled with a wastewater treatment) to produce biofuel and value-added products from microalgae [63]. For these pathways, the techno-economic analysis was found out to be \$5.37, \$4.44, \$4.31, and \$11.13/GGE, respectively. The authors emphasized that for economic viability, the algal biorefinery must produce nonfuel coproducts to reduce capital costs.

Although an integrated biorefinery with combined protein and lipid extraction, followed by thermochemical residue transformation, produces hydrocarbon biofuels along with value-added products, optimization of process conditions and parameters needs to be studied extensively to render microalgae-based hydrocarbon refinery more economical.

13.2.3 Waste-based hydrocarbon biorefinery

Compared to stand-alone activities, biorefinery technologies can also be combined with existing processes to achieve sustainability and enhance the economic aspects. The integration includes process integration, system integration, feedstock and commodity integration, supply-chain integration, and environment integration policy [64]. One choice is to integrate biorefinery concepts in Kraft pulp mills. This allows the possibility of reducing capital costs by utilizing equipment and operating costs by reducing energy, chemical, and raw material costs.

Table 13.6 shows a literature study on the techno-economic study of waste-based hydrocarbon biorefinery. Funkenbusch et al. reported techno-economic analysis for renewable hydrocarbon biofuel production from Kraft pulping lignin by HTL, followed by catalytic upgrading [66]. The lignin from Kraft pulping processing industries was used. This study was focused on three aspects, i.e., the utilization of waste from Kraft pulp processing industries, integration of existing petroleum to produce hydrocarbon biofuels, and conversion of aqueous product of HTL into benzene, toluene, ethylbenzene, and xylene. The reported minimum fuel selling price value of biofuel was in the span of \$3.52–3.86/gal. The MFSP of benzene, toluene, ethylbenzene, and xylene was in the range of \$1.65–2.00/L.

Bioeconomy of hydrocarbon biorefinery processes 375

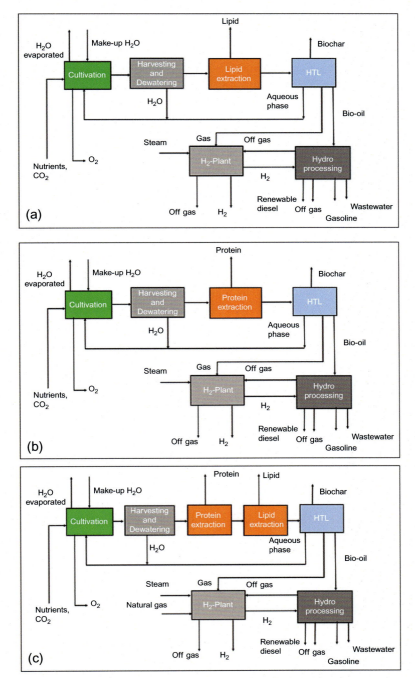

Fig. 13.6 Microalgae-based biorefinery with coproducts using (A) lipid extraction followed by HTL, (B) protein extraction followed by HTL, and (C) both lipid and protein extraction followed by HTL.

Özdenkçi et al. investigated the techno–economic assessment of supercritical water gasification of weak black liquor [68]. The purpose of this attempt was the utilization of black liquor in a Kraft pulp mill for improving the economics of biorefinery. The pulp mill with a plant capacity of around 400 air-dried kilotons of pulp/year produces 10 tons of black liquid per year. The black liquid was used in the supercritical water gasification process. The supercritical water gasification process was designed based on the experimental data of prior work by De Blasio et al. [73]. In this biorefinery, two scenarios of hydrogen production and combined heat and power production were conserved. Also, a type of material of construction of the reactor such as stainless steel 316 and Inconel 625 was studied on the economics of the biorefinery. Considering the case of catalytic hydrothermal production and the cost of production of hydrogen, the Inconel reactor was reported to be more cost-effective than the stainless steel reactor.

Recently, Magdeldin and Järvinen presented a detailed process design and economic evaluation of supercritical water gasification of Kraft black liquor [71]. This study was planned to integrate Nordic softwood pulp mills. The pulp making capacity was 800 air-dried kilotons of pulp per year. The minimum selling price of pulp was found as €637 per air-dried tons of pulp per year which was slightly higher in the reported value (€523 per air-dried tons) in the literature [74]. A techno-economic assessment study was conducted by Gursel et al. for the generation of bio-based aromatics from lignin [75]. Three operational methods were employed: pyrolysis, direct hydrodeoxygenation, and hydrothermal upgrading. The economic assessment study was carried out by estimating operational and capital costs. The plant capacity of 200 kt/year of lignin was used. The products generated during this process are mixed oxygenated aromatic monomers light organics, heavy organics, and char. For these calculations, the year index of 2012 was selected, and Aspen process economic analyzer was used for this study. Bridgewater equation was used for the lignin pyrolysis capital cost estimation [76]. It was reported that the desired product of biofuel and revenue was considerably less for the pyrolysis method and the operating cost was less for pyrolysis due to its simple process compared with the others. Hence, a comprehensive investigation is mandatory to exploit the development of a process that can take advantage of pyrolysis at lower operational costs.

The process design and techno-economic assessment carried out by Abdelaziz et al. were based on the experimental work on the oxidative depolymerization of Kraft lignin [77, 78]. The designing and simulation of the process were carried out in the Aspen Plus v10 software and the process was designed with the possibility of integrating into a prevailing pulp and paper mill. The maximum yield of the products was observed at 160°C and 3 bar oxygen partial pressure. The heat Integration case was also analyzed which was used to recuperate the excess energy and helped in running the existing heat exchangers at lower duties. Metric for Inspecting Sales and Reactant (MISR) method was employed to analyze whether the plant was economically viable or not (if MISR > 1, the process should be analyzed further and if MISR < 1, the process is not economically viable). The

Bioeconomy of hydrocarbon biorefinery processes   377

Table 13.6 Collection of literature data on the techno-economics of waste-based hydrocarbon biorefinery.

Process details	Feedstock	Products	MSFP	References
Conversion process: hydrothermal liquefaction (HTL) Wastewater processing capacity: 133 MTPD; HTL capacity: 100 dry tpd	Sludge waste	Upgraded fuel and bio-crude	\$4.9/GGE for fuel; \$3.8/ GGE for bio-crude	Hallen et al. [65]
Method: HTL Capacity: 400 MTPD of dry lignin Tool: ASPEN model	Lignin from Kraft pulping process	Biofuel, benzene, toluene, ethylbenzene, and xylene	\$3.52–3.86/gal for biofuel \$1.65–2.00 for benzene, toluene, ethylbenzene, and xylene	Funkenbusch et al. [66]
Method: biodiesel production by transesterification Capacity: 120,000 tons of castor and *Eruca sativa* plants per year Tool: Aspen Plus and Aspen process economic analyzer	Castor and *E. sativa*	Biodiesel, biogas, and heat	Average diesel equivalent price of biodiesel was 9% lower than the taxed diesel	Rahimi and Shafiei [67]
Method: supercritical water gasification of black liquor integrated to a Kraft pulp mill Capacity: 400,000 air-dried tons of pulp/year Temperature: 500–700°C and 25 MPa Tool: Aspen Plus	Kraft black liquor	Hydrogen and combined heat and power	Inconel reactor is more profitable than the stainless steel reactor for both energy production and hydrogen production	Özdenkçi et al. [68]
Method: biorefinery annexed to a sugar mill Tool: Aspen Plus	Sugarcane, bagasse, and residue	**(i)** Xylitol and electricity **(ii)** Citric acid and electricity **(iii)** Glutamic **(iv)** Acid and electricity	**(i)** 3000\$/ton **(ii)** 3625\$/ton **(iii)** 680\$/ton	[69]

Continued

378 Hydrocarbon biorefinery

Table 13.6 Collection of literature data on the techno-economics of waste-based hydrocarbon biorefinery—cont'd

Process details	Feedstock	Products	MSFP	References
Method: Biorefinery Capacity: 65 tons/h biomass Operating for 9 months/year Annexed to a sugar mill Tool: Aspen Plus	Sugar mill waste	Main product: sugar Coproducts: electricity and sorbitol	$619/ton of sorbitol	Kapanji et al. [70]
Method: Supercritical water gasification of Kraft black liquor Capacity: 800 kadt pulp/annum Temperature: 600°C and 450°C	Weak black liquor	Pulp product	€637/air-dried tons pulp at 600°C €639/air-dried tons pulp at 450°C	[71]
(a) Fischer-Tropsch gasification (b) CHP gasification (c) Combustion and gasification (d) Slow pyrolysis and combustion (e) Slow pyrolysis and biochar (f) Fast pyrolysis (g) Fast pyrolysis and upgradation (h) HTL (i) HTL and upgradation Tool: Aspen Plus	Poultry litter	(a) Diesel and gasoline (b) Electricity and heat (c) Electricity (d) Bio-oil and electricity (e) Bio-oil, biochar, heat, and electricity (f) Bio-oil, Biochar, electricity, and heat (g) Diesel, gasoline, biochar, electricity, and heat (h) Bio-oil and biochar (i) Diesel, gasoline, and biochar	Revenue in millions (a) Diesel: $24.68, Gasoline: $17.48 (b) $26.91 (c) $13.46 (d) Bio-oil: $11.03, Electricity: $4.28 (e) Bio-oil: $11.03, Biochar: $9.19, Electricity and heat: $4.62 (f) Bio-oil: $22.54, Biochar: $3.12, Electricity and heat: $1.92 (g) Diesel: $31.11, Gasoline: $10.82, Biochar: $1.56, Electricity and heat: $0.96 (h) Bio-oil: $11.56, Biochar: $2.36 (i) Diesel: $25.46, Gasoline: $7.33, Biochar: $2.36	Bora et al. [72]

tpd, tons per day.

total costs incurred as capital and operating costs were calculated using the CAPCOST software and was updated using the CEPCI 2019 and NPV and ROI values were used as the metrics to calculate the viability of the process. A MISR value of (>8) was found with a feedstock cost of (250–700$/tons) of Kraft lignin and hence the process is suitable for further economic analysis. A sensitivity analysis was used to assess the impact of the cost of feedstock and product on the profitability of the process. Even under unfavorable conditions of feedstock and product prices, the ROI attained was still positive, the selling price of Vanillin (one of the aromatics considered to be produced) has an important part in determining the profitability of the process, with an optimum scenario where the Vanillin is sold for $40/kg the ROI shoots beyond 29% for both the scenarios, which is an extremely profitable result.

Another option for developing a biorefinery model is to integrate an existing sugar mill. Kapanji et al. developed the economic feasibility of the biorefinery model using waste biomass from the sugar mill [70]. The proposed biorefinery concept involves the transformation of glucose to sorbitol and gluconic acid, while lignin and hemicellulose from hydrolysis are converted to produce electricity. Sorbital was produced by hydrogenation of glucose at a temperature of 120°C and 70 bar and gluconic acid is produced by oxidation of glucose. They looked at economic feasibility using different pretreatment of dilute acid and steam explosion of biomass. They found that the dilute acid pretreatment scenario is more economically viable than the steam explosion pretreatment methods. The minimum selling price of sorbitol was US$619/ton at a 10% internal rate of return which was 5% below the market price. This biorefinery concept can also be integrated into an existing sugar mill.

Alternative work utilized lignocellulosic biomass from sugarcane including bagasse and harvesting residues as feedstock in a biorefinery to yield biochemicals, such as xylitol, citric acid, and glutamic acid with coproduction of electricity [69]. This biorefinery can also be integrated into the existing sugar mills. The authors investigated several scenarios and found that MFSP as US$3000, 3625, and 680/ton for xylitol and power generation scenario, citric acid and electricity scenario, and glutamic acid and electricity scenario, respectively.

Bora et al. conducted a techno-economic study on poultry waste using various conversion processes [72]. They analyzed the techno-economic viability of different thermochemical methods for the production of valuable products from poultry waste. Pyrolysis (both fast and slow), gasification, and hydrothermal liquefaction methods were employed. The authors inferred from this analysis that the pyrolysis approach is a better option for the economical processing of poultry wastes.

For the comparison of various types of biorefinery, reported MFSP ranges and the status of technology readiness level (TRL) are enlisted in Table 13.7. The results revealed that conversion methods of pyrolysis and HTL can generate bio-oil, electricity, and heat in lower MFSP value. Also, these processes are in higher TRL. Thus, based on the

380 Hydrocarbon biorefinery

Table 13.7 Present status of various pathways for the conversion biomass to hydrocarbon biofuels.

Pathways	Feedstock	MFSP	Status
Sugar-to-hydrocarbons biofuels by catalytic conversion	Lignocellulose	$0.25–1.7/L	TRL3–4
Sugar-to-hydrocarbon biofuels by biological conversion	Lignocellulose	$1–3.8/L	TRL3–4
Bio-crude to hydrocarbon biofuels by pyrolysis	Lignocellulose biomass, and microalgae	$0.5–2/L	TRL6–7
Bio-crude to hydrocarbon biofuels by hydrothermal liquefaction	Lignocellulose, microalgae, sewage sludge, and lignin	0.77–1.2/L	TRL 5

parameters of MFSP and TRL, the pyrolysis and HTL methods are superior to other methods.

13.3 Conclusions and future prospects

The techno-economic analysis of hydrocarbon biorefinery using different feedstock was reviewed in this chapter. The current status of various pathways for the production of hydrocarbon biofuels was also analyzed. The cost of hydrocarbon biofuels production is mostly driven by feedstock and capital costs. Also, to achieve the goal of MFSP of $2.5/GGE proposed by Department of Energy (DOE), USA, as well as reduce uncertainty in various conversion pathways including catalytic, biochemical, and thermochemical, further improvement is needed. Some of the strategies for further improvements are as follows: integrate and optimize the process to reduce process and capital costs; and evaluate coproducts possibilities from waste generated during the various conversion processes, especially the catalytic conversion of lignin to chemicals, the value of the addition of HTL aqueous product. In the case of a biorefinery based on microalgae, the extraction of protein and lipid from microalgae, followed by thermochemical conversion of extracted residue, can improve the economics of the process. Coprocessing of low oxygenate bio-crude to improve the economics of hydrocarbon biofuels can also be tested.

Acknowledgment

Author (JSS) acknowledges MHRD, India, for PhD Scholarship. Author (PR) acknowledges to Department of Science and Technology, India, for DST INSPIRE Faculty program (DST/INSPIRE/2014/ENG-97) and Faculty Research Grand by National Institute of Technology Calicut, India.

References

[1] IEA Bioenergy Task 42 Biorefinery. n.d., Accessed July 7, 2020. https://www.iea-bioenergy.task42-biorefineries.com/en/ieabiorefinery.htm.

[2] European Commission, "A Sustainable Bioeconomy for Europe: Strengthening the Connection between Economy, Society and the Environment Updated Bioeconomy Strategy." n.d. https://doi.org/10.2777/478385.

[3] V. Aristizábal-Marulanda, C.A. Cardona Alzate, Methods for designing and assessing biorefineries: review, Biofuels Bioprod. Biorefin. 13 (3) (2018) 789–808. https://doi.org/10.1002/bbb.1961.

[4] L. Ou, H. Cai, H.J. Seong, D.E. Longman, J.B. Dunn, J.M.E. Storey, T.J. Toops, J.A. Pihl, M. Biddy, M. Thornton, Co-optimization of heavy-duty fuels and engines: cost benefit analysis and implications, Environ. Sci. Technol. (2019). https://doi.org/10.1021/acs.est.9b03690.

[5] G. Zang, J. Jia, S. Tejasvi, A. Ratner, E.S. Lora, Techno-economic comparative analysis of biomass integrated gasification combined cycles with and without CO2 capture, International Journal of Greenhouse Gas Control (2018). https://doi.org/10.1016/j.ijggc.2018.07.023.

[6] G.W. Huber, J.N. Chheda, C.J. Barrett, J.A. Dumesic, Chemistry: production of liquid alkanes by aqueous-phase processing of biomass-derived carbohydrates, Science 308 (5727) (2005) 1446–1450. https://doi.org/10.1126/science.1111166.

[7] M.J. Biddy, R. Davis, D. Humbird, L. Tao, N. Dowe, M.T. Guarnieri, J.G. Linger, et al., The techno-economic basis for coproduct manufacturing to enable hydrocarbon fuel production from lignocellulosic biomass, ACS Sustain. Chem. Eng. 4 (6) (2016) 3196–3211. https://doi.org/10.1021/acssuschemeng.6b00243.

[8] L. Tao, J.N. Markham, Z. Haq, M.J. Biddy, Techno-economic analysis for upgrading the biomass-derived ethanol-to-jet blendstocks, Green Chem. 19 (4) (2017) 1082–1101. https://doi.org/10.1039/c6gc02800d.

[9] J. Byun, J. Han, An integrated strategy for catalytic co-production of jet fuel range alkenes, tetrahydrofurfuryl alcohol, and 1,2-pentanediol from lignocellulosic biomass, Green Chem. 19 (21) (2017) 5214–5229. https://doi.org/10.1039/c7gc02368e.

[10] H. Olcay, R. Malina, A.A. Upadhye, J.I. Hileman, G.W. Huber, S.R.H. Barrett, Techno-economic and environmental evaluation of producing chemicals and drop-in aviation biofuels via aqueous phase processing, Energy Environ. Sci. 11 (8) (2018) 2085–2101. https://doi.org/10.1039/c7ee03557h.

[11] R. Shen, T. Ling, B. Yang, Techno-economic analysis of jet-fuel production from biorefinery waste lignin, Biofuels Bioprod. Biorefin. 13 (3) (2019) 486–501. https://doi.org/10.1002/bbb.1952.

[12] S.H. Mousavi-Avval, A. Shah, Techno-economic analysis of pennycress production, harvest and post-harvest logistics for renewable jet fuel, Renew. Sust. Energ. Rev. 123 (February) (2020) 109764. https://doi.org/10.1016/j.rser.2020.109764.

[13] K.L. Brandt, R.J. Wooley, S.C. Geleynse, J. Gao, J. Zhu, R.P. Cavalieri, M.P. Wolcott, Impact of co-product selection on techno-economic analyses of alternative jet fuel produced with Forest harvest residuals, Biofuels Bioprod. Biorefin. 14 (4) (2020) 764–775. https://doi.org/10.1002/bbb.2111.

[14] D. López Barreiro, W. Prins, F. Ronsse, W. Brilman, Hydrothermal liquefaction (HTL) of microalgae for biofuel production: state of the art review and future prospects, Biomass Bioenergy 53 (June) (2013) 113–127. https://doi.org/10.1016/j.biombioe.2012.12.029.

[15] S.B. Jones, J.E. Holladay, C. Valkenburg, D.J. Stevens, C.W. Walton, C. Kinchin, et al., Production of Gasoline and Diesel from Biomass Via Fast Pyrolysis, Hydrotreating and Hydrocracking: A Design Case, Pacific Northwest National Laboratory, Richland, WA, 2009, PNNL-18284.

[16] R.P. Anex, A. Aden, F.K. Kazi, J. Fortman, R.M. Swanson, M.M. Wright, J.A. Satrio, et al., Techno-economic comparison of biomass-to-transportation fuels via pyrolysis, gasification, and biochemical pathways, Fuel 89 (Suppl. 1) (2010) S29–S35. https://doi.org/10.1016/j.fuel.2010.07.015.

[17] M.M. Wright, D.E. Daugaard, J.A. Satrio, R.C. Brown, Techno-economic analysis of biomass fast pyrolysis to transportation fuels, Fuel 89 (Suppl. 1) (2010). https://doi.org/10.1016/j.fuel.2010.07.029.

[18] T.R. Brown, R. Thilakaratne, R.C. Brown, G. Hu, Techno-economic analysis of biomass to transportation fuels and electricity via fast pyrolysis and hydroprocessing, Fuel 106 (2013) 463–469. https://doi.org/10.1016/j.fuel.2012.11.029.

[19] R. Thilakaratne, M.M. Wright, R.C. Brown, A techno-economic analysis of microalgae remnant catalytic pyrolysis and upgrading to fuels, Fuel 128 (2014) 104–112. https://doi.org/10.1016/j.fuel.2014.02.077.

[20] K. Wang, L. Ou, T. Brown, R.C. Brown, Beyond ethanol: a techno-economic analysis of an integrated corn biorefinery for the production of hydrocarbon fuels and chemicals, Biofuels Bioprod. Biorefin. 9 (2) (2015) 190–200. https://doi.org/10.1002/bbb.1529.

[21] B. Li, L. Ou, Q. Dang, P. Meyer, S. Jones, R. Brown, M. Wright, Techno-economic and uncertainty analysis of in situ and ex situ fast pyrolysis for biofuel production, Bioresour. Technol. 196 (November) (2015) 49–56. https://doi.org/10.1016/j.biortech.2015.07.073.

[22] M.B. Shemfe, S. Gu, P. Ranganathan, Techno-economic performance analysis of biofuel production and miniature electric power generation from biomass fast pyrolysis and bio-oil upgrading, Fuel 143 (March) (2015) 361–372. https://doi.org/10.1016/j.fuel.2014.11.078.

[23] M. Sharifzadeh, C.J. Richard, K. Liu, K. Hellgardt, D. Chadwick, N. Shah, An integrated process for biomass pyrolysis oil upgrading: a synergistic approach, Biomass Bioenergy 76 (May) (2015) 108–117. https://doi.org/10.1016/j.biombioe.2015.03.003.

[24] O. Winjobi, D.R. Shonnard, E. Bar-Ziv, W. Zhou, Techno-economic assessment of the effect of torrefaction on fast pyrolysis of pine, Biofuels Bioprod. Biorefin. 10 (2) (2016) 117–128. https://doi.org/10.1002/bbb.1624.

[25] M. Shemfe, S. Gu, B. Fidalgo, Techno-economic analysis of biofuel production via bio-oil zeolite upgrading: an evaluation of two catalyst regeneration systems, Biomass Bioenergy 98 (March) (2017) 182–193. https://doi.org/10.1016/j.biombioe.2017.01.020.

[26] H. AlMohamadi, S. Gunukula, W.J. DeSisto, M.C. Wheeler, Formate-assisted pyrolysis of biomass: an economic and modeling analysis, Biofuels Bioprod. Biorefin. 12 (1) (2018) 45–55. https://doi.org/10.1002/bbb.1827.

[27] W. Li, Q. Dang, R. Smith, R.C. Brown, M.M. Wright, Techno-economic analysis of the stabilization of bio-oil fractions for insertion into petroleum refineries, ACS Sustain. Chem. Eng. 5 (2) (2017) 1528–1537. https://doi.org/10.1021/acssuschemeng.6b02222.

[28] P.S. Sabaini, A.A. Boateng, M. Schaffer, C.A. Mullen, Y. Elkasabi, C.M. McMahan, N. Macken, Techno-economic analysis of guayule (Parthenium argentatum) pyrolysis biorefining: production of biofuels from guayule bagasse via tail-gas reactive pyrolysis, Ind. Crop. Prod. 112 (February) (2018) 82–89. https://doi.org/10.1016/j.indcrop.2017.11.009.

[29] Y. Yang, J. Wang, K. Chong, A.V. Bridgwater, A techno-economic analysis of energy recovery from organic fraction of municipal solid waste (MSW) by an integrated intermediate pyrolysis and combined heat and power (CHP) plant, Energy Convers. Manag. 174 (October) (2018) 406–416. https://doi.org/10.1016/j.enconman.2018.08.033.

[30] Q. Dang, M.M. Wright, W. Li, Technoeconomic analysis of a hybrid biomass thermochemical and electrochemical conversion system, Energy Technol. 6 (1) (2018) 178–187. https://doi.org/10.1002/ente.201700395.

[31] M. Patel, A.O. Oyedun, A. Kumar, R. Gupta, A techno-economic assessment of renewable diesel and gasoline production from Aspen hardwood, Waste Biomass Valorization 10 (10) (2019) 2745–2760. https://doi.org/10.1007/s12649-018-0359-x.

[32] S. Michailos, A. Bridgwater, A comparative techno-economic assessment of three bio-oil upgrading routes for aviation biofuel production, Int. J. Energy Res. 43 (13) (2019), er.4745. https://doi.org/10.1002/er.4745.

[33] G.V. Brigagão, O. de Queiroz Fernandes Araújo, J.L. de Medeiros, H. Mikulcic, N. Duic, A techno-economic analysis of thermochemical pathways for corncob-to-energy: fast pyrolysis to bio-oil, gasification to methanol and combustion to electricity, Fuel Process. Technol. 193 (2019) 102–113. https://doi.org/10.1016/j.fuproc.2019.05.011.

[34] T.V. Nguyen, L.R. Clausen, Techno-economic analysis of polygeneration systems based on catalytic hydropyrolysis for the production of bio-oil and fuels, Energy Convers. Manag. 184 (March) (2019) 539–558. https://doi.org/10.1016/j.enconman.2019.01.070.

Bioeconomy of hydrocarbon biorefinery processes **383**

[35] I.Y. Mohammed, Y.A. Abakr, R. Mokaya, Integrated biomass thermochemical conversion for clean energy production: process design and economic analysis, Journal of Environmental Chemical Engineering 7 (3) (2019) 103093. https://doi.org/10.1016/j.jece.2019.103093.

[36] G. Bagnato, A. Sanna, Process and techno-economic analysis for fuel and chemical production by hydrodeoxygenation of bio-oil, Catalysts 9 (12) (2019). https://doi.org/10.3390/catal9121021.

[37] Y.-.C. Liu, W.-.C. Wang, Process design and evaluations for producing pyrolytic jet fuel, Biofuels Bioprod. Biorefin. 14 (2) (2020) 249–264. https://doi.org/10.1002/bbb.2061.

[38] D.L. van Schalkwyk, M. Mandegari, S. Farzad, J.F. Görgens, Techno-economic and environmental analysis of bio-oil production from forest residues via non-catalytic and catalytic pyrolysis processes, Energy Convers. Manag. 213 (December 2019) (2020) 112815. https://doi.org/10.1016/j.enconman.2020.112815.

[39] A. Dutta, A. Sahir, E. Tan, Process design and economics for the conversion of lignocellulosic biomass to hydrocarbon fuels thermochemical research pathways with in situ and ex situ upgrading of fast pyrolysis vapors | request PDF, in: Technical Report NREL/TP-5100-62455 PNNL-23823, 2015. https://www.researchgate.net/publication/275465105_Process_Design_and_Economics_for_the_Conversion_of_Lignocellulosic_Biomass_to_Hydrocarbon_Fuels_Thermochemical_Research_Pathways_with_In_Situ_and_Ex_Situ_Upgrading_of_Fast_Pyrolysis_Vapors.

[40] R. Thilakaratne, T. Brown, Y. Li, G. Hu, R. Brown, Mild catalytic pyrolysis of biomass for production of transportation fuels: a techno-economic analysis, Green Chem. 16 (2) (2014) 627–636. https://doi.org/10.1039/c3gc41314d.

[41] T. Kuppens, M. Van Dael, K. Vanreppelen, T. Thewys, J. Yperman, R. Carleer, S. Schreurs, S. Van Passel, Techno-economic assessment of fast pyrolysis for the valorization of short rotation coppice cultivated for phytoextraction, J. Clean. Prod. 88 (February) (2015) 336–344. https://doi.org/10.1016/j.jclepro.2014.07.023.

[42] Y. Zhang, T.R. Brown, G. Hu, R.C. Brown, Techno-economic analysis of monosaccharide production via fast pyrolysis of lignocellulose, Bioresour. Technol. 127 (January) (2013) 358–365. https://doi.org/10.1016/j.biortech.2012.09.070.

[43] T.X. Do, Y.I. Lim, Techno-economic comparison of three energy conversion pathways from empty fruit bunches, Renew. Energy 90 (May) (2016) 307–318. https://doi.org/10.1016/j.renene.2016.01.030.

[44] X. Chen, H. Zhang, R. Xiao, Mobile autothermal pyrolysis system for local biomass conversion: process simulation and techno-economic analysis, Energy Fuel (2018). American Chemical Society https://doi.org/10.1021/acs.energyfuels.7b03172.

[45] Y. Zhu, M.J. Biddy, S.B. Jones, D.C. Elliott, A.J. Schmidt, Techno-economic analysis of liquid fuel production from woody biomass via hydrothermal liquefaction (HTL) and upgrading, Appl. Energy 129 (2014) 384–394. https://doi.org/10.1016/j.apenergy.2014.03.053.

[46] K.F. Tzanetis, J.A. Posada, A. Ramirez, Analysis of biomass hydrothermal liquefaction and biocrude-oil upgrading for renewable jet fuel production: the impact of reaction conditions on production costs and GHG emissions performance, Renew. Energy 113 (December) (2017) 1388–1398. https://doi.org/10.1016/j.renene.2017.06.104.

[47] T.H. Pedersen, N.H. Hansen, O.M. Pérez, D.E.V. Cabezas, L.A. Rosendahl, Renewable hydrocarbon fuels from hydrothermal liquefaction: a techno-economic analysis, Biofuels Bioprod. Biorefin. 12 (2) (2018) 213–223. https://doi.org/10.1002/bbb.1831.

[48] M. Magdeldin, T. Kohl, M. Järvinen, Techno-economic assessment of the by-products contribution from non-catalytic hydrothermal liquefaction of lignocellulose residues, Energy 137 (October) (2017) 679–695. https://doi.org/10.1016/j.energy.2017.06.166.

[49] Y. Nie, X.T. Bi, Techno-economic assessment of transportation biofuels from hydrothermal liquefaction of forest residues in British Columbia, Energy 153 (June) (2018) 464–475. https://doi.org/10.1016/j.energy.2018.04.057.

[50] I. Tews, Y. Zhu, C. Drennan, D. Elliot, L. Snowden-Swan, K. Onarheim, Y. Solantausta, D. Beckman, Biomass Direct Liquefaction Options: TechnoEconomic and Life Cycle Assessment, Pacific Northwest National Laboratory, 2014.

[51] Y. Zhu, K.O. Albrecht, D.C. Elliott, R.T. Hallen, S.B. Jones, Development of hydrothermal liquefaction and upgrading technologies for lipid-extracted algae conversion to liquid fuels, Algal Res. 2 (4) (2013) 455–464. https://doi.org/10.1016/j.algal.2013.07.003.

[52] S. Jones, Y. Zhu, D. Anderson, R. Hallen, D.C. Elliott, E.A. Schmidt, et al., Process Design and Economics for the Conversion of Algal Biomass to Hydrocarbons: Whole Algae Hydrothermal Liquefaction and Upgrading, Pacific Northwest National Laboratory, Richland, WA, 2014, PNNL-23227.

[53] L. Ou, R. Thilakaratne, R.C. Brown, M.M. Wright, Techno-economic analysis of transportation fuels from defatted microalgae via hydrothermal liquefaction and hydroprocessing, Biomass Bioenergy 72 (January) (2015) 45–54. https://doi.org/10.1016/j.biombioe.2014.11.018.

[54] C. Xin, M.M. Addy, J. Zhao, Y. Cheng, S. Cheng, D. Mu, Y. Liu, R. Ding, P. Chen, R. Ruan, Comprehensive techno–economic analysis of wastewater-based algal biofuel production: a case study, Bioresour. Technol. 211 (July) (2016) 584–593. https://doi.org/10.1016/j.biortech.2016.03.102.

[55] A. Juneja, G.S. Murthy, Evaluating the potential of renewable diesel production from algae cultured on wastewater: techno–economic analysis and life cycle assessment, AIMS Energy 5 (2) (2017) 239–257. https://doi.org/10.3934/energy.2017.2.239.

[56] Y. Jiang, S.B. Jones, Y. Zhu, L. Snowden-Swan, A.J. Schmidt, J.M. Billing, D. Anderson, Techno-economic uncertainty quantification of algal-derived biocrude via hydrothermal liquefaction, Algal Res. 39 (2019) 101450. https://doi.org/10.1016/j.algal.2019.101450.

[57] Y. Zhu, S.B. Jones, A.J. Schmidt, K.O. Albrecht, S.J. Edmundson, D.B. Anderson, Techno-economic analysis of alternative aqueous phase treatment methods for microalgae hydrothermal liquefaction and biocrude upgrading system, Algal Res. 39 (May) (2019) 101467. https://doi.org/10.1016/j.algal.2019.101467.

[58] X. Gu, L. Yu, N. Pang, J.S. Martinez-Fernandez, X. Fu, S. Chen, Comparative techno-economic analysis of algal biofuel production via hydrothermal liquefaction: one stage versus two stages, Appl. Energy 259 (2020) 114115. https://doi.org/10.1016/j.apenergy.2019.114115.

[59] M. Kumar, A.O. Oyedun, A. Kumar, A comparative technoeconomic analysis of algal thermochemical conversion technologies for diluent production, Energy Technology 1900828 (2019) 1–15. https://doi.org/10.1002/ente.201900828.

[60] C. Xin, Z. Jinyu, R. Roger, M.M. Addy, S. Liu, D. Mu, Economical feasibility of bio-oil production from sewage sludge through pyrolysis, Therm. Sci. 22 (2018) 459–467. https://doi.org/10.2298/TSCI170921258X.

[61] P. Ranganathan, S. Savithri, Techno–economic analysis of microalgae-based liquid fuels production from wastewater via hydrothermal liquefaction and hydroprocessing, Bioresour. Technol. 284 (July) (2019) 256–265. https://doi.org/10.1016/j.biortech.2019.03.087.

[62] K. DeRose, C. DeMill, R.W. Davis, J.C. Quinn, Integrated techno economic and life cycle assessment of the conversion of high productivity, low lipid algae to renewable fuels, Algal Res. 38 (March) (2019) 101412. https://doi.org/10.1016/j.algal.2019.101412.

[63] J.R. Cruce, J.C. Quinn, Economic viability of multiple algal biorefining pathways and the impact of public policies, Appl. Energy 233–234 (January) (2019) 735–746. https://doi.org/10.1016/j.apenergy.2018.10.046.

[64] P.R. Stuart, M.M. El-Halwagi, Integrated Biorefineries: Design, Analysis, and Optimization—Google Books, CRC Press, 2013. https://books.google.co.in/books/about/Integrated_Biorefineries.html?id=C5IZnQAACAAJ&redir_esc=y.

[65] R. Hallen, L. Snowden-Swan, Y. Zhu, S.B. Jones, D. Elliott, A.J. Schmidt, J. Billing, T.R. Hart, S.P. Fox, G.D. Maupin, Hydrothermal Liquefaction and Upgrading of Municipal Wastewater Treatment Plant Sludge: A Preliminary Techno-Economic Analysis, 2016. https://www.researchgate.net/publication/308643533_Hydrothermal_Liquefaction_and_Upgrading_of_Municipal_Wastewater_Treatment_Plant_Sludge_A_Preliminary_Techno-Economic_Analysis.

[66] L.T. Funkenbusch, M.E. Mullins, L. Vamling, T. Belkhieri, N. Srettiwat, O. Winjobi, D.R. Shonnard, T.N. Rogers, Technoeconomic assessment of hydrothermal liquefaction oil from lignin with catalytic upgrading for renewable fuel and chemical production, Wiley Interdiscip. Rev. Energy Environ. 8 (1) (2018), e319. https://doi.org/10.1002/wene.319.

[67] V. Rahimi, M. Shafiei, Techno-economic assessment of a biorefinery based on low-impact energy crops: a step towards commercial production of biodiesel, biogas, and heat, Energy Convers. Manag. 183 (November 2018) (2019) 698–707. https://doi.org/10.1016/j.enconman.2019.01.020.

[68] K. Özdenkçi, C. De Blasio, G. Sarwar, K. Melin, J. Koskinen, V. Alopaeus, Techno-economic feasibility of supercritical water gasification of black liquor, Energy 189 (2019). https://doi.org/10.1016/j.energy.2019.116284.

[69] H.M.R. Özüdoğru, M. Nieder-Heitmann, K.F. Haigh, J.F. Görgens, Techno-economic analysis of product biorefineries utilizing sugarcane lignocelluloses: xylitol, citric acid and glutamic acid scenarios annexed to sugar mills with electricity co-production, Ind. Crop. Prod. 133 (November 2018) (2019) 259–268. https://doi.org/10.1016/j.indcrop.2019.03.015.

[70] K.K. Kapanji, K.F. Haigh, J.F. Görgens, Techno-economic analysis of chemically catalysed lignocellulose biorefineries at a typical sugar mill: sorbitol or glucaric acid and electricity co-production, Bioresour. Technol. 289 (June) (2019) 121635. https://doi.org/10.1016/j.biortech.2019.121635.

[71] M. Magdeldin, M. Järvinen, Supercritical water gasification of Kraft black liquor: process design, analysis, pulp mill integration and economic evaluation, Appl. Energy 262 (December 2019) (2020) 114558. https://doi.org/10.1016/j.apenergy.2020.114558.

[72] R.R. Bora, Y. Tao, J. Lehmann, J.W. Tester, R.E. Richardson, F. You, Techno-economic feasibility and spatial analysis of thermochemical conversion pathways for Resgional poultry waste valorization, ACS Sustain. Chem. Eng. 8 (14) (2020) 5763–5775. https://doi.org/10.1021/acssuschemeng.0c01229.

[73] C. De Blasio, G. Lucca, K. Özdenkci, M. Mulas, K. Lundqvist, J. Koskinen, M. Santarelli, T. Westerlund, M. Järvinen, A study on supercritical water gasification of black liquor conducted in stainless steel and nickel–chromium–molybdenum reactors, J. Chem. Technol. Biotechnol. 91 (10) (2016) 2664–2678. https://doi.org/10.1002/jctb.4871.

[74] K. Petteri, S. Kaijaluoto, M. Määttänen, Evaluation of future pulp mill concepts—reference model of a modern nordic Kraft pulp mill, Nordic Pulp Paper Res. J. (2014). https://www.researchgate.net/publication/280697430_Evaluation_of_future_pulp_mill_concepts_-_Reference_model_of_a_modern_Nordic_kraft_pulp_mill.

[75] I.V. Gursel, J.W. Dijkstra, W.J.J. Huijgen, A. Ramirez, Techno-economic comparative assessment of novel lignin depolymerization routes to bio-based aromatics, Biofuels Bioprod. Biorefin. 13 (4) (2019) 1068–1084. https://doi.org/10.1002/bbb.1999.

[76] A.V. Bridgwater, Review of fast pyrolysis of biomass and product upgrading, Biomass Bioenergy 38 (March) (2012) 68–94. https://doi.org/10.1016/j.biombioe.2011.01.048.

[77] O.Y. Abdelaziz, A.A. Al-Rabiah, M.M. El-Halwagi, C.P. Hulteberg, Conceptual design of a Kraft lignin biorefinery for the production of valuable chemicals via oxidative depolymerization, ACS Sustain. Chem. Eng. (2020). https://doi.org/10.1021/acssuschemeng.0c02945.

[78] O.Y. Abdelaziz, K. Ravi, F. Mittermeier, S. Meier, A. Riisager, G. Lidén, C.P. Hulteberg, Oxidative depolymerization of Kraft lignin for microbial conversion, ACS Sustain. Chem. Eng. (2019). https://doi.org/10.1021/acssuschemeng.9b01605.

CHAPTER 14

Life-cycle analysis of a hydrocarbon biorefinery

Jasvinder Singh[a,b], Aman Kumar Bhonsle[b,c], and Neeraj Atray[c]

[a]Material Resource and Efficiency Division, CSIR–Indian Institute of Petroleum (CSIR–IIP), Dehradun, India
[b]Academy of Scientific and Innovative Research (AcSIR), Ghaziabad, India
[c]Biofuels Division, CSIR–IIP, Dehradun, India

Contents

14.1 Introduction	388
14.2 Biorefinery configurations	389
14.2.1 Platform-based configuration	389
14.2.2 Feedstock-based configuration	390
14.2.3 Process-based configuration	393
14.2.4 Product-based biorefinery	393
14.3 LCA of biorefineries	396
14.3.1 LCA approach to biodiesel process	399
14.3.2 LCA of a thermal conversion-based biorefinery	401
14.4 Conclusions	403
References	404

Abbreviations

AA	adipic acid
CO	carbon monoxide
CO$_2$	carbon dioxide
FT	Fischer–Tropsch
GHG	greenhouse gases
GJ	Giga Joule
GREET	greenhouse gases, regulated emissions, energy use in transportation
GWP	global warming potential
HTL	hydrothermal liquefaction
IBR	integrated biorefinery
LCA	life-cycle assessment
MJ	Mega Joule
RDB	renewable diesel blend
SA	succinic acid
TEA	techno–economic assessment

Hydrocarbon Biorefinery
https://doi.org/10.1016/B978-0-12-823306-1.00015-7

Copyright © 2022 Elsevier Inc.
All rights reserved.

14.1 Introduction

The global energy security concerns, growing environmental issues, and constant increase in global warming have generated enough impetus for the development of renewable energy. The present largest source of global energy, fossil fuels will not last long due to depleting resources as well as their substantial environmental impact on greenhouse gas (GHG) emissions. The stringent environmental norms are an additional thrust for the development of sustainable and environmentally friendly fuels and chemicals. A continuous search is on worldwide, to identify new and sustainable feedstocks for biofuels as well as green chemicals.

In recent years, biofuels have been identified as a promising substitute for fossil fuels. Various biomasses have been identified for conversion to renewable fuels. However, the exploitation of bio-based feedstocks has certain limitations in terms of their quantum of availability as well as environmental impact. The advanced bioprocesses are thus continuously being developed to convert the various biomass to hydrocarbon biofuels. A large variety of biomass has been identified as a potential source of hydrocarbon biofuels. These feedstocks include the different crops, crop residues, forest wastes, microalgae, other marine vegetation, industrial wastes, and municipal solid wastes. For all of these feedstocks, a variety of processing routes are available. Accordingly, a diverse type of process can be envisaged to convert the biomass to a variety of hydrocarbon biofuels for application as fuels as well as green chemicals. A well-thought-out integration of these processes to generate a diverse array of biofuels and chemicals leads to a biorefinery configuration. Various biorefinery configurations are possible and may be classified based on the base process or the base feedstocks. Some popular biorefinery configurations have been discussed in subsequent subsections.

In the process of converting these biomass feedstocks, the excessive use of the resources may lead to a growing threat to the environment. Rapid industrialization is already putting a constant threat to natural resources, and pollution has become a big concern. *Chemical and Engineering News* issue of July 3, 2017, quotes a research manager at Water Research Foundation, Khiari says that the public won't accept any smell in their water [1]. This indicates a distinguished impact on the water for human use due to the expansion of new process plants. In view of this, it becomes mandatory for the researchers and technologists to develop technologies with minimal impact on the natural resources and the environment. Engineers and scientists are associated with the whole life cycle of a product or process, starting from the stage of discovery to plant commissioning, production, marketing, use phase, and finally the end-of-life treatments. They have to contribute their respective inputs to all subdivisions of the supply chain. Two essential steps for the evaluation of the feasibility and sustainability of a new process are techno–economic assessment (TEA) and life-cycle assessment (LCA). Before a process can be licensed for mass production, its energy economics and environmental impact must be identified.

LCA has emerged out as a very useful tool for the environmental impact assessment of a new process. Basically, it is a standard procedure to estimate the environmental impact of the various stages of a process on the complete process or its end product [2–5]. As part of a complete LCA, a system boundary is first drawn to contain the major process stages, followed by goal and scope definition. The goal and scope determine the objective of LCA analysis. A functional unit is defined to quantify the environmental metrics and identification of energy and material flow to be estimated in terms of this functional unit. The next step is the estimation of the released waste to the environment and its treatment and/or disposal scenario. This whole information is further converted to the standard sets of environmental impact parameters, with the help of standard impact assessment methods, that can be further consolidated to the various impact categories. These impacts finally help to assess the various biorefinery configurations and designs and provide inputs for choosing optimal solutions for setting up sustainable, environment-friendly, and profitable biorefineries. Several studies on LCA of different biorefinery configurations have been reported and published in the literature [6–10]. The effects of the methods of coproduct allocations on the end environment impact have been discussed and reported. It is reported that LCA results could be significantly different for different allocation strategies.

This chapter discusses the various biorefinery configurations that are classified based on the various conversion processes and available feedstocks. The categorized configurations have been discussed in light of LCA analyses presented in the literature. A critical appraisal is further presented with recommendations for future configurations.

14.2 Biorefinery configurations

The main goal of the biorefinery is to develop a various range of products such as biofuels, materials, chemicals, etc. from renewable biomass employing various conversion processes. The main challenges of biorefineries are the efficient production of biofuels and the best utilization of the coproducts involved in the process in economic and environmental terms. The biorefinery system can be divided into four classes: platform based, feedstock based, product based, and process based (Fig. 14.1).

14.2.1 Platform-based configuration

Platforms are an integral part of designing the biorefinery system as they link feedstocks to the final product. Platforms can be explained the same way as the petrochemical industry. The biorefinery has been classified into the various level of platforms such as two, three, four, and five platforms. Two-platform biorefineries using wood chips for FT fuels, electricity, heat, etc. whereas three-platform biorefinery uses a straw to produced fuel and methanol. Four-platform biorefinery uses the wood chip to produce biomethane, biohydrogen, and carbon dioxide. The five-platform biorefinery uses starch crops and straw

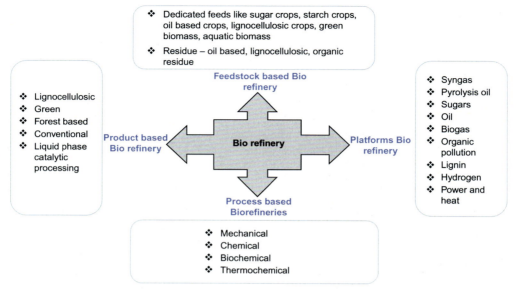

Fig. 14.1 Classification of biorefinery system.

to produce bioethanol, FT fuels, electricity, heat, etc. The other classification based on different platforms is illustrated in Table 14.1.

14.2.2 Feedstock-based configuration

The biorefinery approach uses feedstocks to produce biofuels and other chemicals from renewable biomass. The feedstocks can be from various sectors such as agriculture, industries, forest, algal, and water-based systems. The variety of feedstocks discussed in the literature, viz., carbohydrates lignin-based, triglycerides, and mixed organic residues have been explored for the biorefinery system in the subsequent section.

14.2.2.1 Lignocellulose-based biorefinery

It is type III biorefinery in which lignocellulose biomass is the feedstock. In type III biorefinery, multiple feedstocks are taken and processed through several processes to generate various products. Lignocellulose biomass is composed of three substituents: cellulose (30%–50%), hemicellulose (20%–40%), and lignin (15%–25%). Hemicellulose and cellulose are used in ethanol production and other energy product but not lignin because of its condensed polymer structures. Typical lignocellulosic biomass includes agricultural and forest residues and municipal organic wastes for which pretreatment is done and subsequent purification to generate energy and chemical products [12, 13]. Various pretreatment procedures have been applied because of a variety of lignocellulosic materials such as hardwood, softwood, and grasses and their difference in physical

Life-cycle analysis of a hydrocarbon biorefinery 391

Table 14.1 Biorefinery configurations based on different platform [11].

S. No.	Platform type	Description
1	Syngas platform	Syngas ($CO + H_2$) is generated using a thermal degradation technique in the presence of an oxidant, and the product is used to produce biofuel and chemicals.
2	Sugar platform	Sugars (C_5 and C_6) are generated from sugar, starch, cellulose, etc. using chemical and biological treatments.
3	Pyrolysis oil platform	Pyrolysis of biomass is done to get bio-oil which can be upgraded to petrochemicals.
4	Biogas platform	Biogas is generated through anaerobic digestion of biomass and used to improve energy security.
5	Organic solution platform	The organic solution is part of green biorefinery. The obtained products are biofuel and chemicals.
6	Lignin platform	Lignin and lignocellulosic biomass are used to make high-purity chemicals like vanillin and petrochemicals.
7	Hydrogen platform	Bio-hydrogen is generated using various biochemical and chemical techniques.
8	Oil platform	Oil is pretreated and transesterified to make biodiesel with glycerol as a coproduct. Another technique like catalytic cracking results lubricants and other building-block chemicals.

and chemical properties. Typical processes include acid or basic treatments which are largely done to remove hemicellulose and lignin, respectively [14]. New biological technologies such as genome sequencing and plant breeding are also available along with the mechanical process for lignocellulosic biomass treatment. Various kinds of products (chemical and biofuels) can be obtained from lignocellulosic biomass by processes such as liquefaction, gasification, and pyrolysis. Two routes, viz., thermochemical and biochemical are employed to generate biofuels such as ethanol, hydrogen, methane, etc., and platform chemicals such as levulinic acid, 2,3-butanediol, etc. from biomass [15–17]. Another alternative is an integrated approach in which both processes are used to produce biofuel and value-added coproducts. Lignocellulosic biomass after thermal treatment gives carbonaceous material such as biochar which adds in reducing GHG emission and increasing soil quality, thereby increased environmental benefits.

14.2.2.2 Oilseed biorefinery

In the oilseed biorefinery, the feedstock is tree-born seed oil. In the typical oil composition, it has a series of both saturated and unsaturated carbon chains that are responsible for different chemical and biological properties. Seed oil is taken as feedstock to produce biofuel such as biodiesel using the transesterification process and in another scenario, where catalytic cracking is done to get gasoline, aromatics, and other petrochemicals. Additionally, residual seed biomass is taken and processed through liquefaction, gasification, and pyrolysis. After processing a variety of products such as syngas, biomethanol, biobutanol, and chemical involving sterol, tocol, carotenoid, 1,3-propanediol are obtained. Bioheat and bioelectricity are other products in the biorefinery. Different biomass such as oil palm shell, mixed wood waste, rapeseed grain, and jatropha husk have been identified as potential feedstocks [18–20]. Generating the biofuel and chemicals from the seed oil biorefinery system is a cost-competitive process as compared to the petroleum refinery. As a consequence, the researchers are interested in the new integrated biorefinery. Patti and Arora et al. discussed oil seed biorefinery integrating microwave and nanobiocatalysis on waste pomegranate seeds and found the oil (17.8%) along with protein and abundant dietary fibers [21].

14.2.2.3 Algal biorefinery

It is the third generation biorefinery taking algal biomass as feedstocks into account. Algal biomass is proteins (6%–52%) rich along with lipid (7%–23%) and carbohydrate (5%–23%) as other components. It is the raw material for the generation of a variety of products such as pigments, proteins, vitamins, lipids, polysaccharides as well as biofuels such as biodiesel, biomethane, and biohydrogen [22–24]. Membrane filtration technique and other separation techniques are used for concentrating algal biomass. Still, the sustainable process for concentrating algal biomass for feasibility and biorefinery-related studies is not fully explored yet [25, 26]. The algal biorefinery is comprised of many steps for the generation of biofuel and chemicals. In the first step, biomass is cultivated following different cultivation techniques such as auto-, hetero-, and mixo–trophic cultivation. The biomass is then harvested, followed by other cell extraction treatment if required. Biomass is finally converted to biofuel and valuable chemicals employing thermochemical, biochemical, and chemical conversion processes. Biofuel generation in algal biorefinery is a cost-competitive process, but the other useful coproduct generation from the algal biomass can give new dimensions.

14.2.2.4 Green biorefinery

It is the type III biorefinery system with green biomass as feedstock. The green biomass may be comprised of naturally occurring wet green grass and green crops such as clover, lucerne, immature cereal, and other agricultural residues such as potato leaf and sugar beet leaf. This biomass provides different products such as carbohydrates, lipids, proteins, vitamins, minerals, etc. The type of product obtained in the biorefinery depends on various factors: grassland types, green plant composition, and other environmental and management factors such as temperature, harvesting technique, and nutrition control. In the procedure, green biomass is primarily pretreated to remove the water content and fiber concentration using the hydrothermal and dewatering method. The wet fraction of green biomass is divided into two parts: press juice and press cake. Press juice is nutrient rich and further divided into green juice which gives products such as protein and silage juice which produces lactic acid and amino acid, respectively. Further, residuals of both green juice and red juice can undergo fermentation to generate lactic acid, ethanol, lysine, organic acid, biogas, etc. In another pathway, press cake which is fiber-rich lignocellulose feedstock can be valorized into sugar, second-generation biofuel using thermochemical techniques. Press cake can also be used directly to generate other valuable products such as bio-composite, insulating materials, etc.

14.2.3 Process-based configuration

Biorefinery involves various processes to convert biomass into products. These processes are explained below.

14.2.3.1 Thermochemical processes

The thermochemical process includes gasification, liquefaction, and pyrolysis. The pyrolysis reaction involves the heating of biomass in the absence or limited supply of oxygen to get gas mixtures and pyrolysis oil. Further treatment of pyrolysis oil such as hydrocracking gives diesel and gasoline range blend feedstock. Further, factors such as temperature and heating rate also affect the reaction yield and products. Bio-oil is produced at low temperatures and a high heating rate coupled with short residence time. On the contrary, the fuel gas is produced at high temperature, long residence time, and low heating rate. In the liquefaction process, biomass is heated under syngas and water environment to get various products. In biomass gasification, biomass is heated at high temperatures in the presence of oxygen to produce hydrogen, CO, CO_2, biomethane, and char. The product obtained is syngas which can either be used in the Fischer-Tropsch (FT) process to get liquid hydrocarbon biofuel or higher alcohols.

14.2.3.2 Chemical and biochemical processes

The chemical process involves transesterification reaction and acid or basic hydrolysis. These processes are generally employed in pretreatment steps. The biochemical process involves fermentation, aerobic and anaerobic digestion, and enzymatic reactions to produce the different classes of biofuels.

14.2.3.3 Mechanical process

These techniques are used to extract oil from lignocellulosic and algal biomass which is a crucial step in the biorefinery system. The mechanical process includes physical separation and mechanical fractionation techniques.

14.2.4 Product-based biorefinery

The biorefinery is designed to produce a spectrum of biofuel and chemicals. The conventional biofuels produced from biorefinery are bioethanol, biodiesel, and bio-jet fuel. These are explained in detail in the next section.

14.2.4.1 Biodiesel-based biorefinery

Biodiesel is one of the significant biofuels produced from oilseed with glycerol as a by-product. Various biorefinery such as lignocellulose biorefinery, algal refinery, etc. has been explored for biodiesel production [27]. In the typical biorefinery based on biodiesel, oil is first treated to produce oilseed crops and residue (Fig. 14.2). Oilseed crops underwent filtration to get oil, and oilseed residue is also treated through the physical process such as mechanical pressing to get oil and feeds for biodiesel production. Oil is pretreated to reduce free fatty acid by esterification. The transesterification using short-chain alcohol is done to get biodiesel as the main product whereas glycerine is a by-product [28]. Transesterification is carried out by acid or base catalysis by taking into consideration of free fatty acid amount. Homogeneous and heterogeneous catalysis has been employed for biodiesel production and also has advantages and disadvantages. Solid base catalysis has been found more effective than solid acid catalysis (La/zeolite beta, CaO/SBA-14, Mg-Al–CO$_3$NaOH/c-alumina) in biodiesel production [29–32].

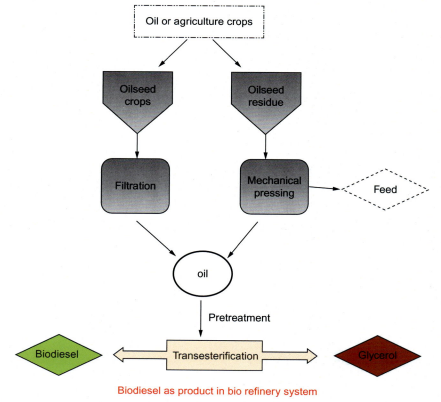

Fig. 14.2 Biodiesel as a product in the biorefinery system.

Microwave, enzymatic–assisted transesterification, or supercritical methanol–based transesterification are the new approaches for biodiesel production. Biodiesel also has been produced in combined oil biorefinery [33].

14.2.4.2 Bioethanol

Bioethanol and short-chain alcohols are the mainstream products in the biorefinery system. Different pathways have been explored in different biorefineries such as lignocellulosic, green, algal–based, oilseed–based, food waste biorefinery, etc. In the lignocellulosic biorefinery, bioethanol is produced from lignocellulosic biomass which involves different steps, viz., pretreatment, hydrolysis, fermentation, and distillation. The pretreatment step involves hemicellulose and cellulose breakage using the steam explosion technique, hot water treatment, and acid or base treatment. The pretreated biomass is enzymatically depolymerized, which yields C_5 and C_6 sugars. Finally, the microbial fermentation process gives bioethanol as a product [34].

In the green biorefinery system, the wet green biomass is fractionated to get press cake which can be a potential lignocellulosic feedstock for bioethanol production [35]. In the oilseed–based biorefinery, biomass is taken, and pretreatment such as a steam explosion is done to get different products. Organosolv pretreatment is the latest advancement in which pretreatment is catalyzed using sodium hydroxide and hydrogen peroxide to get cellulose by removing lignin and hemicellulose. After the pretreatment cellulose is subjected to simultaneous saccharification and fermentation to get aqueous bioethanol which after purification yields pure bioethanol as a product [36, 37]. Algal biomass is also a feedstock for bioethanol production. In a typical process, algal biomass is treated using the cell disruption technique. In the next step, extraction and purification processes are employed to get different fractions such as oil fraction, protein fraction, and carbohydrate fractions. The fermentation of sugar gives bioethanol [22, 23].

14.2.4.3 Bio-jet fuel

Aviation fuel or jet fuel, the kerosene-range product is an essential product of biorefinery. Bio-jet fuel can be produced from both lignocellulosic biomass, oil, and oilseed crop using different thermochemical, chemical, and biochemical processes. Lignin-based bio-jet fuel offers low emissions, favorable energy density, and high-performance characteristics of being used in engines. Biomass has been converted into bio-jet fuel by different processes.

1. Gasification and FT synthesis to bio-jet fuel (GFT-J)
2. Gasification, syngas fermentation to ethanol which is upgraded to bio-jet fuel (SYN-FER-J)
3. Lignocellulose biochemical conversion to ethanol which is upgraded to bio-jet fuel (L-ETH-J)

In the GFT-J process, lignocellulosic biomass along with air and stream is processed in the gasification plant, and syngas is generated. The syngas is further cleaned and subjected to an FT plant to get FT fuels. The hydrogen generated is recovered and used whereas other biofuels are upgraded to bio-jet fuel along with naphtha. In the SYN-FER-J process, the lignocellulosic biomass is sent to the gasification plant for syngas generation. Half of the syngas are used to provide electricity to the plant and the other is sent to the syngas fermenter for fermentation. After the syngas fermentation by using *Clostridium ljungdahlii*, ethanol is formed as the main product along with acetone as a by-product. Ethanol undergoes the dehydration reaction to form ethylene which is then oligomerized to bio-jet fuel, and hydroprocessed to get diesel and jet fuel as a product. In the L-ETH-J process, biomass is pretreated using SO_2 and stream and then undergoes hydrolysis and fermentation step to generate ethanol. The ethanol is dehydrated, oligomerized, and hydroprocessed to produce jet fuel and diesel. In recent advances, catalytic routes are also designed to get jet fuel [38].

Lignocellulosic biomass can be converted to sugar and after fermentation to get a short range of alcohols, a fatty acid, which works as a source for bio-jet fuel [39]. Biofuel is also produced from vegetable oil and sugar cane juice following hydroprocessing and sucrose fermentation. In the sucrose fermentation process, sugarcane is cleaned, and sucrose is extracted, which undergoes the treatment and fermentation process to get bioethanol. The ethanol is dehydrated with the help of generated hydrogen in the plant. After that oligomerization and hydroprocessing yield jet fuel and diesel as the product. In hydroprocessing, vegetable oil is hydrotreated to obtain the heavy range products. The aforementioned are further hydrocracked and separated to produce a range of products such as diesel, naphtha, and bio-jet fuel.

14.3 LCA of biorefineries

LCA studies on many of the biorefinery configurations have been reported in the literature. These studies have been conducted with the targeted feedstocks as well as desired products, and environmental impacts have been reported for the various possible scenarios. The reported studies have mainly been focused on the environmental impact; some unavoidable complementary factors. Economic issues and carbon trade policies, energy directives, and other policy decisions have also been considered to evolve the best biomass valorization route.

Cai et al. have published a comprehensive study on a new integrated biorefinery concept using hemicellulose feedstock and considering the coproduction of hydrocarbon biofuels and bio-based chemicals [10]. They have presented two integrated biorefinery (IBR) designs: one for coproduction of succinic acid (SA) from upgradation of a diverted hemicellulose C_5 sugar stream and the other for coproduction of adipic acid (AA) by upgrading a fraction of lignin. The study was mainly aimed at the enhancement of the economic feasibility of the IBR concept to increase biomass resource utilization efficiency. LCA was conducted at biorefinery configuration as well as the product level. The objective of biorefinery-level LCA was to estimate the global warming potential (GWP) of the production of the targeted final products in that specific configuration, for the total life cycle. The cumulative potential of GHG emission reduction by all products that could displace the conventionally made corresponding item production has been estimated. Product-level LCA was intended to assess the emissions of the complete life-cycle GHG-specific finished product along with the reduction of GHG emissions through displacing the corresponding products with those made by conventional methods.

The authors used process engineering models of SA and AA coproduction scenarios, available in Aspen Plus. System boundary for LCA included the biomass feedstock supply chain, operation of the biorefinery, transportation, and end use of the biorefinery products. LCA data was obtained from the 2016 release of Greenhouse gases, Regulated Emissions, Energy use in Transportation (GREET) model database [40]. The functional unit used for Renewable Diesel Blend (RDB) was taken as 1 MJ energy and for SA and AA for product-specific LCA, 1 US ton was used, respectively. This is because; per MJ basis is acceptable for fuel LCA of RDB per carbon regulation certification and compliance purposes.

Three product handling methods namely system-level, process-level, and displacement method were used in the LCA analysis for the estimation of product-specific GHG. The biorefinery-level GHG emissions were estimated using the aggregated methodology based on product-specific results at a biorefinery level. Reported life-cycle emissions have been calculated along with two steps. The material and energy utilization associated with any conventional processes contributing to both the fuels and chemical products are allocated between those products. Secondly, the life-cycle GHG emissions estimated by GREET include the trails of different resources of material and energy, coupled with net life-cycle GHG emissions calculations are based on material and energy consumption allocations.

Three system-level allocation methods, mass, market value, and energy allocation have been used. Usually, the net energy and material inputs to a process and their associated energy and emission loads may be allocated over all final products, using their respective output ratios, as per their masses, energy contents, or market values. On a mass allocation basis, the underlying assumption is that the equivalent amount of material and energy is required to process the feedstock into an equal amount of various products on a mass basis. It is a widely used method for the products-based LCA [41–43]. The market value approach is based on the assumption of activities and decisions driven by economics. Therefore, the environmental burdens are allocated based on the economic revenue of the individual product [44]. Energy-based allocations are preferred for a system setup aimed at the production of energy-based products, such as biofuels and refinery fuels. For this LCA study of IBR, the energy allocation was not used due to the production of biochemical coproducts that are not fuels. Such allocation methods, which treat the biorefinery as a single block in terms of overall net inputs and total outputs, do not consider details of the process-level inputs ascribed to a specific product/processing string. Nevertheless, it is often used for an LCA analysis due to its simplicity, when the main and co-products demonstrate analogous properties, e.g., both are scaled by mass or both displace the energy products or both possess market value as a commodity, of application. The main setback of this type of method is that it always allocates a small part of the chemical and energy burdens credited to one train (say biochemical production) to the other, i.e., the fuel production process, which in actuality has no requirement for that explicit chemical or energy input.

Contrary to the allocation at the system level, the process-level allocation methods assign the material and energy use of the distinctive processes of the biorefinery to the typical product stream attached to those processes with common unit operations, using the mass, energy, or market value allocation criteria. In other words, the processes connected with both fuels and chemical trains are allocated between the two products. Conversely, the processes associated with only fuel production are allocated to the fuel, whereas those associated with only chemical production are allocated into chemicals [6]. Consequently, the emission related to the allocated consumption of material and energy from all the processes contributing to a particular product is amassed for that product.

Nevertheless, allocation methods, system-level as well as process-level, treat the main product and coproducts the same for the physical property used for the allocation basis, e.-g., the cost or mass. These methods do not differentiate between possible, sometimes vivid, variations in the superiority of services or the functionality provided by the main product and the coproducts.

If a coproduct generated during a process can be used in the process itself, e.g., a recyclable solvent, it can be allocated as an avoided product. But in IBR configuration, there are some by-products that have to be recovered as coproducts. The emissions associated with these products may be handled with the displacement method, during LCA analysis. In the displacement method, also termed as system boundary extension, net associated emissions, and energy inputs of the process are first allocated to the main product. Then the credit is taken for the avoided emissions by displacement of the coproduct with a similar product, which is otherwise made by other conventional processes. This method forms a robust basis for the techno-economic evaluation and sustainability assessment of an IBR configuration, by analysis of the impact of the coproduction of the chemicals on the least fuel selling price, and accordingly adding all operational costs for energy and chemical inputs for the biorefinery collectively. At the same time, coproduct incomes are also generated by the sale of coproducts at its prevailing cost eventually reduce the requisite fuel market price, which makes the biorefinery a profitable enterprise [45].

14.3.1 LCA approach to biodiesel process

Biodiesel, a vital biofuel, can replace fossil fuel-based transportation fuels and their generated emission. The by-product obtained in biodiesel production is glycerol that has potential applications in the production of different compounds such as ethers, alcohols, acids, PHB, sugars, and bio-hydrogen through chemical and biochemical routes [46, 47]. Increased biodiesel production has generated excess glycerol in the last few years. This excess glycerol, which is considered as valuable coproducts earlier, is now considered as a waste stream [48]. The scrutiny of the biodiesel process is therefore required to assess the biodiesel and glycerol impacts on the environment. Keeping in mind, the LCA approach is discussed for the biodiesel production process. In the LCA approach, the goal and scope of the processes are defined along with its system boundary. The different allocation approaches are used as explored in the literature. The performance and emission characteristics of biodiesel are then evaluated and compared to fossil-based diesel for a concluding a sustainable approach to the system taken [49]. The production of biodiesel has been studied by various authors, as can be found in the literature. The life-cycle analysis of biodiesel is assessed on different parameters such as feedstock origin (country based) and its type as the first-, second-, and third-generation feedstock, process based, and reaction condition based.

Nazir studied the LCA of biodiesel produced from first- and second-generation feedstock, palm oil, and jatropha oil in Indonesia [50]. They evaluated the environmental impact in terms of mid-point categories and end point categories and GWP in terms of CO_2 equivalent. The author concludes that biodiesel production from palm oil consumes higher energy than jatropha oil and the highest energy consumption was for the transesterification step. Further, the cultivation process also contributes to the environmental impact. Various studies have been reported the LCA of biodiesel on different feedstocks. Some of the studies are illustrated in Table 14.2.

In similar studies by Srinophakun et al. concluded that the transesterification step has a major environmental impact [55]. Mupondwa et al. studied the camelina oil-derived biodiesel in Canadian prairies and found that it has less GHG emissions and energy consumption as compared to other oilseeds [51]. Kumar et al. did a comprehensive LCA of jatropha biodiesel of Indian origin. They evaluated biodiesel impact in terms of GHG emission and net energy ratio for irrigated and rain-fed scenarios and found a reduction in both values [52].

Schneider et al. studied the LCA of biodiesel from tobacco seeds [53]. They concluded that solaris seeds tobacco-derived biodiesel prevents environmental impact and results were found good based on damage-oriented categories such as human health, ecosystem, and resources. Further, they concluded that fertilizer and energy use contribute highly to environmental impact. Peiro et al. did the LCA and energetic LCA of biodiesel produced from used cooking oil to evaluate environmental impact and exergy input to the system [54]. As per their results, the transesterification caused 68% of the

Table 14.2 Reported studies on biodiesel production from various feedstocks.

S. No.	Feedstock	Inferences	References
1.	Camelina	• Biodiesel was prepared from camelina oil on Canadian prairies. • 1 kg of biodiesel was taken as functional unit at agricultural stage and 1 MJ was used for fuel combustion stage. • GHG emission and carbon footprint from 1 MJ of biodiesel decreased.	[51]
2.	Jatropha	• Biodiesel was synthesized from jatropha oil and environmental impacts were assessed through with and without allocation approach. • Environmental impact was assessed in irrigation and rain fed scenario. • GHG emissions and net energy ratio was evaluated.	[52]
3.	Tobacco seeds	• 1 kg of biodiesel was used as functional unit. • All the process parameters were compared with two impact assessment method, viz., Recipe and Eco-indicator 99. • Total damages were correlated with damage categories.	[53]
4.	Used cooking oil	• One ton of biodiesel was used as functional unit. • LCA and exergetic LCA was done on biodiesel production from used cooking oil. • CO and CO_2 emissions were reduced to 26%.	[54]

total environmental impact, and maximum exergy inputs were from uranium and natural gas for the production of electricity. Lardon et al. studied biodiesel based on microalgae and compared it with first-generation biodiesel and petro-diesel [56]. It was concluded that the algae have excellent potential for biodiesel production but its impact on the environment is yet to be explored. Foteinis and coworkers studied the LCA for biodiesel produced from used cooking oil and compared it with petro-diesel, first- and third-generation biodiesel [57]. The total carbon footprint and environmental impact were lower than first-generation and third-generation biofuels. Comparing with petro-diesel, the environmental impact was three times lesser. Overall, it was more sustainable than other generation biofuels. Piumsomboon et al. reported a comparative LCA study on biodiesel production using conventional supercritical methanol methods [58]. The study concluded that although the supercritical process offers a simple production process and high yield but still not feasible. It generates high environmental owes to the methanol recovery unit, which itself needs high energy. In a different study by Srinophakun et al., the different catalyst, viz., sodium hydroxide and potassium hydroxide were taken for the transesterification process of jatropha oil. They evaluated the effects in terms of damage-oriented categories [55]. Authors concluded that despite sodium hydroxide has low resource depletion, it has a high environmental impact on human health and the ecosystem. In conclusion, LCA can be used to assess the parameters and interpret the results in terms of different impact categories for the sustainability of the system.

14.3.2 LCA of a thermal conversion-based biorefinery

A diverse variety of the processes have been reported for thermal conversion of various biomass to the bio-oil as well as other hydrocarbon biofuels, such as pyrolysis (fast and slow), torrefaction, liquefaction, and gasification. Similar to different processes, the production of hydrocarbon biofuels and chemicals, through the thermal process involves raw products as well as other natural or technical inputs, which results in various degrees of environmental impact. Therefore, several studies have been carried out on the LCA of the different thermochemical processes. Ubando et al. have reported a comprehensive review of the LCA of bioenergy products from microalgal as well as lignocellulosic feedstocks [59]. These two feedstocks form a major share of the biomass, after oil seeds, and are also widely used in the majority of the thermal conversion processes. According to the report of the International Energy Agency (2017), the development of the thermochemical conversion is one of the priority areas to achieve the goals set by their global energy technology roadmap. Therefore, the systematic LCA studies for thermochemical processes also become necessary for sustainability assessment. Some latest studies on LCA of various thermochemical processes are summarized in the following sections.

14.3.2.1 Torrefaction

With the development of the concept of zero residue-based refineries, torrefaction has emerged as an essential process for microalgae and lignocellulose residues. Torrefaction is used to achieve a higher energy density for the possible replacement of fossil coal. A number of studies are reported in the literature on the torrefaction of the residual biomass at different temperatures and residence time. However, no potential research has been published on the LCA of torrefaction of microalgae, which quantifies its environmental effect. However, few studies on the torrefaction of lignocelluloses biomass have been reported [60, 61]. They have performed the comparative LCA on a wood pallet made by torrefaction and compared it with raw wood biomass. McNamee et al. have reported the highest GHG emissions during the drying of wood, followed by torrefaction of biomass, and finally, during transport. Christoforou and Fokaides studied the ecological impact of torrefaction of the olive husk and reported similar observations. They conclude that the energy saving during the drying of biomass contributes more in reduction of overall environmental impact than transportation [62].

Tsalidis et al. studied the environmental payback of cofiring of the torrefied lignocellulosic biomass and coal with the conventional coal power plant [63]. According to them, the cofiring results in the reduction of GWP by 12%, acidification by 7%, and the photochemical oxidation potential by 5%, in comparison to coal. Arteaga-Pérez et al. reported a study on the effect of mixing pure pine pellets with the torrefied pine pellets and cofired with coal [64]. A mixture of 20:80 of torrefied biomass with coal, used in power generation reduced abiotic depletion by 11%, acidification potential by 20%, and human toxicity by 20%. The GWP was reduced by 16% as compared to the pure coal plant. Thus, the LCA of torrefaction has reported the increasing trend of the GWP reduction, along with other categories of environmental impact.

14.3.2.2 Pyrolysis and hydrothermal liquefaction

Slow as well as fast pyrolysis are the most popular processes of thermal conversion of biomass. Recent studies are mainly focused on the pyrolysis of the lignocellulosic biomass feedstocks to generate bio-oil. However, some studies have also been centered on microalgae pyrolysis as well [65]. Grierson et al. and Wang et al. reported the LCA studies on the pyrolysis of the microalgal biomass to estimate the environmental impact of bioenergy products of pyrolysis [66, 67]. They assessed the environmental impact of bio-oil generation from microalgae on a manufacturing scale and have shown a reduction of GWP impact by 40% for electricity generation from the pyrolyzed biomass, as compared to the conventional electricity. Their study indicates the major environmental impacts during the cultivation and drying. Wang et al. expanded their research to estimate the impact of the production of aviation fuel from microalgae bio-oil and reported an 11% improvement in energy input. Limited LCA has been carried out on the pyrolysis of lignocelluloses biomass [66]. Fan et al. reported the GHG emissions of pyrolyzed bio-oil from various forest waste biomass streams, such as logging residues of the poplar and willow [68]. Their study reveals lower environmental impacts as compared to fossil fuels. Roberts et al. calculated the energetic, along with the climate change impact of the switchgrass [69]. Their studies report negative GHG implications in view of CO_2 uptake during the biomass growth phase. A study published by Chan et al. evaluated the ecological impact of bio-oil production from the pyrolysis of empty fruit bunches, relative to hydrothermal liquefaction (HTL) [70].

Liu et al. have reported a study on the LCA of biofuels production by the HTL process. Three different cases of the lab-, bench-, and pilot-scale process were studied to assess the environmental impacts for industrial scale [71]. The results indicate that the GWP impacts of all three cases (lab-, bench-, and pilot-plant) are lower than the regular gasoline. Another route of bio-oil upgradation via HTL is conversion to bio–jet fuel. Fortier et al. have reported the LCA study evaluating the GWP of the microalgae bio–jet fuel. At the same time, HTL is located in a refinery complex, with HTL colocated in the microalgae waste treatment plant [72]. A significant GHG emission reduction of 51.3 kg CO_2e/GJ was observed in the latter. Similar studies were also reported by Connelly and coworkers [73]. Thus, the LCA studies highly recommend HTL for the wet biomass similar to the microalgae, due to the significant reduction of GWP by nearly 50% due to the drying process being energy intensive.

Despite the majority of HTL studies being focused on the various lignocellulosic biomass, there are few studies aimed at the sustainability assessment of producing bioenergy products via the HTL route. Environmental impact assessment of the production of empty fruit bunch oil by liquefaction has been reported by Chan et al. [70]. In a subsequent study, Chen et al. reported the expansion of this work by including palm kernel shell oil [74]. A recent study by Michailos has reported a 50% GWP reduction by producing bio-jet fuel from sugarcane residue, as compared to fossil jet fuel [75].

14.3.2.3 Gasification

LCA of the gasification of biomass is not well reported. The reason may be a few plants operating at an industrial scale. An LCA study by Azadi et al. has published the coproduction of synthesis gas and biodiesel from microalgae feedstock [76]. According to their study, the net GHG emissions of the generated synthesis gas were reduced at least four times using solar drying compared to the conventional drying method.

Many authors have reported LCA studies on the gasification of lignocellulosic biomass. Renó et al. reported negative carbon impact ($-2.28\,kg\ CO_2e/kg_{methanol}$) during the production of methanol from sugarcane biomass, due to a large amount of CO_2 absorbed at the cultivation stage [77]. Similarly, negative carbon impact has been reported by other authors as well in the case of methanol production by wood gasification [78–80]. A reduction in emissions of $0.325–0.653\,kg\ CO_2e/kWh$ for electricity generation was reported by Nuss et al. and $6.72\,gCO_2e/(MW\ of\ H_2)$ was reported by Kalinci et al. for hydrogen production, both by wood gasification [81, 82].

To summarize the GWP impacts of thermal conversion biorefineries, the studies show the relatively high GWP on torrefaction, in the case of cofiring of the torrefied biomass and bio-coal [63, 83]. Negative carbon emissions have also been reported for pyrolysis of corn stover, switchgrass, and yard waste, in view of carbon sequestration through the cultivation stage of the life cycle. Relatively high GWP reduction has been reported in the case of microalgae to fuel conversion at the pilot-scale by Liu et al. [71]. The order of the average GWP in increasing order is as follows.

$$\text{Torrefaction} < \text{Gasification} < \text{Liquefaction} < \text{Pyrolysis}$$

In addition to the GHG emission reduction, the energy contents of the biofuels showed an improvement in the calorific value of the biomass [74].

14.4 Conclusions

The biorefinery concept has emerged out with the advent of a variety of processes for conversion of the biomass feedstocks to useful chemicals and hydrocarbon biofuels.

However, to identify the sustainable routes for biofuels and chemicals from renewable feedstock, LCA is being used effectively. A complete LCA study not only firms up the conceptual design of the biomass conversion processes but also helps in the profitable operation of the various conversion processes. LCA of any biorefinery configuration can be carried out using data generated indigenously, along with the standard life-cycle inventories available commercially. Three main conversion processes, namely biochemical, chemical, and thermochemical, have been reported in the literature. It has been undisputedly accepted by the technologists that these processes have been proved to be effective for GWP reduction and energy-efficient processes. A well-designed biorefinery configuration helps in the maximization of energy recovery from the biomass feedstock and zero waste generation from the constituent conversion processes. The concept of LCA may further be extended to identify the potential of various configurations for contribution to the circular economy and materials recycling. This will further help in the development of zero residue biorefineries which can lead to simultaneous pollution abatement.

References

[1] E. Reichmanis, M. Sabahi, Life cycle inventory assessment as a sustainable chemistry and engineering education tool, ACS Sustain. Chem. Eng. 5 (2017) 9603–9613, https://doi.org/10.1021/acssuschemeng.7b03144.

[2] ISO 14040:2006, 2020.000Z. https://www.iso.org/standard/37456.html. (Accessed 10 July 2020).

[3] IS/ISO 14044, Environmental Management-Life Cycle Assessment-Requirements and Guidelines, Bureau of Indian Standards, 2006.

[4] W.M.J. Hugo, J.A. Badenhorst, E.H.B. van Biljon, Supply Chain Management: Logistics in Perspective, first ed., Van Schaik, Pretoria, South Africa, 2013.

[5] J.A. Cano-Ruiz, G.J. McRae, Environmentally conscious chemical process design, Annu. Rev. Energy Environ. 23 (1998) 499–536, https://doi.org/10.1146/annurev.energy.23.1.499.

[6] Q. Wang, W. Liu, X. Yuan, H. Tang, Y. Tang, M. Wang, J. Zuo, Z. Song, J. Sun, Environmental impact analysis and process optimization of batteries based on life cycle assessment, J. Clean. Prod. 174 (2018) 1262–1273, https://doi.org/10.1016/j.jclepro.2017.11.059.

[7] H.K. Jeswani, S. Hellweg, A. Azapagic, Accounting for land use, biodiversity and ecosystem services in life cycle assessment: impacts of breakfast cereals, Sci. Total Environ. 645 (2018) 51–59, https://doi.org/10.1016/j.scitotenv.2018.07.088.

[8] S. de Jong, K. Antonissen, R. Hoefnagels, L. Lonza, M. Wang, A. Faaij, M. Junginger, Life-cycle analysis of greenhouse gas emissions from renewable jet fuel production, Biotechnol. Biofuels 10 (2017) 64, https://doi.org/10.1186/s13068-017-0739-7.

[9] C.E. Canter, J.B. Dunn, J. Han, Z. Wang, M. Wang, Policy implications of allocation methods in the life cycle analysis of integrated corn and corn stover ethanol production, Bioenerg. Res. 9 (2016) 77–87, https://doi.org/10.1007/s12155-015-9664-4.

[10] H. Cai, J. Han, M. Wang, R. Davis, M. Biddy, E. Tan, Life-cycle analysis of integrated biorefineries with co-production of biofuels and bio-based chemicals: co-product handling methods and implications, Biofuels Bioprod. Biorefin. 12 (2018) 815–833, https://doi.org/10.1002/bbb.1893.

[11] IEA Bioenergy Task 42 Biorefinery, NaN. https://www.iea-bioenergy.task42-biorefineries.com/en/ieabiorefinery.htm. (Accessed 13 July 2020).

[12] M. FitzPatrick, P. Champagne, M.F. Cunningham, R.A. Whitney, A biorefinery processing perspective: treatment of lignocellulosic materials for the production of value-added products, Bioresour. Technol. 101 (2010) 8915–8922, https://doi.org/10.1016/j.biortech.2010.06.125.

[13] F. Cherubini, The biorefinery concept: using biomass instead of oil for producing energy and chemicals, Energy Convers. Manag. 51 (2010) 1412–1421, https://doi.org/10.1016/j.enconman.2010.01.015.

[14] C. Liu, C.E. Wyman, Partial flow of compressed-hot water through corn Stover to enhance hemicellulose sugar recovery and enzymatic digestibility of cellulose, Bioresour. Technol. 96 (2005) 1978–1985, https://doi.org/10.1016/j.biortech.2005.01.012.

[15] S.H. Hazeena, R. Sindhu, A. Pandey, P. Binod, Lignocellulosic bio-refinery approach for microbial 2,3-butanediol production, Bioresour. Technol. 302 (2020) 122873, https://doi.org/10.1016/j.biortech.2020.122873.

[16] P. Lamers, E.C.D. Tan, E.M. Searcy, C.J. Scarlata, K.G. Cafferty, J.J. Jacobson, Strategic supply system design—a holistic evaluation of operational and production cost for a biorefinery supply chain, Biofuels Bioprod. Biorefin. 9 (2015) 648–660, https://doi.org/10.1002/bbb.1575.

[17] L.A. Lucia, D.S. Argyropoulos, L. Adamopoulos, A.R. Gaspar, Chemicals and energy from biomass, Can. J. Chem. 84 (2006) 960–970, https://doi.org/10.1139/v06-117.

[18] M. Predel, Pyrolysis of rape-seed in a fluidised-bed reactor, Bioresour. Technol. 66 (1998) 113–117, https://doi.org/10.1016/S0960-8524(98)00059-5.

[19] P.A. Horne, P.T. Williams, Influence of temperature on the products from the flash pyrolysis of biomass, Fuel 75 (1996) 1051–1059, https://doi.org/10.1016/0016-2361(96)00081-6.

[20] M. Nayaggy, Z.A. Putra, Process simulation on fast pyrolysis of palm kernel shell for production of fuel, Indonesian J. Sci. Technol. 4 (2019) 64, https://doi.org/10.17509/ijost.v4i1.15803.

[21] S. Talekar, A.F. Patti, R. Vijayraghavan, A. Arora, Complete utilization of waste pomegranate peels to produce a hydrocolloid, punicalagin rich phenolics, and a hard carbon electrode, ACS Sustain. Chem. Eng. 6 (2018) 16363–16374, https://doi.org/10.1021/acssuschemeng.8b03452.

[22] A.K. Koyande, P.-L. Show, R. Guo, B. Tang, C. Ogino, J.-S. Chang, Bio-processing of algal biorefinery: a review on current advances and future perspectives, Bioengineered 10 (2019) 574–592, https://doi.org/10.1080/21655979.2019.1679697.

[23] R. Chandra, H.M.N. Iqbal, G. Vishal, H.-S. Lee, S. Nagra, Algal biorefinery: a sustainable approach to valorize algal-based biomass towards multiple product recovery, Bioresour. Technol. 278 (2019) 346–359, https://doi.org/10.1016/j.biortech.2019.01.104.

[24] J. Trivedi, M. Aila, D.P. Bangwal, S. Kaul, M.O. Garg, Algae based biorefinery—how to make sense? Renew. Sust. Energ. Rev. 47 (2015) 295–307, https://doi.org/10.1016/j.rser.2015.03.052.

[25] J. Trivedi, M. Aila, C.D. Sharma, P. Gupta, S. Kaul, Clean synthesis of biolubricant range esters using novel liquid lipase enzyme in solvent free medium, Springerplus 4 (2015) 165, https://doi.org/10.1186/s40064-015-0937-3.

[26] K. Özdenkçi, C. de Blasio, H.R. Muddassar, K. Melin, P. Oinas, J. Koskinen, G. Sarwar, M. Järvinen, A novel biorefinery integration concept for lignocellulosic biomass, Energy Convers. Manag. 149 (2017) 974–987, https://doi.org/10.1016/j.enconman.2017.04.034.

[27] N. Srivastava, M. Srivastava, P.K. Mishra, S.N. Upadhyay, P.W. Ramteke, V.K. Gupta (Eds.), Sustainable Approaches for Biofuels Production Technologies, Springer International Publishing, Cham, Switzerland, 2019.

[28] S. Al-Zuhair, Production of biodiesel: possibilities and challenges, Biofuels Bioprod. Biorefin. 1 (2007) 57–66, https://doi.org/10.1002/bbb.2.

[29] N. Barakos, S. Pasias, N. Papayannakos, Transesterification of triglycerides in high and low quality oil feeds over an HT2 hydrotalcite catalyst, Bioresour. Technol. 99 (2008) 5037–5042, https://doi.org/10.1016/j.biortech.2007.09.008.

[30] Q. Shu, Z. Nawaz, J. Gao, Y. Liao, Q. Zhang, D. Wang, J. Wang, Synthesis of biodiesel from a model waste oil feedstock using a carbon-based solid acid catalyst: reaction and separation, Bioresour. Technol. 101 (2010) 5374–5384, https://doi.org/10.1016/j.biortech.2010.02.050.

[31] G. Arzamendi, I. Campo, E. Arguiñarena, M. Sánchez, M. Montes, L.M. Gandía, Synthesis of biodiesel with heterogeneous NaOH/alumina catalysts: comparison with homogeneous NaOH, Chem. Eng. J. 134 (2007) 123–130, https://doi.org/10.1016/j.cej.2007.03.049.

[32] G. Arzamendi, E. Arguiñarena, I. Campo, S. Zabala, L.M. Gandía, Alkaline and alkaline-earth metals compounds as catalysts for the methanolysis of sunflower oil, Catal. Today 133-135 (2008) 305–313, https://doi.org/10.1016/j.cattod.2007.11.029.

[33] A. Niño-Villalobos, J. Puello-Yarce, Á.D. González-Delgado, K.A. Ojeda, E. Sánchez-Tuirán, Biodiesel and hydrogen production in a combined palm and jatropha biomass biorefinery: simulation, techno-economic, and environmental evaluation, ACS Omega 5 (2020) 7074–7084, https://doi.org/10.1021/acsomega.9b03049.

[34] J.K. Saini, R. Gupta, A. Hemansi, P. Verma, R.S. Gaur, R. Shukla, R.C. Kuhad, Integrated lignocellulosic biorefinery for sustainable bio-based economy, in: N. Srivastava, M. Srivastava, P.K. Mishra, S.-N. Upadhyay, P.W. Ramteke, V.K. Gupta (Eds.), Sustainable Approaches for Biofuels Production Technologies, Springer International Publishing, Cham, Switzerland, 2019, pp. 25–46.

[35] J. McEniy, P. O'Kiely, Developments in grass-/forage-based biorefineries, in: Advances in Biorefineries, Elsevier, 2014, pp. 335–363.

[36] C. Ofori-Boateng, K.T. Lee, Comparative thermodynamic sustainability assessment of lignocellulosic pretreatment methods for bioethanol production via exergy analysis, Chem. Eng. J. 228 (2013) 162–171, https://doi.org/10.1016/j.cej.2013.04.082.

[37] S.Y. Lee (Ed.), Consequences of Microbial Interactions with Hydrocarbons, Oils, and Lipids: Production of Fuels and Chemicals, Springer International Publishing, Cham, 2017.

[38] H. Wang, M.A. Oehlschlaeger, Autoignition studies of conventional and Fischer–Tropsch jet fuels, Fuel 98 (2012) 249–258, https://doi.org/10.1016/j.fuel.2012.03.041.

[39] T. Tian, T.S. Lee, Advanced biodiesel and biojet fuels from lignocellulosic biomass, in: S.Y. Lee (Ed.), Consequences of Microbial Interactions with Hydrocarbons, Oils, and Lipids: Production of Fuels and Chemicals, Springer International Publishing, Cham, Switzerland, 2017, pp. 1–25.

[40] Argonne GREET Model, 2020.000Z. https://greet.es.anl.gov/. (Accessed 10 July 2020).

[41] R. Parker, Implications of high animal by-product feed inputs in life cycle assessments of farmed Atlantic salmon, Int. J. Life Cycle Assess. 23 (2018) 982–994, https://doi.org/10.1007/s11367-017-1340-9.

[42] G. Jungmeier, F. Werner, A. Jarnehammar, C. Hohenthal, K. Richter, Allocation in LCA of wood-based products experiences of cost action E9 part I. Methodology, Int. J. Life Cycle Assess. 7 (2002).

[43] H. Ellingsen, S.A. Aanondsen, Environmental impacts of wild caught cod and farmed Salmon—a comparison with chicken (7 pp), Int. J. Life Cycle Assess. 11 (2006) 60–65, https://doi.org/10.1065/lca2006.01.236.

[44] M. Wang, H. Huo, S. Arora, Methods of dealing with co-products of biofuels in life-cycle analysis and consequent results within the U.S. context, Energy Policy 39 (2011) 5726–5736, https://doi.org/10.1016/j.enpol.2010.03.052.

[45] https://www.osti.gov/biblio/1244312-chemicals-from-biomass-market-assessment-bioproducts-near-term-potential. (Accessed 20 May 2021).

[46] J. Singh, J. Kumar, M.S. Negi, D. Bangwal, S. Kaul, M.O. Garg, Kinetics and modeling study on etherification of glycerol using isobutylene by in situ production from tert-butyl alcohol, Ind. Eng. Chem. Res. 54 (2015) 5213–5219, https://doi.org/10.1021/acs.iecr.5b00079.

[47] C. Varrone, R. Liberatore, T. Crescenzi, G. Izzo, A. Wang, The valorization of glycerol: economic assessment of an innovative process for the bioconversion of crude glycerol into ethanol and hydrogen, Appl. Energy 105 (2013) 349–357, https://doi.org/10.1016/j.apenergy.2013.01.015.

[48] J.R.M. Almeida, L.C.L. Fávaro, B.F. Quirino, Biodiesel biorefinery: opportunities and challenges for microbial production of fuels and chemicals from glycerol waste, Biotechnol. Biofuels 5 (2012) 48, https://doi.org/10.1186/1754-6834-5-48.

[49] E. Gnansounou, A. Dauriat, J. Villegas, L. Panichelli, Life cycle assessment of biofuels: energy and greenhouse gas balances, Bioresour. Technol. 100 (2009) 4919–4930, https://doi.org/10.1016/j.biortech.2009.05.067.

[50] N. Nazir, N. Ramli, D. Mangunwidjaja, E. Hambali, D. Setyaningsih, S. Yuliani, M.A. Yarmo, J. Salimon, Extraction, transesterification and process control in biodiesel production from Jatropha curcas, Eur. J. Lipid Sci. Technol. 111 (2009) 1185–1200, https://doi.org/10.1002/ejlt.200800259.

[51] X. Li, E. Mupondwa, Life cycle assessment of camelina oil derived biodiesel and jet fuel in the Canadian prairies, Sci. Total Environ. 481 (2014) 17–26, https://doi.org/10.1016/j.scitotenv.2014.02.003.

[52] S. Kumar, J. Singh, S.M. Nanoti, M.O. Garg, A comprehensive life cycle assessment (LCA) of Jatropha biodiesel production in India, Bioresour. Technol. 110 (2012) 723–729, https://doi.org/10.1016/j.biortech.2012.01.142.

[53] F.S. Carvalho, F. Fornasier, J.O.M. Leitão, J.A.R. Moraes, R.C.S. Schneider, Life cycle assessment of biodiesel production from solaris seed tobacco, J. Clean. Prod. 230 (2019) 1085–1095, https://doi.org/10.1016/j.jclepro.2019.05.177.

[54] L. Talens Peiró, L. Lombardi, G. Villalba Méndez, X. Gabarrell i Durany, Life cycle assessment (LCA) and exergetic life cycle assessment (ELCA) of the production of biodiesel from used cooking oil (UCO), Energy 35 (2010) 889–893, https://doi.org/10.1016/j.energy.2009.07.013.

[55] U. Kaewcharoensombat, K. Prommetta, T. Srinophakun, Life cycle assessment of biodiesel production from jatropha, J. Taiwan Inst. Chem. Eng. 42 (2011) 454–462, https://doi.org/10.1016/j.jtice.2010.09.008.

[56] L. Lardon, A. Hélias, B. Sialve, J.-P. Steyer, O. Bernard, Life-cycle assessment of biodiesel production from microalgae, Environ. Sci. Technol. 43 (2009) 6475–6481, https://doi.org/10.1021/es900705j.

[57] S. Foteinis, E. Chatzisymeon, A. Litinas, T. Tsoutsos, Used-cooking-oil biodiesel: life cycle assessment and comparison with first- and third-generation biofuel, Renew. Energy 153 (2020) 588–600, https://doi.org/10.1016/j.renene.2020.02.022.

[58] C. Kiwjaroun, C. Tubtimdee, P. Piumsomboon, LCA studies comparing biodiesel synthesized by conventional and supercritical methanol methods, J. Clean. Prod. 17 (2009) 143–153, https://doi.org/10.1016/j.jclepro.2008.03.011.

[59] A.T. Ubando, D.R.T. Rivera, W.-H. Chen, A.B. Culaba, A comprehensive review of life cycle assessment (LCA) of microalgal and lignocellulosic bioenergy products from thermochemical processes, Bioresour. Technol. 291 (2019) 121837, https://doi.org/10.1016/j.biortech.2019.121837.

[60] P. McNamee, P.W.R. Adams, M.C. McManus, B. Dooley, L.I. Darvell, A. Williams, J.M. Jones, An assessment of the torrefaction of North American pine and life cycle greenhouse gas emissions, Energy Convers. Manag. 113 (2016) 177–188, https://doi.org/10.1016/j.enconman.2016.01.006.

[61] P.W.R. Adams, J.E.J. Shirley, M.C. McManus, Comparative cradle-to-gate life cycle assessment of wood pellet production with torrefaction, Appl. Energy 138 (2015) 367–380, https://doi.org/10.1016/j.apenergy.2014.11.002.

[62] E.A. Christoforou, P.A. Fokaides, Life cycle assessment (LCA) of olive husk torrefaction, Renew. Energy 90 (2016) 257–266, https://doi.org/10.1016/j.renene.2016.01.022.

[63] G.-A. Tsalidis, Y. Joshi, G. Korevaar, W. de Jong, Life cycle assessment of direct co-firing of torrefied and/or pelletised woody biomass with coal in the Netherlands, J. Clean. Prod. 81 (2014) 168–177, https://doi.org/10.1016/j.jclepro.2014.06.049.

[64] L.E. Arteaga-Pérez, M. Vega, L.C. Rodríguez, M. Flores, C.A. Zaror, Y. Casas Ledón, Life-cycle assessment of coal–biomass based electricity in Chile: focus on using raw vs torrefied wood, Energy Sustainable Dev. 29 (2015) 81–90, https://doi.org/10.1016/j.esd.2015.10.004.

[65] Q.-V. Bach, W.-H. Chen, Pyrolysis characteristics and kinetics of microalgae via thermogravimetric analysis (TGA): a state-of-the-art review, Bioresour. Technol. 246 (2017) 88–100, https://doi.org/10.1016/j.biortech.2017.06.087.

[66] X. Wang, F. Guo, Y. Li, X. Yang, Effect of pretreatment on microalgae pyrolysis: kinetics, biocrude yield and quality, and life cycle assessment, Energy Convers. Manag. 132 (2017) 161–171, https://doi.org/10.1016/j.enconman.2016.11.006.

[67] S. Grierson, V. Strezov, J. Bengtsson, Life cycle assessment of a microalgae biomass cultivation, bio-oil extraction and pyrolysis processing regime, Algal Res. 2 (2013) 299–311, https://doi.org/10.1016/j.algal.2013.04.004.

[68] J. Fan, T.N. Kalnes, M. Alward, J. Klinger, A. Sadehvandi, D.R. Shonnard, Life cycle assessment of electricity generation using fast pyrolysis bio-oil, Renew. Energy 36 (2011) 632–641, https://doi.org/10.1016/j.renene.2010.06.045.

[69] K.G. Roberts, B.A. Gloy, S. Joseph, N.R. Scott, J. Lehmann, Life cycle assessment of biochar systems: estimating the energetic, economic, and climate change potential, Environ. Sci. Technol. 44 (2010) 827–833, https://doi.org/10.1021/es902266r.

[70] Y.H. Chan, R.R. Tan, S. Yusup, H.L. Lam, A.T. Quitain, Comparative life cycle assessment (LCA) of bio-oil production from fast pyrolysis and hydrothermal liquefaction of oil palm empty fruit bunch (EFB), Clean Techn. Environ. Policy 18 (2016) 1759–1768, https://doi.org/10.1007/s10098-016-1172-5.

[71] X. Liu, B. Saydah, P. Eranki, L.M. Colosi, B. Greg Mitchell, J. Rhodes, A.F. Clarens, Pilot-scale data provide enhanced estimates of the life cycle energy and emissions profile of algae biofuels produced via hydrothermal liquefaction, Bioresour. Technol. 148 (2013) 163–171, https://doi.org/10.1016/j.biortech.2013.08.112.

[72] M.-O.P. Fortier, G.W. Roberts, S.M. Stagg-Williams, B.S.M. Sturm, Life cycle assessment of bio-jet fuel from hydrothermal liquefaction of microalgae, Appl. Energy 122 (2014) 73–82, https://doi.org/10.1016/j.apenergy.2014.01.077.

[73] E.B. Connelly, L.M. Colosi, A.F. Clarens, J.H. Lambert, Life cycle assessment of biofuels from algae hydrothermal liquefaction: the upstream and downstream factors affecting regulatory compliance, Energy Fuel 29 (2015) 1653–1661, https://doi.org/10.1021/ef502100f.

[74] W.-H. Chen, J. Peng, X.T. Bi, A state-of-the-art review of biomass torrefaction, densification and applications, Renew. Sust. Energ. Rev. 44 (2015) 847–866, https://doi.org/10.1016/j.rser.2014.12.039.

[75] S. Michailos, Process design, economic evaluation and life cycle assessment of jet fuel production from sugar cane residue, Environ. Prog. Sustain. Energy 37 (2018) 1227–1235, https://doi.org/10.1002/ep.12840.

[76] P. Azadi, G. Brownbridge, S. Mosbach, O. Inderwildi, M. Kraft, Simulation and life cycle assessment of algae gasification process in dual fluidized bed gasifiers, Green Chem. 17 (2015) 1793–1801, https://doi.org/10.1039/C4GC01698J.

[77] M.L.G. Renó, E.E.S. Lora, J.C.E. Palacio, O.J. Venturini, J. Buchgeister, O. Almazan, A LCA (life cycle assessment) of the methanol production from sugarcane bagasse, Energy 36 (2011) 3716–3726, https://doi.org/10.1016/j.energy.2010.12.010.

[78] M. Carpentieri, A. Corti, L. Lombardi, Life cycle assessment (LCA) of an integrated biomass gasification combined cycle (IBGCC) with CO2 removal, Energy Convers. Manag. 46 (2005) 1790–1808, https://doi.org/10.1016/j.enconman.2004.08.010.

[79] J.S. Luterbacher, M. Fröling, F. Vogel, F. Maréchal, J.W. Tester, Hydrothermal gasification of waste biomass: process design and life cycle assessment, Environ. Sci. Technol. 43 (2009) 1578–1583, https://doi.org/10.1021/es801532f.

[80] S. González-García, D. Iribarren, A. Susmozas, J. Dufour, R.J. Murphy, Life cycle assessment of two alternative bioenergy systems involving Salix spp. biomass: bioethanol production and power generation, Appl. Energy 95 (2012) 111–122, https://doi.org/10.1016/j.apenergy.2012.02.022.

[81] Y. Kalinci, A. Hepbasli, I. Dincer, Life cycle assessment of hydrogen production from biomass gasification systems, Int. J. Hydrog. Energy 37 (2012) 14026–14039, https://doi.org/10.1016/j.ijhydene.2012.06.015.

[82] P. Nuss, K.H. Gardner, J.R. Jambeck, Comparative life cycle assessment (LCA) of construction and demolition (C&D) derived biomass and U.S. northeast forest residuals gasification for electricity production, Environ. Sci. Technol. 47 (2013) 3463–3471, https://doi.org/10.1021/es304312f.

[83] N. Kaliyan, R.V. Morey, D.G. Tiffany, W.F. Lee, Life cycle assessment of a corn Stover torrefaction plant integrated with a corn ethanol plant and a coal fired power plant, Biomass Bioenergy 63 (2014) 92–100, https://doi.org/10.1016/j.biombioe.2014.02.008.

CHAPTER 15

Hydrodeoxygenation of lignin-derived platform chemicals on transition metal catalysts

Shelaka Gupta[a] and M. Ali Haider[b]

[a]Multiscale Modelling for Energy and Catalysis Laboratory, Department of Chemical Engineering, Indian Institute of Technology Hyderabad, Kandi, Telangana, India
[b]Renewable Energy and Chemicals Laboratory, Department of Chemical Engineering, Indian Institute of Technology Delhi, Hauz Khas, Delhi, India

Contents

15.1 Introduction	409
15.2 Transition metals as hydrodeoxygenation catalysts	411
15.2.1 Active site in HDO reaction	411
15.2.2 Support effect in HDO reaction	412
15.2.3 Product selectivity trends for hydrogenolysis products over other products	416
15.3 Rational design of bimetallic alloys for HDO reaction	421
15.4 Oxophilic metals as acidic sites for HDO reaction	424
15.5 Conclusions	425
Acknowledgment	426
References	426

Abbreviations

ASA	amorphous silica alumina
CN	carbon nitrides
DDO	direct deoxygenation
DFT	density functional theory
HDO	hydrodeoxygenation
HYD	hydrogenation–dehydration
KIT-6	mesoporous silica
MKM	microkinetic model
TMB	trimethyl benzene
TON	turnover number

15.1 Introduction

Lignocellulosic biomass is a renewable source of organic carbon which comprises of three major constituents: cellulose, hemicellulose, and lignin [1–4]. Out of these, the usage of

Hydrocarbon Biorefinery
https://doi.org/10.1016/B978-0-12-823306-1.00007-8

Copyright © 2022 Elsevier Inc.
All rights reserved.

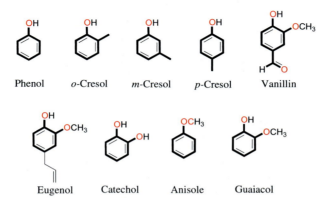

Fig. 15.1 Phenolic compounds obtained from the pyrolysis of biomass.

lignin is limited due to its complex three-dimensional structure comprising phenylpropane units and high thermal stability [5, 6]. Pyrolysis is generally employed for lignin utilization to produce biofuels and value-added chemicals [7]. The pyrolysis oil obtained from the lignin part of biomass comprises of 25–40 wt% of phenolics (e.g., anisole, *m*-cresol, *p*-cresol, phenol, guaiacol, vanillin, etc.) resulting in low heating value and poor chemical and thermal stability thereby making the bio-oil unsuitable for direct use as a liquid fuel (Fig. 15.1) [8]. Hence, the upgradation of lignin-derived bio-oil is essential for producing hydrocarbon biofuels that can be directly used as an alternative to petroleum [9]. Hydrodeoxygenation (HDO) reactions have played an important role in the upgradation of lignin-derived pyrolysis oil into transportation fuels [10]. During the HDO reaction of biomass-derived platform molecules, the C—O bonds present in the platform molecule are selectively cleaved by the addition of H_2, and the oxygen is removed in the form of water [11]. Developing stable catalysts for the HDO reactions which exhibit high conversion and selectivity toward deoxygenated products (aromatics and cyclohexane) at low temperature and pressure conditions has always been a challenge [12]. Aromatics, such as toluene and benzene, are value-added chemicals, while cyclohexane is used as an intermediate to produce oxygen-free biofuel [12].

HDO reaction of phenolic compounds is generally suggested to proceed by three different routes, namely direct deoxygenation (DDO) [13, 14], tautomerization-deoxygenation [15], and hydrogenation-dehydration (HYD) [16], as shown in Fig. 15.2. Out of these, the first two routes mainly yield aromatic products, whereas the last route produces either cycloalkenes or cycloalkanes. DDO pathway proceeds via direct C—O bond hydrogenolysis of the oxygenated molecule without any hydrogenation steps and requires high temperature and low pressures [17]. Also, the DDO pathway requires a lesser amount of external H_2 as it does not involve the ring hydrogenation. This route produces mainly aromatics with higher octane rating, which is suitable for oil blending [10]. In contrast, the HYD pathway is favored at low

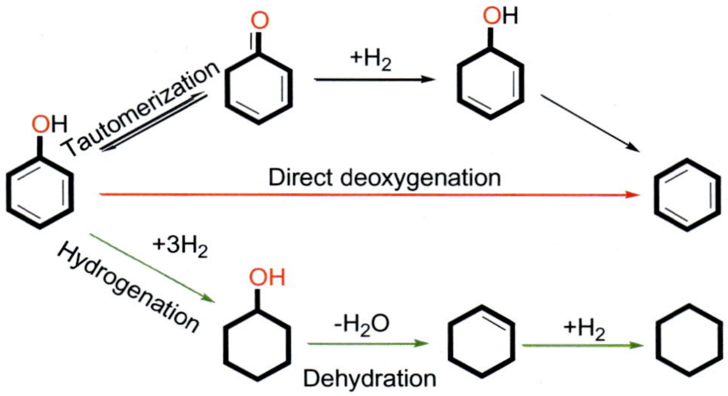

Hydrodeoxygenation of lignin-derived platform chemicals **411**

Fig. 15.2 Different reaction pathways for the hydrodeoxygenation (HDO) reaction of phenolics [15].

temperatures and high-pressure conditions. The reaction pathway during the HDO of phenolic compounds depends on the type of catalysts or reaction conditions, such as temperature and pressure [18].

The overall efficiency of the HDO process of a phenolic compound is governed by numerous parameters, such as the hydrogen pressure, type of reactor, temperature, solvent, catalyst, etc. [19] Among these, the molecular level design of active sites in a heterogeneous catalyst plays an important role in the HDO reactions [20, 21]. Many studies have reported the effect of type of support, metals, or the type of sites on the overall reactivity and selectivity in HDO reactions of phenolics [12]. This chapter provides a brief summary of the catalytic performance of the supported metal and bimetallic catalysts in the HDO reaction of biomass-derived phenolic compounds in order to serve as a guide for developing efficient catalysts for the HDO process.

15.2 Transition metals as hydrodeoxygenation catalysts

15.2.1 Active site in HDO reaction

Transition metal-based catalysts (e.g., Ni, Pt, Pd, Ru, Rh, Fe, Co, Mo, Re, etc.) have been generally employed for HDO reactions of phenolics (Fig. 15.3). The HDO activity and product selectivity over these metal catalysts has been observed to be dependent on

3d	Sc	Ti	V	Cr	Mn	Fe	Co	Ni	Cu	Zn
4d	Y	Zr	Nb	Mo	Tc	Ru	Rh	Pd	Ag	Cd
5d		Hf	Ta	W	Re	Os	Ir	Pt	Au	Hg

Fig. 15.3 Commonly used transition metals for the hydrodeoxygenation (HDO) of phenolics.

two factors, namely the metal oxophilicity and the metal atom coordination [22]. The oxophilicity of a metal is a measure of the M—O bond strength, which depends on the d-band center position relative to the Fermi level and governs the selectivity of DDO products [23]. When the d-band center is close to the Fermi level, the antibonding hybridization orbitals of the metal and the substrate remains unfilled and far from the Fermi level. This increases the M—O bond strength and oxophilicity. It has been further suggested that the lone pair electrons present in the O atom can be easily donated to the empty d orbitals of the early transition metal and hence forms stronger M—O bonds. Experimental and computational studies have suggested that a strong metal–oxygen bond is observed for early transition metals for 3d, 4d, and 5d group elements in the periodic table (Fig. 15.3). As we move upward and toward the left of the periodic table, the d-band center of the elements moves closer to the Fermi level. This trend was evident in the *m*-cresol HDO wherein DFT calculated activation barrier for the *m*-cresol DDO correlated with the metal oxophilicity and followed the trend (Pt < Pd < Rh < Ru < Fe) [22]. In addition to oxophilicity of different metals, product selectivity in the HDO reactions is also found to vary with the size of the metal clusters. It was found that the ratio of low coordination atoms (corners and steps) to the high coordination atoms (terrace or face) changes with the change in the size, which in turn led to the variation in the reaction mechanism and product selectivity [24, 25]. Duong et al. observed that smaller particle size 2.5% Rh/SiO_2 catalyst containing a higher amount of low coordination sites was more active with a turn over number (TON) of about 0.054 compared to 7 wt% Rh/SiO_2 (TON of about 0.032) having larger particles in the HDO reaction of *m*-cresol [22]. This was further supported by the DFT simulations which showed that the low coordination step sites bound O with higher strength (−610 kJ/mol) compared to the flat Rh (111) surface (−575 kJ/mol). Thus the C—O bond dissociation in *m*-cresol was reduced from 164 kJ/mol on Rh (111) to 122 kJ/mol on Rh (533) step sites. It was further ascertained that the d-band center varies with the metal atom coordination. The surface defect sites, such as steps, corners, and edges, having fewer atoms, had sharper d-band. In order to have the same occupancy, d-band center moves toward the Fermi level, thus increasing the reactivity compared to the flat sites with broad d-band center [26–28]. Thus product selectivity and activity in HDO reaction on metal-supported particles can be varied by changing the metal type or the size of the metal catalysts.

15.2.2 Support effect in HDO reaction

The type of support is an important design parameter for developing catalysts for HDO reactions [29]. Different supports, such as silica [30], alumina [31], activated carbon [18], zeolites [32], zirconia [33], etc., have been widely used for carrying out the HDO reactions of phenolics. Out of these, activated carbon is hydrothermally stable and neutral support, which provides higher dispersion of metal [18]. On the other hand, alumina

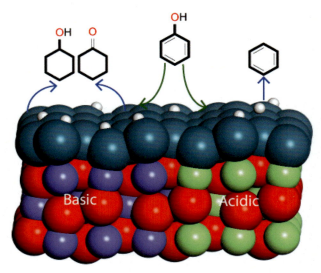

Fig. 15.4 Effect of acidic/basic support on the product selectivity in the hydrodeoxygenation (HDO) reaction of phenolics. The metal atoms (Pd, Pt, Ni, etc.) are displayed in *blue*, O in *red*, H in *white*, basic atoms (Mg, Zr, etc.) in *purple* and acidic atoms (Ce, Si, Al, etc.) in *green*.

is acidic but unstable under hydrothermal conditions. The zeolites have higher diffusion and shape-selective effects. The product selectivity and activity of the catalysts in these reactions have been observed to be dependent on the surface area, pore size, and acidic or basic nature of the supports (Fig. 15.4) [34]. For example, the product distribution in the HDO of phenolics on the Pd-supported catalyst was found to vary with the type of support. Pd/C at 200°C yielded cyclohexanol with 98% selectivity, whereas Lewis acidic supports, such as SiO_2, Al_2O_3, and amorphous silica–alumina (ASA), produced cyclohexanone along with cyclohexanol (Table 15.1, entry 1) [18]. In addition to these acidic supports, Pd-supported catalyst on CeO_2 also produced cyclohexanone (~46.2%), cyclohexane (19%), and cyclohexanol (~34.8%) with 80% conversion of phenol at 180°C (Table 15.1, entry 5) [35]. In contrast, Pd/ZrO_2 selectively produced cyclohexanone (~93%) with very low selectivity toward cyclohexanol (~3.5%) and cyclohexane (~2.7%) (Table 15.1, entry 6) which was similar to the observed selectivity trends on basic support, Pd/MgO (Table 15.1, entry 7) and Pd/TiO_2 catalyst [35]. It was suggested that the difference in acid-base properties of the supports in Pd/ZrO_2 and Pd/CeO_2 catalyst affected the adsorption geometry of phenol on these catalysts, which in turn affect the product selectivity. In the case of Pd/CeO_2, it was suggested that Ce^{4+} ions formed during the reduction of CeO_2 to CeO_{2-x} acted as Lewis acid sites. The phenol was adsorbed in a coplanar mode, which led to the formation of cyclohexanol. In contrast, Pd/ZrO_2 was less acidic and more basic in nature, and the noncoplanar mode of adsorption on these basic sites led to more cyclohexanone. A similar difference in activity and

Table 15.1 Role of support on the catalytic HDO reactions of phenolics.

S. no.	Catalyst	T (°C)	T (h)	P (MPa)	Solvent	Reactant	Conversion (%)	Selectivity (%)	Ref.
1	Pd/C	200	0.5	5	Water	Phenol	100	Cyclohexanone $=2$ Cyclohexanol $=98$	[18]
2	Pd/Al$_2$O$_3$	200	0.5	5	Water	Phenol	82	Cyclohexanone $=28$ Cyclohexanol $=72$	[18]
3	Pd/SiO$_2$	200	0.5	5	Water	Phenol	87	Cyclohexanone $=38$ Cyclohexanol $=62$	[18]
4	Pd/ASA	200	0.5	5	Water	Phenol	96	Cyclohexanone $=27$ Cyclohexanol $=73$	[18]
5	3% Pd/CeO$_2$ (0.5 g)	180	1	0.1	Benzene	Phenol	81.4	Cyclohexane $=19$ Cyclohexanol $=34.8$ Cyclohexanone $=46.2$	[35]
6	3% Pd/ZrO$_2$	180	1	0.1	Benzene	Phenol	62.9	Cyclohexane $=2.7$ Cyclohexanol $=3.5$ Cyclohexanone $=93$	[35]
7	Pd/MgO	180	1	0.1	Benzene	Phenol	77	Cyclohexane $=0.6$ Cyclohexanol $=4.9$ Cyclohexanone $=90.5$	[35]
8	Ni/SiO$_2$	300	16	5	n-Octane	Phenol	100	Benzene $=2.5$ Cyclohexane $=85$ Cyclohexene $=1$ Cyclohexanol $=6$	[40]
9	Ni/Al$_2$O$_3$	300	16	5	n-Octane	Phenol	100	Benzene $=5$ Cyclohexane $=90$ Cyclohexene $=2$	[40]
10	Ni/HZSM-5 (10% Ni)	240	3	4		Phenol	91.8	Benzene $=25$ Cyclohexane $=65$ Cyclohexanol $=5$ Cyclohexene $=4$	[41]

11	Pt/HZSM-5	250	3	4	Decane	Guaiacol	15	Cyclohexane	[39]
12	Pt/mesoporous beta	250	3	4	Decane	Guaiacol	99	Cyclohexane = 26.13 Methylcyclopentane = 13.1 2,4,6 Trimethyl phenol = 8.61 2,3,5,6 Tetramethyl phenol = 8.97 1-Methoxy-4-methyl- 2-(1-methylethyl)-benzene = 6.71	[39]
13	Pt/HBeta	250	3	4	Decane	Guaiacol	97	Cyclohexane = 45.3 Methylcyclopentane = 15.2 2,4,6-Trimethyl phenol = 4.2 2,3,5,6-Tetramethyl phenol = 4.6	[39]
14	Pt/MMZ$_{beta}$	250	3	4	Decane	Guaiacol	34	Cyclohexane = 23.94	[39]
15	Pt/Al-ACM-48	250	3	4	Decane	Guaiacol	64	Cyclohexane = 37.45 1,10-Bicyclohexyl = 13.91	[39]
16	Pt/Si-MCM-48	250	3	4	Decane	Guaiacol	3	1,2 Dimethoxybenzene = 0.08	[39]

selectivity for phenol HDO product was observed for Ni when supported on SiO_2, Al_2O_3, and H–ZSM5 (Table 15.1, entries 8–10) [36,37]. Selectivity toward benzene increased in the order of their acidity Ni-HZSM-5 > Ni/Al_2O_3 > NiSiO$_2$. Thus HZSM-5 catalyst having more acidity easily cleaved the C_{aryl}—O bond, whereas SiO_2 being inert in nature underwent complete ring hydrogenation followed by oxygen removal. However, the high acidity of the support led to unfavorable cracking reactions. Therefore, the optimum acidity of the support was found necessary for the removal of oxygen in highly oxygenated species. The product selectivity trends of m-cresol HDO over Pd-supported catalyst (Pd/SiO_2, Pd/Al_2O_3, or Pd/HBeta) was found to be dependent on the acid density and pore structure of the catalyst [38]. Higher acid sites density and microporous structure of HBeta-based catalyst led to the lowest activity. Similarly, in guaiacol HDO over different Pt-supported catalysts, such as Pt/Beta, Pt/Meso Beta, Pt/MMZ, and Pt/HZSM-5, both conversion and product selectivity varied depending on the type of support. HZSM-5, having the highest acidity, exhibited lowest guaiacol conversion ($\sim$15%) due to the small size of the pores. In contrast, Pt/Meso beta and Pt/HBeta having high acidity showed high conversion (>90%) whereas Pt/MMZ$_{Beta}$ with less number of acidic sites showed low conversion ($\sim$34%) [39]. The main product formed on all these zeolites was cyclohexane (Table 15.1, entries 11–16). However, the Pt/Meso Beta catalyst having larger pores promoted the formation of high molecular weight product (yield $\sim$35.48%), therefore making it less selective toward cyclohexane (26.13%) in spite of higher conversion (>97%) compared to Pt/HBeta (45.3%) [39]. Thus support with an optimum amount of acidic sites as well as sufficient pore size is needed for achieving higher conversion and desired product selectivity in the HDO of phenolics.

15.2.3 Product selectivity trends for hydrogenolysis products over other products

Noble metal (Pt, Ru, Rh, Pd, etc.) catalysts have been extensively studied for HDO reactions. However, due to higher hydrogenation activity of noble metal-based catalyst (Pt > Ru > Pd), lower selectivity for aromatic products is observed as they tend to hydrogenate the substrate ring as well [42]. The combined theoretical and experimental investigation by Lu et al. showed that Pt (111) sites were least active for guaiacol HDO in comparison to Pt corner or step sites or the catalyst support, which accounted for the Pt-based catalyst deoxygenation activity [43]. Therefore, in order to enhance the HDO activity of Pt-based catalyst, supports having good deoxygenation ability should be used. In a recent study, Pt-based supported catalyst showed reactivity in the order Pt/SiO_2 > Pt/Al_2O_3 > Pt/ZrO_2 > Pt/P_{25} > Pt/TiO_2 > Pt/ZrO_2 ~ Pt/CeO_2 [44] (Table 15.2, entry 1). On mild acidic supports, such as ZrO_2, high activity and selectivity toward deoxygenated products were observed for Pt-supported catalyst and can be used as an alternative to the sulfided catalyst, such as CoMo/Al_2O_3 [45]. In contrast to Pt, another noble metal Pd is known for higher hydrogenation ability and has lower

Table 15.2 Role of transition metals on the catalytic HDO reaction of phenolics.

S. no.	Catalyst	T (°C)	T (h)	P (MPa)	Solvent	Reactant	Conversion (%)	Selectivity (%)	Ref.
1	Pt/SiO$_2$	180	5	5	n-Hexadecane	Guaiacol	86	Cyclohexane = 8 Cyclohexanol = 20 Methoxycyclohexane = 10 Methoxycyclohexanol = 51	[44]
2	Pd/CN	50	0.1	0.5	Water	Vanillin	99	2-Methoxy-4-methyl-phenol = 99	[46]
3	Pd/KIT-6	300	0.1	6	Methanol	Vanillin	98	p-Cresol = 94	[47]
4	Pd/HZSM-5	200	6	2	n-Dodecane	m-Cresol	99.9	Methylcyclohexane = 99	[48]
5	Ru/meso-TiO$_2$	200	3	30	—	Anisole	100	Cyclohexane = 26.8 Methoxycyclohexane = 64.6 Benzene = 4.1	[50]
6	Ru/meso Al$_2$O$_3$	200	3	30	—	Anisole	100	Cyclohexane = 46.5 Methoxycyclohexane = 24.3 Cyclohexanol = 21	[50]
7	Ru/meso silica	200	3	30	—	Anisole	100	Cyclohexane = 55.6 Methoxycyclohexane = 21.4 Cyclohexanol = 10.8	[50]
8	Ni/SiO$_2$	300	16	5	n-Octane	Phenol	99	Cyclohexane = 90	[40]
9	Ni/γ-Al$_2$O$_3$	300	16	5	n-Octane	Phenol	100	Cyclohexane = 88	[40]
10	Co/SiO$_2$	300	1	1	—	Guaiacol	100	Aromatics = 53.1	[56]
11	CoN$_x$/NC	200	2	2	n-Dodecane	Eugenol	100	Propylcyclohexanol = 99.9	[57]
12	CoP/SiO$_2$	300	6	3	Octane	Phenol	98	Cyclohexane = 93	[58]
13	Pd/C	350		4	Gas phase	Guaiacol	98.8	Ben/Tol/TMB = 2.7 Phenol = 78.8 Gas phase = 16.2	[49]

Continued

Table 15.2 Role of transition metals on the catalytic HDO reaction of phenolics—cont'd

S. no.	Catalyst	T (°C)	T (h)	P (MPa)	Solvent	Reactant	Conversion (%)	Selectivity (%)	Ref.
14	Pt/C	350		4	Gas phase	Guaiacol	97.1	Ben/Tol/TMB = 30 Phenol = 28 Gas phase = 38.8	[49]
15	Ru/C	350		4	Gas phase	Guaiacol	100	Gas phase = ~100	[49]
16	Fe/C	350		4	Gas phase	Guaiacol	96	Ben/Tol/TMB = ~6.4% o-Cresol = 4.5 Phenol = 72.4 Gas phase (12.7)	[49]
17	Cu/C	350		4	Gas phase	Guaiacol	69.5	Ben/Tol/TMB $\leq$ 1 Phenol = 55 Catechol = 2.5 Gas phase = 10	[49]
18	Mo$_2$C/C	247	14	0.1 MPa	—	Anisole	100	Benzene = ~95	[59]
19	MoO$_3$	320	3	0.1	—	Anisole	78.7	Benzene = 55.5 Toluene = 19.8	[60]
20	MoO$_3$	320	3	0.1	—	Cresol	48.9	Toluene = 9.3	[60]
21	MoO$_3$/ZrO$_2$	320	3	0.1	—	Cresol	78	Toluene = 98.7	[61]
22	MoO$_3$/TiO$_2$	320	3	0.1	—	Cresol	47	Toluene = 97.9	[61]

cost and available in abundance. Pd on inert supports, such as mesoporous silica (KIT-6) or carbon nitrides (CN), is partially active for deoxygenation of phenolic compounds [46,47]. The N atoms present in carbon nitride increase the hydrophilic property in Pd/CN due to which it shows better activity than the catalyst having oxide support (e.g., Al_2O_3, CeO_2, MgO, and TiO_2). Pd/CN was observed to be effective for cleaving the C—O bond (100% conversion and 99% selectivity for 2-methoxy-4-methylphenol) in vanillin and inactive for C_{aryl}—OCH_3 or C_{aryl}—OH bond (Table 15.2, entry 2) [46]. On the other hand, Pd/KIT-6 showed selectivity for both C—O and C_{aryl}—OCH_3 bond cleavage with 98% conversion and 94% selectivity for p-cresol [47] (Table 15.2, entry 3). Thus by combining hydrogenation ability of Pd with acidic supports, such as HZSM-5, high selectivity for methylcyclohexane ($\sim$99%) in HDO reaction of m-cresol (99% conversion) was achieved (Table 15.2, entry 4) [48]. Similar to Pd and Pt, Ru also exhibits good hydrogenation activity. However, Ru produces more ring-opened products during HDO reactions at a high temperature [49]. Ru, when present on different supports, showed different activities depending on the size of Ru particles or the nature of the support. Ru/meso-TiO_2 having the smallest particle size exhibited high conversion and some selectivity toward benzene in the HDO of anisole via direct deoxygenation (DDO) route due to high hydrogen spillover and redox activity of TiO_2 which favored a strong oxygen interaction and enhanced the C—O bond cleavage [50] (Table 15.2, entry 5). On the other hand, Ru/Al_2O_3 or Ru/SiO_2 produced ring-hydrogenated products (Table 15.2, entries 6 and 7) [50]. Ru supported on mixed oxides, such as WO_x-ZrO_2, ZrO_2-La(OH)$_3$, or TiO_2-ZrO_2 [51–53], also exhibited higher conversions and selectivity toward saturated deoxygenated products, such as cyclopentane, cyclohexane, or methyl cyclopentane. In contrast, Rh, which is very expensive, has been reported in lesser studies [54]. DFT studies have shown that the phenolic compounds on Rh-based catalyst are first hydrogenated, followed by C—O bond cleavage to produce deoxygenated products, such as cyclohexane, etc. [55]. Overall higher hydrogenation activity of noble metals when combined with acidic supports, produces ring-saturated deoxygenated products, whereas selectivity toward deoxygenated aromatics remained lower.

On the other hand, base metals, such as Fe, Co, Ni, Mo, etc., have been used in many studies for HDO reaction of phenolics due to their easy availability as well as lower cost in comparison to noble metal catalysts. Similar to noble metals, base metal catalysts containing Ni was found to mainly yield cyclohexanes instead of aromatic hydrocarbons [62]. Many studies have revealed that the reaction pathways and distribution of products on these catalysts are governed by various parameters, such as particle size, type of support, etc. For example, the product selectivity for the HDO reaction of phenol in the presence of Ni/SiO_2 catalyst was found to be size dependent [63]. Optimal deoxygenation activity was observed for Ni catalyst below 300°C, having Ni particle with size varying from 9 to 10 nm [63]. The type of support was further seen to affect the product selectivity. On

Ni/SiO$_2$ catalyst, first cyclohexanol was produced via phenol hydrogenation, which on subsequent dehydroxylation produced cyclohexane [40] (Table 15.2, entry 8). In contrast, in the presence of Lewis acid support, such as γ-Al$_2$O$_3$, benzene was seen as an intermediate suggesting the cleavage of hydroxyl group first followed by hydrogenation to produce cyclohexane [40] (Table 15.2, entry 9). Due to low electrophilicity of Ni, hydrogenation, decarboxylation, and decarbonylation reactions were predominant on Ni-based catalyst compared to the direct C—O bond scission reactions (HDO) [64]. It is further observed that Ni in the presence of other oxophilic metals or supports with deoxygenation ability exhibited higher reactivity for HDO reactions. In contrast, Co-based catalysts were found to be efficient for the removal of oxygen via DDO reactions [56] and for the production of aromatic hydrocarbons (Table 15.2, entry10). Mochizuki et al. studied the mechanism of guaiacol HDO over Co/SiO$_2$ catalyst in which it was shown that the H$_2$ activation first takes place on the Co surface, which leads to hydrogen spillover over the catalyst surface with subsequent C$_{aryl}$—O—CH$_3$ bond hydrogenolysis. Similarly, cobalt phosphide (Co$_X$P$_Y$) [58] having Brønsted sites and cobalt nitride [57] also showed high HDO reactivity for phenol (Table 15.2, entries 11 and 12). Since cobalt shows higher DDO activity as compared to hydrogenation, it is used as a promoter during the HDO of phenolic compound.

In addition to Ni- and Co-, Fe-based catalyst shows high HDO activity but low hydrogenation of aromatic ring [65]. Compared to Pd, Pt, Cu, and Ru, Fe/C was observed to have a higher selectivity for benzene in guaiacol HDO [49]. Phenol was seen as an intermediate in guaiacol vapor phase HDO reaction on all the carbon-supported metal catalysts, such as Fe/C, Ru/C, Cu/C, Pd/C, Pt/C, etc. (Table 15.2, entries 13–17). However, as discussed previously aromatic ring saturation was the major pathway on the noble metal catalysts on which first cyclohexanol and cyclohexanone were formed followed by the formation of gaseous products via ring-opening at a higher temperature (>350°C) (mainly C1 products including CH$_4$, CO$_2$, and CO in the order of 5Ru/C (100%) > 5Pt/C (38.8%) > 5Pd/C (16.2%)) (Table 15.2, entries 13–15) [49]. On the other hand, base metal catalysts although exhibited low activity but showed higher selectivity for benzene, small amounts of cresol, toluene, and trimethyl benzene (TMB) without any ring-opening or ring-saturated products [49]. Olcese et al. [66] suggested that higher oxophilicity of Fe led to direct C—O bond cleavage. However, Fe catalysts suffer a lot of deactivation due to carbon deposition or the presence of oxygenated species, which tends to oxidize Fe surface [66]. Similar to Fe, Mo is also oxophilic in nature and is effective for carrying out HDO processes to produce aromatic hydrocarbons at relatively lower H$_2$ pressures. At ambient pressure and temperature ranging from 147°C to 247°C β-Mo$_2$C catalyst showed high H$_2$ efficiency (<9% selectivity of cyclohexane) and high selectivity for benzene (>90% among C6+ products) for the HDO reaction of anisole [59] (Table 15.2, entry 18). Similarly, the unsupported MoO$_3$ catalyst exhibited high efficiency for the HDO of phenolics (anisole, cresol,

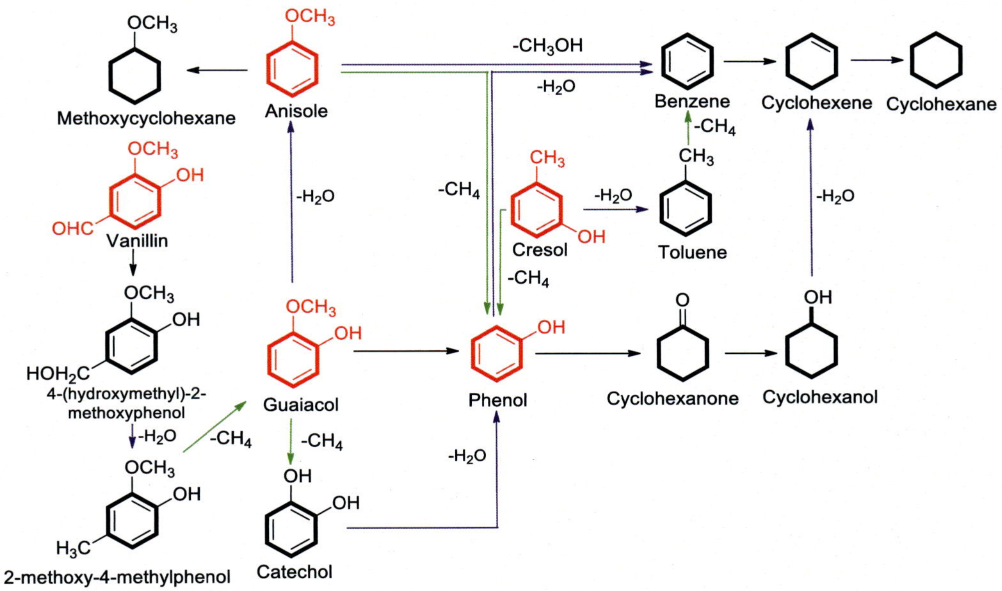

Fig. 15.5 Reaction network for HDO, hydrogenation, and hydrogenolysis of different biomass-derived phenolics represented by *blue*, *black*, and *green arrows* on different supported metal catalyst [67].

guaiacol, or phenol) by cleaving the C_{ar}—O bond (Table 15.2, entries 19 and 20) [60]. It was speculated that the Mo^{5+} present in MoO_3 acted as Lewis acid sites and cleaved the C—O bond present in the reactant and intermediates [60]. Further, ZrO_2 and TiO_2 were found to be optimum supports that stabilized the Mo^{5+} species and promoted the selectivity of MoO_3 with higher hydrocarbon yield ($\sim$98.7% and 97.9%) as compared to unsupported MoO_3 catalyst [61] (Table 15.2, entries 20–22). However, Mo_2C-based catalysts were also deactivated in the presence of water and exhibited lower overall conversion. Thus, the type of metal catalyst affects both the activity and selectivity of the products during HDO reactions (Fig. 15.5).

15.3 Rational design of bimetallic alloys for HDO reaction

When one transition metal is alloyed with another metal, a shift in the reactivity for C—O vs C—C bond cleavage has been observed in the HDO reaction of biomass-derived platform molecules. For example, when NiRe catalyst was used for *m*-cresol HDO, the selectivity toward toluene (i.e., C—O bond scission product) increased to 23.7% compared to monometallic catalyst Re (2.9%) and Ni (1.9%) [68]. Similarly, when metals having high hydrogenation capability, such as Ni, Ru, Pd, Pt, etc., are combined with the metals with high oxophilic strength, such as Co, Mo, Re, or Fe, an improved conversion of phenolics and high selectivity toward aromatic compounds are achieved in the HDO reactions under mild reaction conditions with less hydrogen consumption compared to their monometallic counterparts [69]. This trend was evident in the improved catalytic performance of NiCo compared to monometallic Ni or Co, which in turn was attributed to the increased stability and better dispersion of Ni [70]. The presence of oxophilic promoters improved the HDO activity without using acidic supports and thus prevented side reactions, such as polymerization or coke formation. The addition of oxophilic promoters, such as Fe to Ru/TiO$_2$ was suggested to change the HDO pathway of anisole from HYD to DDO which resulted in higher selectivity toward aromatics such as benzene (>80%) at 1 MPa of H$_2$ pressure and 250°C [71] (Table 15.3, entry 1). Further, it was suggested that H$_2$ dissociation takes place on Ru, whereas Fe increased the interaction between the oxygen present in the compounds with the support TiO$_2$ and assisted C—O bond cleavage. Similarly, PdFe bimetallic catalyst in guaiacol HDO process also favored the direct DDO pathway and enhanced the selectivity of aromatic compounds, such as benzene, trimethylbenzene, and toluene (~83%), as compared to its monometallic counterpart Fe (yield ~43.3%) at 450°C [49] (Table 15.3, entry 2). Combined experimental and theoretical investigations further showed that Pd atoms prevented the oxidation of Fe and promoted water formation due to their enhanced reducibility [72]. On the other hand, NiFe bimetallic catalyst showed high selectivity toward cyclohexane and phenol, where Ni promoted H$_2$ dissociation, and Fe showed higher oxygen affinity [73] (Table 15.3, entries 3 and 4). Similarly, Re-promoted Ni catalyst showed high selectivity toward toluene during m-cresol HDO, which was attributed to both electronic and geometric effects. Re, when present in the vicinity of Ni, lowered the d-band center, which prevented the adsorption of phenyl rings on the surface, thereby preventing C—C bond cleavage [68]. ReRu/MWCNT also exhibited higher selectivity for HDO of guaiacol due to the presence of RuRe clusters on the support [74] (Table 15.3, entry 5). Similarly, PtMo/C bimetallic catalyst was used in the HDO of m-cresol, which exhibited higher selectivity toward methylcyclohexane (~70%) [75] (Table 15.3, entry 6). DFT simulations further showed that Mo present in the catalyst reduced the tautomerization and deoxygenation barrier for m-cresol. Thus instead of ring-hydrogenation as observed over monometallic Pt, tautomerization-deoxygenation was more prevalent on PtMo/C catalyst. Further, it was observed that the addition of the small amount of alkali metals (K or Na) as dopants to the bimetallic catalysts, such as CoMo/Al$_2$O$_3$ or NiMo/Al$_2$O$_3$, led to increased C$_{ar}$—OH bond cleavage which not only improved the desired product selectivity but also led to lesser deactivation [76]. Thus the usage of bimetallic catalyst modulates the product turnover owing to electronic or geometric effects.

Table 15.3 Product selectivity trends for the HDO of phenolics over bimetallic catalysts.

S. no.	Catalyst	T (°C)	T (h)	P (MPa)	Solvent	Reactant	Conversion (%)	Selectivity (%)	Ref.
1	Ru–Fe/ TiO$_2$	250	3	1	Decane	Anisole	98	Benzene = 84 Cyclohexane = 7 Methoxycyclohexane = 2	[71]
2	Pd–Fe/C	450	1	0.04	—	Guaiacol	100	Benzene + toluene + trimethylbenzene = 83	[49]
3	5Ni–1Fe/ CNT	400	2	3	—	Guaiacol	96.8	Cyclohexane = 83.4	[73]
4	1Ni–5Fe/ CNT	400	2	3	—	Guaiacol	47.2	Phenol = 83.3 Cyclohexane = 2.5	[73]
5	Ru–Re/ MWCNT	240	1	2	n-Heptane	Guaiacol	100	Cyclohexane = 80	[74]
6	PtMo/C	250		0.5	—	m-Cresol	100	Methylcyclohexane = 70	[75]
7	NiMo/ASA	310	4	3	Decalin	Phenol	96	Cyclohexane = 83.6 Cyclohexene = 5 Benzene = 3.4 Cyclohexanol = 8	[77]
8	NiFe/SiO$_2$	300		1 atm	—	m-Cresol	13.7	Toluene = 52.6 Methane = 2.2 Transalkylation products = 45.3	[10]
9	NiW/AC	300	—	1.5		Phenol	98	Cyclohexane = 89.9 Benzene = 6.8 Methylcyclopentane = 3.3	[78]
10	NiRe/SiO$_2$	300	1	0.1	—	m-Cresol	47.6	Toluene = 23.7	[68]
11	PtZn/C	325	1	2.75	n-Heptane	Anisole	25	Phenol = 66 Cyclohexane = 14 Cyclohexanol = 13 Cyclohexanone = 3 Methoxycyclohexane = 4	[79]
12	Pd/Fe$_2$O$_3$	300	2	—	—	m-Cresol	>55	Toluene ≥ 80	[80]

424 Hydrocarbon biorefinery

15.4 Oxophilic metals as acidic sites for HDO reaction

From the above discussion, it is evident that monometallic catalysts, such as Ni, Pd, Pt, Ru, Fe, etc., are active. In contrast, the bimetallic catalysts, such as Ni-Mo/ASA (Table 15.3, entry 7) [77], Ni-Fe/SiO$_2$ [10] (Table 15.3, entry 8), Ni-W/AC [78] (Table 15.3, entry 9), Pd-Fe/C [49] (Table 15.3, entry 2), Ni-Re/SiO$_2$ (Table 15.3, entry 10) [68], Pt-Zn/C [79] (Table 15.3, entry 11), Pt-Mo/C [75] (Table 15.3, entry 6), etc., exhibit higher conversion and selectivity for HDO reactions than their monometallic counterparts. Several factors, such as electronic, geometric, synergistic, and bifunctional effects, and the oxophilic nature of secondary metals in bimetallic alloys, contribute to the enhanced performance in HDO reactions.

The bimetallic catalysts consisting of a hydrogenating metal with an oxophilic metal exhibited high activity for HDO reactions irrespective of the type of support. For example, Pd/Fe$_2$O$_3$ catalyst showed enhanced activity for DDO reaction of *m*-cresol (Table 15.3, entry 12) to toluene (55% conversion and >80% selectivity) where Pd interaction with Fe reduced Fe^{2+} and Fe^{3+} oxides and Fe site favored the C—O bond cleavage [80]. Hong et al. by combining experiments and DFT simulations, studied the Pd/Fe synergistic effect in the Pd/Fe$_2$O$_3$ catalyst in the HDO of phenolics wherein they showed that Pd partially donated electrons to the oxide surface. This led to more delocalized electrons present on the surface, which in turn stabilized the reduced Fe oxide surface under a hydrogenation environment. The stabilization of reduced Fe sites and facilitation of water formation on Pd led to an increase in HDO activity on Pd/Fe$_2$O$_3$ catalyst [72]. Similarly, in *m*-cresol selective deoxygenation to toluene on Ni-Re/SiO$_2$ bimetallic catalyst, it was shown that Re breaks the ensembles of Ni and reduces the d-band center thus making Ni—Re sites active for deoxygenation [68]. Similarly, Pt—Fe, Ni—Fe, Pt—Mo, etc. were active in promoting the deoxygenation of phenolics. From all of these studies, one main feature of bimetallic alloys that promotes the deoxygenation of phenolics is the oxophilic nature of the secondary metal. In a recent study by Zhou et al. Ni-based bimetallic single-atom catalyst (M@Ni where M was Sc, Ti, V, Cr, Mn, Fe, Co, Ni, Cu, Zn, etc.) was shown to be effective in the HDO of phenol [81]. The presence of a secondary atom near the Ni surface modified the electronic structure of the catalyst (Ni-M-NI), thereby moving the d-band center toward the Fermi level and increasing the OH binding energy, which increased the C$_{aryl}$—OH as well as C$_{aliphatic}$—OH bond length. The increased bond length was showed to be a good descriptor for predicting C—OH bond scission and was found to be proportional to the oxophilicity of the secondary metal. In order to achieve optimum HDO activity, balanced oxophilicity of the secondary atom should be considered so that the DDO activity could be enhanced with desorption of OH as water [81]. The effect of oxophilic promoters was also observed in the C—O bond hydrogenolysis in cyclic ethers and polyols on Rh-ReO$_x$/C catalyst where high activity and selectivity toward terminal diols was attributed to the hydroxyl group adsorbed on the Re atom present at corner site of the

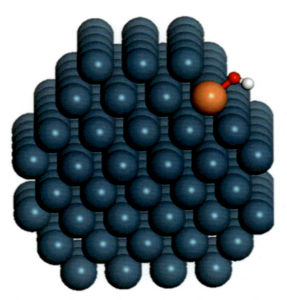

Fig. 15.6 Hydroxyl group adsorbed on the Re atom (*orange*) present at the corner site of Ir (*blue*) nanoparticle [82]. The O atom is displayed in *red* and H in *white color*, respectively.

Rh-Re/C catalyst (Fig. 15.6). The hydroxyl group acted as Brønsted acid due to strong binding of O atom with Re and hence easily donated proton, which helped in the facile ring-opening of cyclic ethers [82]. Thus under the hydrogenation environment, the hydrogenating metal provided the H atom, whereas oxophilic promoter facilitated the C—O bond cleavage.

By utilizing ab initio microkinetic model (MKM), Jalid et al. presented a framework for the bimetallic catalyst design (Co$_3$Ni, Ni$_3$Fe, Co$_3$Ru, Ni$_3$Pt, Ni$_3$Rh, Ni$_3$Cu, Ni$_3$Co, and Co$_3$Fe) for the HDO reactions of phenolics [83]. Ethanol was used as a model compound in this study. It was observed that at 523 K temperature, higher selectivity was obtained for C—C bond cleavage product, i.e., methane. In contrast to Co$_3$Fe and Ni$_3$Fe bimetallic catalyst ethane was obtained as the product. However, a high turnover was achieved for ethane (C—O bond cleavage product) at a low temperature (∼323 K) on all the bimetallic catalysts surface. Thus the reaction conditions played an important role in the bimetallic catalyst design for specific HDO reactions wherein at two different temperatures (523 K and 323 K) different sets of bimetallic catalysts were found to be active [83].

15.5 Conclusions

Many catalytic systems have been explored for the HDO of phenolics to biofuels and value-added chemicals. The reaction pathway and product selectivity during the catalytic HDO reactions are strongly dependent on the catalyst properties. This chapter reviews the effect of the type of support, metals, promoters, and nature of the sites on the activity and product selectivity in HDO of phenolics (anisole, cresol, guaiacol, phenol, etc.). Enhanced interaction between the active sites and reactants can be achieved by using appropriate support. The supports with optimum acidity and pore size improve the rate of HDO reaction with lesser deactivation. Catalyst properties, such as size and type of active sites, influence the catalyst activity and product selectivity. Undercoordinated steps and corner sites in metal catalysts are more active for HDO reactions compared to the majority of terrace sites. The transition metals having the potential of activating both C—O bond and H_2 molecule showed strong activity. In general, a combination of a hydrogenating metal (Pt, Pd, Ni, Ru, etc.) with another metal of higher oxophilic strength (Fe, Mo, Co, Re, etc.) promotes the activity and selectivity toward aromatic products in HDO reactions. Since HDO reactions of phenolics proceed by different mechanisms with the formation of different products, further research is needed to develop efficient catalysts for selective HDO to produce desired hydrocarbons. Metals functionality, the optimum ratio of acid sites, and *meso*porous supports with mild acidity are some of the factors that should be considered while designing catalysts for the HDO reactions.

Acknowledgment

The authors acknowledge the Ministry of Human Resource and Development, Government of India grant number # P1318 and Department of Science and Technology CRG/2019/006176 for financial assistance, Seed grant from Indian Institute of Technology Hyderabad.

References

[1] M. Stöcker, Biofuels and biomass-to-liquid fuels in the biorefinery: catalytic conversion of lignocellulosic biomass using porous materials, Angew. Chem. 47 (2008) 9200–9211.

[2] S.K. Bardhan, S. Gupta, M.E. Gorman, M.A. Haider, Biorenewable chemicals: feedstocks, technologies and the conflict with food production, Renew. Sustain. Energy Rev. 51 (2015) 506–520.

[3] M.I. Alam, et al., Development of 6-amyl-α-pyrone as a potential biomass-derived platform molecule, Green Chem. 18 (2016) 6399–6696.

[4] S. Gupta, R. Arora, N. Sinha, M.I. Alam, M.A. Haider, Mechanistic insights into the ring-opening of biomass derived lactones, RSC Adv. 6 (2016) 12932–12942.

[5] T. Kotake, H. Kawamoto, S. Saka, Mechanisms for the formation of monomers and oligomers during the pyrolysis of a softwood lignin, J. Anal. Appl. Pyrolysis 105 (2014) 309–316.

[6] G. Porwal, et al., Mechanistic insights into the pathways of phenol hydrogenation on Pd nanostructures, ACS Sustain. Chem. Eng. 7 (2019) 17126–17136.

[7] S. Wang, B. Ru, H. Lin, W. Sun, Z. Luo, Pyrolysis behaviors of four lignin polymers isolated from the same pine wood, Bioresour. Technol. 182 (2015) 120–127.

[8] H. Prajitno, R. Insyani, J. Park, C. Ryu, J. Kim, Non-catalytic upgrading of fast pyrolysis bio-oil in supercritical ethanol and combustion behavior of the upgraded oil, Appl. Energy 172 (2016) 12–22.

[9] D. Elliott, C. Historical, Developments in hydroprocessing bio-oils, Energy & Fuels 21 (2007) 1792–1815.

Hydrodeoxygenation of lignin-derived platform chemicals **427**

[10] L. Nie, et al., Selective conversion of m-cresol to toluene over bimetallic Ni-Fe catalysts, J. Mol. Catal. A Chem. 388–389 (2014) 47–55.

[11] S. Gupta, T.S. Khan, B. Saha, M.A. Haider, Synergistic effect of Zn in a bimetallic PdZn catalyst: elucidating the role of undercoordinated sites in the hydrodeoxygenation reactions of biorenewable platforms, Ind. Eng. Chem. Res. 58 (2019) 16153–16163.

[12] W. Jin, L. Pastor-pérez, D. Shen, A. Sepúlveda-Escribano, S. Gu, Catalytic upgrading of biomass model compounds: novel approaches and lessons learnt from traditional hydrodeoxygenation—a review, ChemCatChem 11 (2019) 924–960.

[13] R.C. Nelson, et al., Experimental and theoretical insights into the hydrogen-efficient direct hydrodeoxygenation mechanism of phenol over Ru/TiO2, ACS Catal. 5 (2015) 6509–6523.

[14] Y. Romero, F. Richard, S. Brunet, Hydrodeoxygenation of 2-ethylphenol as a model compound of bio-crude over sulfided Mo-based catalysts: promoting effect and reaction mechanism, Appl. Catal. B Environ. 98 (2010) 213–223.

[15] L. Nie, D.E. Resasco, Kinetics and mechanism of m-cresol hydrodeoxygenation on a Pt/SiO2 catalyst, J. Catal. 317 (2014) 22–29.

[16] Z. Luo, et al., Hydrothermally stable Ru/HZSM-5-catalyzed selective hydrogenolysis of lignin-derived substituted phenols to bio-arenes in water, Green Chem. 18 (2016) 5845–5858.

[17] P.M. De Souza, V. Teixeira, F.B. Noronha, F. Richard, Kinetics of the hydrodeoxygenation of cresol isomers over Ni2P/SiO2: proposals of nature of deoxygenation active sites based on an experimental study, Appl. Catal. B Environ. 205 (2017) 357–367.

[18] C. Zhao, J. He, A.A. Lemonidou, X. Li, J.A. Lercher, Aqueous-phase hydrodeoxygenation of bio-derived phenols to cycloalkanes, J. Catal. 280 (2011) 8–16.

[19] Z. He, X. Wang, Hydrodeoxygenation of model compounds and catalytic systems for pyrolysis bio-oils upgrading, Catal. Sustain. Energy 1 (2013) 28–52.

[20] S. Kim, et al., Recent advances in hydrodeoxygenation of biomass-derived oxygenates over heterogeneous catalysts, Green Chem. 21 (2019) 3715–3743.

[21] S. Gupta, T.S. Khan, M.A. Haider, Mechanistic approaches toward rational design of a heterogeneous catalyst for ring-opening and deoxygenation of biomass-derived cyclic compounds, ACS Sustain. Chem. Eng. 7 (2019) 10165–10181.

[22] N. Duong, Q. Tan, D.E. Resasco, Controlling phenolic hydrodeoxygenation by tailoring metal–O bond strength via specific catalyst metal type and particle size selection, C. R. Chim. 21 (2018) 155–163.

[23] J. Greeley, J.K. Nørskov, A general scheme for the estimation of oxygen binding energies on binary transition metal surface alloys, Surf. Sci. 592 (2005) 104–111.

[24] G. Maire, G. Plouidy, The mechanisms of hydrogenolysis hydrocarbons isomerization on metals, J. Catal. 4 (1965) 556–569.

[25] D.H. Aue, H.M. Webb, M.T. Bowers, S. Barbara, Isomerization on metals. Correlation between metal particle size and reaction mechanisms, J. Am. Chem. Soc. 98 (1976) 856–857.

[26] R.M. Watwe, R.D. Cortright, J.K. Nørskov, J.A. Dumesic, Theoretical studies of stability and reactivity of C2 hydrocarbon species on Pt clusters, Pt (111), and Pt (211), J. Phys. Chem. B 104 (2000) 2299–2310.

[27] J. Kleis, et al., Finite size effects in chemical bonding: from small clusters to solids, Catal. Lett. 141 (2011) 1067–1071.

[28] M. Mavrikakis, B. Hammer, J.K. Nørskov, Effect of strain on the reactivity of metal surfaces, Phys. Rev. Lett. 81 (1998) 2819–2822.

[29] A.J. Foster, P.T.M. Do, R.F. Lobo, The synergy of the support acid function and the metal function in the catalytic hydrodeoxygenation of m-cresol, Top. Catal. 55 (2012) 118–128.

[30] B.S. Angelos, E. Johansson, J.F. Stoddart, J.I. Zink, Mesostructured silica supports for functional materials and molecular machines, Adv. Funct. Mater. 17 (2007) 2261–2271.

[31] B.H. Pines, W.O. Haag, Alumina: catalyst and support. Alumina, its intrinsic acidity and catalytic activity, J. Am. Chem. Soc. 82 (1960) 2471–2483.

[32] X. Zhu, L.L. Lobban, R.G. Mallinson, D.E. Resasco, Bifunctional transalkylation and hydrodeoxygenation of anisole over a Pt/HBeta catalyst, J. Catal. 281 (2011) 21–29.

[33] M.V. Bykova, et al., Environmental Ni-based sol-gel catalysts as promising systems for crude bio-oil upgrading: guaiacol hydrodeoxygenation study, Appl. Catal. B Environ. 113–114 (2012) 296–307.

[34] Y. Ren, B. Yue, M. Gu, H. He, Progress of the application of mesoporous silica-supported heteropolyacids in heterogeneous catalysis and preparation of nanostructured metal oxides, Materials (Basel) 3 (2010) 764–785.

[35] S. Velu, M.P. Kapoor, S. Inagaki, K. Suzuki, Vapor phase hydrogenation of phenol over palladium supported on mesoporous CeO2 and ZrO2, Appl. Catal. A Gen. 245 (2003) 317–331.

[36] Z. Hussain, et al., The conversion of biomass into liquid hydrocarbon fuel by two step pyrolysis using cement as catalyst, J. Anal. Appl. Pyrolysis 101 (2013) 90–95.

[37] S. Sankaranarayanapillai, et al., Catalytic upgrading of biomass-derived methyl ketones to liquid transportation fuel precursors by an organocatalytic approach, Angew. Chem. Int. Ed. Engl. 54 (2015) 4673–4677.

[38] M.S. Zanuttini, B.O.D. Costa, C.A. Querini, M.A. Peralta, Hydrodeoxygenation of m-cresol with Pt supported over mild acid materials, Appl. Catal. A Gen. 482 (2014) 352–361.

[39] E. Hwa, et al., Hydrodeoxygenation of guaiacol over Pt loaded zeolitic materials, J. Ind. Eng. Chem. 37 (2016) 18–21.

[40] X. Zhang, W. Tang, Q. Zhang, T. Wang, L. Ma, Hydrodeoxygenation of lignin-derived phenoic compounds to hydrocarbon fuel over supported Ni-based catalysts, Appl. Energy 227 (2018) 73–79.

[41] X. Zhang, T. Wang, L. Ma, Q. Zhang, T. Jiang, Hydrotreatment of bio-oil over Ni-based catalyst, Bioresour. Technol. 127 (2013) 306–311.

[42] K. Alharbi, E.F. Kozhevnikova, I.V. Kozhevnikov, Hydrogenation of ketones over bifunctional Pt-heteropoly acid catalyst in the gas phase, Appl. Catal. A Gen. 504 (2015) 457–462.

[43] J. Lu, S. Behtash, O. Mamun, A. Heyden, Theoretical investigation of the reaction mechanism of the guaiacol hydrogenation over a Pt (111) catalyst, ACS Catal. 5 (2015) 2423–2435.

[44] M. Hellinger, et al., General catalytic hydrodeoxygenation of guaiacol over platinum supported on metal oxides and zeolites, Appl. Catal. A Gen. 490 (2015) 181–192.

[45] A. Gutierrez, R.K. Kaila, M.L. Honkela, R. Slioor, A.O.I. Krause, Hydrodeoxygenation of guaiacol on noble metal catalysts, Catal. Today 147 (2009) 239–246.

[46] H. Jiang, et al., Efficient aqueous hydrodeoxygenation of vanillin over a mesoporous carbon nitride-modified Pd, RSC Adv. 6 (2016) 69045–69051.

[47] J. Kayalvizhi, A. Pandurangan, Hydrodeoxygenation of vanillin using palladium on mesoporous KIT-6 in vapour phase reactor, Mol. Catal. 436 (2017) 67–77.

[48] J.A. Hunns, et al., Hierarchical mesoporous Pd/ZSM-5 for the selective catalytic hydrodeoxygenation of m-cresol to methylcyclohexane, Catal. Sci. Technol. 6 (2016) 2560–2564.

[49] J. Sun, et al., Carbon-supported bimetallic Pd–Fe catalysts for vapor-phase hydrodeoxygenation of guaiacol, J. Catal. 306 (2013) 47–57.

[50] T. Ngoc, Y. Park, I. Lee, C. Hyun, Enhancement of C-O bond cleavage to afford aromatics in the hydrodeoxygenation of anisole over ruthenium-supporting mesoporous metal oxides, Appl. Catal. A Gen. 544 (2017) 84–93.

[51] A. Adep, I. Kim, L. Zhou, J. Choi, D. Jin, Hydrodeoxygenation of guaiacol on tungstated zirconia supported Ru catalysts, Appl. Catal. A Gen. 543 (2017) 10–16.

[52] M. Lu, et al., Hydrodeoxygenation of guaiacol on Ru catalysts: influence of TiO2-ZrO2 composite oxide supports, Ind. Eng. Chem. Res. 56 (2017) 12070–12079.

[53] G. Xu, et al., Selective hydrodeoxygenation of lignin-derived phenols to alkyl cyclohexanols over a Ru-solid, Green Chem. 18 (2016) 5510–5517.

[54] A. Ng, K. Lup, F. Abnisa, W. Mohd, A. Wan, A review on reactivity and stability of heterogeneous metal catalysts for deoxygenation of bio-oil model compounds, J. Ind. Eng. Chem. 56 (2017) 1–34.

[55] D. Garcia-pintos, J. Voss, A.D. Jensen, F. Studt, Hydrodeoxygenation of phenol to benzene and cyclohexane on Rh (111) and Rh (211) surfaces: insights from density functional theory, J. Phys. Chem. C 120 (2016) 18529–18537.

[56] T. Mochizuki, S. Chen, M. Toba, Y. Yoshimura, Deoxygenation of guaiacol and woody tar over reduced catalysts, Appl. Catal. B Environ. 146 (2014) 237–243.

Hydrodeoxygenation of lignin-derived platform chemicals **429**

[57] X. Liu, L. Xu, G. Xu, W. Jia, Y. Zhang, Selective hydrodeoxygenation of lignin-derived phenols to cyclohexanols or cyclohexanes over magnetic CoNx@NC catalysts under mild conditions, ACS Catal. 6 (2016) 7611–7620.

[58] E.R. Aguado, A.I. Molina, J.A.C.D. Ballesteros, CoxPy catalysts in HDO of phenol and dibenzofuran: effect of P content, Top. Catal. 60 (2017) 1094–1107.

[59] W. Lee, Z. Wang, R.J. Wu, A. Bhan, Selective vapor-phase hydrodeoxygenation of anisole to benzene on molybdenum carbide catalysts, J. Catal. 319 (2014) 44–53.

[60] T. Prasomsri, M. Shetty, M. Karthick, Y.R. Leshkiv, Insights into the catalytic activity and surface modification of MoO3 during the hydrodeoxygenation of lignin-derived model compounds into aromatic hydrocarbons under low hydrogen pressures, Energy Environ. Sci. 7 (2014) 2660–2669.

[61] M. Shetty, K. Murugappan, T. Prasomsri, W.H. Green, Y. Román-Leshkov, Reactivity and stability investigation of supported molybdenum oxide catalysts for the hydrodeoxygenation (HDO) of m-cresol, J. Catal. 331 (2015) 86–97.

[62] H. Shi, J. Chen, Y. Yang, S. Tian, Catalytic deoxygenation of methyl laurate as a model compound to hydrocarbons on nickel phosphide catalysts: remarkable support effect, Fuel Process. Technol. 118 (2014) 161–170.

[63] P.M. Mortensen, J. Grunwaldt, P.A. Jensen, A.D. Jensen, Influence on nickel particle size on the hydrodeoxygenation of phenol over Ni/SiO2, Catal. Today 259 (2016) 277–284.

[64] D. Kubic, L. Kaluza, Deoxygenation of vegetable oils over sulfided Ni, Mo and NiMo catalysts, Appl. Catal. A Gen. 372 (2010) 199–208.

[65] Y. Hong, A. Hensley, J. McEwen, Y. Wang, Perspective on catalytic hydrodeoxygenation of biomass pyrolysis oils: essential roles of Fe-based catalysts, Catal. Lett. 146 (2016) 1621–1633.

[66] R. Olcese, et al., Environmental gas-phase hydrodeoxygenation of guaiacol over iron-based catalysts. Effect of gases composition, iron load and supports (silica and activated carbon), Appl. Catal. B Environ. 129 (2013) 528–538.

[67] R.C. Runnebaum, T. Nimmanwudipong, D.E. Block, B.C. Gates, Catalytic conversion of compounds representative of lignin-derived bio-oils: a reaction network for guaiacol, anisole, 4-methylanisole, and cyclohexanone conversion catalysed by Pt/gamma-Al$_2$O$_3$, Catal. Sci. Technol. 2 (2012) 113–118.

[68] F. Yang, et al., Geometric and electronic effects of bimetallic Ni–Re catalysts for selective deoxygenation of m-cresol to toluene, J. Catal. 349 (2017) 84–97.

[69] C. Li, X. Zhao, A. Wang, G.W. Huber, T. Zhang, Catalytic transformation of lignin for the production of chemicals and fuels, Chem. Rev. 115 (2015) 11559–11624.

[70] N.T.T. Tran, Y. Uemura, S. Chowdhury, A. Ramli, Vapor-phase hydrodeoxygenation of guaiacol on Al-MCM-41 supported Ni and Co catalysts, Appl. Catal. A Gen. 512 (2016) 93–100.

[71] T.N. Phan, C.H. Ko, Synergistic effects of Ru and Fe on titania-supported catalyst for enhanced anisole hydrodeoxygenation selectivity, Catal. Today 303 (2018) 219–226.

[72] A.J.R. Hensley, et al., Enhanced Fe2O3 reducibility via surface modification with Pd: characterizing the synergy within Pd/Fe catalysts for hydrodeoxygenation reactions, ACS Catal. 4 (2014) 3381–3392.

[73] H. Fang, et al., Product tunable behavior of carbon nanotubes-supported Ni- Fe catalysts for guaiacol hydrodeoxygenation, Appl. Catal. A Gen. 529 (2017) 20–31.

[74] K. Bin, J. Lee, J. Ha, H. Lee, D. Jin, Effective hydrodeoxygenation of lignin-derived phenols using bimetallic RuRe catalysts: effect of carbon supports, Catal. Today 303 (2018) 191–199.

[75] A. Robinson, et al., Enhanced hydrodeoxygenation of m-cresol over bimetallic Pt-Mo catalysts through an oxophilic metal-induced tautomerization pathway, ACS Catal. 6 (2016) 4356–4368.

[76] I.D. Mora-Vergara, L. Hernández, E.M. Gaigneaux, S.A. Giraldo, V.G. Baldovino-Medrano, Hydrodeoxygenation of guaiacol using NiMo and CoMo catalysts supported on alumina modified with potassium, Catal. Today 302 (2018) 125–135.

[77] E. Kordouli, B. Pawelec, C. Kordulis, A. Lycourghiotis, J.L.G. Fierro, Hydrodeoxygenation of phenol on bifunctional Ni-based catalysts: effects of Mo promotion and support, Appl. Catal. B Environ. 238 (2018) 147–160.

[78] S. Echeandia, P.L. Arias, V.L. Barrio, B. Pawelec, J.L.G. Fierro, Synergy effect in the HDO of phenol over Ni-W catalysts supported on active carbon: effect of tungsten precursors, Appl. Catal. B Environ. 101 (2010) 1–12.

[79] D. Shi, L. Arroyo-Ramirez, J.M. Vohs, The use of bimetallics to control the selectivity for the upgrading of lignin-derived oxygenates: reaction of anisole on Pt and PtZn catalysts, J. Catal. 340 (2016) 219–226.

[80] Y. Hong, et al., Synergistic catalysis between Pd and Fe in gas phase hydrodeoxygenation of m-cresol, ACS Catal. 4 (2014) 3335–3345.

[81] J. Zhou, W. An, Z. Wang, X. Jia, Hydrodeoxygenation of phenol over Ni-based bimetallic single-atom surface alloys: mechanism, kinetics and descriptor, Catal. Sci. Technol. 9 (2019) 4314–4326.

[82] M. Chia, et al., Selective hydrogenolysis of polyols and cyclic ethers over bifunctional surface sites on rhodium-rhenium catalysts, J. Am. Chem. Soc. 133 (2011) 12675–12689.

[83] F. Jalid, T.S. Khan, F.Q. Mir, M.A. Haider, Understanding trends in hydrodeoxygenation reactivity of metal and bimetallic alloy catalysts from ethanol reaction on stepped surface, J. Catal. 353 (2017) 265–273.

Index

Note: Page numbers followed by *f* indicate figures, *t* indicate tables, and *s* indicate schemes.

A

Ablative plate reactor, 50–51
Ablative reactors model, 50, 51*f*
Acetone, self-aldol condensation, 305–306, 306–307*f*
Acetone–butanol–ethanol (ABE) separation processes, 16–17
Acetyl-CoA carboxylase (ACC), 234, 259–260
Acidic pretreatment, 210, 218–220
Acidity, role of, 111–112, 112*f*
Acids, 26–28
Acyl-ACP reduases (AARs), 245
Adaptive laboratory evolution (ALE), 261
Agricultural wastes, 142
    and residues, 5
Alcohols, 26–28, 246
Alcohol-to-bio-jet fuel, 283
Alcohol-to-jet process, 277
Aldehydes, 26–28
Aldol condensation, 329–331
    of carbonyl compounds, 330*s*
Algae, 234, 235*f*, 240–242, 242*t*
Algal biorefinery, 392
Alkali pretreatment, 209–210, 218–220
    method, 262
Alkanes, 245
    from biomass-derived carbohydrates, 341, 342*f*
    from corncob biomass, 341–343, 342*f*
    synthesis reaction, 340–341, 340*f*
Alkenes, 244–245
Alkyl levulinates, 19–21
Alkylphenols, 347–348
Alpha olefins, 286–287
Alternative jet fuel (AJF), 362
American Society for Testing and Materials (ASTM), 272
Ammonia fiber explosion, 208
Amylopectin, 6
Amylose, 6
Anabaena cylindrica, 264
Anaerobic digestion, 212
Anderson-Shultz-Flory distribution (ASF distribution), 78

Angelica lactone (AL), 310–311, 345–346
Animal manure, 5
Aqueous-phase reforming process, 281, 282*f*
Aqueous phase valorization, 150–152, 151*t*
Aromatic hydrocarbons, 247–248
Aromatization, 287–289
Aspen Plus, 357–358
ASTMD1655, 275
ATP-citrate lyase (ACL), 234
Auto-hydrolysis technique, 208
Aviation fuels. *See* Bio-jet-fuel production
Aviation turbine fuel (ATF), 273–275

B

Bacteria, 237–240, 238–239*t*
Bacterial strains, 11–12
Baker's yeast, 16
BDI-BioEnergy International bioCRACK process, 163–165, 164–165*f*
β-hydroxyaldehyde, 304–305
Bifunctional catalysts, 338
Bifunctional metal–acid catalysts, 65
Bimetallic catalysts, 338, 421–423, 423*t*
Biobutanol, 16–17
Biochemical conversion processes, 28–29
Biochemical processes, 393
Bioconversion, of CO_2, 263–264
bioCRACK oil, 163
Bio-crude-based hydrocarbon biorefinery, 362–369, 364–366*t*
Biocrude properties, and storage, 146, 146*t*
Biodiesel, 15, 24, 98, 328
    based biorefinery, 394–395, 394*f*
    life-cycle analysis, 399–401, 400*t*
Bioeconomy, 356. *See also* Techno-economic analysis (TEA)
Bioethanol, 16, 328, 395
Biofine Corporation, 12
Biofuels, 3, 98, 298, 299*t*
    challenges and opportunities for, 90–94
    fatty acid-derived, 258–260
    isoprenoid-derived, 258
    polyketide-derived, 260*t*, 261

431

432 Index

Biogas, 277
  platform, 391*t*
Bio-jet-fuel production, 278, 395–396
  alcohol-to-bio-jet fuel, 283
  catalytic reforming of sugars, 281–282, 282*f*
  economics of, 289–290
  Fischer-Tropsch synthesis
    (FTS), 280–281, 280–281*f*
  hydro-processing of lipid feedstock, 279–280,
    279*f*
Biological conversion processes, 28–29
Biological pretreatment, 262–263
  lignocellulosic feedstock pretreatment
    techniques, 210–211, 211*t*
  sewage sludge pretreatment techniques, 221–222
    microbial electrolysis cell, 221–222
    temperature-phased anaerobic digestion
     pretreatment (TPAD), 221
Biomass, 3–10, 98, 298
  conversion technologies, 47–48, 48*f*
  gasifiers, 80–81
  lignocellulosic, 299, 321–323
  liquefaction technology, 128
  methanol production from, 182–183, 183*f*
  petrochemical building-block chemicals
    from, 30, 30*f*
  resources, 47–48
Biomass pretreatment technologies, 203–204, 205*f*
  advantages of, 219–220*t*
  challenges, 222–223
  disadvantages of, 219–220*t*
  lignocellulosic feedstock composition
    and, 204–207, 206*f*
    biological pretreatment, 210–211, 211*t*
    chemical pretreatment, 209–210
    internal structure, 206*f*
    physical pretreatment, 207
    physicochemical pretreatment, 208–209
  sewage sludge composition and, 211–213
    biological pretreatment, 221–222
    chemical treatment, 218–221
    mechanical treatment, 213–216
    thermal treatment, 217–218
Biomass Technology Group, 53
Biomass-to-liquid (BTL) plants, 82, 82*t*
Biomass-to-liquid process, 26–28, 280
Bio-oils, 12, 26–28, 276, 362
  hydroprocessing, 369
  techno-economic assessment, 367–368

Bio-oil upgrading techniques, 58–71, 61*f*
  catalyst for hydrodeoxygenation
    process, 64–69, 64*t*, 67*f*
  chemical upgradation technique of, 58–61
  hydrodeoxygenation process, 61–62
  physical upgradation technique of, 58
  reactor configuration, 62–63
  in situ hydrogenation, 64
Bio-olefins
  oligomerization of, 271–272, 278, 283
    catalysts used, 284, 285*t*
    reaction conditions, 286–287
    reaction mechanism, 284–286, 286–287*f*
  sources of, 275
    alcohols, 277
    bio-oils feedstock, 276
    gaseous feedstock, 277–278
    sugar, 276–277
Biorefinery system, 10–13, 11*t*, 388–389
  for biofuels, 13, 13*t*
  classification of, 390*f*
  definition of, 355–356
  feedstock-based configuration, 390
    algal biorefinery, 392
    green biorefinery, 392–393
    lignocellulose-based biorefinery, 390–391
    oilseed biorefinery, 391–392
  for fuel-additives, 18–24, 19*f*, 20*t*
  platform-based configuration, 389–390, 391*t*
  process-based configuration, 393
    chemical and biochemical processes, 393
    mechanical process, 393
    thermochemical processes, 393
  product-based biorefinery, 393
    biodiesel, 394–395, 394*f*
    bioethanol, 395
    bio-jet fuel, 395–396
  techno-economic analysis, 357–358, 359*f*
  for traditional biofuels, 13–18, 13*f*, 14*t*
  types, 357*t*
Bisabolene, 258, 260*t*
Black liquor, 376
Bottom-up approach, 113–114
Bridgewater equation, 376
Brønsted acid sites, 65, 114–115
Brønsted/Lewis acidity ratio, 189
Bubbling fluidized bed reactor, 51, 52*f*
Butamax, 16–17
Butanol, dehydration, 302, 303*f*

Index **433**

C

Camelina biodiesel, 399, 400*t*
CAPCOST software, 376–379
Carbided metal catalysts, 110
Carbon–carbon bond forming reactions, 34–35
Carbon dioxide (CO_2), methanol production from, 183
Carbon-free energy resource, 298
Carbon nitride (CN), 416–419
Carbon Offset and Reduction Scheme for International Aviation, 272
Catalysts, 105–115, 106*f*
  deactivation, 115
  transition metal-based, 411–412, 411*f*
Catalytic cosite model, 65–66
Catalytic cracking *vs.* hydrodeoxygenation (HDO) of triglycerides, 101, 102*t*
Catalytic fast pyrolysis, 33–34, 53–56, 71
Catalytic hydrodeoxygenation process, 61
Catalytic hydrothermolysis, 279*f*, 280
C–C coupling, 305–306, 306*f*, 329
  Aldol condensation, 329–331, 330*s*
  of furfural derivatives, 314–317, 315*f*, 317*f*
  hydroxyalkylation-alkylation (HAA), 331–332, 331*s*
  5-hydroxylmethylfurfural (HMF), 320–321, 320*f*, 322–323*f*
  ketonization, 332–334, 333*s*
  of levulinic acid derivatives, 310–312, 311*f*
  pinacol coupling, 334, 334*s*
C–C dissociation, 337–338, 337*t*
Cellulose, 6–8, 131–134, 132–133*f*, 204–205
  bio/thermochemical transformation of, 301*t*
CGSB-3.22-93, 275
Chemical conversion processes, 26–28
Chemical pretreatment
  lignocellulosic feedstock pretreatment techniques, 209–210
    acid pretreatment, 210
    alkali pretreatment, 209–210
  sewage sludge pretreatment techniques, 218–221
    acidic pretreatment, 218–220
    alkali pretreatment, 218–220
    ozonation, 220–221
Chemical process, 393
Chemical upgradation technique, of bio-oil, 58–61
Chlorella, 241
Cinnamyl alcohol, 246

Circulating fluidized bed reactor, 52, 53*f*
Clemmensen-type cathodic reduction, 305–306, 306*f*
Clostridium acetobutylicum, 16–17, 246
Clostridium autoethanogenum, 246
Clostridium beijerinckii, 16–17
Clostridium tyrobutyricum, 246
Coal, 10–11, 299*t*
  methanol production from, 180–181, 181*f*
Cobalt catalysts, 85
Co-based self-catalytic reactor (Co-SCRs), 90
C–O hydrogenolysis, 336–337, 337*t*
Coke oven gas (COG), conversion of, 182
Combustion processes, 48
Commercial Aviation Alternative Fuels Initiative, 273
Commercialization status, 119–120
Complete hydrodeoxygenation process, 62
Composite catalyst, 111, 111*t*
Coniferyl alcohol, 204–205
Co-(Ni)–MoS model, 65–66
Conventional biorefinery, 357*t*
Cross-condensation
  of furfural derivatives, 316–317, 317*f*
  of levulinic acid, 311–312, 312*f*
Cunninghamella echinulata, 248
Cyanobacteria, 260
  CO_2 to hydrocarbon conversion, 263
Cyclohexane, 412–416, 414–415*t*, 419–420
Cyclopentanone, 314

D

Decarbonylation, 337–338, 337*t*
Decarboxylation, 337–338, 337*t*
Defense standard 91-91, 275
Degree of polymerization (DP), 131
Dehydration, 131–132, 336–337, 337*t*
  reaction, 65
Dehydration-oligomerization approach, 302–304, 303*f*
Dehydrogenation-aldol condensation, 304–305, 304*f*
Deoxyxylulose 5-phosphate (DXP) pathway, 255, 257*f*
Department of Defense Air Force USA, 272
Diesel, 2–3, 98
5,5'-Dihydroxymethyl-furoin (DHMF), 343–344

434 Index

Dimethylallyl pyrophosphate (DMAPP) pathway, 255–256
Dimethyl ether, 17–18
Dimethylformamide (DMF), 318, 320–321
2,5-Dimethylfuran (2,5-DMF), 22–23
2,5-Dimethyltetrahydrofuran (2,5-DMTHF), 23, 318–320
Direct deoxygenation (DDO), 410–411
Direct mechanisms, 184–187
Direct synthesis, 17–18
Displacement method, 398
Diverse microbial systems, for hydrocarbon generation, 236–243
Drop-in fuels
  fermentation fuels, upgradation of
    C–C coupling reactions, 305–306, 306f
    dehydration-oligomerization approach, 302–304, 303f
    dehydrogenation-aldol condensation, 304–305, 304f
  properties of, 301, 301t
  sugar dehydration intermediates, upgradation of, 306–321
Dunaliella salina, 230–231

E

Ecofining process, 120f
Eichhornia crassipes, 279
Electrokinetic disintegration technique, 216
Emerging Fuels Technology, 87
Energy crops, 3–5
Engineered microorganisms, 259t
  hydrocarbon biofuels production, 260t
Esters, 26–28, 247
Ethanol, 153, 299–301
  dehydration, 16, 302, 303f
  Guerbet condensation, 304, 304f
  minimum fuel selling price (MFSP), 360–362
5-Ethoxymethylfurfural (5-EMF), 23
Ethylene oligomerization, 284–286, 286f
Euglena gracilis, 247
European Renewable Energy Council, 3
Extended X-ray absorption fine structure (EXAFS), 110
Extracellular polymeric substances (EPS), 212
ExxonMobil, 194–195, 195f

F

Farnesene, 258, 260t
Farnesol, 260t
Fast pyrolysis process, 26–28, 50–56, 362, 363f
  bio-crude-based hydrocarbon biorefinery, 364–366t
  catalytic, 53–56
  microalgae-based hydrocarbon biorefinery, 369, 370f
  operation conditions, 56–58, 59–60t
  reactor configuration, 50–53
    ablative reactors model, 50, 51f
    bubbling fluidized bed reactor, 51, 52f
    circulating fluidized bed reactor, 52, 53f
    fluidized bed reactor, 51
    rotating cone reactors, 53
Fatty acid
  biosynthesis pathway, 256–257
  derived biofuels, 258–260, 260t
  hydro-processed, 279
Fatty acid methyl esters (FAME) hydrogenolysis, 308–310
Fatty acids synthetase (FAS)
  complex, 233–234, 233f, 256–257
Fatty alcohol, 104–105
Fibers, 204–205
Filtration, 58
Fischer-Tropsch (FT) process, 12
Fischer-Tropsch synthesis (FTS), 25–28, 33, 78, 180, 275, 280–281, 280–281f
  challenges and opportunities for biofuels, 90–94
    automated and unattended operations, 93
    distributed processing, 90–91
    gas loop configuration, 91
    oxygenated fuels, 93–94
    plant modularization, 93
    reactor/catalyst selection, 91–92, 92t
    upgrading and refining steps, 93
  improvement of catalysts, 82–90, 83t
    cobalt catalysts, 85
    fixed-bed multitubular reactor, 86–87
    fluidized-bed reactor, 88
    iron catalysts, 83–84
    microchannel reactors, 89
    reactor design, 85–86, 86t, 87f
    slurry bubble column reactor, 88
    3D printing technology, 89–90
  industrial scale, 78, 79f

operating conditions, 78, 79t
  synthesis gas manufacture, 79–82, 80–81f, 81–82t
Five-platform biorefinery, 389–390
Fixed-bed multitubular reactor, 86–87
Fluid catalytic cracking (FCC) unit, 157
Fluidized bed reactor, 51, 88
Folic acid, 247–248
Forest waste and residue, 5
Fossil fuels, 298, 299t
Four-platform biorefinery, 389–390
Fractionation, 240–241
Free fatty acids, 232
Fulcrum Bioenergy, 82, 87
Fungus, 243, 243t
Furan, 331s, 332
Furanic aldehydes, pinacol coupling, 334, 334s
Furfural, 312–313
  C–C coupling, 314–317, 315f, 317f
  hydrodeoxygenation of, 313–314, 313f
  reductive self-condensation, 343–344, 343s

G

γ-Valerolactone (GVL), 18–19, 21f, 336, 346–347
Gas chromatography, 245
Gasification, 12, 212–213, 280–281
  life-cycle analysis, 403
  processes, 48
  unit, 182–183
Gas loop configuration, 91
Gasoline, 2–3, 98
Gas-to-liquid (GTL) plants, 81–82
GB 6537-2006 Chinese standard, 275
Genetic modification, of microorganisms, 16–17
Gevo, 16–17
GFT-J process, 396
Global bioenergy, 9–10
Global warming potential (GWP), 396–397, 399, 403
Glucose degradation mechanism, 131
Glutamic acid, 11–12
Glycerine, 99–101
Glycerol acetals, 24
GOST 10227, 275
Green Biologics, 16–17
Green biorefinery, 357t, 392–393
Green diesel, 26–28, 99
Green gasoline, 101
Greenhouse gases, 265

Greenhouse gases, Regulated Emissions, Energy use in Transportation (GREET), 397
Guaiacols, 26–28
Guerbet condensation, 304–305, 304f

H

Hasan parameter, 155
HBeta-based catalyst, 412–416
Heating rate, 141, 393
Heavy paraffin conversion, 85
Heavy paraffin synthesis (HPS), 85
Hemicellulose, 8–9, 134, 135f, 204–205
  bio/thermochemical transformation of, 301t
Heterogeneous catalysis, 145–146, 330–331
  role, in hydrocarbon biorefinery, 30–35, 31f
Heterologous pathways, 259t
High-lipid low-protein microalgae, 141–142
High-pressure homogenization, 216, 217f
High-pressure liquefaction, 161–162
High-temperature Fischer-Tropsch
  (HTFT), 82–83
Homogeneous catalysts, 144–145
Host microorganisms, 259t
Hydration reaction, 65
Hydrocarbon biofuels, 24, 99
  biomass-derived compounds
    conversion, 29–30
  from platform chemicals, 29f
  production, 25–26, 25f
  properties of, 27t
Hydrocarbon biofuels, liquefaction production of
  high-pressure liquefaction, 161–162
    Iowa State University and Chevron, 161–162, 161f
  hydrothermal liquefaction (HTL), 129–152,
    158–160, 166–167, 167f
    aqueous phase valorization, 150–152, 151t
    biocrude properties and storage, 146, 146t
    biocrude upgrading, 147–150, 149t
    catalyst selection, 144–146
    feedstock type, 141–143, 143–144t
    Licella CAT-HTR process, 158–159, 159f
    parameters, 138–141
    process fundamentals, 129–130, 130t
    reaction pathways, 130–138, 139f
    steeper energy hydrofaction, 159–160, 160f
  low-pressure liquefaction, 162–165, 162f

436   Index

Hydrocarbon biofuels, liquefaction production of (*Continued*)
  BDI-BioEnergy International bioCRACK process, 163–165, 164–165*f*
  process economics, 165–168, 166*t*
  using organic solvents, 152–158, 153*t*, 161–165, 167–168, 168*f*
    for commercial application, 155–158, 157*t*
    in pure organic compounds, 153–155, 154*f*, 156*f*
Hydrocarbon biorefinery approaches, 24–30, 99, 357–358, 359*f*
  biological and biochemical conversion processes, 28–29
  biomass, 3–10
    availability of, 9–10
    chemistry of, 6–9, 7*f*, 8*t*
    types of, 3–6, 4*t*
  biorefinery, 10–13, 11*t*
    for biofuels, 13, 13*t*
    lignocellulose-based biorefinery, 12
    sugar and starch-based biorefinery, 11–12
    triglyceride-based biorefinery, 12–13
  chemical and thermochemical conversion processes, 26–28
  conversion of biomass-derived compounds to hydrocarbon biofuels, 29–30, 29*f*
  fuel-additives, 18–24, 19*f*, 20*t*
    alkyl levulinates, 19–21
    5-Ethoxymethylfurfural (5-EMF), 23
    γ-valerolactone (GVL), 18–19, 21*f*
    glycerol acetals, 24
    2,5-Dimethylfuran (2,5-DMF), 22–23
    2,5-Dimethyltetrahydrofuran (2,5-DMTHF), 23
    2-Methylfuran (2-MF), 22
    2-Methyltetrahydrofuran (2-MTHF), 22
  heterogeneous catalysis role in, 30–35, 31*f*
    metal catalysts, 32–33
    solid-acid catalysts, 34–35
    solid-base catalysts, 34–35
    zeolite-type catalysts, 33–34
  hydrocarbon biofuels production, 25–26, 25*f*
  petrochemical building-block chemicals from biomass, 30, 30*f*
  traditional biofuels, 13–18, 13*f*, 14*t*
    biobutanol, 16–17
    biodiesel, 15

bioethanol, 16
dimethyl ether, 17–18
Hydrocarbons, 231–232
  biosynthesis of, 244–248
  pool mechanism, 184–187
Hydrocracking process, 62–63
Hydrodenitrogenation (HDN), 148, 149*f*
Hydrodeoxygenation (HDO) process, 12, 147, 157–158, 328, 336–337, 409–410
  bimetallic catalysts, 421–423, 423*t*
  of bio-oil, 33, 61–62, 70–73
  catalysts for, 64–69, 64*t*, 67*f*, 337–338, 337*t*
    bifunctional metal-acid catalysts, 65
    metal catalyst, 64
    nonnoble metal, 69
    oxides-based catalyst, 66
    reduced transition metals, 68
    solid acid support, 65
    sulfide-based catalyst, 65
  dual-stage reactor for, 63*f*
  of fast pyrolytic oil, 72*t*
  of furfural, 313–314, 313*f*
  of 5-hydroxylmethylfurfural, 318–320, 319*f*
  of levulinic acid, 308–310, 309*f*
  oxophilic metals for, 424–425, 425*f*
  oxygenated fuel precursors, 33
  of phenolic compounds, 410–411, 411*f*, 425–426
  reactor configuration of, 62–63
  transition metal-based catalysts, 411*f*
    active site, 411–412
    product selectivity trends, 416–421, 421*f*
    support effect, 412–416, 413*f*, 414–415*t*
  of triglycerides, 32
  of vegetable oil, 32*f*
Hydrodeoxygenation (HDO) of triglycerides
  catalysts, 105–115, 106*f*
    carbided metal catalysts, 110
    composite catalyst, 111, 111*t*
    deactivation, 115
    development for, 114*f*
    mesoporous support, 113–115
    nitride catalysts, 110
    noble metal catalysts, 106
    phosphided catalysts, 110
    reduced transition metal catalysts, 108–109, 108–109*f*
    reducible oxides, 113, 113*f*
    role of acidity, 111–112, 112*f*

sulfided transition metal catalysts, 107
catalytic cracking *vs.*, 101, 102*t*
chemical structure, 99*f*
commercialization status, 119–120
oveview, 122
petroleum feedstock, 116–119, 118*t*
pressure, 115–116
process design and economics, 120–121, 121*f*
reaction mechanism, 101–105, 103–105*f*
structure and composition of, 99–101, 100*t*
temperatures, 115–116
Hydrodesulfurization (HDS) process, 336–337
MoS_2 catalyst in, 66
Hydrogenation, 49, 336–337, 337*t*
Hydrogenation-dehydration (HYD), 410–411
Hydrogen donor solvents, 18–19, 155–156
Hydrogenolysis, 49
Hydrogen platform, 391*t*
Hydrogen pressure, 71
Hydrolysis reaction, 65, 135
Hydroprocessing technology, 116–117, 117*f*
Hydrotalcites, 34–35
Hydrothermal liquefaction (HTL), 26–28, 129–152,
     158–160, 166–167, 167*f*, 212, 240–241,
     362, 363*f*
    bio-crude-based hydrocarbon biorefinery, 368*t*
    life-cycle analysis, 402–403
    microalgae-based hydrocarbon biorefinery, 369,
     370*f*
Hydrothermolysis, 279*f*, 280
Hydrotreatment, 49, 147–148, 157–158
Hydroxyalkylation-alkylation (HAA), 331–332,
    331*s*
5-Hydroxylmethylfurfural (HMF), 11–12,
    317–318
    C–C coupling, 320–321, 320*f*, 322–323*f*
    hydrodeoxygenation of, 318–320, 319*f*
4-hydroxyphenylacetic acid (4HPAA), 247–248
2-(4-hydroxyphenyl) ethanol (4HPE), 247–248
4-hydroxyphenyllactic acid (4HPLA), 247–248

I

Imperial Chemical Industries (ICI), 179
India, 10
Indirect synthesis, 17–18
Induction period, 184–187
Industry and municipality waste, 5
In situ hydrogenation, 64

In situ solvent removal techniques, 16–17
Integrated biorefinery (IBR), 396–397
Intercalation model, 65–66
International Air Transport Association, 271–272
International Energy Agency, 230
International Renewable Energy Agency, 3
Iowa State University, and Chevron, 161–162, 161*f*
Iron catalysts, 83–84
Isoamyl alcohol, 246
Isomerization reactions, 65
Isoparaffin, 281
Isoparaffinic kerosene (IPK), 283
Isopentenols, 260*t*
Isopentenyl pyrophosphate (IPP) pathway, 255–256
Isoprene, 258, 260*t*
Isoprenoid biosynthesis pathway, 255–256, 257*f*
Isoprenoid-derived hydrocarbon
     biofuels, 254, 260*t*
Itaconic acid, 11–12

J

Jatropha biodiesel, 399, 400*t*
Jet A-1, 273
Jet fuel, 2–3, 98. *See also* Bio-jet-fuel production
    composition of, 289*t*
    from renewable feedstocks, 272
    specifications, 273–275
Jet fuel A, 273
Jet fuel TS-1, 273–275

K

Kerosene-type jet fuel, 273
Ketones, 26–28
Ketonization, 332–334, 333*s*
    reaction, 29–30

L

Lactic acid, 11–12
Lauric acid, ketonization, 332–333, 333*s*
L-ETH-J process, 396
Levulinic acid (LA), 11–12, 308
    C-C coupling, 310–312, 311*f*
    cross-condensation, 311–312, 312*f*
    hydrodeoxygenation of, 308–310, 309*f*
Lewis acid sites, 65
Licella CAT-HTR process, 158–159, 159*f*
Life-cycle analysis, 388–389, 396, 403–404
    biodiesel production, 399–401, 400*t*

438 Index

Life-cycle analysis *(Continued)*
  integrated biorefinery (IBR), 398
  products-based, 397–398
  system boundary for, 397
  thermal conversion-based biorefinery, 401
    gasification, 403
    hydrothermal liquefaction, 402–403
    pyrolysis, 402
    torrefaction, 401–402
Light cycle oil (LCO), 157
Lignin, 134, 136*f*
  based bio-jet fuel, 395–396
  derived hydrocarbon biofuels, 347–348, 348*s*
  platform, 391*t*
Lignocellulose-based biorefinery, 12, 390–391
Lignocellulose-based hydrocarbon biofuels,
    299–301
  processes/unit operations, 300*f*
Lignocellulosic biomass, 6–8, 142, 143*f*, 204–205,
    208–210, 299, 321–323, 409–410
  based hydrocarbon biorefinery, 358–369, 359*f*
Lignocellulosic biorefinery, 357*t*
Lignocellulosic feedstock
  composition and pretreatment, 204–207
  pretreatment techniques for, 207–211
Lignocellulosic materials, 262–263
Limonene, 258, 260*t*
Lipids, 135–137, 137*f*, 248
Liquefaction
  process, 129*f*
  in pure organic compounds, 153–155,
    154*f*, 156*f*
  using organic solvents, 152–158, 153*t*, 161–165,
    167–168, 168*f*
Liquid hot water pretreatment, 208–209
Lobry de Bruyn-Alberda van Ekenstein (LBET)
    transformation, 134
Long-chain alkanes, 260*t*
Long-chain alkenes, 260*t*
Long-chain hydrocarbon biofuels
  from furan-based feedstocks, 341–344, 343*s*
  from levulinic acid derivatives, 344, 345*s*
    direct route, 344–345
    via angelica lactone, 345–346
    via γ-valerolactone, 346–347
  from lignin, 347–348, 348*s*
Low-frequency ultrasound technique, 213
Low-lipid high-protein microalgae, 141–142

Low-pressure liquefaction process, 162–165, 162*f*
Low-temperature Fischer-Tropsch (LTFT), 82–83
Lurgi technology, 179

M

Macrococcus caseolyticus, 237
Maillard reactions, 138
Malate dehydrogenase (MDH), 234
Malic enzyme (ME), 234
Marine-based biorefinery, 357*t*
Marine biomass, 6
Mars-van Krevelen mechanism, 66
Mass spectrometry, 245
Mechanical process, 393
Mechanical treatment
  sewage sludge pretreatment techniques, 213–216
    electrokinetic disintegration, 216
    high-pressure homogenization, 216
    microwave irradiation, 213–216
    ultrasonic pretreatment, 213
Membrane filtration, 392
Mesoporous aluminosilicates, 285*t*
Metabolic engineering, 254–255, 259*t*
  fatty acid-derived biofuels, 258–260
  isoprenoid-derived biofuels, 258
  polyketide-derived biofuels, 260*t*, 261
Metal catalysts, 32–33, 64
Metallacycle mechanism, 284–286, 287*f*
Metal oxides, 55
Methane, 245
Methanogens, 237
Methanol, 15, 17–18, 26–28
  production of, 179–183, 180*f*
  synthesis unit, 182–183
Methanol to gasoline (MTG) process, 26–28, 178,
    184–194, 184*s*, 185–186*t*
  crystal size of catalysts and surface properties,
    189–191, 192*t*
  effect of reactor configuration, 193–194, 194*t*
  effect of Si/Al ratio in zeolite, 189, 190*t*
  ExxonMobil, 194–195, 195*f*
  industrial development, 194–196, 196*t*
  overview, 196–197
  process parameters, 191–193, 193*t*
  production of methanol, 179–183, 180*f*
    from biomass, 182–183, 183*f*
    from carbon dioxide (CO_2), 183
    from coal, 180–181, 181*f*

coke oven gas (COG), 182
natural gas, 181–182
reaction mechanism, 184–187, 188f
shape-selectivity of zeolites, 187–188
zeolite-based catalysts, 178, 191, 196–197
Methylcyclopentane, 320, 320f
Methylerythritol phosphate pathway (MEP),
244–245
2-Methylfuran (2-MF), 22, 331s, 332
Methyl isobutyl ketone (MIBK), 310–311, 312f,
318
2-Methyltetrahydrofuran (2-MTHF), 22
Metric for Inspecting Sales and Reactant (MISR)
method, 376–379
Mevalonate (MVA) pathway, 232–233, 255, 257f
MFSP. See Minimum fuel selling price (MFSP)
Michael addition, 335, 335s
Microalgae-based hydrocarbon biorefinery,
369–374, 370f, 371t, 375f
Microbes, hydrocarbons production using
biochemistry of hydrocarbon synthesis in,
231–235
biosynthesis of, 244–248
alcohols, 246
alkanes, 245
alkenes, 244–245
aromatic hydrocarbons, 247–248
esters, 247
lipids, 248
neutral lipids, 248
nitrogen-containing heterocyclic
hydrocarbons, 247–248
phenolic lipids, 248
phospholipid, 248
polycyclic aromatic hydrocarbon, 247–248
diverse microbial systems for, 236–243
algae, 235f, 240–242, 242t
bacteria, 237–240, 238–239t
fungus, 243, 243t
Escherichia coli strains, 235, 236f
overview, 230–231, 249
Microbial electrolysis cell, 221–222
Microbial hydrocarbon synthesis
challenges, 265
metabolic pathways, 255, 256f
fatty acid biosynthesis pathway, 256–257
isoprenoid biosynthesis pathway, 255–256,
257f

polyketide biosynthesis pathway, 257–258
Microchannel reactors, 89
Microcoleus vaginatus, 264
Microcrystalline materials, 55, 55t
Microkinetic model (MKM), 425
Micro/mesoporous zeotypes, 285t
Microorganisms
engineered, 259t
host, 259t
pretreatment, 210–211
Microwave irradiation, 213–216
Mild hydrodeoxygenation process, 62
Million metric ton (MMT), 9–10
Mineral catalyst, 54
Minimum fuel selling price (MFSP), 165–166,
357–358, 380t
algal feedstock and, 373
alternative jet fuel (AJF), 362
bio-crude-based hydrocarbon, 363–366,
364–366t
of ethanol, 360–362
of upgraded bio-oil, 367
for xylitol, 379
Minimum selling price
pulp, 376
sorbital, 379
Mitsubishi technology, 179
Mobil Oil, 28–29
Monolayer model, 65–66
Monometallic catalysts, 424
Monoterpenes, 258
Mortierella isabellina, 248

N

NADPH, 234
Nafion-212, 332
Naphtha-type jet fuel, 273
National Biofuel Policy, 10
Natural gas, 299t
conversion of, 181–182
n-Butanol, 16–17
Neste Oil, 119–120
Net present value (NPV), 357–358
Neutral lipids, 248
compounds, 247
NEXBTL technology, 119–120
Nickel-substituted synthetic mica-montmorillonite,
285t

NiFe bimetallic catalyst, 421–422, 423t
Ni-hybrid mechanism, 284–286
Ni/Mo mole ratio, 108–109
NiRe catalyst, 421–422
Nitride catalysts, 110
Nitrogen-containing heterocyclic hydrocarbons, 247–248
Noble metals, 62–63
    catalysts, 106
Nonnoble metals, 69

O

Oil platform, 391t
Oilseed biorefinery, 391–392
Olefins. *See* Bio-olefins
Oligomer, 283
Oligomerization process, 28, 336, 336s
    of bio-olefins, 271–272, 278, 283
        catalysts used, 284, 285t
        reaction conditions, 286–287
        reaction mechanism, 284–286, 286–287f
    of olefins, 34
Organic solution platform, 391t
Organization of the Petroleum Exporting Countries (OPEC), 240–241
Organized mesoporous alumina (OMA), 114–115
Organosolv pretreatment, 395
Oryx GTL plant, 85
Oxides-based catalyst, 66
Oxophilic metals, 338, 424–425
Oxygenated fuels, 93–94
Ozonation, 220–221

P

Pacific Northwest National Laboratory (PNNL), 62, 148–149
Palladium (Pd), 62–63
Paris Agreement, 290
Particle size, 57–58, 207
p-Coumaryl alcohol, 204–205
PdFe bimetallic catalyst, 421–422, 423t
Pd-supported catalyst, 412–416, 414–415t, 417–418t
Pearl GTL facility, 85
Penicillium cyclopium, 230–231
Pentadecaheptaene, 260t, 261
Pentane, 312–313
1,5-Pentanediol, 313–314
Pentanoic acid, 346–347

Petroleum, 2–3, 299t
    derived liquid fuels, 298
Phanerochaete chrysosporium, 262–263
Phenolic compounds
    from biomass pyrolysis, 409–410, 410f
    hydrodeoxygenation (HDO) reaction, 410–411, 411f
Phenolics, 26–28
    lipids, 248
Phenylacetic acid, 247–248
Phenylalanine (Phe), 247–248
2-phenylethanol, 247–248
Phenyl-lactic acid, 247–248
Phosphided catalysts, 110
Phospholipids, 248
Phosphoric acid, 15
Physical pretreatment, lignocellulosic feedstock pretreatment techniques, 207
Physical separation methods, 11–12
Physical upgradation technique, of bio-oil, 58
Physicochemical pretreatment
    lignocellulosic feedstock pretreatment techniques, 208–209
        ammonia fiber explosion, 208
        liquid hot water pretreatment, 208–209
        steam explosion/auto-hydrolysis, 208
Pichia pastoris, 243
Pinacol coupling, 314–316, 334, 334s
Pinene, 260t, 264
Pittsburgh Energy Research Center (PERC), 128
Plant modularization, 93
Platform-based biorefinery configuration, 389–390, 391t
Platform chemicals, 11, 11t, 35
Pleurotus ostreatus, 262–263
Polar solvents, 58
Polycyclic aromatic hydrocarbon, 247–248
Polyketide-derived biofuels, 260t, 261
Polyketide synthases (PKSs), 257–258
Postsynthetic modification, 113–114
Pressure, 141
Primary synthesis approach, 113–114
Primus Green Energy, 196
Products-based LCA, 397–398
Propane, 260
Proteins, 135–137, 137f
"P-Series fuel,", 22

Index **441**

PtMo/C bimetallic catalyst, 421–422, 423t
Pt-supported catalysts, 412–416, 414–415t,
    417–418t
Pulp, minimum selling price, 376
Pyrolysis processes, 48, 240–241, 362, 363f, 393
    of biomass, 48–49
    life-cycle analysis, 402
    oil platform, 391t
    process, 240–241
Pyruvate carboxylase (PC), 234

R

Reaction mechanism, 101–105, 103–105f,
    184–187, 188f
Reactor configuration
    fast pyrolysis process, 50–53
    of hydrodeoxygenation process, 62–63
Red Mud, 344
Reduced transition metals, 68
    catalysts, 108–109, 108–109f
Renewable Diesel Blend (RDB), 397
Re-promoted Ni catalyst, 421–422
ReRu/MWCNT, 421–422
Residence time, 140, 140t
Return on investment (ROI), 376–379
Reverse Mars-van Krevelen Mechanism,
    66, 67f, 68
Reverse water gas shift reaction (RWGSR), 179
Rh-based catalyst, 416–419
Rhizosolenia setigera, 244–245
Rhodobacter sphaeroides, 258
Rhodotorula minuta, 243
Rhodymenia pseudopalmata, 241
Rice husk, 5
Ring-opening (RO), 339–340
Ring-saturation (RS), 339–340
Robinson annulation, 335, 335s
Rotating cone reactors, 53
    for fast pyrolysis, 54f

S

Sabinene, 260t
Saccharomyces cerevisiae, 246
Sasol Synthol technology, 88
Seed oil, 391–392
Self-condensation
    acetone, 305–306, 306–307f
    of furfural derivatives, 314–316, 315f
    of levulinic acid derivatives, 310–311, 311f

Sewage sludge, 142
    composition and pretreatment, 211–213
    pretreatment techniques for, 213–222,
        214–215t
Shell Middle Distillate Synthesis (SMDS)
    technology, 85
Shikimate pathway, 247–248
Short-chain alkanes, 260t
Silica, 69–70
Silicalite, 33–34
Sinapyl alcohol, 204–205
Single-cell oils (SCO), 248
SiO_2-Al_2O_3 supported Ni_2 P, 285t
Sludge disintegration techniques, 220
Slurry bubble column reactor, 88
Slurry concentration, 140–141
Solid acid, 65
    catalysts, 34–35
Solid-base catalysts, 34–35
Solid phosphoric acid, 285t
Solid wood, 5
Solubilized Chemical oxygen
    demand (SCOD), 216
Solvent, addition of, 58
Sorbital, minimum selling price, 379
Starch, 6
    based biorefinery, 11–12
Steam explosion technique, 208
Steeper energy hydrofaction process, 159–160, 160f
Sterol biosynthesis pathways, 231–232
Succinic acid (SA), 11–12
Sugars, 6
    based biorefinery, 11–12
    based hydrocarbon biorefinery, 360–362, 361t
    fermentation of, 11–12
    platform, 391t
Sulfide-based catalyst, 65
Sulfided transition metal catalysts, 107
Supercritical water gasification, 376
SuperPro Designer, 357–358
Sylvan process, 332
SYN-FER-J process, 396
Syngas, 277
    platform, 391t
Synthesis gas manufacture, 79–82, 80–81f,
    81–82t
Synthetic fuels, 82, 94
Synthetic isoparaffin, 281
System boundary extension, 398

T

Tautomerization-deoxygenation, 410–411
Techno-economic analysis (TEA), 165–166,
  356–357
  bio-crude-based hydrocarbon biorefinery,
    362–369
    using fast pyrolysis, 364–366t
    using hydrothermal liquefaction, 368t
  methodology of, 357, 358f
  sugar-based hydrocarbon biofuels, 360–362, 361t
  waste-based hydrocarbon biorefinery, 374–380,
    377–378t
Temperature, 26–28, 57, 70–71, 138–139, 140f, 191
Temperature-phased anaerobic digestion
  pretreatment (TPAD), 221
Thermal hydrolysis pretreatment technology,
  217–218
Thermal treatment, sewage sludge pretreatment
  techniques, 217–218
Thermochemical biorefinery, 357t
Thermochemical conversion, 401
  processes, 26–28
  technologies, 48, 49f
Thermochemical processes, 212, 393
3D printing technology, 89–90
3-Hydroxy propanoic acid, 11–12
Three-platform biorefinery, 389–390
Tobacco seed biodiesel, 399–401, 400t
Top-down modification, 113–114
Topsoe technology, 179
Torrefaction, life-cycle analysis, 401–402
Total capital investment (TCI), 165–166
Total chemical oxygen demand (TCOD), 216
Trametes versicolor, 262–263
Transesterification, 394–395
  method, 241
  reaction, 15, 15f
Transition metals, as hydrodeoxygenation catalysts,
  411–421
Transportation fuels, 2–3, 24, 49, 71, 98
Triacylglycerols (TAG), 99–101
Triglycerides, 6
  based biorefinery, 12–13
Tryptophan, 247–248
Turn over number (TON), 411–412
Two-platform biorefineries, 389–390
Two platform concept biorefinery, 357t

Tyrosine (Tyr), 247–248
Tyrosol, 247–248

U

Ultrasonication, configurations of, 217f
Ultrasonic pretreatment, 213
US Department of Energy, 245
Used cooking oil, 399–401, 400t
US Technology Corporation, 16–17

V

Vacuum gas oil (VGO), 157, 163–165
Vanillin, 334
  selling price, 376–379
Vanillyl alcohol, 246
Vapor residence time, 50, 58
Velocys Inc., 89

W

Waste-based hydrocarbon biorefinery, 374–380,
  377–378t
Water-gas-shift (WGS) reaction, 145
Water-phase diagram, 131f
Wax esters, 247
  Weight hourly space velocity (WHSV),
    191–193
Wet feedstocks, 141–142
Whole crop biorefinery, 357t
Wood conversion, 128
"World Oil Outlook 2020,", 240–241

X

Xylitol, selling price, 379
Xylose, 134

Y

Yarrowia lipolytica, 258–259
Yineng Bio-Energy Co. Ltd., 64

Z

Zeolites
  effect of Si/Al ratio, 189, 190t
  shape-selectivity of, 187–188
Zeolite-type catalysts, 33–34
ZSM-5
  catalyst, 184, 189–191
  zeolite, 285t, 288

Printed in the United States
by Baker & Taylor Publisher Services